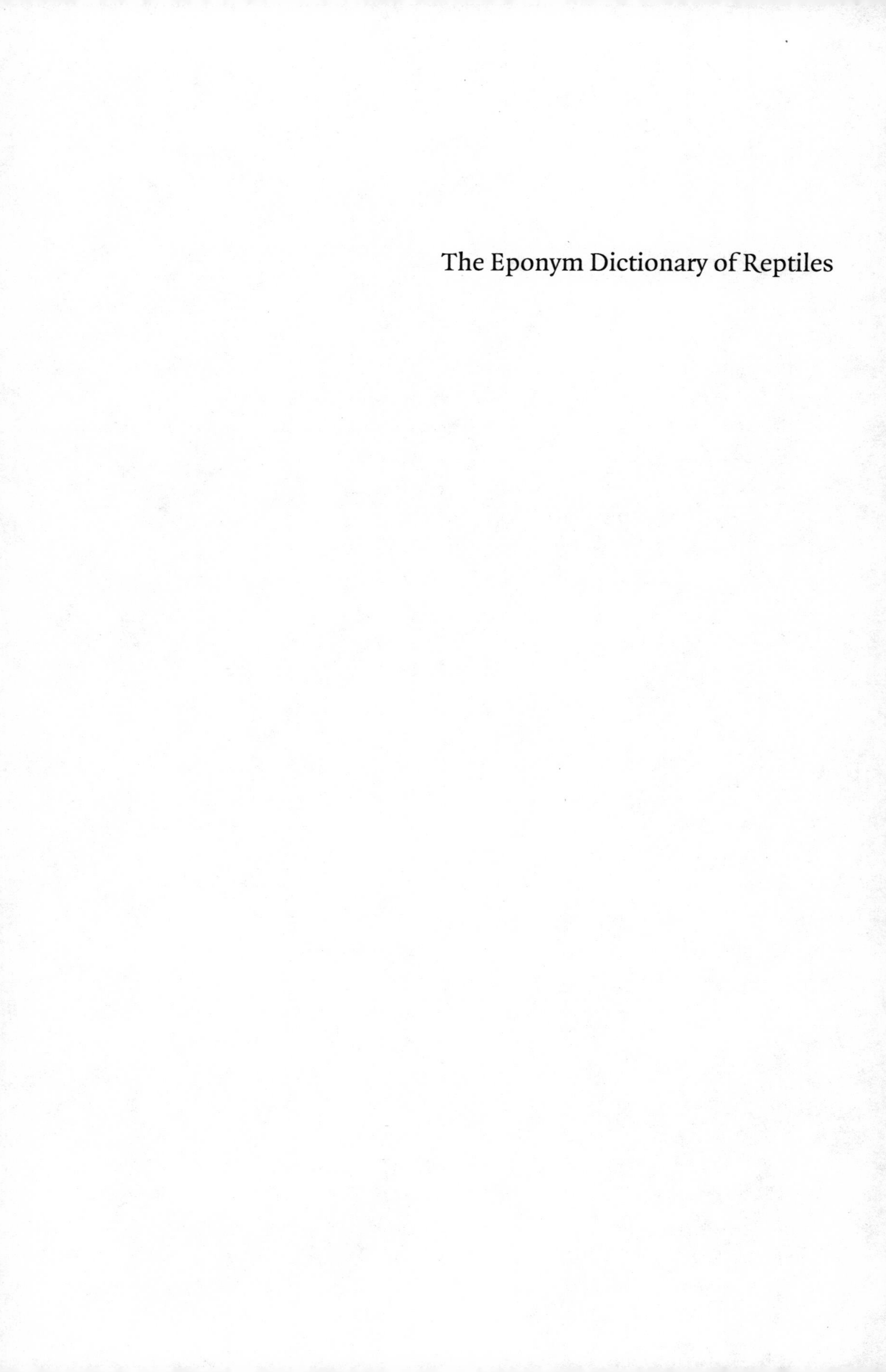

The Eponym Dictionary of Reptiles

The Eponym Dictionary of Reptiles

Bo Beolens
Michael Watkins
Michael Grayson

The Johns Hopkins University Press • Baltimore

 Published 2011
Printed in the United States of America on acid-free paper
9 8 7 6 5 4 3 2 1

The Johns Hopkins University Press
2715 North Charles Street
Baltimore, Maryland 21218-4363
www.press.jhu.edu

Library of Congress Cataloging-in-Publication Data

Beolens, Bo.
The eponym dictionary of reptiles / Bo Beolens, Michael Watkins, and Michael Grayson.
p. cm.
Includes bibliographical references.
ISBN-13: 978-1-4214-0135-5 (hardcover : alk. paper)
ISBN-10: 1-4214-0135-5 (hardcover : alk. paper)
1. Reptiles—Dictionaries. 2. Eponyms—Dictionaries. I. Watkins, Michael, 1940– II. Grayson, Michael. III. Title.
QL640.7.B46 2011
597.903—dc22 2010050260

A catalog record for this book is available from the British Library.

Special discounts are available for bulk purchases of this book. For more information, please contact Special Sales at 410-516-6936 or specialsales@press.jhu.edu.

The Johns Hopkins University Press uses environmentally friendly book materials, including recycled text paper that is composed of at least 30 percent post-consumer waste, whenever possible.

To my late parents, Eric and Phyllis Crombet-Beolens, lost to me during the writing of this eponym dictionary, who encouraged and fostered my lifelong interest in wildlife

Bo Beolens

CONTENTS

PREFACE

Two of us, Bo Beolens and Mike Watkins, wrote *Whose Bird?*, published in 2003. A review of it was written by Nicholas Gould for *International Zoo News*. Gould suggested that there could be a need for similar volumes on other animal orders, and among them he suggested *Whose Mammal?* and *Whose Reptile?* We wish to give credit to him whose suggestion led us to write this book as well as *The Eponym Dictionary of Mammals*, published by the Johns Hopkins University Press in 2009.

We are deeply indebted to the following people and organizations for their generous help with research and, where needed, translations: Kraig Adler, Professor of Biology, Cornell University, USA; Sebastian Aigner, Österr. Akademie der Wissenschaften, Vienna, Austria; Eva M. Albert, Estación Biológica de Doñana (CSIC), Spain; Sylvie Coten-Watkins, Montmorency, France; Patrick Couper, Curator of Herpetology, Queensland Museum, Brisbane, Australia; Paola Gonzalez Abarca, Librarian, Museo Nacional de Historia Natural, Santiago de Chile; Jakob Hallerman, Zoologischen Museum der Universität Hamburg; Notker Helfenberger, St. Gallen, Switzerland; Robert William Henderson, Senior Curator, Herpetology, Milwaukee Public Museum, USA; Stefan Koerber, Ichthyologist, Muelheim, Germany; Anton Maslov, St. Petersburg, Russia; Petr Nečas, Herpetologist, Czech Republic; Giuseppe Osella, Professor of Zoology and Entomology, L'Aquila University, Italy; Geoffrey Patterson, Wellington, New Zealand; Christopher Raxworthy, American Museum of Natural History, New York, USA; Edoardo Razetti, Museum of Natural History, University of Pavia, Italy; John Rose, Assistant Librarian, British Museum of Natural History, London; Jevgeni Shergalin, Carmarthen, Wales; Peter Uetz, University of Delaware, Newark, Delaware, USA; Fernando Videla, Professor, Instituto Argentino de Investigaciones de Zonas Aridas; Manfred Warth, Staatlisches Museum für Naturkunde, Stuttgart, Germany; Charles Watkins, Montmorency, France; Nicholas and Katherine Watkins, Oxford, England; Suzanne Watkins, Bushey Heath, Herts, England; Madame Zouaq, Librarian, Muséum National d'Histoire Naturelle, Paris; Dr. Marco Zuffi, Editor-in-Chief, *Acta Herpetologica Italiana*, Italy.

Our thanks are also due to those people mentioned in the book who amended "their" entries.

INTRODUCTION

Who Is It For?

Vernacular names of animals often contain a person's name (such names are called "eponyms"). We have all heard of Russell's Viper, but how familiar is Uzzell's Lizard? So this book is for both the amateur herpetologist and the student of zoology.

How to Use This Book

This book is arranged alphabetically by the names of the people after whom reptiles have been named. Generally, the easiest way to find your animal is to look it up under the personal name that is apparently embedded in the animal's common or scientific name. We say "apparently" because things are rarely as simple as they seem. In some names, for example, the apostrophe implying ownership is a transcription error; other names may refer to places rather than to people. We have included any names where we think confusion might arise, but we do not promise to have been completely comprehensive in that respect. You should also beware of spelling. Surf the Net, and you may well find animals' names spelled in a number of different ways—that greatest resource is also full of inaccuracies and misinformation. We have tried to include entries on those alternatives that we have come across.

Each biography follows a standard format: First, you will find the name of the person honored. Next, there follows a list of animals named after that person, arranged in order of the year in which they were described. (This list gives common names, scientific names, names of the people who first described each species, and the date of the original descriptions—in that sequence). Alternative English names follow in square brackets and are each preceded by the abbreviation Alt. Different scientific names (where taxonomists disagree) are preceded by the abbreviation Syn. (synonym). Finally, there is a brief biography of that individual. Space constraints limited us to mentioning only one publication written by each person.

To assist you in your search, we have cross-referenced the entries by highlighting (in bold) the names of those describers who also appear elsewhere in the book. Some reptiles are named in different ways after the same person, and we have also tried to marry these up using cross-references.

A person's fame does not correspond to the length of the entry—in fact, often the opposite. A very famous person would have a fairly brief write-up, as he or she would be so well known, and so written-about, that it is necessary for us just to indicate that he or she is the person commemorated.

Sometimes reptiles are named in the vernacular after the finder, the person who wrote the description, or some other person of the latter's choice. When more than one person has thought a species new, the reptiles may get more than one set of names, so it can warrant an entry in several places.

There are a great number of recent namings of fossil animals. As the rate at which fossil remains are discovered and described seems to be increasing exponentially and the disagreement among the paleontologists epidemic, we decided that we would ignore any prehistoric extinctions—in simplistic terms, those occurring before Columbus discovered America.

In the last 100 years many cities and countries have changed names. We have normally used the name by which the subject would have known it, putting in parentheses the name by which it is now known—for example, Salisbury (Harare), Rhodesia (Zimbabwe).

What's in a Name?

Tracking down the provenance of eponymous reptile names, and finding out about the individuals responsible for them, proved to be fraught with difficulties. The final count is 4,130 animals plus 43 genera and including a few where the same animal has been named after two people. The names honor 2,330 individual people, but there are also 99 that sound like people's names but in fact are not, plus 15 indigenous peoples, 5 fictional characters, 2 biblical references, and 34 references to mythology. Additionally, there are entries for 15 names of people whom we have been unable to identify.

Describers and Namers

New species are first brought to the notice of the scientific community in a formal, published description of a type specimen—essentially a dead example of the species—that will eventually be lodged in a scientific collection. The person who describes the species will give it its scientific name, usually in Latin but sometimes in Latinized ancient Greek. Sometimes the "new" animal is later reclassified, and then the scientific name may be changed. This frequently applies to generic names (the first part of a binomial name), but specific scientific names (the second part of a binomial), once proposed, usually cannot be amended or replaced; there are precise and complicated rules governing any such changes. Conventionally, a changed genus is indicated by putting parentheses around the describer's name. An example from *Whose Bird?* illustrates this. The Grey Heron was named *Ardea cinerea* by Linnaeus in 1758, and since that name remains recognized to this day, the bird is officially named *Ardea cinerea* Linnaeus, 1758. Linnaeus also described the Great Bittern as *Ardea stellaris*. However, Bitterns have since been awarded their own genus, and we now call the bird *Botaurus stellaris* (Linnaeus, 1758), the parentheses showing that the name was not the describer's original choice. The scientific names used in this book are largely those used in the Reptile Database, at the time of writing maintained on line by J. Craig Venter Institute. We may have missed a few recently published taxonomic changes, but we have put the name of the original describer after every entry; hence the normal convention regarding parentheses does not apply here. Because alterations to taxonomy have been so radical, and so swiftly changing, we decided we would never get the parentheses around the right entries and so have omitted them entirely.

Although we have used current scientific names as far as possible, these are not always as universal as the casual observer might suppose. There is no "world authority" on such matters.

There are no agreed-upon conventions for English names, and indeed the choice of vernacular names is often controversial. Often the person who coined the scientific name will also have given it a vernacular name, which may not be an English name if the describer was not an English speaker. On the other hand, vernacular names have often been added afterward, frequently by people other than the describers. In this book, therefore, when we refer to an animal having been *named* by someone, we mean that that person gave it the English name in question. We refer to someone as a *describer* when the person was responsible for the original description of the species and hence for its scientific name. As we have said above, it is the describer's name that is given after the scientific name in the biographies.

Currently, there are standardized American English common names lists for reptiles in Europe, North America, and Mexico. At times these names differ from the common names used in the past by describers and in subsequent herpetological literature. We suggest that modern researchers consult the following references when using common names of reptiles, where appropriate:

Crother, B. I. (compiler). 2008. *Scientific and Standard English Names of Amphibians and Reptiles of North America North of Mexico, with Comments Regarding Confidence in Our Understanding*. 6th edition. SSAR Herpetological Circular No. 37, 84 pp.

Liner, E. A., and G. Casas-Andreu. 2008. *Standard Spanish, English, and Scientific Names of the Amphibians and Reptiles of Mexico*. 2nd edition. SSAR Herpetological Circular No. 38, 162 pp.

Stumpel-Rienks, S. E. 1992. *Nomina Herpetofaunae Europapeae. Handbuch der Reptilien und Amphibien Europas*. AULA-Verlag, Wiesbaden.

Animals Named after More Than One Person

Throughout the text you may come across several different names for the same species. In some cases these names are honorifics; for example, Brygoo's Chameleon is the same species as Peyrieras' Chameleon. This peculiarity has sometimes come about through simple mistakes or misunderstandings—such as believing juveniles or females to be a different species from the adult male. In some cases the same animal was found at about the same time in two different places, and only later has it emerged that this is the same animal named twice. Some of these duplications persist even today, with the same reptile being called something different in different places or by different people.

Male or Female

In some cases we know that an animal is named after a man, even although its scientific name is in the feminine. This seems to occur only when a name ends in the letter *a*. Presumably the reason for this is that many singular Latin nouns ending in *a* are feminine: for example, *mensa* means "table" (nothing very feminine about that), and the possessive/genitive case is *mensae*, not *mensai*. There are a number of Latin masculine nouns (e.g. *agricola*, meaning "farmer") that are declined as though they were feminine, the convention being that the feminine form is adopted in such cases. For example, Anchieta's Cobra is named after José Alberto de Oliveira Anchieta, but the scientific name is *Naja anchietae*. This convention has been falling into disuse in recent years. It is quite striking how many modern namings ignore it.

Red Herrings

Further confusion arises from a number of animals that appear to be named after people but upon closer examination turn out to be named after a place that was itself named after a person. We have included these with an appropriate note, as other sources of reference will not necessarily help the enquirer.

Dubious Names

There are a number of vernacular and scientific names that are dubious and have, because of their origin, proved to be impossible to identify or to amplify, or are regarded by many authorities as just plain wrong. These are all omitted from this book, but we have fuller particulars of them for inclusion in any subsequent edition, should any ever become widely recognized by the scientific community.

Vernacular name. When Frank and Ramus published *A Complete Guide to the Scientific and Common Names of Reptiles and Amphibians of the World* (1995), a number of vernacular names suddenly appeared for the first time. They include a number of "personalized" names, such as Stacy's Bachia, a name having absolutely nothing to do with *Bachia trisanale* Cope, 1868, which before 1995 appears not to have had any recognized vernacular name. We suspect that Frank and Ramus named a number of reptiles after their friends and relatives and, perhaps, even after themselves. Ramus' first name is Erica, so perhaps Frank coined Erica's Worm Snake in her honor. Unfortunately Norman Frank is dead, so we can't ask him. In addition to Stacy and Erica, their list includes reptiles named after such unknown personalities as Gail, Gary, Grace, Karen, Norman, and William. These choices of name remain mysterious.

Scientific name. In 1984 and 1985, Richard Wells and Ross Wellington described and named well over 300 species and genera, and 141 of them are named after people. It is, to say the least, an eclectic bunch. Sons, daughters, brothers, girlfriends, and an ex-wife; Prime Ministers Bob Hawke of Australia and Sir Michael Somare of Papua New Guinea; actors Chips Rafferty and Burt Lancaster; folk hero (or villain, depending on your point of view) Ned Kelly; a relative of Ronnie Biggs, the Great Train Robber; the admired Australian poet and writer Henry Lawson; and a genuine world figure, Neil Armstrong, the first man to step onto the moon. Additionally, species are named after colleagues, other herpetologists, retired crocodile hunters, computer experts, and a policeman who gave up the law in favor of selling drinks. At the time of writing their work was still not universally accepted, as opinions had been expressed that there were insufficient grounds for taxonomic changes proposed by them, and they were widely criticized for not publishing "proper" species descriptions. Pending more general acceptance, these names are omitted.

For More Information

Although our work is not intended as a historical document, it may stimulate interest in some of the personalities mentioned. For readers interested in additional information on herpetological history, we suggest consulting the following publications:

Adler, K. 1989. "Herpetologists of the Past." Pp. 5–141 in K. Adler (ed.), *Contributions to the History of Herpetology*. SSAR, Oxford, Ohio.

Adler, K. 2007. "Herpetologists of the Past." Pp. 7–273 in K. Adler (ed.), *Contributions to the History of Herpetology*, vol. 2. SSAR, St. Louis, Missouri.

Bauer, A. M. (ed.). 2003. "Studies in the History of Herpetological Exploration." *Bonner zoologische Beiträge* 52:181–335 (special issue).

Bibliotheca Herpetologica. 1999–2010. Vols. 1–8. International Society for the History and Bibliography of Herpetology.

Rieck, W., G. Hallmann, and W. Bischoff. 2001. *Die Geschichte der Herpetologie und Terrarienkunde im deutschsprachigen Raum. Mertensiella* No. 12, 759 pp.

Rossolimo, O. L., and E. A. Dunayev. 2003. *Herpetologists from Moscow*. Nikolsky Herpetological Society, Moscow. 579 pp. (in Russian, but with short English summaries at the end).

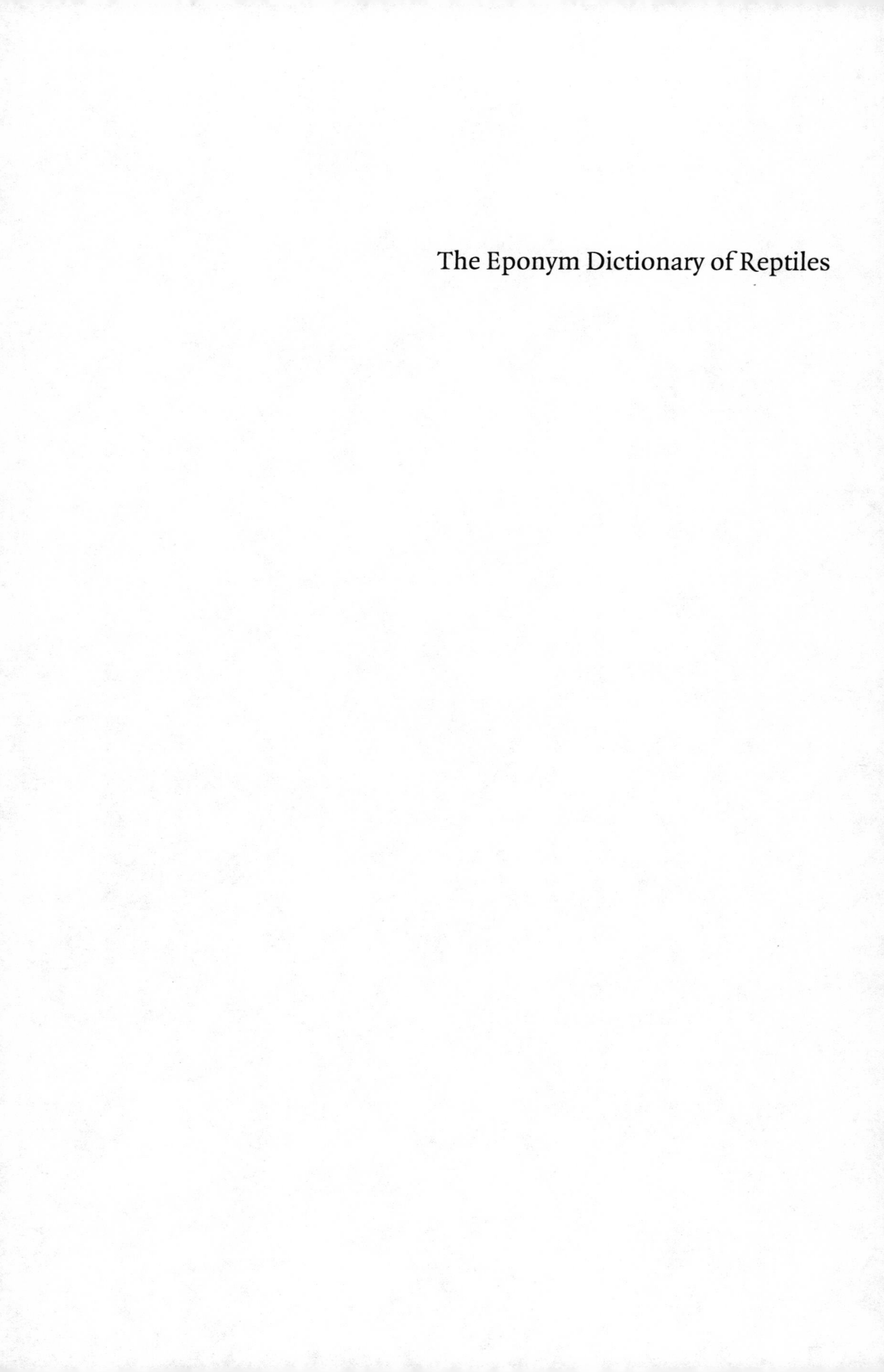

The Eponym Dictionary of Reptiles

A

Aaron

Aaron's Bent-toed Gecko *Cyrtodactylus aaroni* **Günther** and Rösler, 2003

See **Bauer, A. M.**

Abbott, C. G.

Abbott's Banded Gecko *Coleonyx variegatus abbotti* **Klauber,** 1945 [Alt. San Diego Banded Gecko]

Clinton Gilbert Abbott (1881–1946) was born in Liverpool, England, but raised and educated in the USA. He earned his first degree at Columbia University (1903), followed by postgraduate work at Cornell (1914–1915). He was Vice President of the Linnaean Society of New York (1911–1914) and Confidential Secretary of the New York State Conservation Commission (1918–1921). He was put in charge of public education at the San Diego Natural History Museum (1921), becoming the museum's Director (1922–1925), and was President of the San Diego Natural History Society (1923–1925).

Abbott, W. L.

Abbott's Day Gecko *Phelsuma abbotti* **Stejneger,** 1893

Abbott's Anglehead Lizard *Gonocephalus (doriae) abbotti* **Cochran,** 1922 [Alt. Cochran's Forest Dragon]

Dr. William Louis Abbott (1860–1936) was a naturalist and collector. He qualified as a physician at the University of Pennsylvania and as a surgeon at Guy's Hospital, London, but instead of practicing medicine decided to use his wealth for scientific exploration. As a student (1880) he collected in Iowa, North Dakota, Cuba, and Hispaniola (1883). He traveled in East Africa (1887–1889). He studied the wildlife of the Indo-Malayan region, on his ship *Terrapin*, making large collections of mammals for the Smithsonian, and exploring in and around the China Sea (1897–1907). He provided much of the Kenya material in the Smithsonian and wrote *Ethnological Collections in the United States National Museum from Kilima-Njaro, East Africa* (1890–1891). He retired to Maryland but continued his lifelong study of natural history until his death. Several birds and two mammals are named after him.

Abd el Kuri

Abd el Kuri Rock Gecko *Pristurus abdelkuri* **Arnold,** 1986

Abd el Kuri (or Ab-El-Kuri) is a small island near Socotra, off the Horn of Africa.

Abendroth

Abendroth's Bachia *Bachia trisanale abendrothii* **Peters,** 1871

Dr. Ernst Robert Abendroth (1810–1871) was a German zoologist and arachnologist. He published *Über Morphologie und Verwandtschaftsverhältnisse der Arachniden* (1868).

Abingdon

Abingdon Giant Tortoise *Chelonoidis (nigra) abingdonii* **Günther,** 1877 [Formerly *Geochelone elephantopus abingdonii*]

This race of giant tortoise was named after Abingdon (Pinta) Island in the Galapagos. The island was named after James Bertie, First Earl of Abingdon (1653–1699), in 1684. "Lonesome George," probably the last survivor of this taxon, is still alive at the time of writing (December 2010).

Abra

Rainforest Banded Gecko *Cyrtodactylus abrae* **Wells,** 2002

Lyn Abra is a naturalist who worked at the Australian Reptile Park, New South Wales, where she was noted for "milking" funnel-web spiders to use their venom for antivenin production.

Abramov

Reed Snake sp. *Calamaria abramovi* **Orlov,** 2009

Dr. Alexey V. Abramov is a zoologist and herpetologist at the Zoological Institute, Russian Academy of Sciences, St. Petersburg. He specializes in the study of Vietnamese herpetofauna and collected the snake paratype.

Abubaker

False Smooth Snake sp. *Macroprotodon abubakeri* Wade, 2001

M. Aboubakeur Sid-Ahmed is a naturalist from Tlemcen in northwestern Algeria.

Achilles

Achilles' Anole *Anolis achilles* **Taylor,** 1956

Achilles was a Greek hero of the Trojan War and a leading character in Homer's *Iliad*. His mother, Thetis, bathed him in the infernal River Styx to make him immortal but forgot about the heel by which she held him in the current. This became his one weakness, or, in modern parlance, his Achilles' heel. Taylor describes the anole as having "widened heel plates" and gives the etymology, "Referring to the Greek hero, Achilles who had a specialized heel."

Adanson

Adanson's Mud Turtle *Pelusios adansonii* Schweigger, 1812

Michel Adanson (1727–1806) was a French botanist. He was a bookkeeper for Compagnie des Indes in West Africa, mostly Senegal (1748–1754). He collected specimens of all kinds and after returning to France wrote two books, including *Histoire naturelle du Senegal* (1757). He lost his position and income because of the 1789 Revolution, being supported by his servants. He died in penury, leaving a last wish that a wreath representing the 58 plant families he had named be placed on his grave. Other taxa including a bird and many plants are named after him.

Adelyn

Adelyn's (Black-headed) Python *Aspidites melanocephalus adelynensis* Hoser, 2000

Adelyn Hoser (b. 1999) is the describer's elder daughter.

Aderca

Lualaba Worm Lizard *Monopeltis adercae* **Witte,** 1953

Bernard Max Aderca was a Belgian geologist and paleontologist who wrote the sections on those subjects in G. F. de Witte's 1953 report *Reptiles. Exploration du Parc National de l'Upemba* (Belgian Congo). He wrote about the minerals of the area (1950s and 1960s).

Adler

Adler's Anole *Anolis adleri* **H. M. Smith,** 1972

Adler's Spiny Lizard *Sceloporus adleri* H. M. Smith and Savitzky, 1974

Adler's Keelback *Rhabdophis adleri* **Zhao,** 1997

Nicobar Bent-toed Gecko *Cyrtodactylus adleri* **Das,** 1997

Adler's Worm Snake *Leptotyphlops adleri* Hahn and **Wallach,** 1998

Dr. Kraig Kerr Adler (b. 1940) is Professor of Biology at Cornell. He was co-founder and Chairman of the Ohio Herpetological Society (later the Society for the Study of Amphibians and Reptiles). He has written extensively on herpetology and was co-editor of *The New Encyclopedia of Reptiles and Amphibians* (2002). For the last two decades his major research interest has been the sensory basis of long-distance orientation and navigation in reptiles and amphibians.

Adolf Frideric

Forest Lizard genus *Adolfus* **Sternfeld,** 1912

Ituri Chameleon *Kinyongia adolfifriderici* Sternfeld, 1912

Duke Adolf Friedrich Von Mecklenburg (1873–1969) was an explorer and colonial administrator in Africa, and the first President of the German Olympic Committee (1949–1952). He conducted scientific research in the African Rift Valley and crossed Africa from east to west (1907–1908). He led another expedition to Lake Chad and the upper reaches of the Congo River and the Nile in Sudan (1910–1911) and wrote *Vom Kongo zum Niger und Nil*. He was the last Governor of the German colony of Togo (1912–1914). After WW1 he became Vice President of the German Colonial Society for Southwest Africa.

Adonis

Ingram's Litter Skink *Lampropholis adonis* **Ingram,** 1991

In Greek mythology Adonis was a god of vegetation and rebirth, as well as being an ideal of masculine beauty. No reason was specified by Ingram for the choice of name.

Aesculapius

Aesculapian False Coral Snake *Erythrolamprus aesculapii* **Linnaeus,** 1766

Aesculapian Snake *Zamenis longissimus* **Laurenti,** 1768 [Syn. *Elaphe longissima*]

Aesculapius (or Asclepius) was the god of medicine in Greek and Roman mythology. The staff of Asclepius, a rod with a snake wrapped around it, is a symbol of medicine.

Agassiz, A. E.

Agassiz's Anole *Anolis agassizi* **Stejneger,** 1900

Alexander Emanuel Agassiz (1835–1910) was born in Switzerland, moved to the USA with his father, Louis Agassiz (q.v.) (1849), and made a fortune from copper mining. He graduated from Harvard (1855) and took a second degree in engineering and chemistry (1857). In 1859 he joined the U.S. Coastal Survey, later becoming a specialist marine ichthyologist. Until 1866 he was an assistant at the Museum of Comparative Zoology, Harvard, founded by his father. He invested in a copper-mining venture in Michigan, becoming the company's Treasurer. He made the company prosperous, acquired other companies, expanded the conglomerate, and served as its President (1871–1910). He returned to Harvard (early 1870s) to pursue his interests in natural history, giving $500,000 to the Museum of Comparative Zoology and becoming its Curator (1874–1885). He visited copper mines in Peru and Chile and surveyed Lake Titicaca (1875). He helped Wyville Thomson in examining and classifying specimens on the *Challenger* expedition and took part in three dredging expeditions (1877–1880). He published many works on marine zoology, such as *Seaside Studies in Natural History* (1865), co-written with his stepmother, Elizabeth Cary Agassiz.

Agassiz, J. L. R.

Agassiz's Tortoise *Gopherus agassizi* Cooper, 1863 [Alt. Desert Gopher Tortoise]

Burrowing Night Snake *Pseudablabes agassizii* **Jan,** 1863

Agassiz's Green Sea Turtle *Chelonia mydas agassizii* **Bocourt,** 1868 [Alt. Eastern Pacific Green Turtle, Black (Sea) Turtle]

Jean Louis Rudolphe Agassiz (1807–1873) was a Swiss-

American geologist, glaciologist, and zoologist whose speciality was ichthyology. He studied at Zurich, Heidelberg, and Munich, where he qualified as a physician (1830), and in Paris under Cuvier (1831). While still a student he was tasked with working on the Spix and Martius Brazilian freshwater fish collection. He became Professor of Natural History at Lyceum de Neuchâtel (1832). He was the first person to scientifically propose that the Earth had been subject to an ice age and to study ice as a subject, having lived in a special hut built on a glacier in the Alps (1837). He went to the USA in 1846 to study American natural history and geology and to deliver a course of zoology lectures. He visited again in 1848, remaining there for the rest of his life and becoming Professor of Zoology and Geology at Harvard, where he founded and directed the Museum of Comparative Zoology (1859–1873). Latterly he took up studies of Brazilian fishes again and led the Thayer expedition there (1865). He established the Marine Biological Laboratory (1873). Cooper's etymology for the tortoise says that he is "naming this fine tortoise after the celebrated zoologist, whose work on the development, anatomy and classification of American Turtles . . . leaves nothing to be desired in these particulars."

Aguero

Aguero's Anole *Anolis agueroi* Diaz, **Navarro,** and **Garrido,** 1998 [Alt. Cabo Cruz Bearded Anole; Syn. *Chamaeleolis agueroi*]

Dr. Jose de la Cruz Agüero is the Curator of the ichthyological collection at CICIMAR (Centro Interdisciplinario de Ciencias Marinas), La Paz, Baja California Sur, Mexico.

Ahl

Ahl's Anole *Anolis ahli* **Barbour,** 1925

Ahl's Emo Skink *Emoia ahli* **Vogt,** 1932

Dr. Christoph Gustav Ernst Ahl (1898–1945) was an ichthyologist, herpetologist, and aquarist. He served in the artillery during WW1 (1916). He studied natural science at Humboldt-Universität, Berlin (1919–1921), where he was awarded his doctorate, his thesis being on the systematics of a fish family. He was in the Department of Ichthyology and Herpetology, Zoological Museum (1921–1941), becoming Curator of Herpetology (1923). He was editor-in-chief of the magazine *Das Aquarium* (1927–1934). He joined the Nazi party in the 1930s in order to keep his job but was expelled for indiscipline (1939). He was also sacked by the museum (1941), probably because his scientific work was "superficial and careless and his knowledge of the literature poor," rather than because he had been recalled to the Wehrmacht (1939). He fought in Poland and North Africa and was reported as missing in action in Herzegovina (1945). He wrote 170 papers on fishes and amphibians, but most of the names he coined are no longer considered valid.

Ahmad

Fringe-fingered Lizard sp. *Acanthodactylus ahmaddisii* **Werner,** 2004

Professor Ahmad M. Disi is Chair of Biology, University of Jordan, Amman. The etymology honors him "in recognition of his pioneering, continual and prolific contribution to the herpetology of the Hashemite Kingdom of Jordan and of the Levant in general." He has written many scientific papers and longer works including *Amphibians and Reptiles of the Hashemite Kingdom of Jordan: An Atlas and Field Guide* (2001).

Ahsan

Ahsan's Blind Snake *Typhlops ahsanai* **Khan,** 1999

Professor Dr. Ahsanul-Islam (1927–1974) was Principal of the Zoology Department, Government College, Lahore, Pakistan. He taught the describer and initiated his studies of Pakistani herpetology.

Alan

Alan's Skink *Cyclodina alani* Robb, 1970

The original description contains no etymology—a rare omission in a modern scientific description and suggestive of the omission being deliberate. Perhaps this is a private tribute to some friend of Robb's, in which case we speculate that it may be Alan J. Saunders, an official at the Department of Conservation, Wellington, New Zealand. Saunders has been involved in a number of projects involving the Mercury Islands, where Joan Robb also spent time.

Alana

Alana's Menetia *Menetia alanae* **Rankin,** 1979 [Alt. Top End Dwarf Skink]

Ms. Alana Young was working at the Department of Herpetology, Australian Museum, when this skink was described.

Alayón

Alayón's Anole *Anolis alayoni* Estrada and Hedges, 1995

Dr. Giraldo Alayón Garcia is a zoologist who has been Curator of Arachnids at Museo Nacional de Historia Natural, Havana, since 1988. He graduated in biological sciences from the University of Havana (1973), having originally enrolled to study physics (1966). He is co-author of a number of scientific papers and a longer work, *Cuba Natural*. His interests include ornithology and entomology. He is best known outside Cuba as leading an expedition (1986) that claimed to have rediscovered the

Ivory-billed Woodpecker. Other taxa such as arachnid species are named after him.

Albert

Albert's Burrowing Skink *Sepsina alberti* **Hewitt,** 1929

We believe "Albert" may refer to George Albert Boulenger. However, we cannot be sure, as Hewitt first introduces this scientific name in a short footnote, giving no explanation. See **Boulenger.**

Alberti

D'Alberti's Python *Leiopython albertisii* **Peters** and **Doria,** 1878 [Alt. White-lipped Python]

See **D'Albertis.**

Albert Schwartz

Schwartz's Wall Gecko *Tarentola albertschwartzi* Sprackland and Swinney, 1998

See **Schwartz.**

Albuquerque

Albuquerque Ground Snake *Atractus albuquerquei* **Da Cunha** and Nascimento, 1983

Professor Dr. Dalcy de Oliveira Albuquerque (1902–1982) was an entomologist. He was Director, Museu Paraense Emílio Goeldi (1962–1968) and Museu Nacional, Rio de Janeiro (1972–1976). He started the Brazilian national collection of Diptera in Rio de Janeiro (1944). He went on many expeditions within Brazil and other South American countries.

Alcala

Gary's Mountain Keelback *Opisthotropis alcalai* **Brown** and **Leviton,** 1961

Alcala's Wolf Snake *Lycodon alcalai* **Ota** and Ross, 1993

Dr. Angel Chua Alcala (b. 1929) is a biologist and herpetologist who has studied the reptiles and amphibians of the Philippines for 50 years. His major area of interest is marine biology, having founded the Silliman Marine Research Laboratory (1973). His bachelor's degree was awarded by Silliman University, Philippines (1951). His master's (1960) and doctorate (1966) were awarded by Stanford. He is University Research Professor at Silliman, having previously been Professor of Biology there.

Alcock

Alcock's Flying Dragon *Draco norvillii* Alcock, 1895

Alcock's Toad-headed Agama *Phrynocephalus euptilopus* Alcock and Finn, 1897

Major Dr. Alfred William Alcock (1859–1933) joined the Indian Medical Service in 1885, having qualified as a physician at the University of Aberdeen. He was appointed Surgeon-Naturalist to the Indian Marine Service. As a naturalist he was mainly interested in fishes. He was based in Calcutta, dividing his time between the Indian Museum and the survey ship *Investigator.* He published a number of papers on the ichthyology of the Bay of Bengal, after which he returned to England. He wrote *A Naturalist in the Indian Seas* (1902).

Aldrovandi

Aldrovandi's Skink *Eumeces schneideri aldrovandii* **Duméril** and **Bibron,** 1839

Dr. Ulisse Aldrovandi (1522–1605) was an Italian naturalist. He studied law and medicine, graduating as a physician at Bologna (1553). He was accused of heresy before the Inquisition (1549) but was able to clear himself. He was appointed Professor of Philosophy and Lecturer on Botany at Università di Bologna (1551), becoming Professor of Natural History (1561). He became the first Director of Bologna's botanical garden (1568). He was instrumental in the founding of Bologna's public museum. He willed his huge collection of natural history specimens to the Senate of Bologna, but these were gradually distributed among a variety of institutions.

Aleman

Aleman's Snail-eater *Dipsas perijanensis* Aleman, 1953

G. César Alemán was a herpetologist and a curator at Museo de Historia Natural La Salle, Venezuela. His publications on herpetology include "Contribución al estudio de los reptiles y batracios de la Sierra de Perijá" (1953).

Alexandre

Gymnophthalmid lizard genus *Alexandresaurus* **Rodrigues** et al., 1997

Alexandre Rodrigues Ferreira (1756–1815) was born in Bahia, then capital city of the Portuguese colony of Brazil. He went to Portugal and studied law, mathematics, and natural philosophy at Universidade de Coimbra. He was awarded a bachelor's degree (1778) and a doctorate (1779) and worked at the Royal Museum, Lisbon. He returned to Brazil (1783) as head of the "Philosophical Trip for the Captainships of Gran-Pará, Rio Negro, Mato Grosso and Cuiabá." He arrived at Belém, Pará (1783), and started a trip that lasted years. It is estimated that he covered over 35,000 kilometers (22,000 miles), traveling on foot, on horseback, and, most important, on the rivers of the Amazon system. He returned to Lisbon (1793) and worked as an administrator, which included being in charge of the Royal Natural History Collection and the Lisbon Botanical Gardens. Most of his work was taken advantage of by other people, including Étienne Geoffroy

Saint-Hilaire, who was in Lisbon when Napoleon conquered Portugal. It appears that Saint-Hilaire "requested" (almost certainly meaning "looted") part of Ferreira's collection.

Alfaro

Alfaro's Anole *Anolis alfaroi* **Garrido** and Hedges, 1992 [Alt. Small-fanned Bush Anole]

Emilio Alfaro is a zoologist at Museo Nacional de Historia Natural, Havana, and collected the holotype of this lizard. He took part in Penn State / Cuba herpetological expeditions (1989, 1990, and 1994).

Alfred

Forest Skink sp. *Sphenomorphus alfredi* **Boulenger,** 1898

See **Everett.**

Alfred, E.

Alfred's Limbless Skink *Dibamus alfredi* **Taylor,** 1962 [Alt. Taylor's Limbless Skink]

Eric R. Alfred was an ichthyologist and Curator of the Raffles Museum, Singapore (1957–1967). He became Director (1967–1972) following a year's attachment to the British Museum and Rijksmuseum van Natuurlijke Histoire, Leiden. He collected fish species from the Malay Peninsula, and a number of fishes are named after him. He went with the museum's zoological collection when it was moved to the Maritime Museum, Sentosa.

Alfred Schmidt

Campeche Spiny-tailed Iguana *Ctenosaura alfredschmidti* Köhler, 1995

Schmidt's Mastigure *Uromastyx alfredschmidti* Wilms and **Böhme,** 2001 [Alt. Ebony Mastigure]

Blind Snake sp. *Leptotyphlops alfredschmidti* Lehr, **Wallach,** Köhler, and Aguilar, 2002

See **Schmidt, A.**

Alida

Dwarf/Reed Snake sp. *Calamaria alidae* **Boulenger,** 1920

Alida Brooks. The etymology states, "Named in memory of the late Mrs. Brooks, who helped her husband in collecting in Sumatra." See **Brooks, C. J.**

Allan

Allan's Lerista *Lerista allanae* **Longman,** 1937

Mrs. P. C. Allan, according to the etymology, "has presented many interesting specimens to the Queensland Museum." Unfortunately no first names are given in the original description.

Allen

Allen's Cay Iguana *Cyclura inornata* **Barbour** and **Noble,** 1916 [Alt. Allen Cays Iguana; Syn. *C. cychlura inornata*]

Allen's Cay is an island in the Bahamas.

Allen, G. M.

Alpine-meadow Lizard *Adolfus alleni* **Barbour,** 1914

La Guaira Bachia *Bachia heteropa alleni* Barbour, 1914

Dr. Glover Morrill Allen (1879–1942) was a collector, curator, editor, librarian, mammalogist, ornithologist, scientist, taxonomist, teacher, and writer. He was Librarian at the Boston Society of Natural History (1901–1927). He was employed to oversee the mammal collection at the Museum of Comparative Zoology, Harvard (1907), becoming Curator of Mammals (1925–1938) and then Professor of Zoology (1938–1942). He was President of the American Society of Mammalogists (1927–1929). He made many collecting trips (1903–1931), variously to Africa, including the Harvard expedition to Liberia (1926), Australia, the Bahamas, Brazil, and the West Indies. He wrote a great many scientific papers and articles and a number of books. His *Bats* (1939) is still considered a classic. A bat species is named after him.

Allen, J. A.

Striped Swamp Snake *Regina alleni* **Garman,** 1874

Joel Asaph Allen (1838–1921) was a zoologist chiefly interested in mammals and birds. He studied under Louis Agassiz and accompanied him to Brazil (1865). He made a number of field trips in North America, and led an expedition for the Northern Pacific Railroad (1873). He was an Assistant in Ornithology at the Museum of Comparative Zoology, Harvard (1870), and was Curator of the Department of Mammals and Birds, American Museum of Natural History, New York (1885–1921). In addition to naming many species, he made important studies on geographic variation relative to climate. Allen's recognition of "variation within populations and intergradation across geographic gradients" helped to overturn the typological species concept current in the mid-1800s, setting out the principle that intergrading populations should be treated as subspecies instead of separate species. This idea led to the widespread adoption of trinomials by American zoologists, a practice that Allen helped to spread through his editorship of the *Auk* and through the American Ornithologists' Union code of nomenclature. Two birds and 11 mammals are named after him.

Allen, M. J.

Allen's Coral Snake *Micrurus alleni* **Schmidt,** 1936

Morrow J. Allen (1909–1988) collected the holotype of

this snake (1935). He published *A Survey of the Amphibians and Reptiles of Harrison County, Mississippi* (1932).

Allen, N. T.

Allen's Ctenotus *Ctenotus alleni* **Storr,** 1974

Nicholas T. Allen collected the holotype of this skink.

Allison

Allison's Anole *Anolis allisoni* **Barbour,** 1928

See **Armour.**

Allison, A.

Allison's Emo Skink *Emoia guttata* **Brown** and Allison, 1982

Dr. Allen Allison (b. 1950) is a herpetologist who joined the staff of the Bishop Museum, Hawaii (1973), and is currently Assistant Director of Research. He gained his doctorate from the University of California (1979).

Alluaud

Colubrid snake genus *Alluaudina* **Mocquard,** 1894

Northern Flat-tail Gecko *Uroplatus alluaudi* Mocquard, 1894

Southern Leafnose Snake *Langaha alluaudi* Mocquard, 1901

Brygoo's Burrowing Skink *Amphiglossus alluaudi* **Brygoo,** 1981

Charles Alluaud (1861–1949) was an entomologist, botanist, and naturalist who came from a wealthy family. His father was President of the Royal Porcelain Factory, Limoges. He traveled extensively, his trips including scientific expeditions to the Seychelles and Madagascar (1892–1893). A botanical genus is named after him.

Alma

Neotropical House Snake sp. *Thamnodynastes almae* Franco and Ferreira, 2003

Silvia Alma Renata Wilma Lemos Romano-Hoge is a herpetologist and former research scientist at the Herpetology Laboratory, Instituto Butantan, São Paulo, Brazil. She is the widow of Dr. Alphonse Richard Hoge (q.v.), who was Director of Instituto Butantan. She has published widely, including, jointly with her husband, *Poisonous Snakes of the World. Part 1. Check List of the Pit Vipers Viperoidea, Viperidae, Crotalinae* (1978).

Aloysius

Uganda Five-toed Skink *Leptosiaphos aloysiisabaudiae* **Peracca,** 1907

Prince Luigi Amedeo of Savoy, Duke of Abruzzi (1873–1933), was an explorer. He led a polar expedition (1900) and an expedition to explore and climb the Ruwenzori Range in Uganda (1906). They made the first detailed maps of the area and climbed a number of the peaks for the first time, including Mount Stanley. He returned to climb Mount Kenya (1909). He was Commander-in-Chief of the Italian navy (1915–1917). The name *aloysiisabaudiae* is just a fancy, Latinized way of saying Luigi of Sabaudia, which was one of the territories of the ancestors of the House of Savoy. Luigi himself was fond of eponyms, as he named a peak, Luigi de Savoia, after himself. A species of bat is named after him.

Alphonse

Alphonse's Ground Snake *Atractus alphonsehogei* **Cunha** and Nascimento, 1983

See **Hoge.**

Altamirano

Tepalcatepec Skink *Mesoscincus altamirani* **Dugès,** 1891

Federico Altamirano collected the holotype in Mexico.

Alvarez, A.

Argentine Rainbow Boa *Epicrates alvarezi* Abalos, Baez, and Nader, 1964 [Syn. *E. cenchria alvarezi*]

Dr. Antenor Alvarez (1864–1948) was an Argentine physician, politician, and herpetologist. Three years after obtaining his doctorate he became Chairman of the Council of Hygiene in Santiago del Estero Province. He was elected Senator (1898–1902), Deputy (1904), National Senator (1909), and Governor of the province (1912–1916). He was responsible for sanitary and water supply improvements to combat malaria. He wrote *Flora y fauna de la province de Santiago del Estero* (1919).

Alvarez, M.

Alvarez del Toro's Night Lizard *Lepidophyma alvarezi* **H. M. Smith,** 1973

Alvarez's Two-legged Worm Lizard *Bipes alvarezi* H. M. Smith and R. B. Smith 1977 [Junior syn. of *B. canaliculatus* Bonnaterre, 1789, according to some]

Chiapan Stripeless Snake *Coniophanes alvarezi* **Campbell,** 1989

Anole sp. *Anolis alvarezdeltoroi* Nieto Montes de Oca, 1996

Miguel Alvarez del Toro (1917–1996) was a naturalist working mostly in Chiapas State, Mexico. In his early days he worked as a taxidermist at Museo de la Fauna y Flora, Mexico City. He collected birds for the Academy of Natural Sciences, Philadelphia (1938–1939), operating around Mexico City, in Morelos, and Colonia Sarabia, Oaxaca, in the rainforests. He became a taxidermist at Museo de Historia Natural in Tuxtla Gutierrez, Chiapas (1942), collected vertebrates near the city, and surveyed more remote areas. Despite having had no formal education, he was a great teacher, scientist, and conservationist. He taught at Colegio de Ciencias y Artes de Chiapas and at Universidad Nacional Autonoma de Mexico. Among his many honors was an honorary doctorate awarded by Universidad de Chiapas. He was the principal force behind the creation of six protected areas in Chiapas. He published many papers, including "Los reptiles de Chiapas" (1960).

Alvaro

Alvaro's Anole *Pristidactylus alvaroi* **Donoso-Barros,** 1975

Alvaro Donoso-Barros is the describer's son. The etymology says, "La presente designación se hace por mi hijo Alvaro."

Alwis

Alwis' Day Gecko *Cnemaspis alwisi* Wickramasinghe and **Munindradasa,** 2007

Lyn De Alwis (1930–2006) was a zoologist who graduated from the University of Colombo, Sri Lanka, and in 1955 joined the staff of the National Zoological Gardens at Dehiwela. He became the Director of the Department of Wildlife Conservation (1965–1969 and 1978–1983). His reputation as a zookeeper was such that Prime Minister Bandaranaike of Sri Lanka was persuaded by Singapore's premier, Lee Kwan Yew, to second Alwis to Singapore (1970–1972) to oversee the construction of their zoo. He founded the Young Zoologists Association in Sri Lanka. The etymology honors him "for his initiative in igniting a research culture in the country leading to conservation of wildlife resources."

Amaral

Amaral's Burrowing Snake *Apostolepis rondoni* Amaral, 1925

Amaral's Gecko *Gymnodactylus amarali* **Barbour,** 1925

Amaral's Ground Snake *Liophis amarali* **Wettstein,** 1930

Amaral's Boa *Boa constrictor amarali* Stull, 1932

Amaral's Colobosaura *Colobosaura mentalis* Amaral, 1933

Amaral's Brazilian Gecko *Hemidactylus brasiliana* Amaral, 1935

Amaral's Tropical Racer *Mastigodryas amarali* **Stuart,** 1938

Amaral's Whipsnake *Masticophis stigodryas* Amaral, 1943

Amaral's Blind Snake *Leptotyphlops koppesi* Amaral, 1955

Dr. Afranio do Amaral (1894–1982) was a physician, zoologist, and herpetologist and was Director of Instituto Butantan (1919–1921 and 1928–1938). *Time* magazine described him on his arrival in Manhattan (1929) as "the soft-voiced suave herpetologist." He originally trained as a physician at the Medical School of Bahia. His Ph.D. was awarded by Harvard, where he also taught. He was Director of the Antivenin Institute of America (1927) and conducted the first major study of the incidence of snakebite in Texas. He published, with Barbour, "Notes on Some Central American Snakes" (1924).

Amicorum

Skink sp. *Lerista amicorum* **L. A. Smith** and Adams, 2007

Two friends and rivals arrived simultaneously (1964) to see Storr at the Western Australian Museum. Both were convinced they had a new species. Storr dismissed the specimens as *Lerista muelleri* but said that should it ever emerge as a new species, then it would be called *amicorum* after the friends. It did emerge. See the entries for them: **Butler, W. H.,** and **Douglas, A. M.**

Amiet

Cameroon Five-toed Skink *Leptosiaphos amieti* **Perret,** 1973

Professor Jean-Louis Amiet is a Swiss zoologist, herpetologist, entomologist, ecologist, and ichthyologist, formerly at Université de Yaoundé, Cameroon, from where he retired to Europe after 29 years. He wrote *Faune du Cameroun* (1987).

Amoipira

Worm Snake sp. *Typhlops amoipira* **Rodrigues** and Juncá, 2002

Named after the extinct Amoipira Indians, a Tupi tribe that lived in Bahia, Brazil.

Ampueda

Lancini's Sun Tegu *Euspondylus ampuedae* **Lancini,** 1968 [Syn. *Cercosaura ampuedae*]

Ramon Ampueda was a Venezuelan who collected the holotype with Lancini.

Amy

Centralian Rough Knob-tail (Gecko) *Nephrurus amyae* **Couper,** 1994

Amy Couper (b. 1993) is the describer's daughter.

Anan

Anan's Rock Agama *Stellio sacra* **M. A. Smith,** 1935

Professor Dr. Natalia Borisovna Ananjeva (b. 1946) is head of the Department of Herpetology and Vice Director at the Zoological Institute, Russian Academy of Sciences, St. Petersburg, being the first female zoologist member of the academy. She graduated in biology at the Leningrad State University. Just how her name got attached to this

Tibetan lizard is one of those mysteries that makes this kind of research so beguiling. Possibly it happened because of an article written by Ananjeva and others, "*Stellio sacra* (Smith 1935)—A Distinct Species of Asiatic Rock Agamid from Tibet" (1990). As a result you may come across reference to Anan et al., 1990 as the species' original describers, but that is incorrect. An amphibian is named after her. See also **Natalia.**

Anchieta

Anchieta's Desert Lizard *Meroles anchietae* **Bocage,** 1867 [Alt. Namib Sand-diver; Syn. *Aporosaura anchietae*]
Anchieta's Serpentiform Skink *Eumecia anchietae* Bocage, 1870
Anchieta's Chameleon *Chamaeleo anchietae* Bocage, 1872
Anchieta's Spade-snouted Worm Lizard *Monopeltis anchietae* Bocage, 1873
Anchieta's Cobra *Naja anchietae* Bocage, 1879
Anchieta's Dwarf Python *Python anchietae* Bocage, 1887 [Alt. Angolan Python]
Anchieta's Agama *Agama anchietae* Bocage, 1896

José Alberto de Oliveira Anchieta (1832–1897) was an independent naturalist and collector who traveled widely in Africa. He left Lisbon (1857) to join a close friend who had settled in Cabo Verde off the African coast where, although self-taught, he practiced medicine. He nearly died in a cholera outbreak, so he returned to Portugal (1859). He studied medicine in Lisbon but left for Angola before completing his studies. There he collected many natural history specimens that he donated to museums on his return to Portugal, before leaving for the last time (1866). Little is known about the next period of Anchieta's life because most of the museum specimens, as well as his letters to Bocage, disappeared in a fire (1978). We know that he is recorded at various locations (1866–1897) in Angola and Mozambique and that he died, probably from chronic malaria, when returning from a expedition to Caconda. He was responsible for identifying at least 25 new mammals, 46 birds, and as many amphibians and reptiles. Three birds, four mammals, and many other taxa are named after him.

Andaman

Andaman's Cat Snake *Boiga andamanensis* **Wall,** 1909

Named after the Andaman Islands.

Anderson, J.

Anderson's Pit-viper *Trimeresurus andersonii* **Theobald,** 1868 [Syn. *Cryptelytrops andersonii*]
Anderson's Mabuya *Mabuya novemcarinata* Anderson, 1871
Anderson's Bloodsucker *Calotes kakhienesis* Anderson, 1878
Anderson's Ground Skink *Scincella exigua* Anderson, 1878 [Junior syn. of *S. modesta* Günther, 1864]
Anderson's Mountain Keelback *Opisthotropis andersonii* **Boulenger,** 1888 [Alt. Anderson's Stream Snake]
Anderson's Rock Agama *Acanthocercus adramitanus* Anderson, 1896
Anderson's Short-fingered Gecko *Stenodactylus petrii* Anderson, 1896
Anderson's Japalure *Japalura andersoniana* **Annandale,** 1905

Dr. John Anderson (1833–1900) was a naturalist who was Professor of Natural History at Free Church College, Edinburgh. He became Curator of the Indian Museum, Calcutta (1865), and collected for the Trustees. He went on scientific expeditions to Yunnan (1867), Burma (1875–1876), and the Mergui Archipelago (1881–1882). He became Professor of Comparative Anatomy at the Calcutta Medical School (1885), then returned to London (1886). He wrote *A Contribution to the Herpetology of Arabia* (1896). Three birds and three mammals are named after him.

Anderson, S. C.

Anderson's Agama *Trapelus blanfordi* Anderson, 1966
Anderson's Racerunner *Eremias andersoni* **Darevsky** and Szczerbak, 1978

Dr. Steven Clement Anderson (b. 1936) is a herpetologist. He graduated in 1957 and was awarded his M.A. (1962) and doctorate by Stanford (1966), during which time he was employed as a Lab Technician, Instructor, Museum Assistant, and then Assistant Curator of Herpetology, California Academy of Science, until 1967. He held various teaching posts including Assistant, then Associate Professor of Environmental Science, University of the Pacific, until he became Professor of Biology and Environmental Science (1980). He became Professor Emeritus at the Department of Biological Sciences, University of the Pacific, California, on retirement (1996) and is currently a self-employed writer and consultant. His fieldwork has taken him all over the world: to Iran (1958); Australia, New Zealand and Fiji (1984); Kenya (1987); the UK (1990); and Turkey (2000). He qualified as a scuba instructor (1985). He has written many papers and articles including "The Lizards of Iran" (1999). He has also contributed to many longer works, in particular the *Encyclopaedia Iranica*.

Andersson

Andersson's Leaf-toed Gecko *Hemidactylus laticaudatus* Andersson, 1910

Dr. Lars Gabriel Andersson (1868–1951) was a zoologist

and herpetologist at Naturhistoriska Riksmuseet, Stockholm. He studied at Uppsala Universitet (1887–1909), right through to his doctorate. He was an Assistant at the museum (1894–1895 and 1897–1902). He became Professor in Stockholm, took part in the Swedish Australian expedition (1910–1913), and collected many reptiles. He co-wrote *The Swedish Zoological Expedition to British East Africa* 1911.

Andrea

Cuban Lesser Racer *Antillophis andreae* **Reinhardt** and Lütken, 1862

Captain Andrea was a Danish ship's master who collected several examples of this snake in Isla de la Juventud, Cuba, donating them to the museum at Københavns Universitet.

Andrea Z.

Andrea's Keelback *Amphiesma andreae* **Ziegler** and Quyet, 2006

Andrea Ziegler is the wife of Thomas Ziegler, who collected the holotype.

Andreansky

Andreansky's Lizard *Lacerta andreanskyi* **Werner,** 1929 [Syn. *Atlantolacerta andreanskyi*]

Baron Gábor Andreánszky (1895–1967) was a Hungarian-born Viennese botanist and zoologist who on several occasions accompanied Werner to Morocco (where he collected the holotype of this lizard). He served in WW1, which interrupted his education, and he finished his degree after a five-year break. He joined Magyar Természettudományi Múzeum, Budapest (1941), becoming Director (1943–1945). He then became Professor of Plant Morphology and Taxonomy at Budapest University, heading that department until 1953. He took part in several other field trips, including to Sicily, Corsica, Transylvania, and Poland. His greatest work was *Die Flora der sarmatischen Stufe* (1959).

Andreas

Andreas' Racer *Coluber andreanus* **Werner,** 1917

Professor Andreas. Unfortunately Werner gives no more information in his etymology. Andreas collected in Far Province in Persia (Iran) (1905). Later Werner described the collection (1917). This may have been the same Andreas who was Professor at Göttingen (1909). He was considered an Orientalist, having spent six years in Persia studying Oriental languages.

Andresen

Andresen's Snake *Coniophanes andresensis* **Bailey,** 1937

The vernacular name is an apparent misunderstanding of the scientific name, which derives from the Isla San Andres, Lesser Antilles.

Angel

Angel's Kukri Snake *Oligodon macrurus* Angel, 1927
Angel's Mountain Keelback *Opisthotropis praemaxillaris* Angel, 1929
Angel's Petite Gecko *Paragehyra petiti* Angel, 1929
Angel's Dwarf Gecko *Lygodactylus decaryi* Angel, 1930
Keelback (snake) sp. *Rhabdophis angelii* **Bourret,** 1934
Angel's Writhing Skink *Lygosoma angeli* **M. A. Smith,** 1937
Angel's Five-toed Skink *Lacertaspis lepesmei* Angel, 1940
Angel's Gecko *Geckolepis petiti* Angel, 1942
Angel's Chameleon *Furcifer angeli* **Brygoo** and **Domergue,** 1968

Fernand Angel (1881–1950) was a zoologist and herpetologist. He joined Muséum National d'Histoire Naturelle, Paris (1905), as an Assistant Taxidermist for Léon Vaillant and then François Mocquard, working there until his death. The only break in his career was his French army service (1914–1918).

Angelorum

Skink sp. *Pseudoacontias angelorum* Nussbaum and **Raxworthy,** 1995

Angeluc and Angelien Razafimanantsoa are self-taught experts in Madagascan fauna. They have assisted many visiting zoologists; in 1993, for example, they worked for a month on a survey with 19 Earthwatch volunteers. They often work as wildlife guides. The etymology says the species is named "in honor of the twin brothers . . . in recognition of their outstanding contributions to herpetological field research in Madagascar."

Anna

Saint-Barts Blind Snake *Typhlops annae* **Breuil,** 1999

Anna Breuil is the describer's wife, whom he thanked for all the help she had given him during his herpetological expeditions.

Annandale

Annandale's Sea Snake *Kolpophis annandalei* Laidlaw, 1901
Yellow-headed Temple Turtle *Hieremys annandalii* **Boulenger,** 1903
Annandale's Dragon *Mictopholis austeniana* **Annandale,** 1908
Annandale's Leaf-toed Gecko *Cyrtodactylus annandalei* **Bauer,** 2003

Dr. "Thomas" Nelson Annandale (1876–1924) was a zoologist and Superintendent of the Indian Museum, Calcutta. He published a number of scientific papers

(1900–1930), including "Fauna of the Chilka Lake: Mammals, Reptiles, and Batrachians" (1915). He was instrumental in establishing a purely zoological survey, not combined with anthropology, undertaking several expeditions, most notably the Annandale-Robinson expedition that collected in Malaya (1901–1902). A rat is named after him.

Ansorge

Nigeria Leaf-toed Gecko *Hemidactylus ansorgii* **Werner,** 1897
Angolan Link-marked Sand Racer *Psammophis ansorgii* **Boulenger,** 1905
Ansorge's Afrogecko *Afrogecko ansorgii* Boulenger, 1907

Dr. William John Ansorge (1850–1913) was an explorer and collector in Africa in the second half of the 19th century. He wrote *Under the African Sun* (1899). Three birds, four mammals, and many African fish, including one with the splendid common name of Slender Stonebasher, are named after him. His son, Sir Eric Cecil Ansorge (1887–1977), was a lepidopterist.

Anthony

Clarion Island Whipsnake *Masticophis anthonyi* **Stejneger,** 1901
Anthony's Ring-necked Snake *Diadophis punctatus anthonyi* **Van Denburgh** and **Slevin,** 1923 [Alt. Todos Santos Island Ringneck Snake]

Alfred Webster Anthony (1865–1939) was a collector, conservationist, ornithologist, and naturalist. He collected birds, mammals, reptiles, invertebrates, plants, and minerals. His specimens are now in the Carnegie Museum, Pittsburgh; the American Museum of Natural History; and the San Diego Museum of Natural History. He wrote "Field Notes on the Birds of Washington County, Oregon" (1886). His first trip out of the USA was to North Coronado Island, Mexico. He later went up to Alaska during the gold rush, and on a collecting trip to Guatemala, where he met Slevin, who wrote in 1939, "The following paper is based on a collection of snakes made in Guatemala during the spring of 1924 and of 1926, when the author in company with Mr. A. W. Anthony, veteran ornithologist well known for his work on the west coasts of America and Mexico, was making his initial efforts at gathering a representative collection of Guatemalan Birds." Five birds, two mammals, and various other taxa are named after him.

Anton

Anton's Anole *Anolis antonii* **Boulenger,** 1908

Named after San Antonio, Colombia.

Antoni

Skink sp. *Eremiascincus antoniorum* **M. A. Smith,** 1927

This skink is endemic to Timor, Indonesia, where the islanders call themselves "the Antoni."

Anzueto

Anzueto's Arboreal Alligator Lizard *Abronia anzuetoi* **Campbell** and **Frost,** 1993

Roderico Anzueto, a Guatemalan naturalist, collected the holotype.

Aphrodite

Oorida Ctenotus *Ctenotus aphrodite* **Ingram** and **Czechura,** 1990

In Greek mythology, Aphrodite was the goddess of love and beauty.

Appert

Colubrid snake sp. *Liophidium apperti* **Domergue,** 1984

Reverend Dr. Otto Appert (b. 1930) is a Swiss missionary and amateur naturalist in Madagascar. He has written several books and articles such as "Distribution and Biology of the Newtonias (Newtonia, Sylviidae) in the Mangoky Region, Southwest Madagascar" (1997). A bird is named after him.

Arango

Colombian Ground Snake *Atractus arangoi* **Prado,** 1939

Professor Andrés Posada Arango (1839–1923) was a Colombian physician and natural scientist who graduated at Universidad de Medellin (1859). He was active in many fields of study, from anthropology to ichthyology. A number of libraries, botanical gardens, and other institutions in Colombia are named after him.

Arcellazzi

Central Asian Toadhead Agama *Phrynocephalus arcellazzii* **Bedriaga,** 1909

Bedriaga says in his description that the agama is named after T. Arcellazzi and uses the word for "drawer," as in "art." We think he meant the Arcellazzi who was an Italian illustrator of natural history subjects. Much later the Italian postal authorities used a number of his images on postage stamps. We have been unable to find more about him.

Archer

Archer's Post Gecko *Hemidactylus funaiolii* **Lanza,** 1978 [Alt. Kenya Leaf-toed Gecko]

Archer's Post is a town in Kenya.

Ardouin

Yellow Skink *Amphiglossus ardouini* **Mocquard,** 1897

Captain Léon Ardouin (1841–1909) was a plant collector in Madagascar (1896–1897). We believe him to be the "Captain Ardouin" to whom Mocquard refers in the etymology.

Argus

Ocellated Gecko *Sphaerodactylus argus* Gosse, 1850
Mongolian Racerunner *Eremias argus* **Peters,** 1869
Argus Snail-sucker *Sibon argus* **Cope,** 1876
Argus Gecko *Cnemaspis argus* **Dring,** 1979 [Alt. Dring's Rock Gecko]

Argus was a mythological Greek giant who had many eyes (some accounts say 100), which made him a very popular watchman. The reptiles named after him are adorned with *ocelli* (eyelike spots).

Ariadna

Ariadna's Ctenotus *Ctenotus ariadnae* **Storr,** 1969

Ariadna Neumann. The etymology states, "Named after Mrs Ariadna Neumann (Librarian, Western Australian Museum)."

Arias

Jaragua Sphaero *Sphaerodactylus ariasae* **Hedges** and **Thomas,** 2001 [Alt. Jaragua Dwarf Gecko]

Yvonne Arias is a herpetologist, ecologist, and biologist. She is a leading proponent of conservation in the Dominican Republic and President of the Grupo Jaragua, which runs the nature reserve home of this gecko—said to be the world's smallest reptile, at 1.6 centimeters (0.6 inch) long.

Armas

Guantanamo Least Gecko *Sphaerodactylus armasi* **Schwartz** and **Garrido,** 1974

Dr. Luis F. de Armas (b. 1945) is a Cuban zoologist, arachnologist, and herpetologist at Instituto de Zoología Academia de Ciencias de Cuba, Havana.

Armitage

Armitage's Cylindrical Skink *Chalcides armitagei* **Boulenger,** 1920

Captain Sir Cecil Hamilton Armitage (1869–1933), who discovered the skink, was the Governor of the Gambia (1920–1927) and a generous donor to the Zoological Gardens of London. He published "The Tribal Markings and Marks of Adornment of the Natives of the Northern Territories of the Gold Coast Colony" (1924).

Armour

Armour's Anole *Anolis armouri* **Cochran,** 1934 [Alt. Armoured Anole]

Allison Vincent Armour (1863–1941) was a meatpacking millionaire from Chicago. He was a generous sponsor of natural history expeditions, archeological digs, and agricultural research. He regularly cruised the Caribbean on his yacht the *Utowana*, a super-yacht of the era. It was his custom to take parties of guests on board, among whom was Barbour (and all his family). See also **Allison.**

Armstrong, B. L.

Armstrong's Dusky Rattlesnake *Crotalus triseriatus armstrongi* **Campbell,** 1979

Barry L. Armstrong is a herpetologist. He and Campbell collected the snake together (1976) and published a paper, "Geographic Variation in the Mexican Pigmy Rattlesnake, *Sistrurus ravus*, with the Description of a New Subspecies" (1979).

Armstrong, J. C.

Armstrong's Least Gecko *Sphaerodactylus armstrongi* **Noble** and **Hassler,** 1933

John C. Armstrong was Assistant Curator of the American Museum of Natural History (1930s). He wrote a number of articles, including "New Caridea from the Dominican Republic" (1949). The etymology says the gecko is "named in honor of Mr. John C. Armstrong who planned the second expedition and assisted in much of the herpetological collecting."

Arnaldo

Arnaldo's Green Racer *Philodryas arnaldoi* **Amaral,** 1932

Arnaldo França was Amaral's technical assistant.

Arnold, E. N.

Arnold's Rock Gecko *Pristurus minimus* Arnold, 1977
Arnold's Leaf-toed Gecko *Hemidactylus arnoldi* **Lanza,** 1978
Arnold's Fringe-fingered Lizard *Acanthodactylus opheodurus* Arnold, 1980
Arnold's Sand Lizard *Mesalina ayunensis* Arnold, 1980
Arnold's Giant Tortoise *Dipsochelys arnoldi* Bour, 1982

Dr. Edwin Nicholas "Nick" Arnold (b. 1940) is Curator of Herpetology at the Natural History Museum, London. His publications include *Reptiles and Amphibians of Europe* (2003).

Arnold, G.

Arnold's Montane Skink *Proscelotes arnoldi* **Hewitt,** 1932

Dr. George Arnold (1881–1962) was an entomologist who was educated in France and Germany. After qualifying as a Doctor of Science, he joined the Department of Cytology and Cancer Research in Liverpool and, as a hobby, worked on Hymenoptera. He became Curator (1911) and, later, Director of the National Museum in Bulawayo, Southern Rhodesia (Zimbabwe).

Arnoult

Arnoult's Dwarf Gecko *Lygodactylus arnoulti* **Pasteur,** 1964

Jacques Arnoult (1914–1995) was an ichthyologist and herpetologist. He graduated in biology, agriculture, and hydrobiology at Université de Toulouse. He was in charge of zoological research at the Institute of Scientific Research, Madagascar (1951), and became an Assistant in the Department of Reptiles and Fish, Muséum National d'Histoire Naturelle, Paris (1954). He collected frequently in Africa and was notably successful in getting his live specimens to breed in captivity. He was Director of the Aquarium, Monaco (1968–1981).

Arny

Arny's Ringneck Snake *Diadophis punctatus arnyi* **Kennicott,** 1859 [Alt. Prairie Ringneck Snake]

Samuel Arny collected snakes and lizards for Kennicott in the American Midwest (1856–1858).

Ashe

Ashe's Bush Viper *Atheris desaixi* Ashe, 1968 [Alt. Desaix's Bush Viper]

Ashe's Spitting Cobra *Naja ashei* Wüster and **Broadley,** 2007

Mount Nyiro Bearded Chameleon *Kinyongia asheorum* **Necas** et al., 2009

James Ashe (1925–2004) was a British-born farmer and self-taught herpetologist who lived at Watumu near Malindi, Kenya. He kept poisonous snakes to "milk" for their venom. In WW2 he was a British army paratrooper and first visited Africa in 1949. After having been a miner there for a number of years, he settled in Kenya (1956). He worked as a volunteer at the Herpetological Department of the National Museum of Kenya (1960s). The chameleon is named after James and his wife, Sandra.

Asmuss

Iranian Mastigure *Uromastyx asmussi* **Strauch,** 1863

Hermann Martin Asmuss (1812–1859) was a paleozoologist Assistant Professor who was made Assistant Director of the "Cabinet of Natural History" of the University of Tartu (1835). As full Professor of Zoology he became Director (1857–1859). His specialty was the classification of Hemiptera. Strauch was a student at the University of Tartu during Asmuss' time. The etymology misspells his first name and refers to "Dr. Herrmann Asmuss, weiland Professor der Zoologie an der Kais. Universität zu Dorpat."

Astarte

Stony Downs Ctenotus *Ctenotus astarte* **Czechura,** 1986

The ancient Syrians regarded Astarte as a goddess of fertility, sexuality, and war. Czechura admits, "The name was arbitrarily chosen."

Atropos

Berg Adder *Bitis atropos* **Linnaeus,** 1758

Atropos was one of the three Fates in Greek mythology. She is often portrayed with a pair of scissors in her hand, as she was the one who cut the thread of one's life—thus an apt name for a venomous snake.

Auber

Auber's Ameiva *Ameiva auberi* **Cocteau,** 1838

Professor Pedro Alejandro Auber (1786–1843) was a Cuban botanist and naturalist of French origin (he was originally Pierre Alexandre Auber). He taught at the University in Havana, becoming Director of the Havana Botanical Gardens. He also was responsible for building in Cuba the first railway to be constructed in any Spanish-speaking country (1835).

Aubry

Aubry's Flapshell Turtle *Cycloderma aubryi* **Duméril,** 1856

Skink sp. *Tropidoscincus aubrianus* **Bocage,** 1873

Charles Eugène Aubry-Lecomte (1821–1879) was a French civil servant, an administrator in New Caledonia. As an amateur naturalist he collected wherever he was posted and discovered many new reptiles, also making a collection of fishes in Gabon (1850s). He was also among the first to describe the iboga root, now used to treat substance abuse disorders

Auffenberg

Auffenberg's Monitor *Varanus auffenbergi* Sprackland, 1999 [Alt. Peacock Monitor]

Dr. Walter Auffenberg Jr. (1928–2004) was a herpetologist and paleontologist who wrote about monitor lizards. His parents encouraged his early interest, allowing him to hitchhike from Detroit to Florida and Mexico to collect specimens. After graduating from high school he tended the family's orange groves before enlisting in the navy as a paramedic at the end of WW2. He later graduated (1954) and earned a doctorate from the University of Florida (1956). He was Curator of Herpetology, Florida Natural History Museum (1963–1984), after which he was Professor of Zoology until retiring (1991), continuing as Curator Emeritus despite a debilitating stroke (1995). He spent most of 1970 on Komodo Island studying the Komodo Dragon. He wrote more than 130 scientific papers and a number of books, including *Gray's Monitor Lizard* (1988).

Aurelio

Aurelio's Rock Lizard *Iberolacerta aurelioi* Arribas, 1994

Aurelio Arribas was the father of Oscar Arribas, the describer.

Austen

Annandale's Dragon *Mictopholis austeniana* **Annandale,** 1908

Austen's Thick-toed Gecko *Pachydactylus austeni* **Hewitt,** 1923

Lieutenant Colonel Henry Homersham Godwin-Austen (1834–1923) was an army topographer, geologist, surveyor, and ornithologist. He was assigned to several government surveys in northern India, especially the Himalayas (1851–1877). He explored and surveyed the region of the Karakorum around K2 (formerly Mount Godwin-Austen). He wrote *Birds of Assam* (1870–1878). Three birds are named after him, including one he egocentrically named after himself, which is considered bad form.

Avery

Avery's Coral Snake *Micrurus averyi* **Schmidt,** 1939 [Alt. Black-headed Coral Snake]

Sewell L. Avery (1873–1960) was a financier, the head of the Commercial Club of Chicago, and Chairman of Montgomery Ward and Co. (a group of 575 retail outlets in the USA) (1931–1955). He endowed a number of professorial chairs at the University of Chicago, all entitled "Sewell L. Avery Distinguished Service Professor." The holotype of this coral snake was taken during the 1938 Sewell Avery expedition to British Guiana (Guyana), and the original description refers to "Mr. Sewell Avery, Trustee of Field Museum, whose support of the Museum's research interests made possible its discovery."

Ayala

Ayala's Anole *Anolis calimae* Ayala, **Harris,** and **Williams,** 1983

Stephen Charles Ayala (b. 1942) is a biologist and herpetologist on the staff of the Department of Microbiology at Universidad del Valle, Cali, Colombia.

Ayarzaguena

Blind Snake sp. *Typhlophis ayarzaguenai* Señaris, 1998

Dr. José Ayarzagüena Sanz is a Venezuelan zoologist and herpetologist at Museo de Historia Natural La Salle, Caracas, and Director of the Biological Station of El Frio where Orinoco Crocodiles are bred for release to the wild. He co-wrote *Fauna of the Venezuelan Llanos: Notes on Their Morphology and Ecology* (1985). Señaris' etymology says, "In honor of my teacher, Dr. José Ayarzagüena, in gratitude for his friendly teaching."

Azara

Azara's Tree Iguana *Liolaemus azarai* Avila, 2003

Brigadier General Féliz Manuel de Azara (1746–1811) was a soldier, engineer, and naturalist who distinguished himself in various expeditions. He was on the Spanish Commission in South America (1781–1801) sent to try to settle the boundaries between Portuguese and Spanish colonies. There he started to study animals, particularly observing the behavior of quadrupeds. His notes, sent to his brother, the Spanish Ambassador in Paris, are generally acknowledged to be meticulous, but they also contained reports from others that were not as accurate. He wrote *Voyage dans l'Amérique méridionale depuis 1781 jusqu'en 1801* (1801). Five birds, five mammals, and an amphibian are named after him.

B

Bacon

Bacon's Water Skink *Tropidophorus baconi* **Hikida,** Riyanto, and **Ota,** 2003

James Patterson Bacon Jr. (1940–1986) was a zoologist and herpetologist. He was the Curator of Herpetology, San Diego Zoo, and also General Manager of the zoo for the three months before his death.

Baden

Baden's Pacific Gecko *Gekko badenii* Shcherbak and Nekrasova, 1994

Named after Nui Ba Den (Black Lady Mountain), Vietnam, where it is endemic.

Bagual

Tree Iguana sp. *Liolaemus baguali* **Cei** and Scolaro, 1983

Named after the El Bagual Ecological Reserve, Argentina.

Baha El Din

Lacertid lizard sp. *Mesalina bahaeldini* Segoli, Cohen, and **Werner,** 2002

Sherif M. Baha El Din (b. 1960) is an Egyptian zoologist, conservationist, naturalist, and author who lives in Cairo with his American-born wife, Mindy (see **Mindi**). He studied for his first degree in art in Egypt, then for his master's in Urban and Regional Environmental Planning at Virginia Polytechnic, USA, before registering for his doctorate in Nottingham, UK. He is an Associate at the Field Museum. He wrote *Directory of Important Bird Areas in Egypt* (1999).

Baig

Bent-toed Gecko sp. *Cyrtopodion baigii* Masroor, 2008

Dr. Khalid Javed Baig (1956–2006), a Pakistani herpetologist, was Director, Pakistan Museum of Natural History, Islamabad. He co-wrote "A New Species of *Eremias* (Sauria: Lacertidae) from Cholistan Desert, Pakistan" (2006).

Bailey, F. M.

Bailey's Snake *Thermophis baileyi* **Wall,** 1907

Lieutenant Colonel Frederick Markham Bailey (1882–1967) was a British army officer, spy, explorer, and butterfly collector. He was known as "Eric," as his father's first name was also Frederick. "Eric" was born at Lahore (now part of Pakistan, then India) and educated at Sandhurst. He was commissioned (1901) and posted to the Nilgiri Hills, where he met Richard Meinertzhagen, who was convalescing after a bout of fever; they became longtime friends. Bailey was in Tibet (1903–1909), became proficient in Tibetan, and accompanied Younghusband's invasion of Tibet (1904), later traveling alone in then-unknown parts of Tibet and China. During WW1 he fought in France with the Indian Expeditionary Forces and later at Gallipoli. He was sent to Tashkent (1918) to investigate the new Bolshevik government's intentions in relation to India. Unmasked, he had to flee for his life, making his escape by disguising himself as an Austrian prisoner-of-war and joining the Cheka (Russian secret police), where his task was to track down a dangerous British agent, namely himself. He was Political Officer in Sikkim (1921–1928). He later recorded his experiences in *Mission to Tashkent* (1946). A mammal is named after him.

Bailey, J. R.

Bailey's Blind Snake *Leptotyphlops anthracinus* Bailey, 1946

Dr. Joseph Randle Bailey (1913–1998) was Emeritus Professor of Zoology, Duke University. He worked at the Zoological Museum, University of Michigan (1946). He co-wrote "Snakes from the Uplands of the Canal Zone and of Darien" (1939).

Bailey, V. O.

Bailey's Collared Lizard *Crotaphytus collaris baileyi* **Stejneger,** 1890

Vernon Orlando Bailey (1864–1942) was a naturalist and ethnographer. As a young Minnesota farmer he sent many natural history specimens to C. Hart Merriam, then head of the U.S. Biological Survey, which Bailey later joined (1887–1902), eventually becoming its Chief (1889). While with the Survey he undertook many field trips, including six to Texas. He married Florence Augusta Bailey, née Merriam (1899), who also worked for the Survey and accompanied him on some trips. He wrote *The Mammals and Life Zones of Oregon* (1936). Two mammals are named after him.

Baird, J.

Baird's Black-headed Snake *Tantilla bairdi* **Stuart,** 1941

James Baird was a wealthy donor to the University of Michigan, for which reason Stuart named this snake after him, but gave no further details. We think he was the James Baird (1873–1953) who played football for the University of Michigan (1892–1895), became a civil engineer, and founded the company that constructed many famous buildings, including the Lincoln Memorial in Washington, DC.

Baird, S. F.

Baird's Patch-nosed Snake *Salvadora bairdi* **Jan,** 1860

Baird's Rat Snake *Pantherophis bairdi* **Yarrow,** 1880

Spencer Fullerton Baird (1823–1887) was a zoologist who

became such a giant of ornithology that to give a long biography here seems unnecessary. The young Baird became a friend of John James Audubon (1838), sending him specimens. After studying medicine for a time, Baird became Professor of Natural History, Dickinson College, Pennsylvania (1845). He was Assistant Secretary of the Smithsonian (1850–1878), later becoming Secretary. He organized a number of expeditions (1850–1860). Among his publications is *Catalogue of North American Reptiles* (1853). Eight birds and five mammals are named after him.

Baker, A. B.

Baker's Spinytail Iguana *Ctenosaura bakeri* **Stejneger,** 1901

Baker's Worm Lizard *Amphisbaena bakeri* Stejneger, 1904 [Alt. Puerto Rican Gray Amphisbaena]

Arthur B. Baker (1858–1930) went on a collecting expedition to Puerto Rico aboard the *Kitty Hawk* with Stejneger (1899). At the time he was with the U.S. Fish Commission in Puerto Rico and contributed to the report on the natural history specimens collected for the *Smithsonian Institution Annual Report*, 1899. He wrote *A Notable Success in the Breeding of Black Bears* (1904).

Baker, H. B.

Baker's Cat-eyed Snake *Leptodeira bakeri* **Ruthven,** 1936

Dr. Horace Burrington Baker (1889–1971) was a zoologist who received his doctorate at the University of Michigan (1920). After army service (1917–1918), he became first an Instructor (1920) at the University of Pennsylvania, then Assistant Professor (1926), Associate Professor (1928), and Full Professor (1939–1959). He also worked for *Nautilus*, as business manager (1932) and editor (1957–1970). He collected in Mexico (1910 and 1926), Venezuela (1919), and also in the USA, Jamaica, and Puerto Rico. He wrote many articles (1920s–1960s). He collected the holotype of this snake (1922).

Bakewell

Bakewell's Blind Snake *Leptotyphlops bakewelli* **Oliver,** 1937

Father Anderson Bakewell (1913–1999) was a Jesuit priest from St. Louis, Missouri, a lifelong collector of zoological and botanical specimens for U.S. scientific institutions. As a young man he collected snakes, including the blind snake holotype, for the zoologist Richard Marlin Perkins (1905–1986), who became the television presenter of *Wild Kingdom* in the 1960s. Bakewell graduated from St. Louis University (1937) and joined the Society of Jesus (1942). He became a missionary (1947) and worked in Bombay preparing snakebite antivenin and studying cobras, kraits, and vipers. After ordination (1951), he worked in a jungle mission in Bihar, then returned to the USA (1955). In the early 1960s he was Assistant Pastor in Georgetown. He had a parish of over 90,000 square kilometers (35,000 square miles) in Alaska (1967), then was in Santa Fé, New Mexico (1976–1999). In addition to his pastoral duties, he was a mountaineer, the first priest to climb Mount Everest, and a member of the first nonstop round-the-world flight over both poles (1965).

Balfour

Lacertid lizard sp. *Mesalina balfouri* **Blanford,** 1881

Sir Isaac Bayley Balfour (1853–1922) was a Professor of Botany and Regius Keeper of the Edinburgh Botanic Garden (1890–1922). He explored and collected on a number of islands, notably Rodrigues (1874), and made the first botanical study of Socotra (1879–1880). A bird and numerous plants are named after him.

Ballinger

Ballinger's Canyon Lizard *Sceloporus merriami ballingeri* **Lemos-Espinal** et al., 2001

Dr. Royce Eugene Ballinger (b. 1942) is a herpetologist. He was Professor of Biological Sciences, University of Nebraska, Lincoln, retiring in 2006. The description of the lizard specifically mentions how well he acted as a mentor during Lemos-Espinal's doctorate. In the late 1960s he was at Angelo State University, Texas, and was one of those who started that university's herpetological collection. He was Chairman of the Nebraska EPSCoR (Experimental Program to Stimulate Competitive Research) (1993–2003). Among his publications is *How to Know the Amphibians and Reptiles* (1983).

Balzan

Bolivian Ground Snake *Atractus balzani* **Boulenger,** 1898

Dr. Luigi Balzan (1865–1893) was an Italian naturalist who set out on a grand tour of South America (1890), traveling alone by whatever means he could find through Brazil, Paraguay, Peru, and Bolivia. He collected the holotype of this snake. He wrote *Voyage de M E Simon au Venezuela 1887–1888* (1892). A fish is named after him.

Bampfylde

Bampfylde's Writhing Skink *Lygosoma bampfyldei* **Bartlett,** 1895 [Syn. *Riopa bampfyldei*]

Charles Agar Bampfylde (1856–1918) was Resident of the First Division, Sarawak (1896–1903). He returned to England (1903), acting as "Political Agent" there for Rajah Charles Brooke (1903–1912). He was a member of the Sarawak State Advisory Council (1912–1918). He co-wrote *A History of Sarawak under Its Two White Rajahs, 1839–1908* (1909) with Sabine Baring-Gould, who was most famous for having written the hymn *Onward! Christian Soldiers*. A road in Kuching is named after him.

Banao

Sharp-nosed Blindsnake sp. *Acutotyphlops banaorum* **Wallach, Brown,** Diesmos, and Gee, 2007

Named after the Banao, an indigenous tribe in the north of Luzon Island, Philippines.

Baran

Baran's Viper *Vipera barani* **Böhme** and **Joger,** 1984 [Alt. Turkish Viper]

Baran's Dwarf Racer *Eirenis barani* **Schmidtler,** 1988

Colubrid snake sp. *Rhynchocalamus barani* Olgun et al., 2007

Professor Dr. Ibrahim Baran (b. 1940) is a herpetologist at Dokuz Eylül University, Izmir, Turkey. He and Professor Mehemet K. Atatür published *Turkish Herpetofauna (Amphibians and Reptiles)* (1998).

Bárbara

Bárbara's Lizard *Liolaemus barbarae* Pincheira-Donoso and **Núñez,** 2002 [Junior syn. of *L. puna* Lobo and Espinoza, 2004]

Bárbara Hurtado is the wife of the senior describer, Daniel Pincheira-Donoso.

Barbieri

Gecko *Hemidactylus barbierii* Sindaco et al., 2007

Francesco Barbieri (1944–2001), a zoologist, biologist, mathematician, and herpetologist, was Professor of Zoology, University of Pavia. He graduated in biological sciences (1969) and in the same year joined the staff of the Institute of Zoology, Pavia, becoming Assistant Professor (1975) and Associate Professor of Applied Zoology (1980). He was a founding member of the Italian Herpetological Society.

Barbour

Barbour's Ground Skink *Scincella barbouri* **Stejneger,** 1910

Barbour's Eyelid Skink *Eumeces barbouri* **Van Denbergh,** 1912

Barbour's Least Gecko *Sphaerodactyulus torrei* Barbour, 1914

Barbour's Tropical Racer *Mastigodryas bruesi* Barbour, 1914

Barbour's Worm Lizard *Amphisbaena anomala* Barbour, 1914 [Syn. *Aulura anomala*]

Barbour's Montane Pit-viper *Cerrophidion barbouri* **Dunn,** 1919

Hispaniolan Hopping Anole *Anolis barbouri* **Schmidt,** 1919

Barbour's Clawed Gecko *Pseudogonatodes barbouri* **Noble,** 1921

Espiritu Santo Striped Whipsnake *Masticophis barbouri* Van Denburgh and **Slevin,** 1921

Barbour's Water Snake *Rhabdophis barbouri* **Taylor,** 1922

Barbour's Centipede Snake *Tantilla albiceps* Barbour, 1925

Barbour's Tropical Ground Snake *Trimetopon barbouri* Dunn, 1930

Uzungwe Mountain Bush Viper *Adenorhinos barbouri* **Loveridge,** 1930

Barbour's Bachia *Bachia barbouri* **Burt** and Burt, 1931

Striped Caribbean Gecko *Aristelliger barbouri* Noble and Klingel, 1932

Dusky Pygmy Rattlesnake *Sistrurus miliarius barbouri* **Gloyd,** 1935

Barbour's Anaconda *Eunectes barbouri* Dunn and **Conant,** 1936

Marajo Island Rainbow Boa *Epicrates cenchria barbouri* Stull, 1938

Barbour's Galliwasp *Celestus barbouri* **Grant,** 1940

Barbour's Chameleon *Chamaeleo barbouri* Hechenbleichner, 1942

Barbour's Day Gecko *Phelsuma barbouri* Loveridge, 1942

Barbour's Map Turtle *Graptemys barbouri* **Carr** and Marchand, 1942

Cuban Many-ringed Amphisbaena *Amphisbaena barbouri* **Gans** and Alexander, 1962

Barbour's Gecko *Cnemaspis barbouri* **Perret,** 1986

Dr. Thomas Barbour (1884–1946) was an American zoologist who graduated from Harvard (1906) and obtained his doctorate there (1910). He worked at the Harvard Museum (1911–1946), initially as an Associate Curator of Reptiles and Amphibians, rising to Director and Custodian of the Harvard Biological Station and Botanical Garden, Soledad, Cuba (1927). He was Executive Officer in charge of the Barro Colorado Island Laboratory, Panama (1923–1945). During his time at the museum he explored in the East Indies, the West Indies, India, Burma, China, Japan, and South and Central America. He was famously jovial good company and would invite all and sundry to eat and converse. His special area of interest was Central American herpetology. Among his publications is *Checklist of North American Amphibians and Reptiles* (1933), co-written with Stejneger. Two mammals and three amphibians are named after him.

Barboza

Barboza's Leaf-toed Gecko *Hemidactylus bayonii* **Bocage,** 1893

See **Bocage.**

Barker, R. de la Bere

Barker's Sharp-snouted Worm Lizard *Ancylocranium barkeri* **Loveridge,** 1946

Ronald de la Bere Barker (b. 1889), a New Zealander, studied zoology at Canterbury University. He traveled in Australia, Canada, Japan, India, and the Pacific region. He was a soldier under Frederick Selous in the Frontiermens' Battalion in the campaign against the German forces in Tanganyika (Tanzania) (1915–1918). After WW1 he stayed on in East Africa as an administrator, in charge of the Rufiji district of Tanganyika (1923), and wrote *Crocodiles in Tanganyika* (1953). The Worm Lizard was uncovered in a collection of reptiles that he presented to the Museum of Comparative Zoology, Harvard.

Barker, R. W.

Barker's Anole *Anolis barkeri* **Schmidt,** 1939

Dr. Reginald Wright Barker (d. 1969) of Bellaire, Texas, collected natural history specimens in Mexico (including the anole holotype), as well as ethnographical items. He was a paleontologist and consultant to the petroleum industry and published articles on both subjects.

Barnard, H. G.

Yellow-naped Snake *Furina barnardi* **Kinghorn,** 1939

Henry "Harry" Greensill Barnard (1869–1966) was a zoologist, naturalist, and grazier at Coomooboolaroo, Queensland. His father, George, was an entomologist and an oologist whose collection got so large that he built himself a private museum. On his death (1894) it was sold to Lord Rothschild's private museum at Tring, England, now part of the British Natural History Museum. Naturalists and zoologists such as Carl Lumholtz (1883) stayed at Coomooboolaroo. Henry joined a government expedition to explore the Bellenden Ker Range (1888). He companied Albert Stewart Meek, a family friend, on a collecting trip in Northern Queensland for Rothschild (1894). Barnard was at Cape York collecting for a number of different people (1896 and 1899). A bird and a mammal are named after him.

Barnard, K. H.

Barnard's Thick-toed Gecko *Pachydactylus barnardi* **FitzSimons,** 1941

Barnard's Namib Day Gecko *Rhoptropus barnardi* **Hewitt,** 1926 [Alt. Barnard's Slender Gecko]

Dr. Keppel Harcourt Barnard (1887–1964) was a UK-born South African invertebrate zoologist particularly interested in marine crustaceans. He graduated from Cambridge, then studied law, but being more interested in science was for a short time Honorary Naturalist at the Plymouth Marine Biology Laboratory. He worked at the South African Museum, Cape Town (1911–1964), initially as a marine biologist, becoming Assistant Director (1924) and then Director (1946–1956). He undertook many collecting expeditions in South Africa and Mozambique, and in three donkey-trek expeditions in Namibia (1924–1926). A keen mountaineer, he was Secretary of the Mountain Club of South Africa (1918–1945). He wrote on molluscs and entomology. FitzSimons gives no explanation whatsoever of the Thick-toed Gecko's scientific name. We think this is one of those cases where the describer felt that no explanation was necessary, because "everyone would know who he meant"; so we believe we have identified the right man, especially as FitzSimons based his description of the gecko on a series of specimens in the collection of the South African Museum.

Barnes

Barnes' Cat Snake *Boiga barnesi* **Günther,** 1869

Richard Hawksworth Barnes (b. 1831) was born in Colombo. He made a collection of reptiles in Ceylon (Sri Lanka) and sent it to the British Museum. He lived in Kensington, London (1871), and Dorset (1901).

Barnett, Brian

Barnett's Death Adder *Acanthophis barnetti* Hoser, 1988

Brian Barnett (b. 1940) is an Australian herpetologist. He has successfully bred many species of reptile, supplying them to others. He founded the Victorian Herpetological Society (1977) and became its first President. He was instrumental in introducing a workable reptile licensing system in Victoria. The snake's scientific name should use the plural form *barnettorum*, as the dedication specifically includes Brian's wife, Mrs. Lani Barnett, and their sons Taipan, Brett, and Brian.

Barnett, Burgess

Barnett's Lancehead *Bothrops barnetti* **Parker,** 1938

Dr. Burgess Barnett (1888–1944) was Curator of Reptiles, London Zoo (1932–1937). He was appointed Superintendent, Rangoon Zoological Gardens, Burma (Myanmar) (1938), and held that position until his death in Bengal, where he was evacuated on the invasion of Burma by the Japanese army (1941). He collected the lancehead holotype and paratypes in Peru. He published a booklet called *The Terrarium: Tortoises, Other Reptiles, and Amphibians in Captivity*.

Baron

Baron's Green Racer *Philodryas baroni* Berg, 1895 [Alt. Pampas Green Ratsnake, Argentine Longnose Snake]

Named after D. Manuel Barón Morlat, from whom Berg received specimens of this snake from Tucumán, Argentina.

Barrington

Barrington Land Iguana *Conolophus pallidus* **Heller,** 1903 [Alt. Santa Fé Land Iguana]

Barrington Leaf-toed Gecko *Phyllodactylus barringtonensis* **Van Denburgh,** 1912

Barrington is an early name for the Galapagos island Santa Fé.

Barrio

Barrio's Burrowing Snake *Apostolepis barrioi* **Lema,** 1978

Dr. Avelino Barrio (1920–1979) was a Spanish botanist and herpetologist who lived in Argentina. A graduate of the Colegio Nacional de Buenos Aires, he undertook his doctorate at Buenos Aires University (1948–1954). He spent most of his career there and at the Instituto Nacional de Microbiología "Gustav Malbrán," concentrating on the genetics of snakes and anurans. He founded (1969) the Centro Nacional de Investigaciones Iologícas and dedicated much time to the study of venoms. In addition to working on taxonomy, he studied frogs' vocalizations. Thales de Lema named the snake in "homage to a man who was politically persecuted."

Barron

Barron's Kukri Snake *Oligodon barroni* **M. A. Smith,** 1916

P. A. R. Barron was a collector for the Raffles Museum. He collected three specimens of this species in Siam (Thailand), but we know nothing more about him.

Barry Lyon

Skink sp. *Proablepharus barrylyoni* **Couper** et al., 2010

Barry Lyon is the Ranger in charge of the Steve Irwin Wildlife Reserve, northern Cape York Peninsula, Queensland, Australia. He played an important role in collecting the type series of this species and is also honored "for his contributions to wildlife conservation across Cape York Peninsula."

Barth

Guinea Snake-eater *Polemon barthii* **Jan,** 1858

Heinrich Barth (1821–1865) was an explorer and cartographer who joined an expedition (1849–1855) to the Sudan sponsored by the British government, at the end of which only he was left alive. He studied archeology, history, geography, and law at the University of Berlin and spoke Arabic. He was appointed Professor of Geography, University of Berlin (1863). He visited Asia Minor (1858) and the European part of Turkey (1862). His five-volume journal *Travels and Discoveries in North and Central Africa* was published in German and English (1857–1858).

Bartlett

Bartlett's Flying Dragon *Draco affinis* Bartlett, 1894

Edward Bartlett (1836–1908) was a pioneering ornithologist and herpetologist in Borneo. He was the son of a taxidermist and Superintendent of the London Zoo. Edward Bartlett traveled with H. B. Tristram (q.v.) in Palestine (1863–1864) and also collected in Amazonian Peru (1865–1869). He was Curator, Maidstone Museum (1875–1890), leaving to collect in Borneo, becoming Curator of Sarawak Museum (1893–1897). He published "The Crocodiles and Lizards of Borneo in the Sarawak Museum, with Descriptions of Supposed New Species, and the Variation of Colours in the Several Species during Life" (1895).

Bartsch

West Cuban Anole *Anolis bartschi* **Cochran,** 1928

Bartsch's Iguana *Cyclura carinata bartschi* Cochran, 1931

Paul Bartsch (1871–1960) was born in Silesia (then part of Germany, now Poland). His family moved to the USA, where he developed an interest in nature. He worked at the Smithsonian (1896–1942). Among his many expeditions was that to the Philippines (1907–1910) aboard the *Albatross*, and his account of it was published in *Copeia* (1941). He taught at a number of universities and was known as a world authority on molluscs. He also organized the first Boy Scout group in Washington, DC. A bird, the Guam Swiftlet *Aerodramus bartschi*, is named after him.

Basoglu

Basoglu's Racerunner *Eremias suphani* Basoglu and **Hellmich,** 1980

Basoglu's Bow-fingered Gecko *Cyrtodactylus basoglui* **Baran** and Gruber, 1982

Professor Dr. Muhtar Basoglu (1913–1981) was a herpetologist. He graduated in natural sciences at Istanbul University (1936) and became an assistant there (1941). He published widely and traveled extensively, including a visit to Lake Van with Professor Hellmich (1958). He was Professor of Zoology, University of Izmir (1961–1981).

Bastard

Mocquard's Madagascar Ground Gecko *Paroedura bastardi* **Mocquard,** 1900

Eugène Joseph Bastard (1865–1910) was a colonial administrator, naturalist, and paleontologist particularly connected with Madagascan fauna. He traveled and collected in northern Madagascar (1898 and 1900). He also worked at Rijksmuseum van Natuurlijke Histoire, Leiden, alongside Grandidier. He wrote *Exploration au sud de l'Onilahy: Notes, reconnaissances, et explorations* (1899).

Bates, G. L.

Spotted Dagger-tooth Tree Snake *Rhamnophis batesii* **Boulenger,** 1908

George Latimer Bates (1863–1940) was an American ornithologist and botanist who traveled in West Africa (1895–1931), residing in Cameroon for some years. He sent specimens to, inter alia, the Natural History Museum, London, and the Philadelphia Academy. He wrote *Handbook of the Birds of West Africa* (1930). Five birds and four mammals are named after him.

Bates, H. W.

Amazon Basin Emerald Tree Boa *Corallus batesii* **Gray,** 1860

Henry Walter Bates (1825–1892) was an English explorer, entomologist, and naturalist whose first expedition was with Alfred Russel Wallace, who returned to England in 1848 while Bates only returned in 1859, after 11 years in Brazilian Amazonia. From 1864 he worked at Assistant Secretary, Royal Geographical Society, and became a Fellow of both the Linnean and the Royal Society. He gave the first scientific account of mimicry in animals.

Battersby

Battersby's Emo Skink *Emoia battersbyi* **Procter,** 1923

Battersby's Dwarf Boa *Tropidophis battersbyi* **Laurent,** 1949

Battersby's Green Snake *Philothamnus battersbyi* **Loveridge,** 1951

Battersby's Mole Viper *Atractaspis battersbyi* **Witte,** 1959

Colubrid snake sp. *Calamaria battersbyi* **Inger** and **Marx,** 1965

James Clarence Battersby (1901–1993) was a British herpetologist at the Natural History Museum, London (1916–1961), joining (aged 15) as a Boy Attendant for George Albert Boulenger.

Baudin

Baudin's Window-eyed Skink *Pseudemoia baudini* **Duméril** and **Bibron,** 1839 [Alt. Baudin's Emo Skink, Great Bight Cool-Skink; Syn. *Emoia baudini*]

Nicolas Thomas Baudin (1754–1803). Captain Baudin set out (1800) from Le Havre with two corvettes, the *Géographe* and the *Naturaliste*, with a complement of 27 scientific and artistic supernumeraries. *Géographe* crawled back to Le Havre (1804) with a living cargo of 72 birds and other animals, but Baudin had died of tuberculosis in Mauritius. The expedition was a great scientific and artistic success. A species of cockatoo is named after him.

Bauer, A. M.

Bauer's Chameleon Gecko *Eurydactylodes agricolae* **Henkel** and **Böhme,** 2001

Pulau Aur Rock Gecko *Cnemaspis baueri* **Das** and **Grismer,** 2003

Leaf-toed Gecko sp. *Hemidactylus aaronbaueri* Giri, 2008

Bauer's Leaf-toed Gecko *Dixonius aaronbaueri* Ngo and **Ziegler,** 2009

Professor Aaron Matthew Bauer (b. 1961) is a graduate of Michigan State University and the University of California, Berkeley. Since 1988 he has been a member of the Biology Department, Villanova University, Pennsylvania. The binomial *Eurydactylodes agricolae* contains a nice play on words: *agricola* is Latin for "farmer," for which the German is *Bauer*. See also **Aaron.**

Bauer, H. J.

Bauer's Nightsnake *Hypsiglena torquata baueri* **Zweifel,** 1958

Harry J. Bauer (1886–1960) of Los Angeles was a prominent entrepreneur who became President of Edison International (1933). He was co-sponsor of the Puritan-American Museum of Natural History expedition to Baja California (1958). There is a "Harry J. Bauer Collection in the History of Science" at the California Institute of Technology, Pasadena. He also owned a ship that was used for the expedition.

Baur

Striped Mud Turtle *Kinosternon baurii* **Garman,** 1891

Baur's Leaf-toed Gecko *Phyllodactylus baurii* Garman, 1892

Florida Box Turtle *Terrapene carolina bauri* **Taylor,** 1895

George Herman Carl Ludwig Baur (1859–1898) was an osteologist and testudinologist. He studied in Munich (1882–1884), left for America, and was at Yale (1884–1890). He traveled to the Galapagos Islands (1892), where he made an extensive ornithological collection. He was Assistant Professor of Paleontology, University of Chicago (1893–1898). He died having been committed to a lunatic asylum.

Bavay

Bavay's Gecko *Eurydactylodes vieillardi* Bavay, 1869 [Alt. Vieillard's Chameleon Gecko]

Bavay's Giant Gecko *Rhacodactylus chahoua* Bavay, 1869

Bavay's Skink *Lygosoma arborum* Bavay, 1869 [Syn. *Lioscincus nigrofasciolatum* Peters, 1869]

New Caledonian Gecko genus *Bavayia* **Roux,** 1913

Bavay's Keeled Skink *Tropidophorus baviensis* **Bourret,** 1939

Arthur René Jean Baptiste Bavay (1840–1923) was a

pharmacist, sailor, conchologist, and herpetologist. He was attached to the French navy's hospital at Port-Louis (1864). His first love was conchology, and he collected around Suez but also spent time in New Caledonia and wrote the first catalogue of its reptiles: *Catalogue des reptiles de la Nouvelle-Caledonia et description dupuces nouvelles* (1869). He contributed to a publication that dealt with the Andean lakes visited by de Crequi Montfort and Senechal de la Grange in their expedition (1903).

Bavazzano

Somali Leaf-toed Gecko *Hemidactylus bavazzanoi* **Lanza,** 1978 [Alt. Somali Banded Gecko]

Renato Bavazzano was a botanist who worked for the Tropical Herbarium, Università degli Studi di Firenze, for which he made several expeditions to northeast Africa. He visited many countries in Arabia and Africa, including Ethiopia and Somalia with Lanza. He wrote "Contributo alla conoscenza della flora dello Scioa (Etiopia)" (1964).

Baynes

Baynes' Lerista *Lerista baynesi* **Storr,** 1971

Dr. Alexander Baynes (b. 1944) is an Australian mammalogist who is Curator of Palaeontology, Western Australian Museum. His main interest is Quaternary mammals, particularly rodents. He collected the holotype of this skink.

Bayon

Bayon's Skink *Sepsina bayoni* **Bocage,** 1866

Bayon's Mabuya *Trachylepis bayonii* Bocage, 1872

Barboza's Leaf-toed Gecko *Hemidactylus bayonii* Bocage, 1893

Francisco Antonio Pinheiro Bayão was a Portuguese planter and administrator whose family had settled in the Duque de Bragança district of Angola in the 17th century. He was a keen naturalist and collector and sent many specimens to Lisbon, often using Anchieta as an intermediary. A mammal and an amphibian are named after him.

Beale

Beale's Four-eyed Turtle *Sacalia bealei* **Gray,** 1831 [Alt. Beal's/Beale's Eyed Turtle]

Thomas Beale (1775–1842) was an Englishman who lived in Macao and Canton for 40 years from 1792. He was described as a merchant prince and opium mogul. Before the age of 20, he went to China to join his brother's business; he amassed enormous wealth, only to lose it and fall into debt. He disappeared from his house one day, and his decomposed body was found weeks later half-buried on a beach. Beale was known for his hospitality and the magnificence of his mansion in Macao. Gray's original etymology, like many early sources, is brief: he refers only to "Mr. Beale" and mentions "a drawing [of the turtle] communicated by Mr. Reeves." Thus we cannot be 100 percent certain that we have found the right Beale, but we can find no stronger candidate.

Beatty

Beatty's Least Gecko *Sphaerodactylus beattyi* **Grant,** 1937 [Alt. Saint Croix's Sphaero]

Harry Andrew Beatty (1902–1989) was brought up on his father's sugar cane plantation in St. Croix and sent to Massachusetts to be educated. He was studying zoology when financial problems forced him to return home (1919). He worked on a plantation in Puerto Rico (1921). He was encouraged to publish his ornithological notes and was sent to the Dominican Republic to observe bird life (1925). He went to New York to study medical entomology (1929), then returned to St. Croix (1933) as supervisor and biologist in the Health Department dealing with malaria. He was on the Weber expedition to Venezuela (1937). He left the Department of Health (1941) and worked for the Virgin Islands Wildlife Research and Restoration Project. At some point he became a professional collector for a number of American museums—for example, in West Africa for the Peabody Museum (1948–1952) and the Field Museum (1960). He co-wrote "Herpetological Notes on St. Croix, Virgin Islands" (1944).

Beaufort

Beaufort's Forest Skink *Sphenomorphus beauforti* **de Jong,** 1927

Professor Lieven Ferdinand de Beaufort (1879–1968) was a zoologist, ornithologist, and ichthyologist. He was a co-founder of the Dutch Ornithological Association, becoming its President (1924–1956). He undertook his first expedition to the Indo-Australian archipelago (1900) and collected extensively with Weber in New Guinea (1907–1922). He succeeded Weber as Director of the Zoological Museum, Artis Amsterdam (1922–1949). Beaufort co-wrote with Weber the six-volume work *The Fishes of the Indo-Australian Archipelago* (1911–1964). A bird and a mammal are named after him.

Beccari

Beccari's Keeled Skink *Tropidophorus beccarii* **Peters,** 1871

Black Tree Monitor *Varanus beccarii* **Doria,** 1874

Skink sp. *Emoia beccarii* Doria, 1874

Indonesian Brown Skink *Carlia beccarii* Peters and Doria, 1878

Lined Flying Dragon *Draco beccarii* Peters and Doria, 1878

Sumatran Nose-horned Lizard *Harpesaurus beccarii* Doria, 1888

Dr. Odoardo Beccari (1843–1920) was an Italian botanist. He met James Brooke, first Rajah of Sarawak, who enabled him to stay (1865–1868) in Sarawak, Malaya, and Brunei, after which he visited Ethiopia. He also collected in the Celebes, New Guinea, and Sumatra, where he found (1878) the Titan Arum (Corpse Flower), the world's largest flower. Kew successfully grew it from seed, achieving the first flowering of the species in cultivation (1889). Four birds and four mammals are named after Beccari.

Beck

Island Fence Lizard *Sceloporus occidentalis becki* **Van Denburgh,** 1905

Beck's Least Gecko *Sphaerodactylus becki* **Schmidt,** 1919

Cape Berkeley Giant Tortoise *Chelonoidis nigra becki* **Rothschild,** 1901 [Syn. *Geochelone nigra becki*]

Rollo Howard Beck (1870–1950) was a collector who has been described as "the supreme seabird specialist." He collected for the museum of the California Academy of Sciences and then for the American Museum of Natural History. He collected in the Galapagos (1897–1898), Alaska (1911), and New Guinea (1928). On many of his trips he was accompanied by his wife, Ida. One trip lasted 5 years while they explored the South American coast; another in the South Seas lasted nearly 10, when they were part of the Sanford-Whitney expedition. Two birds are named after him.

Becker

Becker's Lichen Anole *Anolis beckeri* **Boulenger,** 1881

Léon Becker (1826–1909) was an arachnologist and gifted painter whose monumental four-volume *Les arachnides de la Belgique* (1882–1896) contained lithographs of almost all the Belgian spiders then known, and is considered a superb example of the scientific and artistic skills of that period.

Beddome

Beddome's Earth Snake *Uropeltis beddomii* **Günther,** 1862

Beddome's Keelback *Amphiesma beddomei* Günther, 1864 [Alt. Nilgiri Keelback]

Beddome's Golden Gecko *Calodactylodes aureus* Beddome, 1870

Beddome's Indian Gecko *Geckoella collagalensis* Beddome, 1870

Beddome's Ornate Gecko *Cnemaspis ornata* Beddome, 1870

Beddome's Skink *Mabuya beddomii* **Jerdon,** 1870

Beddome's Snake-Eye *Ophisops beddomei* Jerdon, 1870

Beddome's Black Shieldtail *Melanophidium punctatum* Beddome, 1871

Beddome's Day Gecko *Cnemaspis beddomei* **Theobald,** 1876

Indian Kangaroo Lizard *Otocryptis beddomii* **Boulenger,** 1885

Beddome's Ground Skink *Scincella beddomi* Boulenger, 1887

Beddome's Ristella *Ristella beddomii* Boulenger, 1887

Beddome's Worm Snake *Typhlops beddomii* Boulenger, 1890

Beddome's Cat Snake *Boiga beddomei* **Wall,** 1909

Beddome's Coral Snake *Calliophis beddomei* **M. A. Smith,** 1943 [Syn. *Maticora beddomei*]

Colonel Richard Henry Beddome (1830–1911) was a keen naturalist, especially interested in the botany of ferns. He joined the army in India (1848) and became Assistant Conservator of Forests in the Madras presidency (1857), eventually becoming Chief Conservator (1860–1882), and a member of the University of Madras (1880). He was first to recognize and describe the great diversity of South Indian herpetology. After returning to England he frequently visited the nearby botanical gardens at Kew. He wrote *Handbook of the Ferns of British India, Ceylon, and the Malay Peninsula* (1892). A mammal and an amphibian are named after him.

Bedriaga

Bedriaga's Skink *Chalcides bedriagai* **Boscá,** 1880

Bedriaga's Fringe-fingered Lizard *Acanthodactylus bedriagai* **Lataste,** 1881

Bedriaga's Rock Lizard *Lacerta bedriagae* Camerano, 1885 [Syn. *Archaeolacerta bedriagae*]

Bedriaga's Wonder Gecko *Teratoscincus bedriagai* **Nikolsky,** 1900

Bedriaga's Toadhead Agama *Phrynocephalus alpherakii* Bedriaga, 1907

Dr. Jacques Vladimir von Bedriaga (or Jacob Vladimirovich Bedriaga or Johann von [Jean de] Bedriaga) (1854–1906) was a herpetologist. He studied at the University of Moscow, but bad health forced him to leave. He went to the University of Jena, Germany, studied comparative anatomy, and after receiving his doctorate (1875) undertook a series of expeditions to Mediterranean countries for his health and to study reptiles. Periodically he visited Russia to study specimens that Prjevalsky was sending back from his expeditions. His health deteriorated further, so he settled in Nice (1881), later moving to Florence, where he died. He wrote *Die Amphibien und Reptilien Griechenlands* (1880), the first monograph on Greek herpetology.

Beetz

Beetz's Tiger Snake *Telescopus beetzi* **Barbour,** 1922

Dr. Paul Friedrich Werner Beetz (fl. 1907–1950) was a German geologist who joined the Deutsche Diamanten Gesellschaft (1908) to explore German South-West Africa (Namibia). He stayed on after WW1, joined Consolidated Diamond Mines, and found diamonds in South-West Africa (1928). He wrote "Geology of South West Angola, between Cunene and Lunda Axis" (1933). He collected the holotype of the snake.

Belcher

Belcher's Sea Snake *Hydrophis belcheri* **Gray,** 1849

Admiral Sir Edward Belcher CB (1799–1877) was an explorer of the Pacific coast of America (1825–1828). He surveyed the coast of Borneo, the Philippines, and Formosa (Taiwan) (1843–1846). He also explored the Arctic (1852–1854), searching for Franklin. He was court-martialed but acquitted (1854) for abandoning three ships during this search. He wrote *The Last of the Arctic Voyages; Being a Narrative of the Expedition in HMS Assistance, under the Command of . . . in Search of Sir John Franklin, during the Years 1852–53–54 with Notes on the Natural History by Sir John Richardson* (1855). Two birds are named after him.

Belding

Belding's Orange-throated Whiptail *Aspidoscelis hyperythrus beldingi* **Stejneger,** 1894

Lyman Belding (1829–1917) was a professional field collector who specialized in birds. He wrote *A Part of My Experience in Collecting* (1900). Five birds and a mammal are named after him.

Bell, E. L.

Bell's Spiny Lizard *Sceloporus undulatus belli* **H. M. Smith,** Chiszar, and **Lemos-Espinal,** 2002

Dr. Edwin Lewis Bell II (1926–2010) received his bachelor's degree from Bucknell University (1948), his master's from Pennsylvania State University (1950), and a doctorate from the University of Illinois (1954). He joined Albright College, Pennsylvania (1954), retiring as Emeritus Professor of Biology and acting as College Archivist.

Bell, T.

Butterfly Lizard *Leiolepis belliana* **Gray,** 1827 [Alt. Common Butterfly Agama, Smooth-scaled Lizard]
Bell's Hinge-back Tortoise *Kinixys belliana* Gray, 1831
Bell's Painted Turtle *Chrysemys picta belli* Gray, 1831 [Alt. Western Painted Turtle]
Bell's Spiny Lizard *Sceloporus bellii* Gray, 1831
Bell's Anole *Leiosaurus belli* **Duméril** and **Bibron,** 1837
Bell's Forest Dragon *Gonocephalus bellii* Duméril and Bibron, 1837
Bell's Skink *Evesia bellii* Duméril and Bibron, 1839
Bell's Turtle *Elseya bellii* Gray, 1844
Bell's Oak Forest Skink *Eumeces lynxe bellii* Gray, 1845 [Syn. *Plestiodon lynxe bellii*]
Tree Iguana sp. *Liolaemus bellii* Gray, 1845

Thomas Bell (1792–1880) was a naturalist and dental surgeon who worked at Guy's Hospital, London (1816–1860). He became Professor of Zoology at King's College (1834) and described many of the reptiles Darwin collected. He was a friend of Gray and President of the Linnean Society. He wrote *History of British Reptiles* (1839).

Belluomini

Ground Snake sp. *Atractus heliobelluomini* Silva Haad, 2004

Professor Dr. Helio Emerson Belluomini of Instituto Butantan, São Paulo, Brazil, is an expert on snake venoms. He co-wrote "Notes on Breeding Anacondas *Eunectes murinus* at Sao Paulo Zoo" (1967).

Belly

Colubrid snake sp. *Alluaudina bellyi* **Mocquard,** 1894

Belly collected the snake with Charles Alluaud. There is a frog *Mantidactylus bellyi* also named after him in 1895 by Mocquard, who gives no details of Belly beyond his family name.

Bennett, D. and C.

PNG Brown White-lipped Python *Leiopython bennettorum* Schleip, 2008 [Originally described as *L. albertisi bennetti* by Hoser, 2000]

Dr. Daniel Bennett is a conservation biologist and herpetologist and Clive Bennett is an Australian Wildlife Officer; they are unrelated. Hoser's description says, "I have once again taken the liberty of naming the subspecies after two people. This includes the UK herpetologist Daniel Bennett, who is perhaps best known for writing a series of books about Monitor lizards. I have also named the subspecies after former NPWS/NSW Wildlife Enforcement Officer Clive Bennett (who coincidentally shares the same name) in recognition of his voluntary conservation work with birds of prey over many years." Clive Bennett played an essential role in having the problem of corruption within his department raised in the New South Wales Parliament and the Independent Commission against Corruption. After failure by these bodies to investigate, Bennett passed his information to Hoser for inclusion in the book *Smuggled—2: Wildlife Trafficking, Crime, and Corruption in Australia* (1996).

Bennett, G.

Bennett's Water Snake *Enhydris bennetti* **Gray,** 1842

Bennett's Two-pored Dragon *Diporiphora bennettii* Gray, 1845 [Alt. Robust Two-line Dragon]

Dr. George Bennett (1804–1893) was a surgeon, botanist, and zoologist. He was surgeon-naturalist on the *Sophia's* voyage to the South Seas and Australia (1834). He returned to Britain but eventually settled permanently in Australia (1836). He was the first Curator and Secretary of the Australian Museum (1835) and an early conservationist, writing (1860), "Many of the Australian quadrupeds and birds are not only peculiar to that country, but are, even there, of comparatively rare occurrence: and such has been the war of extermination recklessly waged against [them], that they are in a fair way of becoming extinct. . . . The Author hopes that what he has been induced to say with reference to this important subject will not be without weight to every thoughtful colonist." He spent 50 years unsuccessfully trying to fully understand monotreme and marsupial biology. He wrote *Gatherings of a Naturalist in Australia—Being Observations Principally on the Animal and Vegetable Productions of New South Wales, New Zealand, and Some of the Austral Islands* (1860). Two birds and a mammal are named after him.

Benson

Benson's Mabuya *Mabuya bensonii* **Peters,** 1867

Stephen Allen Benson (1816–1865) was President of Liberia (1856–1864). Benson was born in Maryland, USA, to freeborn African-American parents. In 1822 his family moved to the newly created country of Liberia. Benson became a successful businessman and private secretary to Thomas Buchanan (Liberia's last white governor). After Liberian independence, he became a judge. He became Vice President (1853), then President. He obtained diplomatic recognition by the USA (1862), and in that year he visited Europe. Previous administrations emphasized the superiority of the freed slaves and their western customs, but Benson sought collaboration with local people, learning several languages. He retired to a plantation. Peters' description implies that the type specimen of this skink was received directly from Benson.

Bent

Bent's Mastigure *Uromastyx benti* **Anderson,** 1894

James Theodore Bent (1852–1897) was an explorer, archeologist, and author who graduated from Oxford (1875). He made many foreign trips accompanied by his wife, Mabel Hall-Dare, who was a photographer. They traveled in Asia Minor, making archeological excavations there and in Persia (1889–1890), then made their first trip to East Africa, investigating ruins in Abyssinia (now Ethiopia) (1893). They also traveled in the Hadramut region of southern Arabia (1893–1894). He contracted malaria (1897) and returned to London, where he died. Among his publications is *The Ruined Cities of Mashonaland* (1892), and his wife published *Southern Arabia, Soudan, and Sakotra* (1900).

Bequaert

Bequaert's Green Snake *Philothamnus bequaerti* **Loveridge,** 1951

Dr. Joseph Charles Bequaert (1886–1982) was a Belgian botanist, entomologist, and malacologist. He graduated with a doctorate in botany from the University of Ghent (1906) and worked for the colonial government in the Belgian Congo (1910–1915). He moved to the USA (1916), becoming a U.S. citizen (1921). He was a Research Assistant, American Museum of Natural History (1917–1922), then worked at Harvard (1923–1956), initially teaching entomology at Harvard Medical School and finally becoming Professor of Zoology, Museum of Comparative Zoology. In retirement he became Professor of Biology, University of Houston (1956–1960), and Visiting Entomologist, University of Arizona. Among other works he co-wrote *The Mollusks of the Arid Southwest* (1973).

Beraducci

Maasai Girdled Lizard *Cordylus beraduccii* **Broadley** and Branch, 2002

Beraducci's Pygmy Chameleon *Rhampholeon beraduccii* Mariaux and **Tilbury,** 2006 [Alt. Mahenge Pygmy Chameleon]

Joe Beraducci is an Italian collector who works for the Mountain Birds and Trophies Snake Farm and Reptile Centre, Arush, Tanzania. He collected the lizard holotype (1999).

Berber

Berber's Skink *Eumeces algeriensis* **Peters,** 1864 [Alt. Algerian Skink]

The name refers to the Berber people of North Africa.

Berdmore

Berdmore's Water Skink *Tropidophorus berdmorei* **Blyth,** 1853

Oriental Leaf-toed Gecko *Leiurus berdmorei* Blyth, 1853 [Junior syn. of *Hemidactylus bowringii* Gray, 1845]

Captain Thomas Matthew Berdmore (1811–1859) amassed, with Theobald (q.v.), a significant collection of specimens that he presented to the Asiatic Society of Bengal (1856). Several fish, an amphibian, and two mammals are named after him.

Berengere

San Andrés Mabuya *Mabuya berengerae* Miralles, 2006

Bérengère Miralles is the wife of the describer, Aurélien Miralles of Muséum National d'Histoire Naturelle, Paris.

Berger

Berger's Cape Tortoise *Homopus bergeri* Lindholm, 1906

Dr. Arthur Berger (1871–1947) was a German physician who spent his life as a hunter, explorer, traveler, and zoologist. He traveled in the Arctic (1900) and Africa (1908–1909) and visited both India and the USA.

Berghof

Berghof's Day Gecko *Phelsuma berghofi* Krüger, 1996

Hans-Peter Berghof is a German herpetologist, an expert on *Phelsuma* (day geckos), and has spent much time in, and written about, Madagascar.

Berlandier

Berlandier's Tortoise *Gopherus berlandieri* **Agassiz,** 1857 [Alt. Texas Tortoise]

Jean Louis Berlandier (1805–1851) was a Belgian botanist who went to Mexico (1826) to work as a collector. The Mexican government employed him (1827–1828), and he stayed on, marrying a local woman, dividing his time between a pharmaceutical business and collecting botanical specimens. The Mexican government employed him again (1834) to serve as an interpreter to General Arista and to be in charge of the hospitals at Matamoros during the Mexican War. He was drowned while trying to cross the San Fernando River.

Bernad

Bernad's Coral Snake *Micrurus bernadi* **Cope,** 1887 [Alt. Blotched Coral Snake]

Dr. Santiago Bernad was a French physician who practiced in Zacualtipan, Mexico, where he collected both extant and fossil animals. He provided Cope with the coral snake holotype. (We have also found references to this species with the misprint/typo *bernardi.*)

Bernard

Bernard's Dwarf Gecko *Lygodactylus bernardi* **FitzSimons,** 1958 [Alt. FitzSimon's Dwarf Gecko]

Bernard Evelyn Buller Fagg (1915–1987) was an archeologist and anthropologist. After graduating from Cambridge he worked for the British colonial administration in Jos, Nigeria (1939), and spent many years studying antiquities of the Nok culture. He was a museum curator, becoming head of the Nigerian National Museum (1957) and Curator, Pitt Rivers Museum, Oxford (1958). He excavated (1958) at Inyanga in Southern Rhodesia (Zimbabwe).

Bernier

Bernier's Striped Snake *Dromicodryas bernieri* **Duméril** and **Bibron,** 1854

Chevalier Alphonse Charles Joseph Bernier (1802–1858) was a naval surgeon, botanist, and collector who spent time in Madagascar (1831–1834). He took 198 specimens back to France, where they were catalogued (1835). Three birds are named after him.

Berthold

Berthold's Bush Anole *Polychrus gutturosus* Berthold, 1846

Jan's Banded Snake *Simoselaps bertholdi* **Jan,** 1859

Berthold's Graceful Brown Snake *Rhadinaea lateristriga* Berthold, 1859

Berthold's Worm Lizard *Leposternon infraorbitale* Berthold, 1859

Berthold's False Fer-de-Lance *Xenodon bertholdi* Jan, 1863 [Junior syn. of *X. rabdocephalus* **Wied,** 1824]

Dr. Arnold Adolph Berthold (1803–1861) was a physiologist and pioneer endocrinologist who was Professor of Zoology, Göttingen University, Germany, but continued to practice medicine, as his university salary was poor. His work on hormones was virtually ignored for half a century, but he is now recognized as the founder of experimental endocrinology. He succeeded Blumenbach as Curator of the university's museum and made significant contributions to herpetological systematics. He was the co-discoverer with Bunsen (of "Bunsen burner" fame) of the antidote for arsenic poisoning (1834). He published 15 papers on herpetology (1840–1850). He died of typhus.

Betsch

Blanc's Leaf Chameleon *Brookesia betschi* **Brygoo, Blanc,** and **Domergue,** 1974

Professor Jean-Marie Betsch is an entomologist who has worked in many parts of the tropics and is now Director of Muséum National d'Histoire Naturelle, Paris.

Beu

Blind Snake sp. *Liotyphlops beui* **Amaral,** 1924

T. Beu collected the holotype in São Paulo, Brazil.

Beyer

Beyer's Sphenomorphus *Sphenomorphus beyeri* **Taylor,** 1922

Henry Otley Beyer (1883–1966) was an ethnologist who spent much of his career in the Philippines. His bachelor's degree was from Iowa State College (1904) and his master's in chemistry from the University of Denver, Colorado (1905). He joined the Philippines Ethnological Survey (1905), being based in Manila until he returned to the USA (1908) and did a postgraduate program in ethnology at Harvard. He joined the Philippine Bureau of Science as an ethnologist (1909) and carried out fieldwork with a number of indigenous peoples (1910–1915). He was one of the founders of the Department of Anthropology, University of the Philippines (1914), becoming head of that department (1925). He founded the university's museum and Institute of Archaeological Ethnology and presented it with his own collections of artifacts. During the Japanese occupation in WW2, he was interned for two years but fortunately Tado Kano, a Japanese ethnologist, helped him to save his papers and collections. He retired in 1947 but remained associated with the university and was made Emeritus Professor of Anthropology (1954). He died in the Philippines.

Beyschlag

Sumatra Forest Dragon *Gonocephalus beyschlagi* **Boettger,** 1892

Fritz Beyschlag was on a plantation at Langkat, near Deli, Sumatra, where the holotype of this lizard was collected. He also sent other zoological specimens to Naturmuseum Senckenberg, Frankfurt (1891–1893).

Bezy

Bezy's Night Lizard *Xantusia bezyi* **Papenfuss,** Macey, and Schulte, 2001

Dr. Robert Lee Bezy (b. 1941) is Curator Emeritus of Herpetology, Natural History Museum of Los Angeles County. He studied at the University of Arizona. He co-wrote "Reproduction in the Island Night Lizard, *Xantusia riversiana*" (1974).

Bibron

Bibron's Mabuya *Mabuya bibronii* **Gray,** 1838
Chameleon sp. *Chamaeleo bibroni* Martin, 1838 [Junior syn. of *C. oweni* Gray, 1831]
Bibron's Skink *Macroscincus coctei* **Duméril** and Bibron, 1839 EXTINCT [Alt. Cocteau's Skink, Cape Verde Giant Skink]
Bibron's Whiptail *Cnemidophorus lacertoides* Duméril and Bibron, 1839
Bengal Black Monitor *Varanus bibronii* **Blyth,** 1842 [Junior syn. of *V. bengalensis* **Daudin,** 1758]
Bibron's Iguana *Diplolaemus bibronii* **Bell,** 1843
Bibron's Tree Iguana *Liolaemus bibronii* Bell, 1843
Pacific Boa *Candoia bibroni* Duméril and Bibron, 1844
Bibron's Blind Snake *Typhlops bibronii* **Andrew Smith,** 1846
Bibron's Gecko *Pachydactylus bibroni* Andrew Smith, 1846
Bibron's Burrowing Asp *Atractaspis bibroni* Andrew Smith, 1849 [Alt. Bibron's Mole Viper]
New Guinea Giant Softshell Turtle *Pelochelys bibroni* **Owen,** 1853
Bibron's False Coral Snake *Oxyrhopus doliatus* Duméril, Bibron, and **Duméril,** 1854
Bibron's Coral Snake *Calliophis bibroni* **Jan,** 1858
Fathead Anole *Enyalius bibronii* **Boulenger,** 1885

Gabriel Bibron (1806–1848) was a zoologist and herpetologist. He worked closely with André Marie Constant Duméril at Muséum National d'Histoire Naturelle, Paris. Duméril was most interested in dissection and left the naming of species largely to Bibron, technically only his assistant. Bibron resigned (1845), later dying of tuberculosis.

Bica

Anole sp. *Anolis bicaorum* Köhler 1996

Named after the members of the Bay Island Conservation Association (BICA), Honduras.

Bigmore

Stuart Bigmore's Python *Broghammerus reticulatus stuartbigmorei* Hoser, 2003

Stuart Bigmore is an amateur herpetologist, reptile keeper, and breeder. The etymology for this race of Reticulated Python reads, "Named after Stuart Bigmore of Victoria, Australia for his contributions to herpetology over two or more decades, in particular varanid taxonomy." He is a leading light in the Victorian Association of Amateur Herpetologists.

Bignell

Forest Skink sp. *Sphenomorphus bignelli* **Schmidt,** 1932

Charles Robert Bignell (1892–1964) was a planter at Ysabel, Solomon Islands. He and his wife had a reputation for being interested in zoology and as being hospitable to visiting collectors. William M. Mann had stayed with them (1912), as did Schmidt on his visit. He stated that he had collected specimens of a gecko species (*Lepidodactylus lugubris*) "from the walls of the plantation house of Mr. C. R. Bignell."

Birula

Tzarewsky's Toadhead Agama *Phrynocephalus birulai* **Tzarevsky,** 1927

Dr. A. A. Bialynicky-Birula (1864–1938) was Director of the Zoological Museum, Leningrad (St. Petersburg). Various other taxa are named after him.

Bischoff

Wall Gecko sp. *Tarentola bischoffi* **Joger,** 1984
Bischoff's Beaked Snake *Rhinotyphlops episcopus* Franzen and **Wallach,** 2002

Wolfgang Bischoff is a herpetologist working at the University of Ulm, Germany. He often writes with Joger, the gecko's describer. The Beaked Snake's binomial *episcopus* means "bishop" or, in German, *Bischoff.*

Blainville

Blainville's Horned Lizard *Phrynosoma blainvillii* **Gray,** 1839 [Alt. San Diego Horned Lizard]

Henry Marie Ducrotay de Blainville (1777–1850) was a zoologist and anatomist. He came late to science, having lost his father (d. 1783) and his mother to imprisonment. He started attending lectures (1802) at Collège de France, later qualifying as a physician (1808). He became one of Cuvier's bitterest rivals over animal classification, nevertheless succeeding Cuvier to the Comparative Anatomy Chair, Muséum National d'Histoire Naturelle and in Collège de France. He had much influence establishing skeletal evolution as one determinant of classification. He separated amphibians from reptiles on the basis of their lack of scales. He wrote *Cours de physiologie generale et comparée* (1829). A bird and five mammals are named after him.

Blair, H. S.

Blair's Bachia *Bachia blairi* **Dunn,** 1940

Henry Sterling Blair was General Manager of the Chiriqui Land Company, Armuelles, Panama, where his employer (United Fruit Company) had a banana plantation. He was manager of the Almirante division (1919) and was clearly interested in herpetology, as he donated a collection of reptiles from Costa Rica to the Museum of Comparative Zoology (1916). The etymology states that Blair's "assistance and hospitality has been deeply appreciated by many workers, including ourselves."

Blair, W. F.

Blair's Kingsnake *Lampropeltis alterna blairi* Flury, 1950

Dr. William Frank Blair (1912–1984) was a zoologist who specialized in the hybrid zones of mammals. Later he taught herpetology at the University of Texas, becoming Professor of Biology (1955). He was awarded his bachelor's degree by the University of Tulsa (1934), his master's by the University of Florida (1935), and his doctorate by the University of Michigan (1938), where he remained as a Research Associate before service in the U.S. Army Air Corps (1942–1946). After WW2 he joined the faculty of the University of Texas. He started the Texas Natural History Collections (1946), concentrating on mammals, amphibians, and reptiles. He retired as Professor Emeritus (1982). He wrote *The Rusty Lizard: A Population Study* (1960). The annual W. Frank Blair Eminent Naturalist Award was established in his honor.

Blake

Blake's Anadia *Anadia blakei* **Schmidt,** 1932

Emmet "Bob" Reid Blake (1908–1997) was an ornithologist. By age 15 he was so good at capturing reptiles that he became known as "Snaky" Blake. After graduating (1928), he roller-skated 1,500 kilometers (900 miles) to join up as a part-time graduate student at the University of Pittsburgh. His first expedition was to the unexplored Rio Negro Venezuela–Brazil border under the auspices of the *National Geographic.* The Field Museum employed him as a collector (1931) in Venezuela, where he took many specimens. He returned (1933) to Pittsburgh to take his M.S. and was invited on a Field Museum trip to Guatemala. He joined the Field Museum's staff (1935–1973), collecting in the neotropics (1935) and becoming Assistant Curator of Birds (1936), and was actively collecting in Mexico during the 1940s and 1950s, retiring as Emeritus Curator (1973). He wrote *Manual of Neo-tropical Birds* (1977). Two birds are named after him.

Blakeway

Blakeway's Mountain Snake *Plagiopholis blakewayi* **Boulenger,** 1893

Lieutenant Blakeway resigned from the British army (1861) and collected reptiles in Toungyi, South Shan States (Myanmar) (1890s).

Blanc, C. P.

Blanc's Dwarf Gecko *Lygodactylus blanci* **Pasteur,** 1967
Blanc's Leaf Chameleon *Brookesia betschi* **Brygoo, Blanc,** and **Domergue,** 1974
Malagasy Night Snake sp. *Ithycyphus blanci* Domergue, 1988

Emeritus Professor Charles Pierre Blanc (b. 1933) worked at the Laboratory of Zoogeography, Université Paul Valéry, Montpellier, France. He has described other species with Pasteur, and has made many trips to Madagascar. He wrote the *Reptiles, Sauriens, Iguanidae* section in the series *Faune de Madagascar* (1977).

Blanc, F.

Dwarf Gecko sp. *Lygodactylus blancae* **Pasteur,** 1995

Françoise Blanc is a geneticist and biologist at the Laboratory of Zoogeography and Genetics, Université Paul Valéry, Montpellier, France, where she is a Professor. She has worked on pearl oysters.

Blanc, H.

Blanc's Psammodromus *Psammodromus blanci* **Lataste,** 1880

Dr. Henri Blanc (1859–1930) was a zoologist who was Director of the Zoological Museum, Lausanne. He took his initial degree at Stuttgart (1877) and received his doctorate from Freiburg University (1880). He worked at the natural history museums of Kiel and Berlin (1880–1883), returning to Lausanne (1883) to the Faculty of Medicine, becoming Professor of Science (1891). He founded the Department of Zoology and Comparative Anatomy (1890) and remained as Director until retiring (1929). He was also Director of the Zoological Museum of the Canton of Vaud (1904).

Blanc, M.

Blanc's Fringe-fingered Lizard *Acanthodactylus blanci* **Doumergue,** 1901

Doumergue's etymology states that "M Blanc" sent the lizard holotype to him from Tunisia; it is likely that M stands for "Monsieur," rather than being an initial. We cannot further identify this individual.

Blanchard

Blanchard's Helmet Skink *Tribolonotus blanchardi* Burt, 1930

Blanchard's Milk Snake *Lampropelitis triangulum blanchardi* Stuart, 1935

Blanchard's Earth Snake *Geophis blanchardi* Taylor and **H. M. Smith,** 1939

Western Smooth Green Snake *Opheodrys vernalis blanchardi* Grobman, 1941

Dr. Frank Nelson Blanchard (1888–1937) graduated from Tufts University (1913) and taught zoology at an agricultural college (1913–1916). After a zoology fellowship and a doctorate from the University of Michigan (1919) and working briefly at the Smithsonian under Stejneger, he returned to the University of Michigan (1920) to teach zoology, becoming Assistant Professor (1926), then Associate Professor (1934). He is best known for his study of *Lampropeltis* and *Diadophis* snakes. He traveled in New Zealand, Australia, and Tasmania (1927–1928). He was Vice President of the American Society of Ichthyologists and Herpetologists (1936–1937). He first major work was *A Revision of the King Snakes: Genus Lampropeltis* (1921).

Blanding

Blanding's Turtle *Emydoidea blandingii* **Holbrook,** 1838

Blanding's Cat Snake *Boiga blandingii* **Hallowell,** 1844 [Alt. Blanding's Tree snake]

Dr. William Blanding (1772–1857) was an American physician, chemist, ornithologist, amateur herpetologist, and numismatist whose hobby was manufacturing "ancient" coins. He was interested in archeology and native culture and was an active member of the Philadelphia Academy of Natural Sciences. Holbrook's etymology states, "This animal was first observed by Dr. William Blanding, of Philadelphia, an accurate Naturalist, whose name I have given to the species."

Blanford

Blanford's Mabuya *Mabuya innotata* Blanford, 1870

Blanford's Rock Agama *Psammophilus blanfordanus* **Stoliczka,** 1871

Blanford's Greater Spider Gecko *Agamura cruralis* Blanford, 1874

Blanford's Middle-toed Gecko *Mediodactylus heterocercus* Blanford, 1874

Blanford's Semaphore Gecko *Pristurus rupestris* Blanford, 1874

Blanford's Bridle Snake *Dryocalamus davisonii* Blanford, 1878

Blanford's Pipe Snake *Cylindrophis lineatus* Blanford, 1881

Blanford's Rock Gecko *Pristurus insignis* Blanford, 1881

Blanford's Spotted Water Snake *Enhydris maculosa* Blanford, 1881

Blanford's Flying Lizard *Draco blanfordii* **Boulenger,** 1885

Blanford's Ground Gecko *Bunopus blanfordii* **Strauch,** 1887

Blanford's Snake Skink *Ophiomorus blanfordi* Boulenger, 1887

Worm Snake sp. *Typhlops blanfordii* Boulenger, 1889

Blanford's Worm Snake *Leptotyphlops blanfordi* Boulenger, 1890

Blanford's Toadhead Agama *Phrynocephalus blanfordi* **Bedriaga,** 1909

Blanford's Fringe-fingered Lizard *Acanthodactylus blanfordii* Boulenger, 1918

Anderson's Agama *Trapelus blanfordi* **Anderson,** 1966

William Thomas Blanford (1832–1905) was a geologist and zoologist. He studied at the Royal School of Mines (1852–1854) and at Freiberg, Saxony, before joining the Indian Geological Survey (1854). He undertook a geological survey of Burma (Myanmar) (1860) and was appointed Deputy Superintendent. He surveyed in Bombay (1862–1866) and was then attached to the Abyssinian expedition (1867). He wrote the mammals section of *The Scientific Results of the Second Yarkand Mission: Mammalia* (1879). Ill health forced early retirement, and he returned to England (1881). In retirement he edited works for the government on Indian fauna and was President of the Royal Geographical Society (1888–1890). Six birds and six mammals are named after him.

Bleck

Bleck's Kukri Snake *Oligodon waandersi* **Bleeker,** 1860

This is an apparent transcription error for Bleeker's Kukri Snake.

Bleeker

Bleeker's Dwarf Snake *Calamaria margaritophora* Bleeker, 1860

Bleeker's Forest Dragon *Gonocephalus megalepis* Bleeker, 1860

Bleeker's Kukri Snake *Oligodon waandersi* Bleeker, 1860

Dr. Pieter Bleeker (1819–1878) was an ichthyologist and army surgeon commissioned (1841) by the Dutch East India Company. He was apprenticed to an apothecary (1831–1834) and became interested in anatomy and zoology. He qualified as a surgeon at Haarlem (1840) and went to Paris, working in hospitals while attending Blainville's lectures. He was stationed in the Dutch East Indies (Indonesia) (1842–1860) and acquired 12,000 specimens of fishes, most of which are today in Rijksmuseum van Natuurlijke Histoire, Leiden. He returned to Holland (1860), taking his collection with him, which was sold after his death. He wrote over 500 scientific papers of which only about 15 were on herpetology. He wrote a monumental work, *Atlas ichthyologique des Indies Orientales Neerlandaises* (1862–1877).

Blochmann

Zaire Three-toed Skink *Leptosiaphos blochmanni* **Tornier,** 1903

Professor Friedrich Johann Wilhelm Blochmann (1858–1931) was a German zoologist, microbiologist, and entomologist who studied ants. He was a member of the faculty at Eberhard Karls Universität Tübingen (1925).

Blomhoff

Asiatic Pit-viper *Gloydius blomhoffi* **H. Boie,** 1826 [Alt. Mamushi]

Jan Cock Blomhoff (1779–1853), who collected the viper holotype, was a trader who was manager (1817–1824) of the trading colony set up by the Dutch East India Company at Deshima Island in the harbor at Nagasaki, Japan.

Blyth

Blyth's Earth Snake *Rhinophis blythii* **Kelaart,** 1853

Blyth's Reticulate Snake *Blythia reticulata* Blyth, 1854

Skink sp. *Eumeces blythianus* **Anderson,** 1871

Edward Blyth (1810–1873) was a zoologist and author. He was Curator of the Museum, Asiatic Society of Bengal (1842–1864). Arthur Grote said of him, "Had he been a less imaginative and more practical man, he must have been a prosperous one. . . . All that he knew was at the service of everybody. No one asking him for information asked in vain." His monographs were collected and published posthumously under the title *The Natural History of Cranes* (1881). Nineteen birds and three mammals are named after him.

Bocage

Bocage's Chameleon *Chamaeleo quilensis* Bocage, 1866

Bocage's Mabuya *Trachylepis binotata* Bocage, 1867 [Alt. Ovambo Tree Skink; Syn. *Mabuya binotata*]

Bocage's Sand Lizard *Pedioplanis benguelensis* Bocage, 1867

Bocage's Serpentiform Skink *Euprepes binotatus* Bocage, 1867

Bocage's Wall Lizard *Podarcis bocagei* **Seoane,** 1884

Bocage's Blind Snake *Leptotyphlops rostratus* Bocage, 1886

Bocage's Mole Viper *Atractaspis dahomeyensis* Bocage, 1887

Skink sp. *Trachylepis bocagii* **Boulenger,** 1887

Bocage's Horned Adder *Vipera heraldica* Bocage, 1889 [Alt. Angolan Adder]

José Vicente Barboza du Bocage (1823–1907) was Director, Museu de História Natural de Sintra, Portugal, which is now named in his honor. He became known as the father of Angolan ornithology and wrote *Ornithologie d'Angola*. Six birds and five mammals are named after him. See also **Barboza.**

Bock

Bock's Ground Snake *Atractus bocki* **Werner,** 1909

Carl Alfred Bock (1849–1932) was a Norwegian naturalist and ethnologist. He traveled in the Dutch East Indies (Indonesia), Thailand, and Laos (1878–1883). He published *The Head-Hunters of Borneo* (1881). Various London museums and the Ethnographic Museum, Oslo, hold most of his collection.

Bocourt

Bocourt's Water Snake *Enhydris bocourti* **Jan,** 1865

Bocourt's Arboreal Alligator Lizard *Abronia vasconcelosii* Bocourt, 1871

Ecuadorian Coral Snake *Micrurus bocourti* Jan, 1872

Bocourt's Anole *Anolis baccatus* Bocourt, 1873

Bocourt's Emerald Lizard *Sceloporus smaragdinus* Bocourt, 1873

Bocourt's Spiny Lizard *Sceloporus acanthinus* Bocourt, 1873

Bocourt's Ameiva *Ameiva edracantha* Bocourt, 1874

Bocourt's Dwarf Iguana *Enyalioides heterolepis* Bocourt, 1874

Anole sp. *Anolis bocourtii* **Cope,** 1876

Bocourt's Skink *Phoboscincus bocourti* **Brocchi,** 1876 [Alt. Bocourt's Eyelid Skink]

Bocourt's Redback Coffee Snake *Ninia sebae punctulata* Bocourt, 1883
Bocourt's Agama *Agama bocourti* Rochebrune, 1884
Bocourt's Snail-eater *Dipsas viguieri* Bocourt, 1884
Bocourt's Spiny Lizard *Sceloporus occidentalis bocourtii* Boulenger, 1885 [Alt. Coast Range Fence Lizard]
Bocourt's Tropical Racer *Masticodryas dorsalis* Bocourt, 1890
Bocourt's Black-headed Snake *Tantilla bocourti* **Günther,** 1895
Bocourt's Snake-eater *Polemon bocourti* **Mocquard,** 1897
Bocourt's Ground Snake *Atractus bocourti* Boulenger, 1904

Marie Firmin Bocourt (1819–1904) was a zoologist and artist. He followed his father, who engraved copper plates for Muséum National d'Histoire Naturelle, Paris. He became a preparator for Bibron (1834). He was officially designated "Museum Painter" (1854). He was sent to Siam (Thailand) (1861), where he made an important collection that he took back to Paris. He visited Mexico and Central America (1864–1866). He co-wrote, with Duméril and Mocquard, *Études sur les reptiles et les batraciens* (1870).

Boddaert

Boddaert's Tropical Racer *Mastigodryas boddaerti* Sentzen, 1796

Dr. Pieter Boddaert (1730–1796) was a physician, naturalist, zoologist, ornithologist, and physiologist who lectured on natural history at the University of Utrecht (1793) and corresponded regularly with Linnaeus.

Boelen

Boelen's Python *Morelia boeleni* **Brongersma,** 1953 [Syn. *Liasis boeleni*]

K. W. J. Boelen was the Government Surgeon at Enarotali, Irian Jaya, Indonesia (1950s and 1960s). He wrote *Doctor at the Wissel Lakes* (1954), describing the life and habits of the mountain Papuans, who had been recently discovered (1936), and their reactions to the modern world. Brongersma wrote, "My sincere thanks are due to K. W. J. Boelen, M.D., government surgeon at Enarotali, for his successful efforts to procure a specimen of this interesting species for our museum."

Boeseman

Boeseman's Reed Snake *Calamaria boesemani* **Inger** and **Marx,** 1965

Dr. Marinus Boeseman (1916–2006) was an ichthyologist, Department of Zoology, Rijksmuseum van Natuurlijke Histoire, Leiden, becoming Curator of Fishes (1947–1981). His master's degree was awarded by Universiteit Leiden (1941). He was in the Dutch resistance during WW2 and was arrested (1943) but survived imprisonment at Dachau, though for years his health was so badly affected that he could not work. He collected fishes in El Salvador (1953) and traveled in New Guinea (1954–1955). He caught polio (1957) and suffered from a permanent disability in his right arm, but still took part in collecting expeditions to, inter alia, Surinam and Trinidad.

Boettger, C. R.

Boettger's Lizard *Gallotia caesaris* Lehrs, 1914

Caesar Rudolf Boettger (1888–1976) was a noted malacologist and the nephew of Dr. Oskar Boettger (q.v.).

Boettger, O.

Boettger's Keelback *Rhabdophis callistus* **Günther,** 1873
Boettger's Worm Snake *Typhlops mucronatus* Boettger, 1880
Puerto Rican Boa *Piesigaster boettgeri* **Seoane,** 1881 [Junior syn. of *Epicrates inornatus* Reinhardt, 1843]
Nosy Bé Flat-tailed Gecko *Uroplatus boettgeri* **Fischer,** 1884
Boettger's Sipo *Chironius flavolineatus* Boettger, 1885
Boettger's Worm Lizard *Leposternon boettgeri* **Boulenger,** 1885
Boettger's Mabuya *Mabuya boettgeri* Boulenger, 1887
Boettger's Chameleon *Calumma boettgeri* Boulenger, 1888
Boettger's Tortoise *Testudo boettgeri* Mojsisovics, 1889 [Syn. *T. hermanni boettgeri*]
Boettger's Girdled Lizard *Zonosaurus boettgeri* **Steindachner,** 1891
Boettger's Wall Gecko *Tarentola boettgeri* Steindachner, 1891
Boettger's Dwarf Racer *Eirenis punctatolineata* Boettger, 1892
Bearded Pygmy Chameleon *Rhampholeon boettgeri* **Pfeffer,** 1893
Boettger's Kentropyx *Kentropyx paulensis* Boettger, 1893
Boettger's Ground Snake *Atractus boettgeri* Boulenger, 1896
Boettger's Two-headed Snake *Micrelaps boettgeri* Boulenger, 1896 [Alt. Desert Black-headed Snake]
Boettger's Snail-eater *Dipsas boettgeri* **Werner,** 1901
Boettger's Anole *Anolis boettgeri* Boulenger, 1911
Boettger's Whorl-tailed Iguana *Stenocercus boettgeri* Boulenger, 1911
Boettger's Ground Skink *Scincella boettgeri* **Van Denburgh,** 1912
Boettger's Day Gecko *Phelsuma v-nigra* Boettger, 1913
Boettger's Dwarf Gecko *Lygodactylus heterurus* Boettger, 1913

Boettger's Madagascar Snake *Heteroliodon torquatus* **C. R. Boettger,** 1913
Boettger's Emo Skink *Emoia boettgeri* **Sternfeld,** 1918
Oscar Boettger's Tortoise *Testudo oscarboettgeri* Lindholm, 1929

Professor Dr. Oskar Boettger (or Böttger) (1844–1910) was a German zoologist who specialized in herpetology and malacology. He went to the School of Mines, Freiberg, intending to become a mining engineer, but political unrest in Germany prevented him from finding work. He returned to his studies and undertook a doctorate in paleontology at the University of Würzburg (1869). He joined the Senckenberg Museum, Frankfurt (1870), as a paleontologist, becoming Curator of Herpetology (1875). He was unpaid, and he supported himself by taking teaching posts in Offenbach and Frankfurt. He was severely agoraphobic and hardly left his house (1876–1894), but as an avid philatelist the prospect of a rare stamp would get him out. He always asked his correspondents to send him postage stamps of the countries they visited.

Bogadek

Bogadek's Blind Skink *Dibamus bogadeki* **Darevsky,** 1992

Father Anthony Bogadek is a Salesian priest in Hong Kong, where he went in 1949 to study philosophy. He studied theology in England (1954–1958) and after being ordained (1958) returned to Hong Kong. He studied biochemistry and biology at University College Dublin (1964–1968) and returned to Hong Kong to teach biology, with herpetology as his speciality, at St. Louis School. Since retiring from teaching (1994) he has continued to run the school's biology laboratory. He co-wrote *Hong Kong Amphibians and Reptiles* (1986).

Bogert

Bogert's Gecko *Bogertia lutzae* **Loveridge,** 1941
Bogert's Zebra-tailed Lizard *Callisaurus draconoides bogerti* **Martin,** 1943
Bogert's Garter Snake *Thamnophis bogerti* **Hartweg,** 1944
Bogert's Rock Gecko *Afroedura bogerti* Loveridge, 1944
Tucson Banded Gecko *Coleonyx variegates bogerti* **Klauber,** 1945
Bogert's Monitor *Varanus bogerti* **Mertens,** 1950
Bogert's Emo Skink *Emoia bogert* **Brown,** 1953
Bogert's Arboreal Alligator Lizard *Abronia bogerti* Tihen, 1954
Keeled Lava Lizard *Tropidurus bogerti* **Roze,** 1958
Bogert's Blind Dart Skink *Typhlacontias bogerti* **Laurent,** 1966
Bogert's Coral Snake *Micrurus bogerti* Roze, 1967
Bogert's Boa *Exiliboa placata* Bogert, 1968 [Alt. Oaxacan Dwarf Boa]
Oaxacan Graceful Brown Snake *Rhadinaea bogertorum* **Myers,** 1974
Guatemalan Bearded Lizard *Heloderma horridum charlesbogerti* **Campbell** and Vannini, 1988
Rat Snake genus *Bogertophis* Dowling and Price, 1988
Bogert's Garter Snake *Thamnophis bogerti* Rossman and Burbrink, 2005

Charles Mitchill Bogert (1908–1992) was a zoologist and herpetologist. He worked as a technician with the Los Angeles City Schools' Division of Nature Study (1928). He was a guide, Rocky Mountain National Park (1930–1932), then a ranger naturalist, Grand Canyon (1932–1934). His bachelor's (1934) and master's (1936) degrees were awarded by the University of California, where he worked as a teaching assistant (1934–1936). He was Assistant Curator, Herpetology Department, American Museum of Natural History (1936–1940), becoming Associate Curator (1941) and Curator (1943), retiring as Emeritus Curator. He did much work on the fauna of Mexico, where he felt at home, and made recordings of indigenous folk music that were later commercially released. He conducted research (1948–1959) in Central America and Bimini Island (Bahamas). He was President of the American Society of Ichthyologists and Herpetologists (1952–1954). Among his publications, he co-wrote *The Gila Monster and Its Allies* (1956). He had a stroke (1988), feared having to go into a nursing home, and committed suicide. *Rhadinaea bogertorum* is named after Bogert and his wife, Martha.

Böhme

Böhme's Five-toed Skink *Leptosiaphos ianthinoxantha* Böhme, 1975
Böhme's Pit-viper *Gloydius halys boehmei* **Nilson,** 1983
Böhme's Gecko *Tarentola boehmei* **Joger,** 1984 [Alt. Morocco Wall Gecko]
Böhme's Mountain Gecko *Alsophylax boehmei* Shcherbak, 1991
Böhme's Ethiopian Mountain Snake *Pseudoboodon boehmei* **Rasmussen** and **Largen,** 1992
Böhme's Butterfly Lizard *Leiolepis boehmei* **Darevsky** and Kupriyanova, 1993
Böhme's Two-horned Chameleon *Kinyongia boehmei* Lutzmann and **Necas,** 2002
Golden-spotted Tree Monitor *Varanus boehmei* Jacobs, 2003
Skink sp. *Lygosoma boehmei* Ziegler et al., 2007
Boehme's Water Skink *Tropidophorus boehmei* Nguyen et al., 2010

Dr. Wolfgang Böhme (b. 1944) is a German zoologist and herpetologist. He studied zoology, botany, and paleontology for his doctorate at the University of Kiel (1971).

He became Curator of Herpetology and Deputy Director, Zoological Department, Zoologisches Forschungsmuseum Alexander Koenig, Bonn (1971), and Professor of Zoology, Friedrich William University, Bonn (1988). In 1985 he saw an apparently unknown lizard in a television program and so discovered the Yemen Monitor.

Boie

Boie's Many-tooth Snake *Sibynophis geminatus* H. Boie, 1826
Boie's Dwarf Snake *Calamaria virgulata* H. Boie, 1827
Boie's Ground Snake *Atractus badius* F. Boie, 1827
Boie's Keelback *Rhabdophis spilogaster* F. Boie, 1827
Boie's Kukri Snake *Oligodon bitorquatus* F. Boie, 1827
Boie's Rough-sided Snake *Aspidura brachyorrhos* F. Boie, 1827
Boie's Smooth Snake *Gongylosoma baliodeirus* F. Boie, 1827
Boie's Whip Snake *Ahaetulla prasina* F. Boie, 1827
Boie's Sea Snake *Enhydrina schistosa* **Daudin,** 1803
Boie's Day Gecko *Cnemaspis boiei* **Gray,** 1842

There were two German naturalists called Boie; they were brothers. Heinrich Boie (1794–1827) was an explorer and zoologist. He studied under Blumenbach at Göttingen and worked as an assistant to Temminck at Rijksmuseum van Natuurlijke Histoire, Leiden. After Kuhl died, Boie replaced him, went to the Dutch East Indies, and was at Buitenzorg (Bogor), Java, when he died. He wrote some papers with his brother Friedrich (1789–1870), who was a lawyer, herpetologist, ornithologist, and entomologist. Both brothers described reptiles, but Heinrich, after whom the Day Gecko may be named, was more interested in exploring.

Boivin

Velvet Gecko sp. *Blaesodactylus boivini* **Duméril,** 1856

Louis Hyacinthe Boivin (1808–1852) was a botanist and traveler who collected in Madagascar, the Comoro Islands, Réunion, the Canary Islands, and the coasts of Africa for Muséum National d'Histoire Naturelle, Paris.

Bojer

Bojer's Skink *Gongylomorphus bojerii* Desjardins, 1831

Wenzel Bojer (1800–1856) was a Czech naturalist. He collected in tropical Africa. He was Director, Royal Botanic Gardens, Pamplemousses, Mauritius (1948–1949). A bird is named after him.

Bolson

Bolson Tortoise *Gopherus flavomarginatus* Legler, 1959
Bolson Night Lizard *Xantusia bolsonae* **Webb,** 1970

"Bolson" is not a personal name but rather a denotes a flat desert valley surrounded by mountains.

Bolyer

Round Island Boa genus *Bolyeria* **Gray,** 1842

We believe the genus may be named after Friedrich Boie (1789–1870) (q.v.), who originally named this boa *Eryx multocarinata* (1827). Gray later elevated it to a new single-species genus (1842).

Bonnal

Pyrenean Rock Lizard *Iberolacerta bonnali* Lantz, 1927

M. le Comte de Bonnal lived at Montgaillard in the Pyrenees. He was a collector of herpetological specimens.

Bons

Chameleon sp. *Brookesia bonsi* **Ramanantsoa,** 1980

Jacques Bons (b. 1933) is a herpetologist at Université Paul Valéry, Montpellier, France, whose main interest is North African herpetology. He wrote *A Checklist of the Amphibians and Reptiles of Western Sahara* (2000).

Boonsong

Boonsong's Stream Snake *Parahelicops boonsongi* **Taylor** and Elbel, 1958 [Syn. *Opisthotropis boonsongi*]

Dr. Boonsong Lekagul (1907–1992) was a physician, biologist, conservationist, and herpetologist. He qualified as a physician (1933), graduating from Chulalongkorn University, Bangkok. He established the Bangkok Bird Club (1962) and helped to really launch nature conservation in Thailand. His work in lobbying for legislation resulted in a National Parks Act (1961). He wrote "Monitors *(Varanus)* of Thailand" (1969). A bat is named after him.

Bora

Day Gecko sp. *Phelsuma borai* **Glaw, Köhler,** and **Vences,** 2009

Parfait Bora is a Malagasy herpetologist at University of Antananarivo, Madagascar. He was described by Glaw et al. as their "student, colleague and friend . . . who captured the holotype and was of invaluable help during several expeditions in Madagascar." He co-wrote "Which Frogs Are out There? A Preliminary Evaluation of Survey Techniques and Identification Reliability of Malagasy Amphibians" (2008).

Borcke

Guyana Kentropyx *Kentropyx borckiana* **Peters,** 1869

Heinrich Friedrich von Borcke, Count of Kleve (1776–1825), of Hueth Castle, Emmerich, founded both an academy for training draftsmen and a natural history collection; he had bought Albertus Seba's collection. He spent time in Düsseldorf (1806) working to modernize the school board and reorganize Duisburg University. Napoleon made him Prefect of Département du Rhin

(1809) and Conseilleur d'Etat (1812). After Napoleon's defeat, he became a commissioner for Prussia. Merrem, who visited him early in the19th century to study the collection's reptiles, reported that von Borcke presented a number of specimens to the Berlin Museum. Peters, who formally described this lizard, remarked that the original specimen in the Borcke collection had been lost.

Borda

Guerreran Leaf-toed Gecko *Phyllodactylus bordai* **Taylor,** 1942 [Alt. Desert Leaf-toed Gecko]

Don Jose de la Borda (1699–1778) was a Spaniard of French descent. Legend has it that he was riding (1716) in the hills of Taxco, Mexico, when he spotted a rich silver vein. Many places there are named after him, and he built Santa Prisca Cathedral, in which his son served as a priest. Taylor wrote, "The species is named for Joseph le Borde (or Borda), the fabulously wealthy silver miner of Taxco."

Borell

Borell's Worm Lizard *Amphisbaena borelli* **Peracca,** 1897

See **Borelli.**

Borelli

Borelli's Marked Gecko *Homonota borellii* **Peracca,** 1897

Borelli's Worm Lizard *Amphisbaena borelli* Peracca, 1897 [Syn. *Cercolophia borelli*]

Dr. Alfredo Borelli (1858–1943) was an ornithologist who worked at Museo Regionale di Scienze Naturali di Torino (1900–1913). He explored and collected in Argentina and Paraguay (1893–1896). A bird is named after him.

Borri

Sand Boa sp. *Eryx borrii* **Lanza** and Nistri, 2005

Dr. Marco Borri, Zoological Department, Natural History Museum, Università degli Studi di Firenze, is in charge of the section dealing with invertebrate zoology, having previously dealt with marine zoology. The etymology says Borri was an "irreplaceable companion and valuable collaborator of the authors during several expeditions in Italy and abroad."

Börner

Börner's Day Gecko *Phelsuma minuthi* Börner, 1980

Achim-Rudiger Börner (b. 1955) is a German zoologist and herpetologist. He works closely with Walter W. Minuth (q.v.), named in the binomial. They co-wrote "On the Taxonomy of the Indian Ocean Lizards of the *Phelsuma madagascariensis* Species Group *(Reptilia, Gekkonidae)*" (1984).

Bornmüller

Bornmüller's Viper *Vipera bornmuelleri* **Werner,** 1898 [Alt. Lebanon Viper]

Dr. Joseph Friedrich Nicolaus Bornmüller (1862–1948) was a botanist. He studied horticulture at Potsdam in the early 1880s. He was Director of the Herbarium, Weimar (1903–1938), and undertook many trips including one east of Turkestan (1913) with Fedchenko, the Russian botanist. He worked in Macedonia, at that time occupied by German forces (1917–1918). The University of Jena conferred his honorary doctorate. He wrote *Repertorium specierum novarum regni vegetabilis* (1938). After his death his personal collection was sold to Museum für Naturkunde Berlin.

Bosc

Bosc's Monitor *Varanus exanthematicus* Bosc, 1792 [Alt. Savannah Monitor]

Bosk's Fringe-fingered Lizard *Acanthodactylus boskianus* **Daudin,** 1802

Louis Augustin Guillaume Bosc (1759–1828), a naturalist and botanist, was President of the French Natural History Society (1790). After his friend Mme. Roland was guillotined (1793), he had to hide in the Forest of Montmorency, returning to Paris after Robespierre's fall. After the coup d'état (1799) he could only support himself by mass-producing articles for scientific periodicals. He became inspector of the gardens of Versailles and publicly owned nurseries. He often worked with Daudin.

Bosca

Eritrea Longtail Lizard *Latastia boscai* **Bedriaga,** 1884

Eduardo Boscá y Casanoves (1844–1924) was a Spanish herpetologist. He catalogued and described much herpetological fauna of the Iberian Peninsula (1870s–1880s). He described the Spanish Cylindrical Skink *Chalcides bedriagai* (1880), so Bedriaga returned the compliment. He wrote *Catalogue des reptiles et amphibies de la péninsule Ibérique et des iles Baléares* (1880). Several amphibians are named after him.

Boschma

Boschma's Flying Dragon *Draco boschmai* Henning, 1936

Carpentaria Whip Snake *Rhinoplocephalus boschmai* **Brongersma** and Knaap Van Meeuven, 1961

Professor Dr. Hilbrand Boschma (1893–1976) was a zoologist, herpetologist, and expert on crustaceans. His dissertation was on the neck skeleton of crocodiles, but he turned his attention to invertebrates at Rijksmuseum van Natuurlijke Histoire, Leiden (1922), where he became Director (1933–1958). He and Brongersma, the snake describer and his successor, shared a number of expeditions (1920s and 1930s) to Surinam and the Dutch East Indies (Indonesia). Other taxa, including crustaceans, are named after him.

Boshell

Boshell's Forest Racer *Dendrophidion boshelli* **Dunn,** 1944

Dr. Jorge Boshell-Manrique was a physician and epidemiologist who became Director of Instituto Samper Martínez. He was Director of the School of Hygiene, Bogotá, Colombia (1950s). He made noteworthy contributions to the study of yellow fever and wrote a paper on mosquitoes as carriers (1946). He was a keen amateur zoologist, studying mammals and reptiles in their own right and as possible disease carriers. The library at Universidad de Los Llanos is named after him.

Botta

Rubber Boa *Charina bottae* **Blainville,** 1835

Paolo Emilio (Paul-Emile) Botta (1802–1870) was an Italian explorer, archeologist, and physician. Early in his career (1827) he spent a year as ship's surgeon and naturalist and made shore expeditions in California with the ship's captain. He worked in Arabia (1832–1846), excavating near Nineveh, the ancient Assyrian capital (1842–1845). He wrote *Notes on a Journey in Arabia and Account of a Journey in Yemen* (1841). A bird and three mammals are named after him.

Bottego

Somali Agama *Agama bottegi* **Boulenger,** 1897

Bottego's Cylindrical Skink *Chalcides bottegi* Boulenger, 1898

Vittorio Bottego (1860–1897) was an explorer and artilleryman, a skilled horseman who wanted adventure, so he went to Eritrea (1887). He set out with Captain Matteo Grixoni on a journey of exploration (1892–1893) from Berbera, following the Giuba River to its source. After Grixoni had left, Bottego reached Daua Parma, discovered the Barattieri waterfalls, and finally reached Brava. The expedition lost 35 men en route. He set off again (1895) under the auspices of the Italian Geographical Society with a contingent of 250 local troops. He tried crossing Ethiopia, was offered a truce but refused it, and was killed in the fighting. The Ethiopian King kept Bottego's men imprisoned for two years, and only after their release did word of Bottego's fate reach the Italian colonial regime. A shrew is named after him.

Bottom

Northern Hill Death Adder *Acanthophis bottomi* Hoser, 1998

Robert Bottom is an Australian investigative journalist and author on organized crime in Australia. Bottom investigated (mid-1980s) corruption among wildlife officials in New South Wales and in the police force in Victoria before any other journalist dared report it.

Boucard

Boucard's Horned Lizard *Phrynosoma orbiculare boucardii* **Duméril** and **Bocourt,** 1870

Adolphe Boucard (1839–1905) was a French naturalist who worked in Mexico for more than 40 years, collecting hummingbirds for science and the fashion trade. He moved to London (1890) and later to the Isle of Wight. He wrote *The Hummingbird* (1891), saying that "now-a-days the mania of collecting is spread among all classes of society . . . a collection of humming-birds should be the one selected by ladies. It is as beautiful and much more varied than a collection of precious stones and costs much less." Seven birds are named after him.

Bouet

Bouet's Worm Snake *Leptotyphlops boueti* **Chabanaud,** 1917

Chabanaud's Fringe-fingered Lizard *Acanthodactylus boueti* Chabanaud, 1917

Mali Agama *Agama boueti* Chabanaud, 1917

Dr. Georges Bouet (1869–1957) was a physician and ornithologist who worked in Madagascar (1900–1904) and in French West Africa (1906–1930). He wrote *Oiseaux de l'Afrique tropicale* (1955).

Bougainville

Bougainville's Scaly-toed Gecko *Lepidodactylus mutahi* **Brown** and **Parker,** 1977

Named after the island of Bougainville.

Bougainville, H.

Bougainville's Lerista/Skink *Lerista bougainvillii* **Gray,** 1839 [Alt. South-eastern Slider]

Gray probably had Hyacinthe Yves Philippe Potentien, Baron de Bougainville (1781–1846), who visited Australia (1825), in mind, but he provided no etymology.

Boulenger

Water Cobra genus *Boulengerina* Dollo, 1886

Boulenger's Dwarf Iguana *Enyalioides palpebralis* Boulenger, 1883

Boulenger's Forest Dragon *Gonocephalus interruptus* Boulenger, 1885

Boulenger's Indian Gecko *Geckoella albofasciatus* Boulenger, 1885

Boulenger's Pricklenape *Acanthosaura cruciger* Boulenger, 1885

Boulenger's Tree Snake *Sibynomorphus ventrimaculatus* Boulenger, 1885

Boulenger's Agama *Agama boulengeri* **Lataste,** 1886
Boulenger's Burrowing Skink *Scelotes anguineus* Boulenger, 1887
Boulenger's Emo Skink *Emoia mivarti* Boulenger, 1887
Boulenger's Gecko *Cnemaspis boulengerii* **Strauch,** 1887
Boulenger's Legless Skink *Typhlosaurus vermis* Boulenger, 1887
Boulenger's Odd-scaled snake *Achalinus rufescens* Boulenger, 1888
Boulenger's Tree Agama *Dendragama boulengeri* **Doria,** 1888
Boulenger's Bronzeback Tree Snake *Dendrelaphis bifrenalis* Boulenger, 1890
Boulenger's Keelback *Amphiesma parallela* Boulenger, 1890
Boulenger's Snake-eyed Skink *Morethia boulengeri* **Ogilby,** 1890 [Alt. Boulenger's Morethia]
Boulenger's Keelback *Xenochrophis asperrimus* Boulenger, 1891
Boulenger's Tree Lizard *Anisolepis grilli* Boulenger, 1891
Boulenger's Forest Snake *Compsophis boulengeri* **Peracca,** 1892
Boulenger's Wedge-snouted Skink *Chalcides boulengeri* **Anderson,** 1892
Boulenger's Wolf Snake *Lycophidion meleagre* Boulenger, 1893
Stejneger's Spiny Lizard *Sceloporus (clarkii) boulengeri* **Stejneger,** 1893
Boulenger's Night Snake *Hypsiglena torquata affinis* Boulenger, 1894
Boulenger's Tropical Snake *Liophis coralliventris* Boulenger, 1894
Boulenger's Garter Snake *Elapsoidea boulengeri* **Boettger,** 1895
Boulenger's Lipinia *Lipinia miotis* Boulenger, 1895
Boulenger's Rock Agama *Acanthocercus zonurus* Boulenger, 1895
Boulenger's Sand Racer *Psammophis pulcher* Boulenger, 1895
Boulenger's Worm Lizard *Leposternon boulengeri* Boettger, 1895
Boulenger's Dasia *Dasia subcaerulea* Boulenger, 1896
Boulenger's Earth Snake *Uropeltis myhendrae* Boulenger, 1896
Boulenger's Ground Snake *Atractus boulengeri* Peracca, 1896
Boulenger's Slender Snake *Tachymenis affinis* Boulenger, 1896
Boulenger's Tree Iguana *Liolaemus boulengeri* **Koslowsky,** 1896
Boulenger's Bow-fingered Gecko *Cyrtodactylus loriae* Boulenger, 1897
Boulenger's Mole Viper *Atractaspis boulengeri* **Mocquard,** 1897
Boulenger's Scaly Lizard *Sceloporus asper* Boulenger, 1897
Boulenger's Water Snake *Enhydris matanensis* Boulenger, 1897
Rhinoceros Snake *Rhynchophis boulengeri* Mocquard, 1897
Boulenger's Green Anole *Anolis chloris* Boulenger, 1898
Boulenger's Sandveld Lizard *Nucras emini* Boulenger, 1898
Boulenger's Snail-eater *Dipsas ellipsifera* Boulenger, 1898
Boulenger's Blind Snake *Leptotyphlops macrurus* Boulenger, 1899
Boulenger's Tree Skink *Amphiglossus frontoparietalis* Boulenger, 1899
Uganda Savannah Lizard *Nucras boulengeri* **Neumann** 1900
Boulenger's Brown Tree Snake *Dipsadoboa duchesnii* Boulenger, 1901
Boulenger's Lava Lizard *Ophryoessoides scapularis* Boulenger, 1901
Boulenger's Sun Tegu *Euspondylus spinalis* Boulenger, 1901
Boulenger's False Coral Snake *Oxyrhopus marcapatae* Boulenger, 1902
Boulenger's Least Gecko *Sphaerodactylus scapularis* Boulenger, 1902
Boulenger's Lightbulb Lizard *Riama hyposticta* Boulenger, 1902
Boulenger's Centipede Snake *Tantilla alticola* Boulenger, 1903
Boulenger's Cape Tortoise *Homopus boulengeri* **Duerden,** 1906
Boulenger's Bush Anole *Polychrus liogaster* Boulenger, 1908
Boulenger's Largescale Lizard *Ptychoglossus brevifrontalis* Boulenger, 1908
Boulenger's Odd-scaled Snake *Achalinus formosanus* Boulenger, 1908
Boulenger's Dwarf Skink *Afroblepharus tancredi* Boulenger, 1909
Boulenger's Writhing Skink *Lygosoma productum* Boulenger, 1909
Northern Eyelash Boa *Trachyboa boulengeri* Peracca, 1910
Boulenger's Mabuya *Mabuya boulengeri* **Sternfeld,** 1911
Boulenger's Pipe Snake *Cylindrophis boulengeri* **Roux,** 1911
Boulenger's Pygmy Chameleon *Rhampholeon boulengeri* **Steindachner,** 1911
Boulenger's Anadia *Anadia vittata* Boulenger, 1913
Manda Flesh-pink Blind Snake *Leptotyphlops boulengeri* Boettger, 1913
Boulenger's Feylinia *Chabanaudia boulengeri* **Chabanaud,** 1917

Boulenger's Anole *Anolis apollinaris* Boulenger, 1919
Boulenger's Limbless Skink *Scolecoseps boulengeri* **Loveridge,** 1920
Boulenger's Racerunner *Pseuderemias erythrosticta* Boulenger, 1920
Boulenger's Slug-eater *Pareas boulengeri* **Angel,** 1920
Boulenger's Short-legged Skink *Brachymeles boulengeri* **Taylor,** 1922
Southern Sharpnose Snake *Xenoxybelis boulengeri* **Procter,** 1923
Tai-yong Keelback *Amphiesma boulengeri* Gressitt, 1937

George Albert Boulenger (1858–1937) was a Belgian-British herpetologist at the British Museum, London. He graduated from university in Brussels (1876) and worked at Muséum des Sciences Naturelles, Brussels, until moving to London (1880) and taking British nationality (1882). His output was prodigious: nearly 2,600 species described, including 872 reptiles, and 877 scientific papers. He was also a violinist and polyglot. He retired (1920) to grow and study roses. Confusingly, two species of snake have been given the vernacular name Boulenger's Keelback.

Boulton

Boulton's Namib Day Gecko *Rhoptropus boultoni* **Schmidt,** 1933 [Alt. Boulton's Slender Gecko]

Wolfrid Rudyerd Boulton (1901–1983) was the Curator of Birds at the Field Museum, Chicago. He collected in West Africa, Angola, and the Kalahari Desert (1931–1946). Two birds are named after him.

Bourcier

Bourcier's Saphenophis Snake *Saphenophis boursieri* **Jan,** 1867 [Syn. *Liophis boursieri*]

Jules Bourcier (1797–1873) was French Consul to Ecuador (1849–1850), a collector, and a naturalist. He specialized in hummingbirds and has one named after him.

Bourgeau

Bourgeau's Anole *Anolis bourgeaei* **Bocourt,** 1873 [Alt. Bourgeae's Ghost Anole; Junior syn. of *A. laeviventris*]

Eugène Bourgeau (1813–1877) worked in the Botanic Garden, Lyons, before becoming a collector of botanic specimens for a French botanical society. He traveled in Spain (1847–1848), the Canary Islands (1855), and North Africa (1856). The British government appointed him botanist to the Palliser expedition (1857–1860) to the Canadian Northwest. He was the first botanist to examine the Rocky Mountains south of Athabaska Pass, and the prairie south of the North Saskatchewan River. He collected in Mexico (1865), where he probably collected the anole holotype. He later became Curator of the Webb Herbarium, Paris. Bourgeau's Pepperweed is named after him.

Bourquin

Bourquin's Burrowing Skink *Scelotes bourquini* **Broadley,** 1994

Dr. Ortwin "Orty" Bourquin (b. 1942) is a South African biologist, naturalist, and herpetologist who worked for the Natal Parks Board for 30 years. After retirement (2002) he relocated and settled in Columbus, Montana, USA, where he has identified previously unknown species. He co-wrote *The South African Tortoise Book* (1988).

Bourret

Bourret's Blind Skink *Dibamus bourreti* **Angel,** 1935
Bourret's Emo Skink *Emoia laobaoense* Bourret, 1937
Bourret's Ground Skink *Scincella ochracea* Bourret, 1937
Bourret's Odd-scaled Snake *Achalinus ater* Bourret, 1937
Bourret's Pit-viper *Protobothrops jerdonii bourreti* **Klemmer,** 1963
Bourret's Box Turtle *Cuora galbinifrons bourreti* **Obst** and **Reimann,** 1994
Bourret's Cat Snake *Boiga bourreti* Tillack, **Ziegler,** and Quyet, 2004

René Leon Bourret (1884–1957) was a French zoologist. He undertook a comprehensive herpetological survey of Vietnam before WW2 and studied Indochinese fauna (1922–1942). He wrote *Les tortues de l'Indochine* (1941), the first detailed monograph to deal with all the chelonians of Southeast Asia. Two amphibians are named after him.

Bouton

Snake-eyed Skink *Cryptoblepharus boutonii* Des Jardin, 1831

Louis Sulphice Bouton (1799–1878) was a French botanist who worked in Mauritius, where *Trochetia boutoniana* is the national flower. He sent many specimens to Kew. He was a founding member (1826) and Secretary (1866) of the Mauritius Royal Society of Arts and Sciences and co-founder of the Mauritius Herbarium, which was established to develop better strains of sugar cane.

Bouvier

Cape Verde Leaf-toed Gecko *Hemidactylus bouvieri* **Bocourt,** 1870
Anole sp. *Anolis bouvierii* Bocourt 1873 [Junior syn. of *A. ortonii* **Cope,** 1868]

Aimé Bouvier (d. 1919) was a French collector and zoologist who became Secretary of the French Zoological Society (1876). He, with other committee members, was forced to resign (1880) when it was discovered that about 5,000 francs of the society's funds had gone missing. He was the author of *Les mammifères de la France*. Other taxa, including two birds and a mammal, are named after him.

Bovall

Coral Mimic Snake *Rhinobothryum bovallii* **Andersson,** 1916

Carl Erik Alexander Bovallius (or Bowallius) (1849–1907) was Associate Professor of Zoology at Uppsala Universitet, but gave up teaching (1897) and traveled in South America, founding a rubber plantation in Trinidad (1901). He wrote on crustaceans, in papers such as "A New Isopod from the Swedish Arctic Expedition of 1883 Described" (1885), and a crustacean genus is named after him.

Bowring

Bowring's Gecko *Hemidactylus bowringii* **Gray,** 1845

Bowring's Writhing Skink *Lygosoma bowringii* **Günther,** 1864 [Alt. Bowring's Supple Skink]

John Charles Bowring (1820–1893) was a Hong Kong businessman and an amateur naturalist who presented the skink holotype to the Natural History Museum, London. The gecko is named after either him or his father, Sir John Bowring (1792–1872), who was a Member of Parliament (1835–1839 and 1844–1849) before becoming British Consul, Canton (Guangzhou) (1849–1853), and Governor, Hong Kong (1854–1859). The elder Bowring was a hyper-polyglot, understanding 200 languages and speaking 100 of them.

Boyd

Boyd's Forest Dragon *Hypsilurus boydii* **Macleay,** 1884

John Archibald Boyd (1846–1926) was an English-born professional natural history collector who emigrated to Australia (1857) and worked for the Australian Museum. He lived in Fiji (1865–1882) and then on a Queensland sugar plantation.

Boyle

Boyle's Beaked Blind Snake *Rhinotyphlops boylei* **FitzSimons,** 1932 [Syn. *Typhlops boylei*]

This snake was collected during the Vernay-Lang Kalahari expedition (1930). The original description makes no comment on the etymology of *boylei*. We think that Boyle might have been a member of that expedition but have not been able to identify him. One candidate is Howarth S. Boyle, an American zoologist and ornithologist. He accompanied Leo E. Miller on his South American expedition (1915) for the American Museum of Natural History.

Boyle, C. E.

Boyle's Kingsnake *Lampropeltis getula boylii* **Baird** and **Girard,** 1853 [Syn. *L. g. californiae*]

Dr. Charles Elisha Boyle (1821–1870) was a physician and amateur naturalist. After working for a printer and teaching, he graduated from medical college (1847). He caught "gold rush fever" (1848–1849), joining the Columbus and California Industrial Association party as its physician and traveling the Oregon Trail to California. He practiced as a physician in California; collected, particularly herpetological specimens, for the Smithsonian; and with a friend built a boat, which they sailed home via Cape Horn (1850–1852). During the American Civil War (1860–1865) he was a surgeon captain in the army, and then returned to private practice. He was fluent in 32 languages and "gave much of his time and practice to the poor . . . and as a result never amassed much money and died poor himself." A mammal and an amphibian are named after him.

Braack

Braack's Dwarf Leaf-toed Gecko *Goggia braacki* **Good, Bauer,** and Branch, 1996

Dr. Harold H. Braack is a South African conservationist who was the Warden of the Kruger National Park. He set up the Addo Elephant Park in the Eastern Cape and the Rigtersveld National Park in the Western Cape. He has two doctorates. He co-wrote "*Kinixys spekii* Gray, 1863, Speke's Hinged Tortoise—Breeding and Feeding" (2005).

Bracciani

Bracciani's Worm Snake *Leptotyphlops braccianii* **Scortecci,** 1929 [Alt. Scortecci's Blind Snake]

Luigi Bracciani was an Italian explorer who was on the Corni-Calciati-Bracciani expedition to Eritrea (1922–1923), which collected the snake holotype. He was in charge of planning routes while others collected specimens.

Braconnier

Braconnier's Short Skink *Pygomeles braconnieri* **Grandidier,** 1867

Short-tailed Horned Lizard *Phrynosoma braconnieri* **Duméril,** 1870

Slender Gecko sp. *Rhoptropus braconnieri* Thominot, 1878

Séraphin Braconnier was a French naturalist who was employed in the Herpetology and Ichthyology Department, Muséum National d'Histoire Naturelle, Paris, for over 25 years. He collected in New Caledonia and sold a number of specimens to Museum für Naturkunde Berlin.

Bradfield

Bradfield's Dwarf Gecko *Lygodactylus bradfieldi* **Hewitt,** 1932

Bradfield's Namib Day Gecko *Rhoptropus bradfieldi* Hewitt, 1935

R. D. Bradfield (1882–1949) was a South African farmer,

naturalist, and collector who spent most of his life in Namibia. He has three birds named after him.

Braestrup

Colubrid snake sp. *Crotaphopeltis braestrupi* **Rasmussen,** 1985

Dr. Frits Wimpffen Braestrup (1906–1999) was Curator, Zoological Museum, Københavns Universitet, and was an expert on earwigs. He traveled in East and southern Africa. He wrote "Remarks on Faunal Exchange through the Sahara" (1947). An amphibian is named after him.

Brain

Brain's Legless Skink *Typhlosaurus braini* **Haacke,** 1964 [Alt. Haacke's Blind Legless Skink]

Dr. Charles Kimberlin Brain (b. 1931) is a Zimbabwean paleontologist whose main interests are Australopithecines and the taphonomy of caves (how deposits in caves were created and fossilized). His doctorate is in geology (1958). He was Director of the Transvaal Museum (1965–1991), retired (1996), but is still Curator Emeritus and Honorary Professor of Zoology at the University of the Witwatersrand, Johannesburg. He supervised a 30-year-long excavation of the Swartkans Cave, Sterkfontein Valley, producing, as a sample, 240,000 fossils from a most diverse fauna.

Branderhorst

Branderhorst's Turtle *Elseya branderhorsti* Ouwens, 1914

Dr. Bastiaan Branderhorst (b. 1880) was a physician in the Dutch East Indies army (1906). He joined a detachment that was exploring (1907–1910) Dutch New Guinea (West Papua). He collected botanical specimens (1907–1908) and was on the Anglo-Dutch Committee to determine boundaries in Borneo (1912–1913). He retired from the army (1924) and then was employed by the Dutch East Indian Public Health Service. He was appointed physician of the Pengalengan Society for Nursing, Western Java (1925).

Brandt

Lacertid lizard sp. *Iranolacerta brandtii* **De Filippi,** 1863

Johann Friedrich (Fedor Fedorovich) von Brandt (1802–1879) was a German naturalist, surgeon, and pharmacologist who emigrated to Russia in 1831. He explored Siberia and was founding Director of the Zoological Museum, Academy of Science, St. Petersburg. He and De Filippi visited Persia (now Iran) together (1862). He wrote on many subjects and was co-author of *Medical Zoology* (2 vols., 1829–1833). Many taxa are named after him, including five mammals.

Bransford

Bransford's Anole *Anolis bransfordii* **Cope,** 1874 [Junior syn. of *A. limifrons* Cope, 1871]

Dr. John F. Bransford (1846–1911) was an assistant U.S. naval surgeon (1872–1890) on the Nicaragua and Panama Canal surveys (1872–1888). He made three separate herpetological collections: in Nicaragua (1875 and 1885) and in Panama (1875). He was recalled to the colors (1898) for the Spanish-American War, and he retired again (1901) with the rank of Surgeon.

Brauer

Brauer's Skink *Janetascincus braueri* **Boettger,** 1896

Dr. August Bernhard Brauer (1863–1917) was a German zoologist, herpetologist, and ichthyologist. He graduated from Humboldt-Universität, Berlin, in natural sciences (1885) and took his doctorate there (1892). He collected in the Seychelles (1897) and was on the *Valdivia* expedition (1898), describing (1908) the fish they collected. He became a Professor at Berlin University (1905) and Director of the university's Zoological Museum (1906). He was appointed Professor of the Zoological University, Berlin (1914). A rodent is named after him.

Braun

Braun's Mabuya *Mabuya brauni* **Tornier,** 1902 [Alt. Ukinga Montane Skink; Syn. *Trachylepis brauni*]

Rudlolf H. Braun (b. 1908) was a German collector in Angola and southern Africa. He wrote *Beitrage zur Biologie der Vögel von Angola* (1930). Other taxa including a bird and an amphibian are named after him.

Brazil

Brazilian Bird Snake *Rhachidelus brazili* **Boulenger,** 1908

Brazil's Woodland Racer *Drymoluber brazili* **Gomes,** 1918

Brazil's Lancehead *Bothrops brazili* **Hoge,** 1954

Dr. Vital Brazil Mineiro da Campanha (1865–1950) was a Brazilian physician, immunologist, and scientist who developed the serum for use against snakebite from the *Crotalus*, *Bothrops*, and *Micrurus* genera. His names were chosen by his father to honor the city, the state, and the country in which he was born. He is regarded as one of the most important Brazilian scientists ever and has been honored in many ways, including being portrayed on a banknote.

Bredl

Bredl's Carpet Python *Morelia bredli* **Gow,** 1981 [Alt. Centralian Python]

Josef "Joe" Bredl (1948–2007) was a German-born Australian herpetologist. He created the Edward River

Crocodile Farm and was the proprietor of Renmark Reptile Park, South Australia. His younger brother, Rob, is known as the "Barefoot Bushman."

Breedlove

Breedlove's Anole *Anolis breedlovei* **H. M. Smith** and **Paulson,** 1968

Dr. Dennis E. Breedlove (b. 1939) is a botanist who also collected herpetological specimens. His doctorate was awarded by Stanford (1968). He is Curator Emeritus, Botany Department, California Academy of Sciences. He wrote *Introduction to the Flora of Chiapas* (1981).

Breitenstein

Borneo Short-tailed Python *Python breitensteini* **Steindachner,** 1881 [Alt. Borneo Blood Python]

Dr. Heinrich Breitenstein (1848–1930) was a German physician who served with the Dutch East Indies army for 21 years. While in Borneo he collected herpetofauna that Steindachner purchased for Naturhistorisches Museum Wien. He published his memoirs as *21 Jahre in Indien; Aus dem Tagebuchen eines Militärarztes* (1899).

Bremer

Herradura Anole *Anolis bremeri* **Barbour** 1914

Dr. John Lewis Bremer (1874–1959) worked at the Harvard Medical School, where he became Hersey Professor of Anatomy (1931). He took his bachelor's degree (1896) and his M.D. (1901) at Harvard and was an Instructor in Histology and Embryology (1902–1906) and a Demonstrator of Histology (1906–1912). He became an Assistant Professor (1912) and an Associate Professor (1915–1931). Barbour wrote that Bremer "has often most kindly aided me while upon collecting trips." He liked catchy titles for his publications, such as *Microscopic Evidences of Absorption in the Large Intestine.*

Brenchley

Ordos Racerunner *Eremias brenchleyi* **Günther,** 1872

Julius Lucius Brenchley (1816–1873) was an English traveler who collected the racerunner (lizard) holotype. He was in the Solomon Islands (1865) and collected over 1,000 objects, many of them part of the Brenchley Collection in the British Museum. He was a major benefactor of the museum at Maidstone, England. He wrote *The Cruise of the Curacoa among the South Sea Islands* (1865). A bird is named after him.

Brenner

Brenner's Racerunner *Pseuderemias brenneri* **Peters,** 1869

Richard Brenner (1833–1874) was on von der Decken's last African expedition to Somaliland (Somalia) (1865). They were attacked and some members killed, and the survivors had a difficult journey to Zanzibar. He collected in Somaliland (1866–1867) for Museum für Naturkunde Berlin and covered a large area that was previously unmapped. He visited Aden (1868), then returned to Somalia (1869), where he collected the lizard holotype. He was appointed as Austro-Hungarian Consul in Aden (1871). He returned to Zanzibar (1872), where he died.

Bresslau

Bresslau's Bachia *Bachia bresslaui* **Amaral,** 1935

Dr. Ernst Ludwig Bresslau (1877–1935) was a physician and zoologist. He became a naval surgeon and visited Brazil (1904), returned to study marsupials (1913), and left again (1914). He became head of the Zoology Department, Institute of Physician-Therapeutical Research, Georg Speyer Haus, Frankfurt (1920). He was also Professor and Director, Zoological Institute, Universität zu Köln. He visited Brazil again (1929). Being Jewish, he lost his university jobs in Nazi Germany (1933), so left to become the first Director, Department of Zoology, University of São Paulo, Brazil (1934).

Bresson

Bresson's Splendid Cat-eyed Snake *Leptodeira splendida bressoni* **Taylor,** 1938

Michoacán Slender Blind Snake *Leptotyphlops bressoni* Taylor, 1939

Don Julio Raymond Bresson collected the type specimens of both these snakes.

Breuil

St. Lucia Threadsnake *Leptotyphlops breuili* Hedges, 2008

Michel Breuil is a herpetologist at Muséum National d'Histoire Naturelle, Paris, who was honored "for his contribution to the herpetology of the Lesser Antilles."

Breyer

Breyer's Whip Lizard *Tetradactylus breyeri* **Roux,** 1907 [Alt. Breyer's Long-tailed Seps]

Waterberg Girdled Lizard *Cordylus warreni breyeri* **Van Dam,** 1921

Dr. Hermann Gottfried Breyer (1864–1923) was a Dutch naturalist, botanist, and physician who collected in Transvaal and Mozambique (1890–1910). He qualified at Universiteit van Amsterdam, moving to South Africa to teach. He became a Trustee of the Staatsmuseum, Pretoria, serving as its Curator until 1897. He became Curator (1901) and then Director (1913–1921) of the renamed Transvaal Museum and Zoological Gardens.

His name is sometimes spelled Breijer, and his son, J. W. F. Breijer, was also a botanical collector.

Bridges

Bridges' Ameiva *Ameiva bridgesii* **Cope,** 1869

Dr. Robert Bridges (1806–1882) was a Professor of Chemistry. Cope wrote in his etymology of "my friend Robert Bridges, M.D., Professor of Chemistry in the Philadelphia College of Pharmacy, and an active member of the Academy of Natural Sciences."

Briggs

Briggs' Centipede Snake *Tantilla briggsi* Savitzky and **H. M. Smith,** 1971

Dr. William T. Briggs is an academic. The etymology reads, "The name *briggsi* is applied in honor of Dr. William T. Briggs, Dean of the College of Arts and Sciences, in recognition of his generous support of herpetological research at the University of Colorado."

Broadley

Broadley's Lance Skink *Acontias litoralis* Broadley and **Greer,** 1969

Lake Turkana Hinged Terrapin *Pelusios broadleyi* **Bour,** 1986 [Alt. Turkana Mud Turtle]

Broadley's Writhing Skink *Lygosoma lanceolatum* Broadley, 1990

Broadley's Dwarf Gecko *Lygodactylus broadleyi* **Pasteur,** 1995

Broadley's Flat Lizard *Platysaurus broadleyi* Branch and Whiting, 1997

Broadley's Garter Snake *Elapsoidea broadleyi* Jakobsen, 1997

Broadley's Worm Snake *Leptotyphlops broadleyi* **Wallach** and Hahn, 1997

Broadley's Bush Viper *Atheris broadleyi* **Lawson,** 1999

Dr. Donald George Broadley (b. 1932) is a specialist in East African herpetology and Curator of Herpetology, Natural History Museum of Zimbabwe, Bulawayo, where his wife, Sheila, is also a herpetologist. He gained his doctorate from the University of Natal (1966). Among his publications is "On the Status of *Simocephalus riggenbachi* Sternfeld 1910" (2007).

Brocchi

Stone Skink *Paracontias brocchii* **Mocquard,** 1894

Paul Louis Antoine Brocchi (1838–1898) was a herpetologist who worked at Muséum National d'Histoire Naturelle, Paris, with Milne-Edwards. Mocquard gives no explanation of his choice of *brocchii*, but this candidate seems very likely. He wrote "Sur quelques batraciens raniformes et bufoniformes de l'Amérique" (1877). An amphibian is named after him.

Broghammer

Python genus *Broghammerus* Hoser, 2004

Stefan Broghammer is a German herpetologist and breeder of pythons. He wrote *Ball Pythons: Habitat, Care, and Breeding* (2004).

Brongersma

Red Blood Python *Python brongersmai* Stull, 1938

Brongersma's Lipinia *Lipinia venemai* Brongersma, 1953

Brongersma's Reed Snake *Calamaria brongersmai* **Inger** and **Marx,** 1965

Brongersma's Pit-viper *Trimeresurus brongersmai* **Hoge,** 1969

Brongersma's Lobulia *Lobulia brongersmai* **Zweifel,** 1972

Brongersma's Tree Skink *Glaphyromorphus brongersmai* **Storr,** 1972

Brongersma's Worm Snake *Typhlops brongersmianus* **Vanzolini,** 1972

Brongersma's Helmet Skink *Tribolonotus brongersmai* **Cogger,** 1973

Brongersma's Emo Skink *Emoia brongersmai* **Brown,** 1991

Dr. Leo Daniel Brongersma (1907–1994) was an author and zoologist. His doctorate was from Universiteit van Amsterdam (1934). He began studying herpetology as an Assistant, Artis Amsterdam Zoological Museum. He lectured at Universiteit Leiden and was Curator of Reptiles and Amphibians, then Director, Rijksmuseum van Natuurlijke Histoire, Leiden, until his retirement (1972). He explored in New Guinea and Surinam. Among his publications is *European Atlantic Turtles* (1972). Three amphibians are named after him.

Brooke

Brook's House Gecko *Hemidactylus brookii* **Gray,** 1845

Brooke's Keeled Skink *Tropidophorus brookei* Gray, 1845

Brooke's Sea Snake *Hydrophis brookii*, **Günther** 1864

Sir James Brooke (1803–1868), the first White Rajah of Sarawak, is believed to have been the model for the eponymous hero of Joseph Conrad's novel *Lord Jim*. He worked for the Honourable East India Company and was near-fatally wounded in the Anglo-Burmese War (1825). He was inspired by the example of Sir Stamford Raffles and resolved to emulate him. He helped put down a rebellion and, with the blessing of the Sultan of Brunei, became Governor and Rajah of Sarawak (1841). He successfully suppressed the locals' propensity for piracy and headhunting.

Brookes

Chameleon genus *Brookesia* **Gray,** 1865

Joshua Brookes (1761–1833), a British naturalist and anatomist, taught anatomy in London and founded the Brookesian Museum of Comparative Anatomy. Gray examined his private museum.

Brooks, C. J.

Brooks' Wolf Gecko *Luperosaurus brooksi* **Boulenger,** 1920

Brooks' Nose-horned Lizard *Thaumatorhynchus brooksi* **Parker,** 1924

Cecil Joslin Brooks (1875–1953) was a collector, mainly botanical, in Borneo and Sumatra early in the 20th century. He was a metallurgical chemist employed by Borneo Company for gold exploitation in Sarawak (1900–1910), where he collected, helped by Hewitt, Curator of the Sarawak Museum. He was employed at a goldmine in Sumatra (1912–1923), traveled subsequently in the Dutch East Indies, and then sailed, via Australia and New Zealand, to Europe (1924). In England he studied his collections and the butterflies in the British Museum. He wrote mainly on ferns, and several are named after him, as is a mammal.

Brooks, W. S.

Brooks' Kingsnake *Lampropelitis getula brooksi* **Barbour,** 1919

Brooks' Ctenotus *Ctenotus brooksi* **Loveridge,** 1933

Winthrop Sprague Brooks (1887–1965) was a collector and zoologist who was Custodian of Bird's Eggs and Nests, Harvard Museum of Comparative Zoology (1928–1934). He spent time in Eastern Siberia and Alaska (1913–1914) on the Harvard University polar bear–hunting expedition led by John Eliot Thayer; and he spent several months collecting birds in the Falkland Islands (1917). He traveled in Australia and collected the holotype of the ctenotus (1926).

Broom

Broom's Small Skink *Proablepharus tenuis* Broom, 1896 [Alt. Northern Soil-crevice Skink]

Broom's Blind Snake *Ramphotyphlops broomi* **Boulenger,** 1898 [Alt. Faint-striped Blind Snake; Syn. *Austrotyphlops broomi*]

Dr. Robert Broom (1866–1951) was a physician and paleontologist. He qualified as a doctor, receiving his Doctorate of Science from the University of Glasgow (1905). He was Professor of Zoology and Geology, Victoria College, Stellenbosch, South Africa (1903–1910), and later Keeper of Vertebrate Palaeontology, South Africa Museum, Cape Town, and on the staff, Transvaal Museum, Pretoria, as an Assistant in Paleontology (1934). He devoted all his later years to the study of early hominids and proposed the Australopithecinae subfamily (1946). He published *On the Origin of Lizards* (1925).

Brougham

Brougham's Earth Snake *Uropeltis broughami* **Beddome,** 1878

See **Guppy.**

Brown, B. C.

Brown's Coral Snake *Micrurus browni* **Schmidt** and **H. M. Smith,** 1943

Dr. Bryce Cardigan Brown. See **Brown (Family).**

Brown (Family)

Brown's Bunchgrass Lizard *Sceloporus scalaris brownorum* **H. M. Smith** et al., 1997

Dr. Bryce Cardigan Brown (1920–2008), Director Emeritus, Strecker Museum, and Professor Emeritus in Biology, Baylor University, Texas, and his wife, Lilian, and their five children, Alton, Brent, Carol, Leo, and Roy, are all included in the dedication. Bryce graduated in zoology at the University of Texas (1942), joined the U.S. Army Air Corps, and served in India. After WW2 he returned to Texas, finishing his master's at Texas A&M University (1948). The University of Michigan awarded his doctorate (1955). He was President, Texas Herpetological Society (1946), and was hired as Curator of the Strecker Museum, becoming Director (1966–1981). Among his publications is *An Annotated Checklist of the Reptiles and Amphibians of Texas* (1950). The Brown family collected the lizard holotype (1961).

Brown, H.

Saddled Leaf-nosed Snake *Phyllorhynchus browni* **Stejneger,** 1890

Herbert Brown (1848–1913) moved to Tucson, Arizona (1873), to prospect in the mountains and nearly died of thirst, as well as surviving a number of narrow escapes from Apaches. He worked as journalist, editor, and newspaper proprietor in Tucson, was President of the Audubon Society of Arizona, and was Clerk to the Superior Court of Pima County.

Brown, W. C.

Brown's Short-legged Skink *Brachymeles samarensis* Brown, 1956

Brown's Gecko *Gekko athymus* Brown and **Alcala,** 1962

Brown's Wolf Gecko *Luperosaurus browni* **Russell,** 1979

Brown's Mabuya *Mabuya indeprensa* Brown and Alcala, 1980

Brown's Scaly-toed Gecko *Lepidodactylus browni* Pernetta and Black, 1983

Brown's Emo Skink *Emoia aurulenta* Brown and **Parker,** 1985

Skink sp. *Sphenomorphus tagapayo* R. Brown, McGuire, Fewer, and Alcala, 1998

Walter Creighton Brown (1913–2002) was a herpetologist who specialized in the herpetofauna of the South Pacific and the Philippines. He served in New Guinea during WW2. He received his doctorate from Stanford (1955) and went to Silliman University, Philippines, as a Fulbright Professor of Sciences. Brown was Alcala's mentor throughout the latter's postgraduate zoological education in the Philippines and at Stanford. In the skink description the authors say they "name this new species of skink in honor of Walter C. Brown. . . . The specific appellation . . . is chosen from the Tagalog term tagapayo, meaning a wise and trusted friend, advisor, or mentor."

Brues

Barbour's Tropical Racer *Mastigodryas bruesi* **Barbour,** 1914

Dr. Charles Thomas Brues (1879–1955) was a zoologist and entomologist. He gained his bachelor's (1901) and master's (1902) degrees from the University of Texas before moving to Columbia University for a year. He worked for the Department of Agriculture (1904–1905). He joined the staff of the Natural History Museum of Milwaukee (1905) before moving to Harvard (1909), where he and Barbour were members of the Cambridge Entomological Club. He was Associate Curator of Insects, Harvard Museum of Comparative Zoology, and later Professor of Entomology, retiring in 1947.

Bruijn

Bruijn Forest Dragon *Hypsilurus bruijnii* **Peters** and **Doria,** 1878

Anton August Bruijn (d. 1885) was a Dutch *plumassier* (feather merchant). He exhibited many natural history specimens during the Colonial Trade Exhibition in Amsterdam (1883). Others have described him as a botanist, explorer, and zoologist and use a different initial. It could be that there were two men of the same name, perhaps brothers, operating in the same area. Two birds and two mammals are named after him.

Brunet

Brunet's Anole *Anolis bruneti* Thominot, 1887 [Junior syn. of *A. fuscoauratus* D'Orbigny 1837]

M. Brunet gave the holotype to Muséum National d'Histoire Naturelle, Paris, but no further details are known.

Brussaux

Mocquard's African Ground Snake *Gonionotophis brussauxi* **Mocquard,** 1889

Eugène Brussaux was an anthropologist who was active in West and Central Africa in the late 19th and early 20th centuries. He took part in the de Brazza mission (1886–1891). He took photographs of the borders of Chad and Cameroun on the Moll expedition to Cameroons (1905–1907) to establish the countries' borders. He wrote "Notes sur la race Baya" (1908).

Brygoo

Brygoo's Leaf Chameleon *Brookesia antoetrae* Brygoo and **Domergue,** 1971

Brygoo's Chameleon *Calumma peyrierasi* Brygoo, **Blanc,** and Domergue, 1974 [Alt. Peyrieras' Chameleon]

Brygoo's Burrowing Skink *Amphiglossus alluaudi* Brygoo, 1981

Colubrid snake genus *Brygophis* Domergue, 1988

Brygoo's Girdled Lizard *Zonosaurus brygooi* **Lang** and **Böhme,** 1990

Brygoo's Pygmy Chameleon *Brookesia brygooi* **Raxworthy** and Nussbaum, 1995

Colonel Dr. Edouard-Raoul Brygoo (b. 1920) originally trained as a physician and became Professor of Zoology (Reptiles and Amphibians), Muséum National d'Histoire Naturelle, Paris (1977). Among his publications is *Les types de lacértidés (reptiles, sauriens) du Muséum National d'Histoire Naturelle* (1988).

Buch

Dalat Dwarf Snake *Calamaria buchi* **Marx** and **Inger,** 1955

Father Buch (b. 1865) was a French catholic missionary from the Vincentian order in China and Indochina (1906–1952). He collected Lepidoptera specimens and sent them to many museums, including the Field Museum. He collected the holotype of the snake.

Buchard

Buchard's Gecko *Cyrtodactylus buchardi* David, Teynié, and Ohler, 2004

Michel Buchard is a businessman in Clermont-Ferrand, France. Since 1998 he has given generous support to the study of natural history in general and to herpetology and entomology in particular.

Buchwald

Buchwald's Scaly-eyed Gecko *Lepidoblepharis buchwaldi* **Werner,** 1910

Otto von Buchwald (1843–1934) was an engineer, anthropologist, ethnographer, and philologist who spoke at least 12 languages and understood more. He was a soldier in Germany and fought against Denmark and Austria. He left Germany for Peru (1869), where he studied early civilizations, leaving for California (1887) but never arriving, as he stopped off in Ecuador and was persuaded to stay to study indigenous dialects. He worked as an engineer and made a number of expeditions into the Amazonian forests. He was a leader of the German community in Guayaquil and was blacklisted by the Ecuadorian government (1916). He wrote articles in English and German on archeology and natural history. He was crippled by a fall from his horse (1923) and used a cane for walking thereafter. He died of a stroke.

Buckley

Buckley's Teiid *Alopoglossus buckleyi* **O'Shaughnessy,** 1881 [Alt. Smooth-bellied Shade Lizard]

Clarence Buckley (fl. 1839–1889) was a collector who made several expeditions to Ecuador (1880s). He collected over 10,000 specimens of many plant and animal taxa in a small area of Ecuador, sending them to various institutions and scientists but mainly to the British Natural History Museum. A number of insect species are named after him.

Budak

Budak's Skink *Ablepharus budaki* Göcmen, Kumlutap, and Topunodlu, 1996

Dr. Abidin Budak (b. 1943) is a herpetologist and (since 1990) a Professor at the Ege University Zoology Department, Izmir, Turkey, by which university his bachelor's degree (1968), his master's (1972), and his doctorate (1974) were awarded.

Buechner

Kaschar Racerunner *Eremias buechneri* **Bedriaga,** 1906

Eugen A. Büchner was a Russian zoologist of German descent who specialized in mammals and birds. He was co-editor of a journal produced by the museum where he worked, *Annuaire Musée Zoologique de l'Académie des Sciences de St. Petersbourg*.

Buerger

Buerger's Tree Iguana *Liolaemus buergeri* **Werner,** 1907

Professor Dr. Otto Bürger (b. 1865) was a traveler and collector for Zoologisches Forschungsmuseum Alexander Koenig, Bonn. He explored in Colombia (1896–1897), from Barranquilla to the Orinoco, for the Academy of Science, Göttingen. Werner worked on Bürger's herpetological collection (1899–1916). Bürger was in Valparaiso in Chile (1907) and wrote *Die Robinson Insel* (1909) about the Juan Fernandez archipelago, which Defoe used as the setting for *Robinson Crusoe*.

Buergers

Buergers' Forest Snake *Toxicocalamus buergersi* **Sternfeld,** 1913

Burgers' Emo Skink *Papuascincus buergersi* **Vogt,** 1932

Theodore Joseph Bürgers (1881–1954) was a physician and zoologist who participated in the German Sepik expedition in New Guinea for the museum of Humboldt-Universität, Berlin. He became Professor of Hygiene and Bacteriology in Dusseldorf (1923) and later at Georg-August-Universität Göttingen. Three birds are named after him.

Bulel

Gecko sp. *Lepidodactylus buleli* Ineich, 2008

We do not know who or what "Bulel" is or was, and if M. Ineich of Muséum National d'Histoire Naturelle, Paris, has his way, we never will. The etymology reads, "The specific epithet *buleli* is given by the author as a reference to a personal and private story and has no particular signification related to the species, its characteristics, geographical origin, or biology."

Buller

Buller's Spiny Lizard *Sceloporus bulleri* **Boulenger,** 1894

Dr. Audley Cecil Buller (1853–1894) was a collector of mammals and reptiles. He collected the holotypes of several taxa including the spiny lizard. He traveled 1,500 kilometers (1,000 miles) collecting for the American Museum of Natural History across the Sierra de Nayarit and ranges of the Sierra Madre to Zacatecas, then the least known area of Mexico. Two mammals are named after him.

Bunker

Bunker's Earless Lizard *Holbrookia maculata bunkeri* **H. M. Smith,** 1935

Charles Dean Bunker (1870–1948) was a zoologist. On the advice of the family physician, his childhood was mostly spent outdoors, which led to his interest in natural history. He collected specimens, gave them to a local taxidermist in exchange for lessons on how to preserve skins, and went to work at the University of Kansas (1895) as a taxidermist. After an interlude at the University of Oklahoma, he was at the Museum of Natural History, University of Kansas (1904–1942), becoming Assistant Curator of Birds and Mammals (1907), Assistant Curator in Charge (1909), and finally Curator (1912), in which position he served until he retired. He developed innovative techniques for cleaning bones, using dermestid beetles, the larvae of which were already known for their ability to clean bones precisely and without damage. His fame, however, lies chiefly in his teaching ability, the achievements of many of his students, and his delight in their success. A mammal is named after him.

Bunty

Bunty's Dwarf Gecko *Lygodactylus grandisonae* **Pasteur,** 1962 [Alt. Kenyan Dwarf Gecko]

See Alice Georgie Cruikshank Grandison.

Burbidge

Plain-backed Kimberley Ctenotus *Ctenotus burbidgei* **Storr,** 1975

Dr. Andrew A. Burbidge is a zoologist and Research Fellow, Department of Conservation and Land Management, Western Australia. He is Chairman, Australasian Marsupial and Monotreme Specialist Group. He wrote *Threatened Animals of Western Australia* (2004). A mammal is named after him.

Burchell

Burchell's Sand Lizard *Pedioplanis burchelli* **Duméril** and **Bibron,** 1839

William John Burchell (1781–1863) was an explorer-naturalist. He went to the Cape of Good Hope (1810) and undertook a major exploration of the interior of South Africa (1811–1815), traveling over 7,000 kilometers (4,400 miles) through largely unexplored country. He returned to London (1815) to work on his collections, spent two months in Lisbon (1825), and proceeded to Brazil, where he collected extensively, then returned again to England (1830). He became increasingly reclusive and ill, eventually committing suicide. He wrote the two-volume *Travels in the Interior of Southern Africa* (1822–1824). Among the taxa he described is the White Rhinoceros, and he has six birds and a mammal named after him.

Burden

Snake-eyed Skink sp. *Cryptoblepharus burdeni* **Dunn,** 1927

William Douglas Burden (1898–1978) was a wealthy, adventurous man and a Trustee of the American Museum of Natural History. He undertook an expedition to Komodo (1927) to collect Komodo Dragons. Dunn was the expedition's herpetologist. Burden related his adventures to filmmaker Merian Cooper. Cooper changed the dragon to a gigantic ape, added a beautiful heroine, and released the resulting picture as *King Kong*.

Burgeon

Skink sp. *Panaspis burgeoni* **Witte,** 1933

Louis Burgeon (1884–1974) was an entomologist who was Director, Zoology Department, the Royal Museum for Central Africa, Tervuren. He collected in the Belgian Congo (Zaire) (1917–1918) and collected herpetofauna in Ruwenzori, Kivu, and Tanganyika (Tanzania) (1930s). He co-wrote *Les insects du Congo Belge* (1950).

Burmeister

Burmeister's Anole *Pristidactylus scapulatus* Burmeister, 1861

Professor Karl Hermann Konrad Burmeister (1807–1892) was an ornithologist who was Director of the Institute of Zoology, Martin-Luther-Universität Halle-Wittenberg, Germany (1837–1861), for which he made large collections during two expeditions: Brazil (1850–1852) and the La Plata region, Argentina (1857–1860). He lived in Argentina (1861–1892), being founding Director, Museo Nacional, Buenos Aires, until retirement (1880). He was in the Prussian civil service but won his release by using the inventive excuse that a persistent stomach complaint was caused by arsenic emissions in the museum and by the drinking water in Halle, which had a high sulphate content. He wrote *Reise nach Brasilien* (1853). A bird and a mammal are named after him.

Burt

Canyon Spotted Whiptail *Aspidoscelis burti* **Taylor,** 1938

Dr. Charles Earle Burt (1904–1963) was a herpetologist. He took his bachelor's degree at Kansas State Agricultural College, and his master's degree (1927) and doctorate (1930) at the University of Michigan. He worked at the American Museum of Natural History (1929–1930). He taught at Trinity College, Waxahachie, Texas (1930–1931), and at Southwestern College, Winfield, Kansas (1932–1944). He then became owner and manager of Quivira Specialties Co. of Topeka, Kansas (suppliers of such useful items as live toads as food for hog-nosed snakes), and taught at Kansas State College. His wife, May Danheim Burt, was a teacher of home economics but was just as interested as he in herpetology, and they co-wrote several articles and papers (see **Danheim**). Burt wrote *A Key to the Lizards of the US and Canada* (1936). He died of cancer.

Burton, E.

Burton's Legless Lizard *Lialis burtonis* **Gray,** 1835

Burton's Nessia *Nessia burtonii* Gray, 1839 [Alt. Gray's Snake Skink, Three-toed Snake-Skink]

Major Edward Burton (1790–1867) was an army surgeon. He was stationed at Chatham, England (1829–1837), and wrote *A Catalogue of the Collection of Mammalia and Birds in the Museum at Fort Pitt, Chatham* (1838). He wrote a paper on fishes that Cuvier had described.

Burton, R.

Burton's Carpet Viper *Echis coloratus* **Günther,** 1878 [Alt. Arabian Saw-scaled Viper]

Captain Sir Richard Francis Burton (1821–1890) was a

famous explorer, linguist, devotee of erotica, and author who was British Consul in Trieste (1872–1890). Günther's original paper mentions "Burton's *Gold-mines of Midian*," indicating that the author of that work, Richard F. Burton, was the same person who collected the reptile. A mammal and a bird are named after him.

Bury

Bury's Worm Snake *Leptotyphlops burii* **Boulenger,** 1905 [Alt. Arabian Blind Snake]

George Wyman Bury (1874–1920) was a naturalist, explorer, political officer, and Arabist. His career came to an early end due to an unjust charge of corruption. He was closely associated with the "Arab Revolt." He wrote *The Land of Uz* (1911). A bird is named after him.

Busack

Busack's Fringe-fingered Lizard *Acanthodactylus busacki* Salvador, 1982

Stephen Dana Busack (b. 1944), a zoologist and herpetologist, was Director of Research and Collections, North Carolina Museum of Natural Sciences. His bachelor's degree was from Cornell (1967), his master's from George Mason University (1977), and his doctorate from the University of California, Berkeley (1985). He was a Field Associate in Herpetology (1971–1994) at the Carnegie Museum of Natural History, where since 1995 he has been a Research Associate, Amphibians and Reptiles. He studied the herpetofauna of Cadiz, Spain. He and Salvador were close friends and colleagues.

Bush

Bush's Hooded Snake *Suta spectabilis bushi* **Storr,** 1988

Bush's Pygmy Monitor *Varanus bushi* Aplin, Fitch, and King, 2006

Brian Gordon Bush (b. 1947) is a herpetologist in Western Australia, where he manages a company, Snakes Harmful and Harmless. He was originally trained in New South Wales as an electrician but wanted something different, moved to Western Australia (1976), and worked for 10 years as a fencer and windmill mechanic on the Esperance sandplain. He wrote, with Brad Maryan, *Reptiles and Frogs in the Bush: Southwestern Australia* (2007).

Butler, A. L.

Butler's Four-clawed Gecko *Gehyra butleri* **Boulenger,** 1900 [Alt. Butler's Dtella]

Butler's Wolf Snake *Lycodon butleri* Boulenger, 1900

Butler's Black-and-yellow Burrowing Snake *Chilorhinophis butleri* **Werner,** 1907 [Alt. Butler's Two-headed Snake]

Forest Skink sp. *Sphenomorphus butleri* Boulenger, 1912

Arthur Lennox Butler (1873–1939) was a zoologist. He became Curator, Selangor Museum, Malaya, and then Superintendent of Game Preservation in Sudan. A bird is named after him.

Butler, A. W.

Butler's Garter Snake *Thamnophis butleri* **Cope,** 1889

Amos William Butler (1860–1937) was an ornithologist. He was a founding member of the Brookville (Indiana) Society of Natural History (1881). His bachelor's degree (1894) and master's (1900) were awarded by Indiana University. The State of Indiana employed him, first in the Department of Geology and Resources as an ornithologist (1896–1997) and then as Secretary to the Board of State Charities (1897–1923). He wrote *Birds of Indiana* (1891).

Butler, W. H.

Butler's Black Snake *Pseudechis butleri* **L.A.Smith,** 1942

Butler's Morethia *Morethia butleri* **Storr,** 1963

Australian Earless Lizard *Tympanocryptis (parviceps) butleri* Storr, 1977

Butler's Snake-eyed Skink *Notoscincus butleri* Storr, 1979

Butler's Scalyfoot *Delma butleri* Storr, 1987

Skink sp. *Glaphyromorphus butlerorum* Aplin, **How,** and Boeadi, 1993

Dr. William Henry "Harry" Butler (b. 1930) trained as a teacher but worked for corporate and government bodies as an environmental consultant and collector (1963), undertaking a major study of Western Australian fauna. He collected 14 species of mammals new to science. He presented the popular ABC television series *In the Wild* (1976). He was awarded an honorary Doctorate of Science by the Edith Cowan University, Perth (2003). The skink is named after him and his wife. See **Margaret B.**

Büttikofer

Büttikofer's Forest Skink *Sphenomorphus buettikoferi* Lidth de Jeude, 1905

Büttikofer's Glass Lizard *Ophisaurus buettikoferi* Lidth de Jeude, 1905

Dr. Johann Büttikofer (1850–1927) was a Swiss zoologist with the Dutch Borneo expedition. He studied biology at Universität Bern and went to work at Rijksmuseum van Natuurlijke Histoire, Leiden (1879), becoming Curator (1884–1897). He undertook two expeditions to Liberia (1879–1882 and 1886–1887). He accompanied Nieuwenhuis to Borneo (1893–1894). He was Director of the Rotterdam Zoo (1897–1924) before retiring to Bern. He wrote *Mededeelingen over Liberia* (1883). Three mammals and two birds are named after him.

Büttner

Büttner's Mabuya *Mabuya buettneri* **Matschie,** 1893 [Syn. *Trachylepis buettneri*]

Dr. Oskar Alexander Richard Büttner (1858–1927) was a botanist and explorer in Africa who was on the German expedition to the Congo Basin (1884–1886). He became the first Professor of African Languages in Berlin. He published *Reise durch die Deutschen Kolonien* (1910).

Bynoe

Bynoe's Gecko *Heteronotia binoei* **Gray,** 1845

Benjamin Bynoe (1804–1865) was a naval surgeon. He was Assistant Surgeon on HMS *Beagle* (1831–1836). His superior, Robert McCormick, was angry that Darwin, instead of himself, was treated as the ship's naturalist, and he resigned from the expedition (1832). Bynoe was promoted to Surgeon, served in that position for the rest of that voyage, and was given the same position on the *Beagle*'s third voyage (1837–1843). He was a great success as a naturalist and collector. Bynoe Harbour in Australia is named after him.

Byrne

Byrne's Gecko *Diplodactylus byrnei* Lucas and Frost, 1896 [Alt. Gibber Gecko, Pink-blotched Gecko; Syn. *Lucasium byrnei*]

P. M. "Paddy" Byrne was an Australian telegraph official living in frontier settlements hundreds of kilometers from the nearest Europeans. He met the anthropologist and biologist Baldwin Spencer during the Horn scientific expedition to central Australia (1894). He corresponded with Spencer (1894–1925) about the Aboriginal people and the unusual flora and fauna of the region. A mammal is named after him.

C

Caesar

Caesar's African Water Snake *Grayia caesar* **Günther,** 1863

Günther does not name a particular Caesar in the etymology but mentions the beauty of the snake, so we believe the name reflects the impressiveness of the creature—as in the "king of snakes."

Cagle

Cagle's Map Turtle *Graptemys caglei* Haynes and McKown, 1974

Dr. Fred Ray Cagle (1915–1968) was an American herpetologist. He was an Assistant Professor, Tulane University (1940). He served in the U.S. Army Air Corps (1943–1945), reaching the rank of Captain, then returned to Tulane (1946). He was Director of Zoology until 1958. He was also editor-in-chief of *Copeia* (1955–1959). He worked for UNESCO (1961–1963). He wrote over 40 research titles on turtles (1937–1955) and described four turtles new to science.

Caiden

Sun Tegu sp. *Euspondylus caideni* Köhler, 2003

Caiden Christopher Vlasimsky (b. 2003) is the son of Stan Vlasimsky, a Texas business consultant who supports BIOPAT—an organization that raises funds for conservation by selling the rights to name species. Other members of the family have taxa named after them through this method.

Cairo

Cairo Blind Snake *Leptotyphlops cairi* **Duméril** and **Bibron,** 1844

Named after the city of Cairo.

Caligula

Montane Sun Skink *Lampropholis caligula* **Ingram** and **Rawlinson** 1981

Gaius Julius Caesar Augustus Germanicus (A.D. 12–41), more commonly known as Caligula, was the third Julio-Claudian Roman Emperor. He was known for his extreme extravagance, eccentricity, depravity, cruelty, and despotism. His own guards killed him.

Campan

Campan's Chameleon *Furcifer campani* **Grandidier,** 1872

Dominique Campan was a French resident of Madagascar. His uncle was Laborde (q.v.). He was involved in Malagasy politics, inherited half his uncle's fortune, and became French Consul but was threatened with death and forced to leave Tananarive (1882).

Campbell, J.

Campbell's Skink *Emoia campbelli* **Brown** and **Gibbons,** 1986 [Alt. Montane Emo Skink]

John Campbell, a geologist, was a consultant to the Monasavu Hydro Electric Scheme in Fiji. He collected the skink holotype.

Campbell, J. A.

Campbell's Toadheaded Viper *Bothrocophias campbelli* Freire-Lascano, 1991

Campbell's Alligator Lizard *Abronia campbelli* Brodie and Savage, 1993

Campbell's Galliwasp *Diploglossus legnotus* Campbell and Camarillo, 1994

Campbell's Galliwasp *Diploglossus ingridae* Werler and Campbell, 2004

Anole sp. *Anolis campbelli* Köhler and E. N. Smith, 2008

Dr. Jonathan Atwood Campbell (b. 1947) is Professor of Biology, University of Texas, Arlington. His main area of expertise is the herpetofauna of Guatemala. He was awarded his master's degree at UTA (1977), then went to the University of Kansas for his doctorate (1983). He returned to UTA as Curator of Herpetology and Assistant Professor (1983), progressing to Associate Professor (1988) and Full Professor (1993). Both *Diploglossus ingridae* and *D. legnotus* have had the vernacular name Campbell's Galliwasp applied to them. See also **Jonathan.**

Campbell, W. A.

Campbell's Girdled Lizard *Cordylus campbelli* **Fitz-Simons,** 1938 [Alt. Campbell's Spiny-tailed Lizard]

William A. Campbell (1880–1962) was a businessman and a big game hunter in South Africa where he managed estates in Natal (1906–1962). He was a member of the National Parks Board.

Canedi

Scrocchi's Ground Snake *Atractus canedii* **Scrocchi** and **Cei,** 1991

Dr. Arturo Adolfo Canedi of Universidad Nacional de Jujuy, Argentina, is primarily a mammalogist. He graduated as a Doctor of Veterinary Sciences (1972). He has contributed to articles on vicunas, pumas, and jaguars.

Canh

Cahn's Gecko *Gekko canhi* Rösler et al., 2010

Dr. Le Xuan Canh is Professor at and Director of the Institute of Ecology and Biological Resources, Hanoi. His doctorate was awarded by the National University of the Soviet Union, Moscow.

Cann

Cann's Snake-necked Turtle *Chelodina canni* **McCord** and Thomson, 2002

John Robert Cann (b. 1938) is an Australian herpetologist with a long association with the Australian Museum, Sydney. He and his brother, George, run a public exhibition of lizards and snakes in the Sydney area called La Peruse. They took over the 50-year-old business from their parents, who were both experts at handling venomous snakes. John wrote Australian Freshwater Turtles (1968) and is a serious researcher and author, while George runs the business day to day.

Canquel

Tree Iguana sp. *Liolaemus canqueli* **Cei,** 1975

Canquel is the name of the plateau in Argentina where this lizard is found.

Cantor

Cantor's Black-headed Snake *Sibynophis sagittarius* Cantor, 1839

Cantor's Kukri snake *Oligodon cyclurus* Cantor, 1839

Cantor's Wolf Snake *Lycodon atropurpureus* Cantor, 1839

Cantor's Rat Snake *Ptyas dhumnades* Cantor, 1842

Cantor's Pit-viper *Trimeresurus cantori* **Blyth,** 1846 [Syn. *Cryptelytrops cantori*]

Cantor's Dwarf Reed Snake *Pseudorabdion longiceps* Cantor, 1847

Cantor's Water Snake *Cantoria violacea* **Girard,** 1857

Cantor's Giant Softshell Turtle *Pelochelys cantori* Gray, 1864

Cantor's Small-headed Sea Snake *Microcephalophis cantoris* **Günther,** 1864 [Alt. Günther's Sea Snake; Syn. *Hydrophis cantoris*]

Indian Fringe-fingered Lizard *Acanthodactylus cantoris* Günther, 1864

Eastern Trinket Snake *Orthriophis cantoris* **Boulenger,** 1894 [Syn. *Elaphe cantoris*]

Dr. Theodore Edward Cantor (1809–1860) was an amateur zoologist and Superintendent Physician of the European Asylum, Bhowanipur, Calcutta. This was part of the Honourable East India Company's Bengal Medical Service. He was interested in tropical fish, and around 1840 the King of Siam gave him some Bettas, commonly known as fighting fish. He published an article about them that led to "Betta fever," the popular craze in Victorian England for keeping such fish. He wrote 16 herpetological papers (1836–1848), including descriptions of several reptiles and *Catalogue of Reptiles Inhabiting the Malayan Peninsula and Islands, Collected or Observed by Theodore Cantor, Esq., M.D. Bengal Medical Service* (1847). Two mammals are named after him.

Cao Van Sung

Gecko sp. *Cyrtodactylus caovansungi* **Orlov** et al., 2007

Professor Dr. Cao Van Sung (d. 2002), of the University of Hanoi, was Emeritus Director of the Institute for Ecology and Biological Resources, Hanoi. He was a zoologist and biologist who received his training at Russian universities. A mammal is named after him.

Captain

Captain's Wood Snake *Xylophis captaini* Gower and Winkler, 2007

Ashok Captain (b. 1960) is an Indian herpetologist, based at Pune, where he was awarded his bachelor's degree in zoology (1982). He was a leading bicycle racer (1977–1989). He worked as a senior instructor at nature camps for children (1990–1998). He co-wrote, with Whitaker, *Snakes of India: The Field Guide* (2004).

Capuron

Madagascar Chameleon *Calumma capuroni* **Brygoo, Blanc,** and **Domergue,** 1972

Malagasy Tree Snake sp. *Stenophis capuroni* Domergue, 1994

René Capuron (1921–1971) was a French botanist and forester who lived in Madagascar for nearly 25 years, during which he made 154 collecting trips. He was a Principal Inspector, Département d'Eaux et Forêts (1948–1971). He wrote *Essai d'introduction à l'étude de la flore forestiére de Madagascar* (1957). He contracted a grave illness, returned to France, and died.

Carbonell, J.

Carbonell's Wall Lizard *Podarcis carbonelli* Perez-Mellado, 1981

J. Carbonell is the describer's wife. In an all too brief etymology Perez-Mellado writes, "El nombre está dedicado a mi mujer J. Carbonell"—not only failing to give her name but also not using the feminine Latin suffix; we suggest that the name should be *Podarcis carbonellae.*

Carbonell, L.

Roze's Green Racer *Philodryas carbonelli* **Roze,** 1957 [Junior syn. of *P. olfersii herbeus* **Wied,** 1825]

Dr. Luis Manuel Carbonell Parra (b. 1924) was on Roze's expedition (1951–1952) to discover the source of the Orinoco. He graduated as a physician (1948) and was Professor of Biology and Pathology at a number of

Venezuelan universities, becoming President, Academia de Ciencias Físicas, Matemáticas, y Naturales de Venezuela.

Carla

Barbados Threadsnake *Leptotyphlops carlae* Hedges, 2008

Dr. Carla Ann Hass is an American herpetologist and Program Coordinator, Biology Department, Pennsylvania State University. In private life she is the wife of the describer, Dr. S. Blair Hedges. This snake is believed to be the smallest in the world.

Carlet

Colubrid snake sp. *Stenophis carleti* **Domergue,** 1994

The original description gives no etymology, so we cannot be sure, but a likely contender is Dr. Jean Carlet who is a bacteriologist at Hospital of the Pasteur Institute, Paris. Domergue, who described this reptile, also worked at that institute.

Carlos Todd

See **Todd.**

Carnaby

Carnaby's Snake-eyed Skink *Cryptoblepharus carnabyi* **Storr,** 1976 [Alt. Carnaby's Wall Skink]

Keith Carnaby (1910–1994) was an entomologist whose collection is housed by the British Natural History Museum. The town Boyup Brook, Western Australia, has the Carnaby Collection of Beetles and Butterflies, which is regarded as second in excellence only to that of the British Museum. He published *Jewel Beetles of Western Australia* (1986). A bird is named after him.

Carp

Carp's Barking Gecko *Ptenopus carpi* **Brain,** 1962 [Alt. Namib Chirping Gecko]

Bernhard Carp (1901–1966) was a Dutch-born Cape Town businessman and naturalist. He sponsored many collecting expeditions, particularly to Namibia, by the Zoological Museum, Universiteit van Amsterdam. He bought important mammal collections in South Africa and donated them to that museum. Political considerations made it difficult, but the university's board eventually accepted the very important collection, which included rare skins and skulls. He wrote *Why I Chose Africa*. A bird is named after him.

Carpenter, C. C.

Carpenter's Anole *Anolis carpenteri* Echelle, Echelle, and **Fitch,** 1971

Dr. Charles Congden Carpenter (b. 1921) is a noted herpetologist. He gained his master's degree (1947) and doctorate (1951) from the University of Michigan. His earlier studies had been interrupted by service with the U.S. Army Medical Corps (1943–1946). He worked in Michigan (1946–1952) and then went to the University of Oklahoma (1953–1987), first as Assistant Professor, becoming Associate Professor (1959) and Professor (1966). Since 1988 he has been Professor Emeritus of Zoology and Curator Emeritus of Herpetology, Sam Noble Oklahoma Museum of Natural History.

Carpenter, G. D. H.

Liwale Two-headed Snake *Chilorhinophis carpenteri* **Parker,** 1927

Carpenter's Chameleon *Kinyongia carpenteri* Parker, 1929

Dr. Geoffrey Douglas Hale Carpenter (1882–1953) originally qualified as a physician and specialized in tropical medicine, as this allowed him to pursue his interest in natural history (particularly entomology) while earning a living in East Africa. He became Hope Professor of Entomology, Oxford, and is particularly remembered for his work on Ugandan butterflies and for research into the causes of sleeping sickness. He wrote *A Naturalist in East Africa* (1925).

Carr

Carr's Snail-sucker *Sibon carri* **Shreve,** 1951

Carr's Chameleon *Rhampholeon platyceps carri* **Loveridge,** 1953

Dr. Archibald "Archie" Fairly Carr Jr. (1909–1987) was a conservationist who was Professor of Zoology, University of Florida, having previously studied there. His Ph.D. was the first in zoology awarded by that institution; his thesis was entitled "A Contribution to the Herpetology of Florida" (1940). He was especially interested in turtles, becoming one of the world's leading experts on sea turtles. The Dr. Archie Carr National Wildlife Refuge, south of Melbourne, Florida, was established and named in his honor. He wrote several books, including *So Excellent a Fishe* (1967).

Carrau

Giant Hispaniolan Galliwasp *Celestus carraui* Incháustegui, **Schwartz,** and **Henderson,** 1985

Dr. José Antonio Carrau was a contemporary of Incháustegui at Museo de Historia, Universidad Autónoma de Santo Domingo (1980s).

Carrion

Parker's Ground Snake *Atractus carrioni* **Parker,** 1930

Parker's Whorl-tailed Iguana *Stenocercus carrioni* Parker, 1934

Professor Clodoveo Carrión Mora (1883–1957) was the

Ecuadorian natural scientist of the 20th century. He was a paleontologist and naturalist who came from a literary family. Recognizing that he had aptitude for the sciences but none for letters, he traveled to England and studied at universities in Manchester and London, emerging after 10 years as an engineer. Having returned to Ecuador he became Professor of Natural Sciences, Colegio Bernardo Valdevisio. He collected the holotypes of both these reptiles.

Carter

Carter's Rock Gecko *Pristurus carteri* **Gray,** 1863

Henry Carter appears to be the same Dr. Henry Carter who was a great friend of the explorer, Sir Richard Francis Burton. Gray described him as "well known for his researches on the Foraminifera, Sponges, and Microscopic Vegetables of India." He collected the gecko holotype.

Carteret

Carteret's Skink *Gongylus cartereti* **Duméril** and **Bibron,** 1839 [Junior syn. of *Emoia cyanogaster* Lesson, 1830]

Named after Carteret Harbour, New Ireland (Papua New Guinea), where the holotype was collected.

Carvalho, A. L.

Carvalho's Worm Lizard *Amphisbaena carvalhoi* **Gans,** 1965

Carvalho's Coral Snake *Micrurus lemniscatus carvalhoi* **Roze,** 1967

Carvalho's Mabuya *Mabuya carvalhoi* **Rebouças-Spieker** and **Vanzolini,** 1990

Carvalho's Slider *Trachemys adiutrix* Vanzolini, 1996 [Alt. Brazilian Slider]

Gymnophthalmid lizard sp. *Colobosauroides carvalhoi* Soares and Caramaschi, 1998

Dr. Antenor Leitao de Carvalho (1910–1985) was a Brazilian herpetologist and ichthyologist, specialized in frogs. He became a pilot in the merchant marine (1927–1932), and whenever in port (Rio), he volunteered to help out at the museum. He became Field Collector for the Museu Nacional, Rio de Janeiro (1933), undertaking a number of expeditions in Brazil. He became Curator of Herpetology (1941) and, eventually, the museum's Vice Director. He collected many different taxa specimens from all over Brazil (1930s and 1940s). He wrote "A Preliminary Synopsis of the Genera of American Microhylid Frogs" (1954). Several amphibians are named after him.

Carvalho, C. M.

Carvalho's Gecko *Gymnodactylus carvalhoi* **Vanzolini,** 2005

Dr. Celso Morato de Carvalho is a Brazilian herpetologist. He was awarded his bachelor's degree in biology by Universidade Estadual Paulista, Botacatu (1978), and his doctorate in ecology by the National Institute for Amazonian Research, Manaus (1992), where he now works. He was Chief, Center for Research in Aquatic Biology, Roraima (1985–1995), and was Researcher, Universidade Federal de Sergipe (1995–2005). He is an old friend, colleague, and field companion of Vanzolini.

Casamiquela

Tree iguana sp. *Liolaemus casamiquelai* Avila et al., 2010

Dr. Rodolfo Magín Casamiquela (1932–2008) was an Argentine vertebrate palaeontologist, anthropologist, and expert on Patagonia. He was a researcher at Consejo Nacional de Investigaciones Científicas y Técnicas of Argentina.

Casilda

Casilda's Anole *Anolis casildae* Arosemena, Ibanez, and de Sousa, 1991

The description contains no etymology, and we do not know to what it refers.

Castelnau

Northern Velvet Gecko *Oedura castelnaui* Thominot, 1889

Francis Louis Nompar de Caumont, Comte de Laporte de Castelnau—in brief, Comte de Castelnau—(1810–1880) was a career diplomat and naturalist who was born in London, studied natural science in Paris, and then led a French scientific expedition to study the lakes of Canada, the USA, and Mexico (1837–1841). He led the first expedition (1843–1847) to cross South America from Peru to Brazil, following the watershed between the Amazon and the Río de la Plata systems. Soon after his return to France he undertook another long voyage of exploration. Following this he took several diplomatic posts. He lived in Melbourne (1864–1880), being Consul-General (1862) and then French Consul (1864–1877). A bird is named after him.

Castroviejo

Agama sp. *Agama castroviejoi* Padial, 2005

Dr. Javier Castroviejo Bolibar (b. 1940) is a Spanish zoologist and ecologist who created a number of biological stations, including one in the Doñana, Spain. He is President of the Spanish Committee of UNESCO's Man and Biosphere project. He wrote *Premières donées sur l'écologie hivernale des vertébrés de la Cordillière Cantabrique* (1970). A hare is named after him.

Catesby

Catesby's Snail-eater *Dipsas catesbyi* Sentzen, 1796
Catesby's Pointed Snake *Uromacer catesbyi* **Schlegel,** 1837

Mark Catesby (1683–1749) was an English naturalist, artist, and traveler. He made two journeys to the Americas (1712–1719 and 1722–1726). He refers to the American colonies as the Carolinas, which was how 18th-century England thought of them before the War of Independence. During his travels Catesby observed that birds migrate, and he published his observations in *On the Passage of Birds* (1747). This discovery was entirely contrary to the then prevailing view that birds hibernated in caves or at the bottom of ponds in the winter. He used to ship his snake specimens back to England in jars of rum; sometimes the sailors drank the rum and ruined his specimens. He observed the similarity in the features of the Native Americans and peoples of Asiatic origin and was the first person to hypothesize the existence in the distant past of a land bridge between Asia and the Americas. He wrote *The Natural History of Carolina, Florida and the Bahama Islands: Containing the Figures of Birds, Beasts, Fishes, Serpents, Insects and Plants* (1731–1743), which was consulted by Lewis and Clark during their expedition (1804–1806). Four birds and an amphibian are named after him.

Caziani

Tree iguana sp. *Liolaemus cazianiae* Lobo, Slodki, and Valdecantos, 2010

Sandra Caziani (1961–2005) was a teacher and ecologist who graduated from Universidad de Buenos Aires. She was honored for her great knowledge of the central Andes and its fauna and "particularly for her friendship, wisdom, and courage."

Cecilia

Brilliant South American Gecko *Gonatodes ceciliae* **Donoso-Barros,** 1965

Cecilia Donoso-Barros is the daughter of the Chilean herpetologist Dr. Roberto Donoso-Barros (1922–1975). She was seriously injured in the car accident in which her father was killed.

Cege

Worm Lizard sp. *Amphisbaena cegei* Montero, Safadez, and Alvarez, 1997

Carl Gans ("C. G."); *cegei* is a mock-Latin form of the initials. See **Gans.**

Cei

Cei's Tree Iguana *Liolaemus ceii* **Donoso-Barros,** 1971
Cei's Marked Gecko *Homonota andicola* Cei, 1978
Cei's Mountain Lizard *Phymaturus punae* Cei, **Etheridge,** and Videla, 1985
Cei's Ground Snake *Liophis ceii* **Dixon,** 1991
Mountain Lizard sp. *Phymaturus ceii* Scolaro et al., 2007

Professor Dr. José Miguel Cei (1918–2007) was a biologist, ecologist, zoologist, and herpetologist at Universidad Nacional de Córdoba, Argentina. He wrote *Batracios de Chile* (1962).

Celia

Canasi Dwarf Boa *Tropidophis celiae* Hedges, Estrada, and Diaz, 1999

Celia Puerta de Estrada is the wife of one of the describers, Alberto R. Estrada.

Censky

Censky's Ameiva *Ameiva corax* Censky and Paulson, 1992

Dr. Ellen Joan Censky (b. 1955) is Senior Vice President for Museum Programs, Milwaukee Public Museum. Her bachelor's degree in zoology was awarded by the University of Wisconsin, Milwaukee (1979), and her doctorate by the University of Pittsburgh (1994). She worked at the Carnegie Museum of Natural History (1979–1998); was Director, Connecticut State Museum of Natural History, University of Connecticut (1998–2003); and became Professor of Zoology at the University of Oklahoma and Director, Sam Noble Museum of Natural History (2003). She co-wrote "Revision of the Ameiva (Reptilia: Teiidae) of the Anguilla Bank, West Indies" (1992).

Cerberus

Water Snake genus *Cerberus* **Cuvier,** 1829

Cerberus was the three-headed dog who guarded the entrance to Hades in Greek mythology.

Ceron

Ceron's Glass Lizard *Ophisaurus ceroni* Holman, 1965

Carlos Cerón (d. 1999) and Miguel Cerón (d. 2006) were Mexican naturalists who lived in Cuautlapan, Veracruz Province, and who assisted herpetologists in the field for many years. Holman named the lizard after Miguel but mentions both men in the description.

Chabanaud

Chabanaud's Fringe-fingered Lizard *Acanthodactylus boueti* Chabanaud, 1917
Chabanaud's Mabuya *Mabuya breviparietalis* Chabanaud, 1917
Skink genus *Chabanaudia* **Witte** and **Laurent,** 1943

Dr. Paul Chabanaud (1876–1959) was a French ichthyologist and herpetologist. He took his first degree at Poitiers (1897). He volunteered his services at Muséum National d'Histoire Naturelle, Paris (1915), under Louis Roule, who asked him to identify herpetological specimens and sent him on a scientific expedition to French West Africa (1919). He traveled to Senegal and Guinea before walking 1,200 kilometers (750 miles) through southern Guinea and Liberia, returning to France in 1920, when he became a Preparator of Fishes at the museum with a special interest in flatfish. He took his doctorate at the Sorbonne (1936). He wrote 40 papers on herpetology (1915–1954).

Chabaud

Colubrid snake sp. *Liophidium chabaudi* **Domergue,** 1984

Dr. Alain G. Chabaud (b. 1923) is a French scientist, now retired, and an expert on nematodes. He qualified as a physician in 1947, becoming a Doctor of Science in 1954. He became Professor, Muséum National d'Histoire Naturelle, Paris (1960). Among other taxa, the parasite *Plasmodium chabaudi* is named after him.

Chaitzam

Chaitzam's Ameiva *Ameiva chaitzami* **Stuart,** 1942

Stuart gives a brief, and slightly odd, etymology: "Dedicated to Chaitzam, the mountain lord who dominates the lower Cahabón Valley." We are unsure what he was referring to.

Challenger

Challenger's Skink *Saproscincus challengeri* **Boulenger,** 1887 [Alt. Challenging Shade Skink]

Named after the ship HMS *Challenger*, which made several extensive collecting expeditions, during one of which the skink holotype was taken. The alternative common name seems to have been coined on the basis of a misunderstanding.

Chamisso

Chilean Green Racer *Philodryas chamissonis* Wiegmann, 1835

Adelbert von Chamisso (1781–1838), also known as Louis Charles Adelaide de Chamisso, was a botanist, poet, naturalist, and philologist. He was on Von Kotzebue's *Rurik* expedition (1816–1823). He was a page at the Prussian court, served in the army, and became Keeper of the Royal Botanical Gardens. He wrote *Reise um die Welt mit der Romanzoffischen Entdeckungs-Expedition* (1836), and many of his literary works survive; his sentimental poetic cycle *Frauenliebe und Leben* (1830) was set to music by Schumann. A bird is named after him.

Champion, G. C.

Panamanian Earth Snake *Geophis championi* **Boulenger,** 1894

George Charles Champion (1851–1927) was an entomologist who specialized in Coleoptera. He was taken on as a collector by Frederick DuCane Godman and Osbert Salvin and went to Guatemala (1879). He spent four years collecting in Central America and had a collection containing 15,000 insect species when he returned to England. He stayed in Godman and Salvin's employ and saw through the printing process their 52-volume work *Biologia Centrali-Americana;* he wrote a number of its sections. Champion and Godman collected the holotype of the snake.

Champion, I.

Leaftail Gecko sp. *Phyllurus championae* Schneider, **Couper,** Hoskin, and **Covacevich,** 2000

Irene Champion is a Resource Ranger with the Queensland Parks and Wildlife Service, Mackay. She focused the attention of one of the authors on Cameron Creek and Black Mountain as a phyto-geographically interesting area, possibly pointing to the presence of unusual fauna. She wrote "Round Worms *(Ophidascaris robertsii)* in Pythons, Their Treatment, and Some Potential Problems" (1994).

Chan-ard

Chan-ard's Mountain Reed Snake *Macrocalamus chanardi* David and **Pauwels,** 2004

Chan-ard's Water Snake *Enhydris chanardi* **Murphy** and **Voris,** 2005

Khun Tanya Chan-ard is a herpetologist who is Curator, National Science Museum, Pathumthani, Thailand. He co-wrote *Amphibians and Reptiles of Peninsular Malaysia and Thailand* (1999).

Chaney

Chaney's Bunchgrass Lizard *Sceloporus chaneyi* **Liner** and **Dixon,** 1992

Professor Emeritus Dr. Allan Harold Chaney (1923–2009) of Texas A&M University, Corpus Christi, is an ornithologist and herpetologist. He wrote *Keys to the Vertebrates of Texas* (1982).

Chanhome

Gecko sp. *Cyrtodactylus chanhomeae* **Bauer, Sumontha,** and **Pauwels,** 2003

Dr. Lawan Chanhome of the Queen Saovabha Memorial Institute, Thai Red Cross Society, Bangkok, is an active contributor to the study of Thailand's venomous snakes and is Chief Veterinary Surgeon and manager of the snake breeding facilities of the Red Cross in Bangkok.

Chapin

Rough-scaled Lizard sp. *Ichnotropis chapini* **Schmidt,** 1919

Grey Chameleon *Chamaeleo chapini* **Witte,** 1964

Central African Mud Turtle *Pelusios chapini* **Laurent,** 1965

Dr. James Paul Chapin (1889–1964) was an American ornithologist and co-leader of the Lang-Chapin expedition, which made the first comprehensive biological survey of the Belgian Congo (1909–1915). He was President of the Explorers' Club (1949–1950). Witte wrote, "Among the African reptiles that I examined at the American Museum of Natural History in 1963, I discovered three specimens . . . that represent a species previously unrecognized. The three individuals were collected more than 30 years ago by my friend the late Dr. James P. Chapin. . . . I deeply regret that Dr. Chapin did not live to see this paper in print." He wrote *Birds of the Belgian Congo* (1932), which largely earned him the award of the Daniel Giraud Elliot Gold Medal that year. Eight birds and two mammals are named after him.

Chapman

Chapman's Stumptail Chameleon *Rhampholeon chapmanorum* **Tilbury,** 1992

Elisabeth "Betty" G. Chapman (d. 1985) and James "Jim" D. Chapman (d. 2001) were an English married couple. They researched chameleons and other reptiles and were plant collectors in Africa. He was a forestry expert who worked for the Oxford Forestry Institute and made extensive botanical collections in both Nigeria and Malawi. He worked as a forestry officer in Malawi (then called Nyasaland) in the early 1950s, transferring to Nigeria (1973) and back to Malawi (1980).

Charito

Ground Snake sp. *Atractus charitoae* Silva Haad, 2004

Charito de Silva is the describer's wife. He says that he named the snake for her in recognition of her dedication, patience, and love in sharing 40 years of investigation of the Colombian Amazon snake fauna. Charito is perhaps his pet name for her, as elsewhere in the same paper he refers to his wife as Rosario Collazos de Silva.

Charles Bogert

Guatemalan Beaded Lizard *Heloderma horridum charlesbogerti* **Campbell** and Vannini, 1988

See **Bogert.**

Charles Myers

Anole sp. *Anolis charlesmyersi* **Köhler,** 2010

See **Myers, C. W.**

Chasen

Chasen's Pit-viper *Garthius chaseni* **M. A. Smith,** 1931

Frederick Nutter Chasen (1896–1942) was an English zoologist. He was appointed Assistant Curator of the Raffles Museum in 1921, becoming Director in 1932. He was a well-known authority on Malaysian birds and mammals and co-authored many scientific publications on these topics. Chasen perished at sea when fleeing Singapore during WW2. Four birds are named after him.

Chazalia

Helmethead Gecko *Geckonia chazaliae* **Mocquard,** 1895

This gecko is named after a yacht. The collector of the holotype was the French naturalist Raymond Comte de Dalmas (1862–1930), whose yacht, the *Chazalie*, was grounded on the sands off Cap Blanc, Mauritania.

Chazeau

Chazeau's New Caledonian Skink *Caledoniscincus chazeaui* **Sadlier, Bauer,** and Colgan, 1999

Dr. Jean Chazeau is a zoologist who has been at Laboratoire de Zoologie Appliquée at ORSTOM Nouméa since 1977, currently as Director. He initiated ORSTOM's program "Faunistic Characteristics of the Non-anthropogenic Forests and Maquis." He co-wrote *La Nouvelle Calédonie, vestige du continent de Gondwana* (1998).

Cheesman

Cheesman's Lipinia *Lipinia cheesmanae* **Parker,** 1940

Miss Lucy Evelyn Cheesman (1881–1969) wanted to train as a veterinary surgeon, but in her day the restrictions on the education of women precluded it. Instead she became an entomologist, explorer, and traveler who made a number of expeditions to the Galapagos Islands, New Guinea, the New Hebrides, and the Solomon Islands (1924–1936). She worked for many years as a volunteer at the Natural History Museum, London, and was the first female Curator at the London Zoo. She wrote *The Cyclops Mountains of Dutch New Guinea* (1938). She collected the holotype of this skink.

Cheke

Northern Day Gecko *Phelsuma chekei* **Börner** and **Minuth** 1984 [Syn. *P. abbotti chekei*]

Anthony S. Cheke is a writer who concentrates on the birds and herpetofauna of the Indian Ocean area. He worked in plant and animal ecology in the UK, Mauritius, and Thailand and led the British Ornithological Union expedition to the Mascarene Islands (1973). Since 1982 he has run the Inner Bookshop and Dodo Books, Oxford. He wrote "Lizards of the Seychelles," which appeared in *Biogeography and Ecology of the Seychelles Islands* (1984).

Chelazzi

Somali Garter Snake *Elapsoidea chelazzii* **Lanza,** 1979

Guido Chelazzi is Professor of both Zoology (1987) and Ecology (1990) at Università degli Studi di Firenze. He co-wrote "Experiments on the Homing Behaviour of Caged Pigeons" (1972).

Chen, B.

Chen's Bamboo Viper *Trimeresurus stejnegeri chenbihuii* **Zhao,** 1995

Professor Bihui Chen is a Chinese herpetologist at Anhui Normal University. He edited *The Amphibian and Reptilian Fauna of Anhui* (1991).

Chen, Y.-H.

Mountain Keelback sp. *Opisthotropis cheni* **Zhao,** 1999

Chen Yuan-Hui collected the holotype.

Cherchi

Longtail Lizard sp. *Latastia cherchii* Arillo, Balletto, and Spano, 1967

Maria Adelaide Cherchi (1927–1985) was an Italian herpetologist who worked with Arillo on a study of Russian tortoises. She was a member of the Institute of Biology, Università degli Studi di Genova. With Spano she co-wrote "Una nuova specie di *Tropiocolotes* del Sud Arabia spedizione Scortecci nell'Hadramaut 1962" (1963). The scientific name is an example of sloppy nomenclature; when referring to a woman, the binomial should use the feminine genitive: *cherchiae.*

Cherlin

Cherlin's Saw-scaled Viper *Echis megalocephalus* Cherlin, 1990

Dr. Vladimir Alexandrovich Cherlin (b. 1951) is a Russian molecular biologist, zoologist, and herpetologist. He earned his doctorate at Leningrad State University, after which he worked at the Zoological Institute of the Academy of Sciences of the USSR. He worked in Tashkent (Uzbekistan) (1983–1989) at the Zoological Park and Physiological Institute. He returned to Leningrad (1990) to the Institute of Evolutionary Physiology. He became a Director of a small bioscience company that produced snake venom, but it went bankrupt when the state withdrew financial support. He was unemployed from 1995 to 1998, and since then has worked in one of the Russian Orthodox churches. He wrote "Taxonomic Revision of the Snake Genus *Echis* (Viperidae). II. An Analysis of Taxonomy and Description of New Forms" (1990).

Chernov

Chernov's Skink *Ablepharus chernovi* **Darevsky,** 1953

Chernov's Snake Skink *Ophiomorus chernovi* **Anderson** and **Leviton,** 1966

Dr. Sergius Alexandrovich Chernov (1903–1964) was a Russian herpetologist. He wrote "Herpetological Fauna of Armenian SSR and Nakhichevan ASSR" (1939).

Cherrie

Cope's Brown Forest Skink *Sphenomorphus cherriei* **Cope,** 1893

George Kruck Cherrie (1865–1948) was an American naturalist and ornithologist who accompanied Theodore Roosevelt on a trip to Brazil (1913). He was Assistant Curator, Department of Ornithology, Field Museum (1890s). He also collected in Costa Rica (1894–1897) and, with his wife, Stella, in Colombia (1898). Roosevelt described him (1914) as "[an] efficient and fearless man; and willy-nilly he had been forced at times to vary his career by taking part in insurrections. Twice he had been behind . . . bars in consequence, on one occasion spending three months in a prison of a certain South American state, expecting each day to be taken out and shot. In another state he had, as an interlude to his ornithological pursuits, followed the career of a gun-runner, acting as such off and on for two and a half years. The particular revolutionary chief whose fortunes he was following finally came into power, and Cherrie immortalized his name by naming a new species of ant-thrush after him—a delightful touch, in its practical combination of those not normally kindred pursuits, ornithology and gun-running." Cherrie wrote *Ornithology Orinoco* (1916). Four birds and a mammal are named after him.

Chevert

Fitzroy Island Gecko *Nactus cheverti* **Macleay,** 1878

Named after the ship *Chevert*, which Sir William John Macleay bought and used for his expedition to New Guinea (1875).

Children

Children's Python *Antaresia childreni* **Gray,** 1842

John George Children (1777–1852) was a British entomologist and scientist with an interest in electricity, on which he published various notes (1808–1813). He became a Fellow of the Royal Society (1807) and subsequently became Secretary of the society. He visited Pennsylvania (1802) and traveled in Spain and Portugal (1808–1809), recording in his diary details of the Peninsular War as well as observations of minerals and the like. He worked in the British Museum (1816–1840). The mineral childrenite is named after him.

Chiszar

Chiszar's Arboreal Alligator Lizard *Abronia chiszari* **H. M. Smith** and **R. B. Smith,** 1981

Dr. David Alfred Chiszar (b. 1944) is a Professor of Psychology at the University of Colorado and an expert on snake behavior. He was awarded both his bachelor's degree (1966) and doctorate (1970) by Rutgers University. He works closely with Hobart Smith, with whom he co-wrote "Distributional and Variational Data on the Frogs of the Genus *Rana* in Chihuahua, Mexico, Including a New Species" (2003).

Christian

Christian's Scaly-toed Gecko *Lepidodactylus christiani* **Taylor,** 1917

Lieutenant Ralph L. Christian of the U.S. Army was a member of an expedition to Canlaon, the Philippines. He helped in the collecting of samples but suffered a fall on the mountain and was badly cut. Taylor, the describer, was with him at the time. It was obviously an exciting expedition; see **McNamara** for Taylor's comments on how he nearly drowned.

Christina

Christina's Lerista *Lerista christinae* **Storr,** 1979

Dr. Christine Davidge was attached to the School of Environmental and Life Sciences, Murdoch University, Murdoch, Australia. She collected the holotype of this skink (1977). She wrote "A Census of a Community of Small Terrestrial Vertebrates" (1979).

Christine

Burrowing Snake sp. *Apostolepis christineae* **De Lema,** 2002

Dr. Christine Strüssmann is a herpetologist. She graduated in veterinary medicine at Universidade Federal do Rio Grande do Sul (1982) and was awarded her master's degree in ecology by the University of Campinas (1992) and her doctorate in zoology by Pontifícia Universidade Católica do Rio Grande do Sul (2003). She is an Associate Professor at Universidade Federal de Mato Grosso do Sul and a Research Assistant at the Pantanal Research Center of Empresa Brasileira de Pesquisa Agropecuaria. She co-wrote *Serpentes do Pantanal* (2005).

Christophe

King Christophe Anole *Anolis christophei* **Williams,** 1960

Named after the Citadel of King Christophe, Cap Haitien, Haiti.

Christy

Christy's Snake-eater *Polemon christyi* **Boulenger,** 1903
Christy's Water Cobra *Boulengerina christyi* Boulenger, 1904
Christy's Banded Snake *Chamaelycus christyi* Boulenger, 1919

Dr. Cuthbert Christy (1863–1932) qualified as a physician at Edinburgh. In the early 1890s he traveled in the West Indies and South America, subsequently joining the army as a doctor. He was in northern Nigeria (1898–1900) and later in Uganda and the Congo (1902–1903). During WW1 he served in Africa and Mesopotamia (Iraq). After the war he explored in the Sudan, Nyasaland (Malawi), and Tanganyika (Tanzania) and was a member of a League of Nations commission enquiring into slavery and forced labor in Liberia. He was at some time Director of the Congo Museum, Tervuren, Belgium. He was on a zoological expedition to the Congo (1932) when he was gored by a buffalo and killed. A mammal is named after him.

Chris Wild

Chris Wild's Snake-eyed Skink *Lacertaspis chriswildi* **Böhme** and **Schmitz,** 1996

Christopher Wild has helped a number of expeditions collecting in the biological hotspots of southwest Cameroon. He wrote "Notes on the Rediscovery and the Congeneric Associations of the Pfeffer's Chameleon *Chamaeleo pfefferi* (Tornier 1900) (Sauria: Chamaeleonidae) with a Brief Description of the Hitherto Unknown Female of the Species" (1993). He collected the skink holotype.

Chu

Pampas Snake sp. *Phimophis chui* **Rodrigues,** 1993

Dr. Tien Hsi Chu is an expert on fibroblast cultures, working at the Department of Biology, Universidade Federal do Maranhão, São Luís, Brazil. He co-wrote "Chromosomal Characterization and Comparative Cytogenetic Analysis of Two Species of *Proechimys* (Echimyidae, Rodentia) from the Caatinga Domain of the State of Bahia, Brazil" (1992).

Citerni

Speedy Leaf-toed Gecko *Hemidactylus citernii* **Boulenger,** 1912

Captain Carlo Citerni (1873–1918) was an explorer. He took part in Bottego's second expedition (1895) to Lake Rudolph. He led the expedition to mark the border between Italian Somaliland and Ethiopia (1910–1911). He co-wrote *L'Omo. Viaggio d'esplorazione nell'Africa Oriental* (1899). Many Italian cities and towns have a "Via Carlo Citerni" named in his memory.

Clark, H. C.

Balsas Armed Lizard *Ctenosaura clarkii* **Bailey,** 1928
Clark's Coral Snake *Micrurus clarki* **Schmidt,** 1936
Clark's Forest Racer *Dendrophidion clarkii* **Dunn,** 1939
Clark's Ground Snake *Atractus clarki* Dunn and Bailey, 1939
Clark's Water Snake *Hydromorphus clarki* Dunn, 1942 [Junior syn. of *H. concolor* **Peters,** 1859]

Dr. Herbert Charles Clark (1877–1960) was the first Director of Gorgas Memorial Laboratory, Panama (1929–1954). United Fruit Company previously employed him as Director of Laboratories and Preventive Medicine. He also organized an annual census of the snake population of Panama (1929–1953).

Clark, J. H.

Clark's Spiny Lizard *Sceloporus clarkii* **Baird** and **Girard,** 1852

Salt Marsh Snake *Nerodia fasciata clarkii* Baird and Girard, 1853

Lieutenant John Henry Clark (1830–1885) was an American surveyor, naturalist, and collector. He was a student at Dickinson College (ca. 1844). He was a zoologist on the U.S./Mexican Border Survey (1850–1855). Under the auspices of the Smithsonian, he conducted the Texas Boundary Survey (1860). A bird is named after him.

Clark, R. and E.

Clark's Toad-headed Agama *Phrynocephalus clarkorum* **Anderson** and **Leviton** 1967

Clark's Lizard *Darevskia clarkorum* **Darevsky** and Vedmederja, 1977

Dr. Richard J. Clark is a zoologist, herpetologist, and ornithologist retired (1998) from York College, Pennsylvania, where he was Professor of Biology. He now lives in Prescott, Arizona. He took his bachelor's (1959) and masters' (1963) degrees at the State University of New York and was awarded his doctorate by Cornell (1970). His wife, Erica D. Clark, is also a zoologist. They work together and co-wrote "Report on a Collection of Amphibians and Reptiles from Turkey" (1973).

Clay

Clay's Dragon *Ctenophorus clayi* **Storr,** 1966 [Alt. Black-shouldered Ground-Dragon]

Brian T. Clay (fl. 1950–2004) was a forestry official in East Africa who moved to Australia (1960), where he worked as a field assistant for the Zoology Department, University of Western Australia. He studied there, becoming a biologist and herpetologist, and eventually ran the university's marsupial research center near Perth until he retired to Geographe Bay, Western Australia. He wrote a paper, "Observations on the Breeding Biology and Behaviour of the Long-necked Tortoise, *Chelodina oblonga*" (1981). Storr wrote that Clay "helped me collect the holotype and much other material in the North-West Division."

Clelia

Colubrid snake genus *Clelia* **Fitzinger,** 1826

Clelia comes from the Latin name Cloelia. In Roman legend Cloelia was a maiden who was given to an Etruscan invader as a hostage but managed to escape by swimming across the Tiber. The snakes of the genus *Clelia* are also known as mussuranas.

Clench

Peninsula Least Gecko *Sphaerodactylus clenchi* **Shreve,** 1968

Dr. William J. Clench (1897–1984) was an entomologist and malacologist. He received a master's degree from Harvard (1923). He then studied molluscs at University of Michigan but left before completing his doctorate to become custodian of collections at Kent Scientific Museum, Michigan, later becoming Curator of Molluscs, Museum of Comparative Zoology, Harvard (1926–1966). The University of Michigan belatedly recognized his achievements and awarded him his doctorate, and made him an honorary Doctor of Science as well (1953). In retirement he became adjunct Professor of Zoology, Ohio State University, and continued his fieldwork on freshwater molluscs.

Cleopatra

Cleopatra's Asp *Naja haje* **Linnaeus,** 1758 [Alt. Egyptian Cobra]

Cleopatra (69–30 B.C.) was Queen of Egypt. For the full story we recommend Shakespeare's *Antony and Cleopatra.*

Clifton

Clifton's Lizard-eater *Dryadophis cliftoni* **Hardy,** 1964 [Syn. *Mastigodryas cliftoni*]

Percy L. Clifton collected the snake holotype. He collected in the state of Jalisco, Mexico, for the University of Kansas (1962–1967).

Cloete

Cloete's Girdled Lizard *Cordylus cloetei* Mouton and Van Wyk, 1994

Jos Cloete owns the farm—De Hoek, in South Africa—where this species was discovered. It was named after him "in appreciation of his hospitality and keen interest in nature."

Cochran

Cochran's Forest Dragon *Gonocephalus abbotti* Cochran, 1922 [Alt. Abbott's Anglehead Lizard]

Cochran's Haitian Lizard *Wetmorena haetiana* Cochran, 1927

Cochran's Curlytail Lizard *Leiocephalus vinculum* Cochran, 1928

Cochran's Caribbean Gecko *Aristelliger cochranae* **Grant,** 1931 [Alt. Cochran's Croaking Gecko, Navassa Gecko]

Cochran's Neusticurus *Neusticurus cochranae* **Burt** and Burt, 1931

Cochran's Gianthead Anole *Anolis caudalis* Cochran, 1932

Spiny Lizard sp. *Sceloporus cochranae* **H. M. Smith,** 1936 [Junior syn. of *S. cupreus* Bocourt, 1873]
Cochran's Least Gecko *Sphaerodactylus cochranae* **Ruibal,** 1946
Cochran's Sipo *Chironius cochranae* **Hoge** and Romano, 1969

Dr. Doris Mable Cochran (1898–1968) was a zoologist and herpetologist. She was educated in Washington, DC, and after graduating from high school worked for the War Department. She took evening classes at George Washington University, where she obtained her master's degree in science (1921). She received a master's degree in zoology from Johns Hopkins University (1928) and a doctorate from the University of Maryland (1933). She also trained as, and became, a highly skilled scientific illustrator. She worked at the Smithsonian (1919–1968), first as an assistant to Leonhard Stejneger in Division of Reptiles and Amphibians, becoming Assistant Curator, Reptiles and Amphibians (1927), Associate Curator (1942), and Curator (1956). After Stejneger's death (1943) she became acting head of the division. Her main interest was the herpetofauna of the West Indies and tropical America, where she made field trips (1935 and 1962–1963) to work with the Brazilian herpetologists Adolpho and Bertha Lutz. She wrote *The Herpetology of Hispaniola* (1941).

Cocteau

Cocteau's Skink *Macroscincus coctei* **Duméril** and **Bibron,** 1839 EXTINCT [Alt. Bibron's Skink, Cape Verde Giant Skink]

Dr. Jean Théodore Cocteau (1798–1838) was a physician and a noted scientist. He wrote *Études sur les scincoides* (1836). The skink, unseen since 1898, is presumed extinct.

Cofan

Duellman's Dwarf Iguana *Enyalioides cofanorum* **Duellman,** 1973

Named after the Cofan Indians of Ecuador.

Cogger

Northern Spotted Velvet Gecko *Oedura coggeri* Bustard, 1966
Cogger's Ctenotus *Ctenotus coggeri* **Sadlier,** 1985
Cogger's Island Skink *Geomyersia coggeri* **Greer,** 1982
Cogger's Sea Snake *Hydrophis coggeri* **Kharin,** 1984
Cogger's Emo Skink *Emoia coggeri* **Brown,** 1991
Northern Sun Skink *Lampropholis coggeri* **Ingram,** 1991
Skink genus *Coggeria* **Couper, Covacevich,** Marsterson, and **Shea,** 1996

Dr. Harold "Hal" George Cogger (b. 1935) is a herpetologist who spent his entire career at the Australian Museum, Sydney. He was Curator, Reptiles and Amphibians (1960–1975), and Deputy Director (1976–1995) of the museum. He was a Conjoint Professor at the University of Newcastle, New South Wales (1997–2001). He has traveled widely, visiting most parts of Australia and New Guinea, Japan, Indonesia, and many islands in the western Pacific. He wrote *Reptiles and Amphibians of Australia* (1975).

Cole

Cole's Racerunner *Cnemidophorus pseudolemniscatus* Cole and Dessauer, 1993
Cole's Night Lizard *Xantusia jaycolei* **Bezy,** Bezy, and Bolles, 2008

Dr. Charles James "Jay" Cole is a herpetologist whose doctorate was awarded by the University of Arizona (1969). He has spent most of his career at the American Museum of Natural History, where he was Curator-in-Charge and Curator (Herpetology). He is now Curator Emeritus and acts as a leader of expeditions to show reptiles and amphibians to paying customers. He worked with Dessauer, with whom he co-wrote "Unisexual and Bisexual Whiptail Lizards of the *Cnemidophorus lemniscatus* Complex (Squamata: Teiidae) of the Guiana Region, South America, with Descriptions of New Species" (1993).

Colee

Colee's Racerunner *Cnemidophorus pseudolemniscatus* **Cole** and Dessauer, 1993

We think this must be a transcription error for Cole's Racerunner (see above). There are very few references to Colee, but the spelling does occur in print.

Coleman

Coleman's Bunchgrass Lizard *Sceloporus samcolemani* **H. M. Smith** and Hall, 1974

Dr. Sam Coleman wrote data-processing programs for Smith's work on Mexican herpetofauna.

Collett

Collett's Black Snake *Pseudechis colletti* **Boulenger,** 1902
Collett's Ctenotus *Ctenotus colletti* Boulenger, 1896

Dr. Robert Collett (1842–1913) was a Norwegian zoologist and ichthyologist. He worked at Christiania (Oslo) Museum (1871–1913), first as an Assistant Curator, then Curator (1874) and Director (1892), and was Professor of Zoology at the university from 1884. A mammal and a bird are named after him.

Colley

Colley's Iguana *Cyclura collei* **Gray,** 1845 [Alt. Jamaican Iguana]

Sometimes said to be named after its preferred location,

hills—*collei* being Latin for "of the hill." Gray does not identify who Colley was but refers to the iguana as Colley's Iguana, so he must have had someone in mind. For a time it was believed extinct, not having been seen since 1948, but it was refound ca. 1990 and is being bred in captivity to build up its numbers for eventual release back into the wild.

Collie

Narrow-breasted Snake-necked Turtle *Chelodina colliei* **Gray,** 1856

Lieutenant Dr. Alexander Collie (1793–1835) was a physician, naturalist, and explorer. He was the naval surgeon and naturalist on an expedition (1825–1828) led by Captain Frederick Beechey on HMS *Blossom*, which made significant zoological findings during the voyage from Chile to Alaska. Collie collected many specimens that did not survive the return journey to England in good condition, but he made some colored drawings of birds he thought were new and also took extensive notes. Collie later went to Perth as a colonial administrator. A mammal and a bird and a town and a river in Western Australia are named after him.

Colliver

Nubbinned Fine-lined Slider *Lerista colliveri* **Couper** and **Ingram,** 1992

Frederick Stanley "Stan" Colliver (1908–1991) worked as a Scientific Officer at the Geology Department, University of Queensland, from 1948; his chief interests were minerals, fossils, and shells. He was renowned for having encyclopedic knowledge. He was an Honorary Associate of the Queensland Museum and a co-founder of the Anthropological Society of Queensland (1948). He donated his collection of around 400,000 mollusc shells and 5,000 geological specimens to the Queensland Museum. A fossil mammal that he discovered (1939) near Geelong is named after him.

Coloma Roman

Gymnophthalmid lizard sp. *Riama colomaromani* **Kizirian,** 1996

Professor Dr. Luis Aurelio Coloma Román (b. 1962) is a zoologist and herpetologist. He is Curator, Herpetology Section, Museum of Zoology, Pontificia Universidad Católica del Ecuador. He took his initial degrees at universities in Ecuador and followed up with a doctorate from the University of Kansas.

Colosi

Colosi's Cylindrical Skink *Chalcides colosii* **Lanza,** 1957

Giuseppe Colosi (1892–1975) was a zoologist, specializing in crustaceans, who was Director, Zoological Institute, Università degli Studi di Firenze (1940–1962). He taught at Università degli Studi di Torino (1920–1924).

Columbus

San Salvador Blind Snake *Leptotyphlops columbi* **Klauber,** 1939

Christopher Columbus (1451–1506) was a Genoese seaman, navigator, entrepreneur, and explorer. So much has been written elsewhere that anything we might write would be redundant.

Conant

Florida Cottonmouth *Agkistrodon piscivorus conanti* **Gloyd,** 1969

Conant's Milk Snake *Lampropelitis triangulum conanti* **Williams,** 1978

Conant's Garter Snake *Thamnophis conanti* **Rossman** and Burbrink 2005

Roger Conant (1909–2003) was a herpetologist, author, and conservationist. While still a teenager he took a job at the Philadelphia Zoo to earn money to help his widowed mother. The result was a lifelong passion for reptiles. He was Curator of Reptiles at the Toledo Zoo, Ohio (1929–1935), leaving to return to the Philadelphia Zoo, where he eventually became Director (1967–1973). In addition to fieldwork in the USA and Mexico, he visited Asia and Africa. He wrote and presented a 15-minute weekly educational program for a Philadelphia radio station called *Let's Visit the Zoo* (1936–1969). His wife was the animal artist and photographer Isabelle de Peyster Hunt (1901–1976), and together they created *A Field Guide to Reptiles and Amphibians of the United States and Canada East of the 100th Meridian* (1958). In retirement at Albuquerque, New Mexico, he continued to conduct research and write; he was Director Emeritus, Philadelphia Zoo, and an Adjunct Professor, University of New Mexico. He died of cancer.

Conrad

Conrad's Worm Snake *Typhlops conradi* **Peters,** 1874

Captain Paul Conrad (1836–ca. 1873) was a German from Bremen who was in the East Indies (1870–1873).

Conradt

Matschie's Dwarf Gecko *Lygodactylus conradti* **Matschie,** 1892

Leopold Conradt (fl. 1875–1910) sent specimens from Usambara, German East Africa (1890s), and was on Dr. Hans Meyer's Usambara expedition (1888). He made a special study of the snails and ants of the Usambara Mountains. He was in Togo and Cameroon (both then in German West Africa) in the early 20th century. By 1910 he was collecting in Mexico.

Conrau

Cameroon Dwarf Gecko *Lygodactylus conraui* **Tornier,** 1902

Gustav Conrau (d. 1899), a German trader, colonial recruiting agent, explorer, and collector in Cameroon, was always known by his Bali nickname, Manjikwara. He was killed by local tribesmen in Bangwa. He has been described as being in self-imposed exile. He was the first European to encounter the Bangwa people (1898) and to meet Chief Fontem Asunganyi of whose influence and wealth he had heard while looking for ivory and plantation labor. Conrau and Asunganyi took a liking to each other. Conrau "admired the dignified bearing of the young chief." They exchanged gifts. Asunganyi was eager that a German trading "factory" should be established at his capital, Azi, so he let Conrau take away 70 men to work on the plantations in the south. When Conrau returned a year later without these men, the people feared they were dead. Asunganyi and his councillors detained Conrau in Fontem until he arranged for their return. One night, according to the Bangwa, Conrau attempted to escape. Pursued by his captors, Conrau panicked, shot wildly at the Bangwa, and, wounded by a spear and down to his last bullet, shot himself to avoid the torture he feared. The Germans appear to have accepted this version of Conrau's suicide, which was reinforced by the account given by his servants. However, some people suggested that he may have been shot by his adversaries. Whatever the truth, his head was removed and carried to Fontem's palace as a war trophy; the prepared scalp was even worn at a celebration at nearby Fotabong by the chief's mother. The Germans, as a consequence of Conrau's death, sent two military expeditions against Fontem.

Constanza

Constanza's Tree Iguana *Liolaemus constanzae* **Donoso-Barros,** 1961

Constanza Donoso-Barros is the author's eldest daughter.

Cook

Cook's Pallid Anole *Anolis cooki* **Grant,** 1931

Dr. Melville "Mel" Thurston Cook (1869–1952) was a botanist and entomologist who became Director, Insular Experiment Station, Rio Piedras, part of Universidad de Puerto Rico. He joined the faculty of Rutgers University (1910) to teach botany. He was a Professor and was seconded to help teach at New Jersey College for Women (1918). He spent some months in Puerto Rico with a party of his students (1926). He wrote *Snake Killing Mongooses a Plague to Porto Rico* (1929).

Cook(e)

Cook's Tree Boa *Corallus cookii* **Gray,** 1842

Gray's description merely gives the name "Edw. Cooke, Esq." We believe this to be Edward William Cooke (1811–1880), an English artist who was most famed as a marine painter. He came from a family of engravers and grew up surrounded by art and artists. He was also interested in natural history and was made a Fellow of the Geological Society and the Zoological Society. The vernacular name should be "Cooke's," but the *e* is usually dropped in error.

Cope

Cope's Kukri Snake *Rhynchocalamus phaenochalinus* Cope, 1860

Cope's Rustyhead Snake *Amastridium veliferum* Cope, 1860

Cope's Yellowbelly Snake *Coniophanes fissidens proterops* Cope, 1860

Cope's Blunthead Tree Snake *Imantodes cenchoa leucomelas* Cope, 1861

Cope's Gopher Snake *Pituophis lineaticollis lineaticollis* Cope, 1861

Cope's Parrot Snake *Leptophis depressirostris* Cope, 1861

Cope's Vine Snake *Oxybelis brevirostris* Cope, 1861

Cope's Worm Lizard *Amphisbaena fenestrata* Cope, 1861

Cope's Ameiva *Ameiva bifrontata* Cope, 1862

Cope's Bachia *Bachia pallidiceps* Cope, 1862

Cope's False Coral Snake *Pliocercus euryzonus* Cope, 1862

Cope's Galliwasp *Celestus stenurus* Cope, 1862

Cope's Lava Lizard *Ophryoessoides caducus* Cope, 1862

Cope's Mabuya *Mabuya frenata* Cope, 1862

Cope's Stripeless Snake *Coniophanes lateritius* Cope 1862

Cope's Tropical Snake *Lygophis flavifrenatus* Cope, 1862

Cope's Antilles Snake *Antillophis parvifrons* Cope, 1863

Cope's Forest Snake *Taeniophallus poecilopogon* Cope, 1863

Cope's Island Racer *Dromicus funereum* Cope, 1863

Cope's Largescale Spiny Lizard *Sceloporus utiformis* Cope, 1863

Northern Scarlet Snake *Cemophora coccinea copei* **Jan,** 1863

Cope's False Chameleon *Anolis porcus* Cope, 1864

Cope's Forest Ground Skink *Scincella gemmingeri* Cope, 1864

Cope's Rough-sided Snake *Aspidura copii* **Günther,** 1864

Cope's Smooth Anole *Anolis damulus* Cope, 1864

Cope's Vera Cruz Anole *Anolis cymbops* Cope, 1864

Cope's Centipede Snake *Tantilla calamarina* Cope, 1866
Cope's Rat Snake *Elaphe triaspis* Cope, 1866 [Alt. Green Rat Snake; Syn. *Senticolis triaspis*]
Cope's Least Gecko *Sphaerodactylus copei* **Steindachner,** 1867
Cope's Arboreal Alligator Lizard *Abronia aurita* Cope, 1868
Cope's Ground Snake *Liophis chrysostoma* Cope, 1868
Cope's Parrot Snake *Leptophis cupreus* Cope, 1868
Cope's Scaly-toed Gecko *Lepidodactylus pusillus* Cope, 1868
Cope's Snail Sucker *Sibon anthracops* Cope, 1868
Cope's Tropical Racer *Mastigodryas pulchriceps* Cope, 1868 [Alt. Cope's Whipsnake]
Cope's Forest Snake *Taeniophallus persimilis* Cope, 1869 [Syn. *Echinanthera persimilis, Rhadinaea persimilis*]
Cope's Scaly-toed Gecko *Lepidodactylus pusillus* Cope, 1869
Cope's Striped Snake *Coniophanes piceivittis* Cope, 1869
Cope's Thread Coral Snake *Leptomicrurus scutiventris* Cope, 1869
Cope's Snail-eater *Dipsas copei* Günther, 1872
Sepsina Skink *Sepsina copei* **Bocage,** 1873
Cope's Coffee Snake *Ninia psephota* Cope, 1875
Cope's Graceful Brown Snake *Rhadinaea pachyura* Cope, 1875
Cope's Neusticurus *Neusticurus eopleopus* Cope, 1876
Cope's Alligator Lizard *Mesaspis monticola* Cope, 1877
Cope's Racerunner *Aspidoscelis guttatus immutabilis* Cope, 1878 [Syn. *Cnemidophorus guttatus immutabilis*]
Cope's Mountain Meadow Snake *Adelophis copei* **Dugès,** 1879
Cope's Leopard Lizard *Gambelia copeii* **Yarrow,** 1882
Cope's Leposoma *Leptosoma parietale* Cope, 1885
Gymnophthalmid lizard sp. *Alopoglossus copii* **Boulenger,** 1885
Cope's Rain Forest Cat-eyed Snake *Leptodeira frenata* Cope, 1886
Cope's Burrowing Snake *Apostolepis vittatus* Cope, 1887
Cope's Earth Snake *Adelphicos quadrivirgatus visoninus* Cope, 1887
Cope's Spinytail Lizard *Lacerta echinata* Cope, 1887
Cope's Brown Forest Skink *Sphenomorphus cherriei* Cope, 1893
Cope's Forest Racer *Dendrophidion paucicarina* Cope, 1893
Cope's Tropical Ground Snake *Trimetopon pliolepis* Cope, 1894
Cope's Blind Snake *Liotyphlops anops* Cope, 1899
Cope's Skink *Eumeces copei* **Taylor,** 1933
Cope's Black-striped Snake *Coniophanes imperialis copei* **Hartweg** and **Oliver,** 1938

Edward Drinker Cope (1840–1897) was a zoologist and paleontologist who studied under Baird at the Smithsonian (1859), at the British Museum, London, and at Jardin des Plantes, Paris (1863–1867). He was Professor of Comparative Zoology and Botany, Haverford College (1864–1867), and was appointed Curator, Philadelphia Academy of Natural Sciences (1865). He was the paleontologist on the Wheeler survey (1874–1877) west of the 100th meridian in New Mexico, Oregon, Texas, and Montana. He was a Professor at the University of Pennsylvania—Geology and Mineralogy (1889–1895) and Zoology and Comparative Anatomy (1895–1897). From 1878 he was Senior Naturalist for the periodical *American Naturalist*, which he owned. He wrote *Systematic Arrangement of the Extinct Batrachia, Reptiles and Aves of North America* (1869–1870). In his will he asked that his body should be used as the holotype of *Homo sapiens*, but his skeleton was found to be unsuitable because of disease; it was rumored that he died of syphilis, but the cause is more likely to have been prostatitis complicated by self-medication using formalin and belladonna.

Corfield

Corfield's Fringed Gecko *Luperosaurus corfieldi* Gaulke, Roesler, and Brown, 2007

Charles Corfield (b. 1959) is an English-born American businessman who has worked and invested in the software and technology sector for 20 years. He enjoys rock and ice climbing and competes in ultra-marathons. He worked as a project manager on the National Geographic Society / Boston Museum of Science resurvey of Everest, and participated in the rescue of climbers involved in the 1996 Everest disaster chronicled in Jon Krakauer's book *Into Thin Air*. The describers say in their etymology, "We are pleased to name the new species for Charles Corfield for supporting biodiversity research and nature conservation in the Philippines by taking patronage of this endemic Philippine species."

Coronado

Guerreran Centipede Snake *Tantilla coronadoi* **Hartweg,** 1944

Salvador Coronado was a Mexican who collected with Hartweg, who said in his description, "The species is named in honor of my Mexican friend, Señor Salvador Coronado of the Departmento Pesca y Maritima." He was in charge of the Fish Culture Station near Mexico City and sent fish specimens to dealers and scientists. His collecting career spanned at least three decades (1930s–1950s).

Cortez

Cortez's Horned Lizard *Phrynosoma orbiculare cortezii* **Duméril** and **Bocourt,** 1870

Hernán Cortés de Monroy y Pizarro (1485–1547) was a Spanish conquistador whose expedition to Mexico (1519–1521) brought about the collapse of the Aztec empire.

Cottrell

Cottrell's Mountain Lizard *Tropidosaura cottrelli* **Hewitt,** 1925

John Awdry Cottrell (b. 1904) was a South African naturalist and ornithologist who collected the lizard holotype (1925). He wrote *Black Eagle Fly Free* (1970).

Couch

Couch's Garter Snake *Thamnophis couchii* **Kennicott,** 1859 [Alt. Western Aquatic Garter Snake]

Couch's Spiny Lizard *Sceloporus couchii* **Baird,** 1859

Darius Nash Couch (1822–1897) was a U.S. Army officer—a General during the Civil War. He was also an explorer who took a leave of absence to lead a zoological expedition in Mexico. He was a Lieutenant (1846) in Mexico and fought at the Battle of Buena Vista (1847). He returned to Washington (1854), resigned his commission (1855), and became a merchant and manufacturer in New York and Massachusetts. On the outbreak of the American Civil War he rejoined the army as a Colonel. He offered to resign on grounds of ill-health (1863), but was persuaded to stay on by being promoted to Major General. He was in charge of all the ceremonies associated with the consecration of the National Cemetery, Gettysburg (1865), the occasion of Abraham Lincoln's Gettysburg Address. After the Civil War he again resigned from the army and was Collector, Port of Boston (1866–1867), President of a Virginia mining and manufacturing concern (1867–1877), and an administrator in Connecticut (1877–1884). He has two birds named after him.

Coulanges

Colubrid snake sp. *Brygophis coulangesi* **Domergue,** 1988

Dr. Pierre Coulanges is a French epidemiologist who joined Institut Pasteur de Madagascar (1973) and was Director during a period when plague reappeared in Madagascar. He wrote "La peste à Tananarive, de son apparition en 1921 à sa résurgence en 1979" (1989).

Couper, J. H.

Couper's Snake *Drymarchon couperi* **Holbrook,** 1842 [Alt. Eastern Indigo Snake; Syn. *D. corais couperi*]

James Hamilton Couper (1794–1866) was a planter in Georgia who experimented with new methods of agriculture. He was born in Scotland but emigrated to America and graduated from Yale (1814). He built a modern sugar mill (1829) at Hopeton in Georgia, then switched production from sugar to rice (1838). He presented a number of papers to the Academy of Natural Sciences, Philadelphia. Plantation owners' way of life was destroyed by the Civil War. Although he personally was against secession, his five sons all fought in the Confederate army; two of them were killed. This was too much for Couper, and he died, financially and mentally broken.

Couper, P.

Plain-backed Sunskink *Lampropholis couperi* **Ingram,** 1991

Patrick Couper's Python *Broghammerus reticulatus patrickcouperi* Hoser 2003

Patrick J. Couper (b. 1958) is zoologist and geographer who was born in New Zealand and migrated to Australia (1981). He started as a volunteer at the University of New England, moving to the Queensland Museum, Brisbane, as a preparator in 1984 and becoming Curator of Herpetology in 1993. His main research interest is the taxonomy of geckos from the rainforests of eastern Australia.

Covacevich

Clouded Gecko *Oedura jacovae* **Couper,** Keim, and Hoskin 2007

Jeanette Adelaide Covacevich (b. 1945) was, until her retirement (2002), Senior Curator, Reptiles and Amphibians, Queensland Museum. She remains a prominent figure in conservation in Queensland. The describers of the gecko note that she is recognized for her efforts to preserve the unique character of North Stradbroke Island. She wrote *The Snakes of Brisbane* (1980).

Coventry

Coventry's Window-eyed Skink *Niveoscincus coventryi* **Rawlinson,** 1975 [Alt. Southern Forest Cool-Skink; Syn. *Pseudemoia coventryi*]

Coventry's Spinytail Skink *Egernia coventryi* **Storr,** 1978 [Alt. Eastern Mourning Skink]

Albert John Coventry (b. 1936) worked for 47 years at Museum Victoria as a herpetologist and is now Emeritus Curator of Herpetology. He is a former President of the Australian Society of Herpetologists.

Cowles

Angolan Coral Snake *Aspidelaps lubricus cowlesi* **Bogert,** 1940

Cowles' Prairie Lizard *Sceloporus cowlesi* **Lowe** and **Norris,** 1956 [Alt. Southwestern Fence Lizard; Syn. *S. undulatus cowlesi*]

Dr. Raymond Bridgeman Cowles (1896–1975), who spent

most of his career at the University of California, Los Angeles (1929–1963), was born in Zululand, South Africa, to American medical missionary parents. He spent his early years wandering in the bush with young Zulu friends and so became entirely fluent in their language. He took his bachelor's degree at Pomona College (1921) and was an instructor at Cornell (1921–1925), where he took his doctorate (1928). He was in Africa (1925–1927) and joined the University of California, becoming Professor of Vertebrate Zoology (1947) and retiring as Professor Emeritus (1963). He wrote the classic work "A Preliminary Study of the Thermal Requirements of Desert Reptiles" (1944).

Cox

Cox's Sphenomorphus *Sphenomorphus coxi* **Taylor,** 1915

Dr. Alvin J. Cox was the Director, Philippines Bureau of Science, Manila, ca. 1910.

Cradock

Cradock Thick-toed Gecko *Pachydactylus geitje* Sparrman, 1778

Sir John Cradock (1759–1839), later created First Baron Howden, had a successful career as a soldier (1777–1810). He was Governor of Cape Province (1811–1814) and founded a town (1813) named Cradock after him. We cannot be sure, but we think the gecko's vernacular name comes from the town and not the man.

Crane

Forest Skink sp. *Sphenomorphus cranei* **Schmidt,** 1932

Cornelius Crane (1905–1962) was a philanthropist, amateur archeologist, and anthropologist. He sponsored and led the Crane Pacific expedition, under the auspices of the Field Museum, which collected widely in the Pacific (1928–1929). Crane used his own yacht, *Illyria*, for the trip. Schmidt was scientific leader of the expedition. Crane, a wealthy Chicago manufacturer of bathroom equipment, sponsored further expeditions over several decades.

Cranwell

Cranwell's Tree Iguana *Liolaemus cranwelli* **Donoso-Barros,** 1973

Dr. Jorge A. Cranwell was in charge of the Herpetology Section, Museo Argentino de Ciencias Naturales "Bernardino Rivadavia," Buenos Aires. He wrote "Para la herpetologia de Misiones" (1943). An amphibian is named after him.

Creaser

Creaser's Mud Turtle *Kinosternon creaseri* **Hartweg,** 1934

Edwin Phillip Creaser (1907–1981) was based at the University of Michigan. He was a researcher into freshwater Decapoda. He was part of the Carnegie Institution's 1936 investigation of the cenotes of Yucatan (Mexico), which researched the hydrography and zoology of the area. He wrote "A Note on the Food of the Box Turtle" (1940).

Crego/Cregoe/Cregoi

Crego/Cregoe/Cregoi's [Blind] Legless Skink *Typhlosaurus cregoi* **Boulenger,** 1903

John P. Cregoe presented the holotype. He collected plants in South Africa on behalf of Kew.

Crenn

Amphiglossus (skink) sp. *Amphiglossus crenni* **Mocquard,** 1906

Dr. Louis Crenn presented the holotype of this lizard to Mocquard. He wrote "Notes d'ophtalmologie sur Madagascar" (1910).

Cristian

Tree Iguana sp. *Liolaemus cristiani* **Núñez,** 1991

Cristian Simón Abdala is a herpetologist at the Institute of Herpetology, Fondación Miguel Lillo, Tucumán, Argentina. He co-wrote "Description of a New Patagonian Lizard Species of the *Liolaemus silvanae* Group (Iguania: Liolaemidae)" (2006).

Crocker

Crocker's Sea Snake *Laticauda crockeri* **Slevin,** 1934

Charles Templeton Crocker (1885–1948) was a member of a California family that made its money from railways, having invested in the first transcontinental American railroad. He was more interested in exploring the South Pacific than in the South Pacific Railroad Company and had a beautiful yacht built, the *Zaca*, that he used as a floating base for a number of "Templeton Crocker expeditions." These started (1930) with a voyage to Fiji, then (1932) to the Galapagos Islands and (1933) to the Solomon Islands. He published *The Cruise of the Zaca* (1933), and he made a film called *People and Dances of Oceania*.

Croizat

Horton's Mabuya *Mabuya croizati* **Horton,** 1973

Leon Croizat (1894–1982) was a botanist, biogeographer, and evolutionist born in Turin. He was proud of his ancestors, who included Mme. Roland (guillotined during the French Revolution) and Lamartine (a famous 19th-century French poet). He served in the Italian army (1914–1919), after which he returned to Università degli Studi di Torino and graduated in law (1920). He hated the Fascists and emigrated to the USA (1923), where he took any job going. He painted in watercolors, with modest financial success until the Wall Street Crash (1929) killed his market. He tried his luck in Paris, but without success, and returned to New York. Merrill, then Director of the Arnold Arboretum, Harvard, hired Croizat (1937) as a technical assistant. His drawings were said to be unbelievably accurate. He fell out with Harvard over a paper he had written and was sacked (1946); the paper was published through a rival, but equally reputable, outlet. He moved to Caracas, Venezuela, and held a number of academic positions there (1947–1952). He was the botanist on the Franco-Venezuelan expedition to the sources of the Orinoco (1950–1951). He resigned his academic positions (1953) to work full time on biological problems. In 1976 he and his wife became the first Directors of Jardin Botanico Xerofito, Coro, which they had founded (1970). It is now named after him.

Crombie

Wall Gecko sp. *Tarentola crombiei* Diaz and Hedges, 2008

Ronald Ian Crombie (b. 1949) was a herpetologist at the Division of Amphibians and Reptiles, National Museum of Natural History, Washington, DC. He collected on the tiny island of Catalinita, off the coast of the Dominican Republic (1975). He wrote *Herpetological Publications of the National Museum of Natural History* (1994). See also **Dermal.**

Cropani

Cropan's Boa *Xenoboa cropanii* **Hoge,** 1953 [Syn. *Corallus cropanii*]

Ottorino de Fiori, Baron de Cropani, was an Italian who became Professor of Geology and Paleontology, Universidade de São Paulo, Brazil (1935). After 1953 he became Director, Instituto Vulcanològico, Catania, Sicily.

Crosse

Crosse's Beaked Snake *Rhinotyphlops crossii* **Boulenger,** 1893

West African File Snake *Mehelya crossii* Boulenger, 1895

Dr. William Henry Crosse (1859–1903) was a traveler and collector who operated in Nigeria. He wrote "Notes on the Malarial Fevers Met with on the River Niger, West Africa" (1892). He once said, "The ideal traveller is a temperate person, with a sound constitution, a digestion like an ostrich, a good temper, and no race prejudice." A mammal is named after him.

Crotalus

New Guinea Death Adder *Acanthophis crotalusei* Hoser, 1998

This is not named after a person, but indirectly after the rattlesnake genus *Crotalus*. (The name is derived from the Greek word *krotalon*, which means "rattle" or "castanet.") The death adder is named after Hoser's pet dog, Crotalus; we assume the dog was named after the snake genus.

Cuming

Mindanao Water Monitor *Varanus cumingi* **Martin,** 1838

Cuming's Blind Snake *Ramphotyphlops cumingii* **Gray,** 1845 [Alt. Philippine Blind Snake]

Cuming's Flap-legged Gecko *Luperosaurus cumingi* Gray, 1845 [Alt. Philippine Wolf Gecko]

Cuming's Sphenomorphus *Sphenomorphus cumingi* Gray, 1845

Balsas Anole *Anolis cumingii* Peters, 1863

Asian Giant Softshell Turtle *Pelochelys cumingii* Gray, 1864

Cuming's Eared Skink *Mabuya cumingi* **Brown** and **Alcala,** 1980

Hugh Cuming (1791–1865) was an English naturalist and conchologist, often described as the "Prince of Collectors." He was a sailmaker living in Valparaiso, Chile, before changing professions to became a collector, starting in the Neotropics (1822–1826 and 1828–1830). He collected in Polynesia (1827–1828) and the East Indies (1836–1840). He preceded Darwin in having collected in the Galapagos (1829). His shell collection is housed in the Linnean Library, London. Other taxa, including a mammal, are named after him.

Cumming

Top End Death Adder *Acanthophis cummingi* Hoser, 1998

Fia Cumming is a political reporter. In 1998 she was working for the Sydney *Sun-Herald* newspaper. She became the first journalist to write on corruption within the New South Wales Parks and Wildlife Service regarding stolen reptiles (1981). Hoser stated that there would probably have been no overhaul of wildlife laws in New South Wales had Cumming not exposed this scandal. Hoser is clearly a man who has, to quote Ben Jonson, "small Latin and less Greek," as he used the masculine genitive form in the binomial.

Cunha

Cunha's Brazilian Lizard *Placosoma cipoense* Cunha, 1966

Cunha's Teiid *Colobosauroides cearensis* Cunha, Lima-Verde, and Lima, 1991

Worm Lizard sp. *Amphisbaena cunhai* **Hoogmoed** and Avila-Pires, 1991

Osvaldo Rodrigues da Cunha (1916–2007) was a zoologist and herpetologist at Museu Paraense Emilio Goeldi, Belém, Brazil. He wrote *O naturalista Alexandre Rodrigues Ferreira: Uma analise comparativa de sua viagem filosofica (1783–1793) pela Amazonia e Mato Grosso com a de outros naturalistas posteriores* (1989). See also **Osvaldo.**

Cunningham

Cunningham's Skink *Egernia cunninghami* **Gray,** 1832

Allan Cunningham (1791–1839) was an explorer and botanist. He collected in Brazil (1814–1816) and sailed from Rio to Sydney (1816) to explore with Oxley in Australia. He served as ship's botanist under Phillip Parker King's command (1817–1822), twice circumnavigating Australia, in HMS *Mermaid* (1817–1820) and in HMS *Bathurst* (1821–1822), including visits to Timor and Mauritius. He explored New South Wales, including the site of Canberra (1822–1826). He visited New Zealand (1826–1827). Cunningham returned to England (1831) with all his specimens, including, as Gray wrote, "three new animals brought from New Holland by Mr Cunningham." He went again to Australia (1837) as Government Botanist but resigned in disgust (1838) on finding that his job was to run a kitchen garden and grow vegetables for government officials. He is buried in the Royal Botanic Gardens, Sydney.

Curle

Northern Leaf-toed Gecko *Hemidactylus curlei* **Parker,** 1942

Parker did not identify the Curle after whom he named the gecko. We speculate that it was Colonel A. T. Curle (d. 1981) of the King's African Rifles, who served in British East Africa in the 1920s and 1930s. He was alternately a consular official and a political officer in Somaliland (Somalia), where the gecko is found. He was later attached to the British Embassy, Addis Ababa. His interests were anthropology and archeology as well as natural history. He wrote *When a Drought Blights Africa* (1929).

Curror

Curror's Skink *Feylinia currori* **Gray,** 1845 [Alt. Western Forest Limbless Skink, Western Forest Feylinia]

J. Curror RN presented the holotype to J. E. Gray. A "J. Curror," who we believe may have been the same man, presented some botanical specimens from the Congo to G. R. Gray (1856).

Curtiss

Curtiss' Galliwasp *Celestus curtissi* **Grant,** 1951

Anthony Curtiss was a collector and amateur natural historian. He collected in Haiti and Morocco and also in what is now Pakistan (1942–1944). For a time he lived in Port-au-Prince, Haiti (1946). He wrote *A Short Zoology of Tahiti in the Society Islands* (1938).

Cuvier

Cuvier's Dwarf Caiman *Paleosuchus palpebrosus* Cuvier, 1807 [Alt. Smooth-fronted Caiman]

Cuvier's Legless Skink *Typhlosaurus caecus* Cuvier, 1817

Cuvier's Anole *Anolis cuvieri* **Merrem,** 1820

Cuvier's Bachia *Bachia cuvieri* **Fitzinger,** 1826

Cuvier's Three-toed Skink *Hemiergis decresiensis* Cuvier, 1829

Cuvier's Madagascar Skink *Oplurus cuvieri* **Gray,** 1831

Georges Léopold Chrétien Frédéric Dagobert Baron Cuvier (1769–1832)—better known by his pen name Georges Cuvier—was a naturalist, one of the scientific giants of his age. He believed that paleontological discontinuities were evidence of sudden and widespread catastrophes—that is, that extinctions can happen suddenly. He is also famed for having stayed in a top government post, as Permanent Secretary, Academy of Sciences, through three regimes, including Napoleon's. Audubon said of a bird he collected (1812), "I named this pretty and rare species after Baron Cuvier, not merely by way of acknowledgment for the kind attentions which I received at the hands of that deservedly celebrated naturalist, but as a homage due by every student of nature to one unrivalled in the knowledge of General Zoology." Cuvier wrote *Tableau élémentaire de l'histoire naturelle des animaux* (1798). Six birds and three mammals are named after him.

Cyclops

Cyclops Emo Skink *Emoia cyclops* **Brown,** 1991

Named after the Cyclops Mountains in New Guinea.

Czeblukov

Fine-spined Sea Snake *Hydrophis czeblukovi* **Kharin,** 1984

Vladimir P. Czeblukov is a Russian herpetologist. He and Kharin work together and co-wrote "A New Revision of Sea Kraits of Family Laticaudidae Cope, 1879 (Serpentes: Colubroidea)" (2006).

Czechura

Czechura's Skink *Saproscincus czechurai* **Ingram** and **Rawlinson,** 1981

Gregory Vincent Czechura (b. 1953) is a herpetologist and ornithologist and "senior information officer" at the Queensland Museum, Brisbane. He wrote "The Rare Scincid Lizard, *Nannoscincus graciloides*: A Reappraisal" (1981). He is very skeptical about reported sightings of "flying saucers."

D

Dahl, A.

Dahl's Whip Snake *Coluber najadum dahlii* **Fitzinger,** 1826 [Syn. *Platyceps najadum dahlii*]

Dr. Anders Dahl (1751–1789) was a Swedish botanist and a student of Linnaeus. He entered Uppsala Universitet (1770) and, after graduating, worked in Gothenburg as Curator of a private natural history museum. He qualified as a physician (1786) at Christian-Albrechts-Universität zu Kiel and started teaching botany and medicine at Åbo Akademi, Finland (1787). The original description has no etymology so we cannot be entirely certain, but we think we have the right man. The dahlia is named after him.

Dahl, G.

Dahl's Toad-headed Turtle *Batrachemys dahli* Zangerl and **Medem,** 1958 [Syn. *Mesoclemmys dahli*]

Professor George Dahl (1905–1979) was a Swedish biologist and ichthyologist who visited Colombia (1936–1939). He returned there in 1948, settling in Sincelejo, Sucre Province, where he worked at Liceo Bolivar and where a foundation is now named after him at Universidad de Sucre. He was at Institute de Ciencias Naturales, Bogotá, Colombia (1961). He collected the turtle holotype.

Dahl, K. T. F.

New Britain Keelback *Tropidonophis dahlii* **Werner,** 1899

Karl Theodor Friedrich Dahl (1856–1929) was a zoologist. He studied at Leipzig, Freiburg, Berlin, and Kiel. He traveled to the Baltic States and to the Bismarck Archipelago, New Guinea. His main interests were spiders and biogeography.

Dahl, S. K.

Dahl's Rock Lizard *Darevskia dahli* **Darevsky,** 1957

Sergei Konstantinovich Dahl was a Russian zoologist. He wrote *Fauna of the Armenian SSR. Volume 1, Vertebrates* (1954). A mammal is named after him.

D'Albertis

D'Albertis' Python *Leiopython albertisii* **Peters** and **Doria,** 1878 [Alt. White-lipped Python]

Red-bellied Short-necked Turtle *Emydura albertisii* **Boulenger,** 1888 [Syn. *E. subglobosa* Krefft, 1876]

Cavaglieri Luigi Maria D'Albertis (1841–1901) was a botanist, ethnologist, and zoologist. He was in New Guinea (1871–1877), where, using a steamboat, he explored and charted the Fly River, venturing further than any European had before. He is reported to have "collected" a number of human skulls and the recently severed head of an elderly woman. His behavior toward the local people probably contributed considerably to their hostility to later European explorers. Four birds and a mammal are named after him.

Damel

Damel's Marsh Snake *Hemiaspis damelii* **Günther,** 1876

Edward Dämel (1821–1900) was a collector, mainly of arachnological and Pacific entomological specimens. He was particularly active between 1860 and 1874. He collected in Queensland and New South Wales (1867–1874) for the Godeffroy Museum, Hamburg. An amphibian is named after him.

Dame-Marie

Dame-Marie Least Gecko *Sphaerodactylus zygaena* **Schwartz** and **Thomas,** 1977

Dame-Marie is a town in Haiti.

Damian

Earth Snake sp. *Geophis damiani* **Wilson,** McCranie, and **Williams,** 1998

Damian Almendarez is a friend, and valued companion in the field, of the describers.

D'Anchieta

See **Anchieta.**

Danford

Danford's Lizard *Lacerta danfordi* **Günther,** 1876

Charles G. Danford (1843–1928) was a geologist, paleontologist, zoologist, artist, traveler, and explorer. He was in Asia Minor (Turkey) in 1875–1876 and 1879. He co-wrote "Taxonomic Status and Distribution of *Apodemus mystacinus*" (1877). The Danford Iris was named after his wife, who introduced it to England.

Danheim

San Jose Island Blue-throated Whiptail *Aspidoscelis danheimae* **Burt,** 1929

May Danheim Burt. See **Burt.**

Daniel

Daniel's Tropical Racer *Mastigodryas danieli* **Amaral,** 1935

Daniel's Keelback *Helicops danieli* Amaral, 1937

Daniel's Anole *Anolis danieli* **Williams,** 1988

Daniel's Largescale Lizard *Ptychoglossus danieli* **Harris,** 1994

Brother Daniel Gonzalez Patiño (1909–1988) was a Colombian monk who joined Museo de Historia Natural, Instituto de La Salle, Bogotá (1937). He was Director of the museum in the 1980s. His major interests were botany, herpetology, and mineralogy. Two amphibians are named after him.

Daniel, J. C.

Daniel's Bloodsucker *Bronchocela danieli* **Tiwari** and Biswas, 1973

Jivanayakam Cyril (always known as "J. C.") Daniel (b. 1927) is an Indian naturalist particularly interested in herpetology. He joined the Bombay Natural History Society as an assistant to Dr. Salim Ali (1950), eventually becoming its Director and serving in that capacity until his retirement (1991), when he was elected an Honorary Member; he now is Honorary Secretary. He is executive editor of the society's journal and initiated the *Hornbill* magazine. He wrote *The Book of Indian Reptiles* (1983). An amphibian is named after him.

Darevsky

Skink sp. *Eutropis darevskii* Bobrov, 1992

Darevsky's Viper *Vipera darevskii* Vedmederja, **Orlov,** and Tunyev, 1986

Lacertid lizard genus *Darevskia* Arribas, 1997

Dr. Ilya Sergeevich Darevsky (1924–2009) was a herpetologist who often worked with Orlov. He was appointed Curator, Department of Herpetology, St. Petersburg (then Leningrad) (1962). He headed the Laboratory of Ornithology and Herpetology there (1976–1996) and discovered parthenogenesis in reptiles. He co-wrote *The Reptiles of Northern Eurasia: Taxonomic Diversity, Distribution, Conservation Status* (2006).

Darlington

Darlington's Anole *Anolis darlingtoni* **Cochran,** 1935

Darlington's Galliwasp *Celestus darlingtoni* Cochran, 1939

Forest Skink sp. *Sphenomorphus darlingtoni* **Loveridge,** 1945

Darlington's Least Gecko *Sphaerodactylus darlingtoni* **Shreve,** 1968

Dr. Philip Jackson Darlington Jr. (1904–1983) was an evolutionary biologist, zoogeographer, and beetle taxonomist. He collected in Colombia, Puerto Rico, Haiti, Cuba, New Guinea, Australia, and Tierra del Fuego. He was Assistant Curator of Insects, Museum of Comparative Zoology, Harvard (1940), where he later became Curator, and Professor of Zoology. He took a year off (1956–1957) to live out of the back of a truck with his family in Australia. He wrote *Zoogeography: The Geographical Distribution of Animals* (1957). A bat is named after him.

D'Armandville

Darmandville Bow-fingered Gecko *Cyrtodactylus darmandvillei* **Weber,** 1890

Father Cornelis J. F. le Cocq d'Armandville (1846–1896) was a Dutch Jesuit missionary in the Dutch East Indies (Indonesia). He was stationed in East Ceram. He landed at Kapaur (1894), becoming the first Christian missionary in that part of New Guinea. He was due to return to Java for medical treatment (1896) but went on a trip to New Guinea. The ship returned without him, the crew stating he was dead, but there were no clear facts or coherent story as to how he died. A college in West Papua (Indonesia) is named after him.

Darvaz

Skink sp. *Ablepharus darvazi* Yeriomchenko and Panfilov, 1990

Named after the Darvaz mountains, Tajikistan.

Darwin, C.

Darwin's Ringed Lizard *Amphisbaena darwini* **Duméril** and **Bibron,** 1839

Darwin's Iguana *Diplolaemus darwinii* **Bell,** 1843

Darwin's Tree Iguana *Liolaemus darwinii* Bell, 1843

Darwin's Gecko *Gymnodactylus darwini* **Gray,** 1845

Darwin's Marked Gecko *Homonota darwinii* **Boulenger,** 1885

Darwin's Leaf-toed Gecko *Phyllodactylus darwini* **Taylor,** 1942

Darwin's Wall Gecko *Tarentola darwini* **Joger,** 1984

Charles Robert Darwin (1809–1882) was the prime advocate, together with Wallace, of natural selection as the way in which speciation occurs. To quote from his most famous work, *On the Origin of Species by Means of Natural Selection* (1859): "I have called this principle, by which each slight variation, if useful, is preserved, by the term Natural Selection." Darwin was the naturalist on HMS *Beagle* on its scientific expedition around the world (1831–1836). In South America he found fossils of extinct animals that were similar to extant species. On the Galapagos Islands he noticed many variations among plants and animals of the same general type as those in South America. On his return to London he conducted research on his notes and specimens. Out of this study grew several related theories: evolution did occur; evolutionary change was gradual, taking thousands or even millions of years; the primary mechanism for evolution was a process called natural selection; and the millions of species alive today arose from a single original life form through a branching process called "speciation." Four mammals, three amphibians, and several birds (including those famous finches) are named after him.

Darwin, Port

Port Darwin Sea Snake *Hydrelaps darwiniensis* **Boulenger,** 1896 [Alt. Darwin's Sea Snake]

Darwin's Ground Skink *Glaphyromorphus darwiniensis* **Storr,** 1967

These two reptiles are named after the city and port of Darwin, Northern Territory, Australia.

Dary

Dary's Burrowing Snake *Adelphicos daryi* **Campbell** and Ford, 1982

Mario Dary Rivera (1928–1981) was a Guatemalan conservationist and biologist. He was President of Universidad de San Carlos de Guatemala, where he had founded the School of Biology (1973). He was instrumental in establishing the *biotopo del quetzal* (essential to the survival of the Quetzal) in the Guatemalan cloud forest (1976). His actions were not universally popular. Some considered him to be a subversive, and he was assassinated by a death squad (1981).

Das

Agamid lizard sp. *Oriotaris dasi* Shah and **Kästle,** 2002 [Syn. *Japalura dasi*]

Gecko sp. *Cnemaspis indraneildasii* **Bauer,** 2002

Dr. Indraneil Das (b. 1964) is an Indian herpetologist now based at University of Malaysia, Sarawak, where he is an Associate Professor (1998). After early education in India, he received his doctorate in animal ecology from Oxford. He wrote many books, including *Biogeography of the Reptiles of South Asia* (1996).

Datz

Anole sp. *Anolis datzorum* Köhler, Ponce, Sunyer, and Batista, 2007

Erika and Walter Datz were patrons of the sciences. The etymology thanks "Erika Datz and her late brother Walter Datz, for their support of biodiversity research in Germany through the Erika and Walter Datz Foundation."

Daudin, F. M.

Daudin's Leaf-toed Gecko *Afrogecko porphyreus* Daudin, 1802

Daudin's Bronzeback *Dendrelaphis tristis* Daudin, 1803

Daudin's Coral Snake *Micrurus psyches* Daudin, 1803 [Alt. Carib Coral Snake]

Daudin's Vine Snake *Xenoxybelis argenteus* Daudin, 1803

Daudin's Giant Tortoise *Dipsochelys daudinii* **Duméril** and **Bibron,** 1835 EXTINCT

François Marie Daudin (1776–1803) was a French zoologist who was interested in herpetology and ornithology. He wrote *Histoire naturelle, générale et particulière des reptiles* (8 vols., 1802–1803). He was a tragic figure. A childhood disease left his lower limbs paralyzed; his books were not commercially successful; and he and his wife, Adèle, who illustrated his work, lived (though born into a wealthy family) and died in poverty, she first and he a few months later, of tuberculosis.

Daudin, J.

Gecko sp. *Gonatodes daudini* Powell and **Henderson,** 2005

Jacques Daudin (b. 1925) is a naturalist, author, and conservationist who has lived on the island of Union, the Grenadines, for some 30 years and has over 30 adopted children. He is Honorary Chairman of the Eastern Caribbean Coalition for Environmental Awareness. He co-wrote "An Annotated Checklist of the Amphibians and Terrestrial Reptiles of the Grenadines with Notes on Their Local Natural History and Conservation" (2007).

Dave Wake

Skink genus *Davewakeum* **Heyer,** 1972

See **Wake.**

David

See **Père David.**

Davis

Davis' Leaf-toed Gecko *Phyllodactylus davisi* **Dixon,** 1964

Dr. William Bennoni Davis (1902–1995) was a zoologist. He took his doctorate at the University of California, Berkeley. He studied the mammals of Texas for over 50 years and is regarded as the father of Texan mammalogy. He was later Professor Emeritus, Department of Wildlife and Fisheries Sciences, Texas A&M University. He established the Texas Cooperative Wildlife Collection (1938) and was President of the American Society of Mammalogists (1955–1957). He wrote the early editions of *The Mammals of Texas.* Three mammals are named after him.

Davison

Blanford's Bridle Snake *Dryocalamus davisonii* **Blanford,** 1878 [Sometimes given as "Bridal" Snake]

William Ruxton Davison (d. 1893) was an ornithologist and curator, Raffles Museum, Singapore (1887–1893). He co-wrote "A Revised List of the Birds of Tenasserim" (1878). A bird is named after him.

Day

Agama sp. *Laudakia dayana* **Stoliczka,** 1871

Dr. Francis Day (1829–1889) was an ichthyologist, zoologist, and a Fellow of both the Zoological and the Linnean Society. He qualified in medicine (1851), joining the Madras Medical Service (1852) and serving in the second Burmese war. He retired from being Inspector General of Fisheries in India and Assistant Surgeon General (1876). He sold a large collection of fish specimens to the Australian Museum for £200 (1883). Normally such an important collection would have gone to the British Museum, but Day couldn't abide Dr. Albert Günther, then Keeper of Zoology there. A mammal is named after him.

De Borre

Deborre's Casquehead Iguana *Laemanctus longipes deborrei* **Boulenger,** 1877

Charles François Paul Alfred Preudhomme de Borre (1833–1905) was a Belgian engineer who became an entomologist. He was a curator at Muséum des Sciences Naturelles, Brussels. He lived in Geneva for the last 10 years of his life.

Decary

Angel's Dwarf Gecko *Lygodactylus decaryi* **Angel,** 1930
Decary's Pygmy Chameleon *Brookesia decaryi* Angel, 1930 [Alt. Spiny Leaf Chameleon]
Rock Skink *Amphiglossus decaryi* Angel, 1930

Raymond Decary (1891–1973) was a colonial administrator in Madagascar (1916–1944). He was a zoologist, botanist, geologist, and ethnographer and was interested in anything and everything to do with Madagascar, contributing over 40,000 specimens of Malagasy flora to the Paris herbarium. He qualified in law (1912). He was seriously wounded in the Battle of the Marne (1914) and was unable to return to active service. He went to Madagascar (1916) as an officer in the Reserve to release fully fit officers for active service in France. He trained as a colonial administrator (1921), returning to Madagascar (1922). He undertook seven scientific expeditions on the island (1923–1930) and became Director of Scientific Research, Madagascar (1937). He was again in the French army in Madagascar (1939–1944), returning to France after the Liberation. Demobilized (1945), he retired into private life to continue his research. He wrote *Malagasy Fauna* (1950).

Decorse

Mocquard's Worm Snake *Typhlops decorsei* **Mocquard,** 1901

Dr. Gaston-Jules Decorse (1873–1907) was an army doctor interested in ethnography and linguistics. He traveled in Madagascar, where he collected botanical specimens (1898–1902), and went on a French expedition to Lake Chad (1902–1904). He co-wrote *Rabah et les arabes du Chari* (1905).

De Coster

De Coster's Garter Snake *Elapsoidea sundevallii decosteri* **Boulenger,** 1888
De Coster's Spade-snouted Worm Lizard *Monopeltis decosteri* Boulenger, 1910

Juste De Coster was the Belgian Consul, Delagoa Bay, where he collected insects and other specimens (ca. 1889) and sent them to the South African Museum.

Decres

Tawny Crevice-dragon *Ctenophorus decresii* **Duméril** and **Bibron,** 1837

This reptile is named after L'Île de Decrès, the French name for Kangaroo Island, off South Australia.

De Filippi

Snouted Night Adder *Causus defilippii* **Jan,** 1863
Elburz Lizard *Darevskia defilippi* Camerano, 1877 [Syn. *Lacerta defilippi*]

Filippo de' Filippi (1814–1867) was a doctor, traveler, and zoologist. He succeeded Bonelli as Professor of Zoology at Museo Regionale di Scienze Naturali di Torino. He was a supporter of Darwin's evolutionary theory and gave a seminal lecture on it (1864): "L'uomo e le scimmie" (Man and the Apes). Filippi accompanied the Duke of Abruzzi's expedition to Alaska. He led the scientific team on an expedition (1862) to explore Persia (Iran). He was on board *Magenta* and in charge of the scientific aspects of that vessel's circumnavigation, but he died of cholera while the vessel was in Hong Kong and was replaced by Giglioli. who published the expedition results. A bird is named after him. See also **Filippi.**

Degen

Degen's Water Snake *Crotaphopeltis degeni* **Boulenger,** 1906

Edward J. E. Degen (1852–1922) was born in Basel and died in London. He collected reptiles, mammals, and fishes in East Africa (ca. 1895–1905). He was in Abyssinia (Ethiopia) in 1902. After leaving Africa he worked as an articulator/taxidermist at the Natural History Museum, London.

Degenhardt

Degenhardt's Scorpion-eating Snake *Stenorrhina degenhardtii* **Berthold,** 1846

Degenhardt was a German collector in New Grenada in the 1840s. He made a collection of herpetofauna that he sent to Berthold at Georg-August-Universität Göttingen. History records nothing more about him. New Grenada was a country in northern South America that was later split up, so where he collected is not entirely certain, but it is believed to have been in what is now the western part of Colombia.

Degerbol

Degerbol's Blind Snake *Leptotyphlops borrichianus* Degerbol, 1923

Professor Dr. Magnus Anton Degerbøl (1895–1977) was

an expert on Quaternary zoology who was Chief Curator, Zoological Museum, Copenhagen. He was awarded his doctorate by Københavns Universitet (1921). He then joined the museum staff, becoming Curator of Mammals (1924). Before WW2 his main interest was the fauna of Greenland. After WW2 he was on the Danish Central Africa expedition (1947) and developed exhibitions of African savannah animals. He wrote *Acta Arctica* (1943).

De Grys

Sierra Leone Worm Lizard *Cynisca degrysi* **Loveridge,** 1941

Pedro de Grys (de Grijs) worked at the Zoological Museum, Universität Hamburg. Loveridge wrote in his etymology, "Herr P. de Grys of the Hamburg Museum who so kindly lent me the specimen for study."

Deharveng

Deharveng's Blind Snake *Cyclotyphlops deharvengi* In Den Bosch and Ineich, 1994

Dibamid lizard sp. *Dibamus deharvengi* Ineich, 1999

Dr. Louis Deharveng is an entomologist, agriculturist, speleologist, and Director of a research unit at Muséum National d'Histoire Naturelle, Paris. He was on the 2006 Santo expedition to the island of Espiritu Santo.

Deignan

Deignan's Tree Skink *Lankascincus deignani* **Taylor,** 1950

Herbert "Bert" Girton Deignan (1906–1968) was a Fellow of the John Simon Guggenheim Memorial Foundation (1952) and worked for the U.S. National Museum (1938–1962). He graduated from Princeton (1928), then taught English at a school at Chiang Mai, Siam (Thailand) (1928–1932). He was an associate of Wetmore, who helped him get a temporary job at the Smithsonian (1933). Deignan then worked at the Library of Congress (1934–1935), returning to his old job in Thailand, combining teaching with collecting birds for Wetmore (1935–1937). He returned to the USA (1938) and worked at the Smithsonian until 1962, except for his service (1944–1946) in southern Asia as an agent of the Office of Strategic Services (forerunner to the CIA). He retired to live in Switzerland. He wrote *Holotypes of Birds in the United States National Museum* (1961). A bird is named after him.

De Jong

De Jong's Japalure *Japalura nasuta* de Jong, 1930 [Junior syn. of *Aphaniotis ornata* Lidth de Jeude, 1893]

Dr. Jan Kornelis de Jong (1895–1972) was a herpetologist at the Zoological Museum, Buitenzorg, Java (1930). He wrote *Reptiles from Dutch New Guinea* (1927).

DeKay

DeKay's Snake *Storeria dekayi* **Holbrook,** 1836 [Alt. Northern Brown Snake]

Dr. James Ellsworth DeKay (1792–1851) was a zoologist. He studied at Yale (1807–1812) but did not graduate. He studied medicine at Edinburgh, qualifying in 1819. He returned to the USA, married, and then traveled with his father-in-law to Turkey as a ship's physician (1831–1832). He returned to America, forsaking medicine to study natural history, and was on the Geological Survey of New York from 1835. He wrote *The Zoology of New York* (1842–1849).

Delacour

Delacour's Mountain Snake *Plagiopholis delacouri* **Angel,** 1929

Dr. Jean Theodore Delacour (1890–1985) was a French-American ornithologist renowned for discovering, keeping, and breeding some of the rarest birds in the world. He was born in Paris and died in Los Angeles. He created (1919–1920) the zoological gardens at Clères and donated them to Muséum National d'Histoire Naturelle, Paris (1967). He undertook several expeditions to Indochina, particularly Vietnam, and collected specimens, especially pheasants. He co-wrote *Birds of the Philippines* (1946). Among the taxa named after him are three birds and three mammals.

De la Fuente

Guamuhaya Anole *Anolis delafuentei* **Garrido** 1982

Marcelo S. de la Fuente is an Argentine herpetological paleontologist. He was at Museo de La Plata, Universidad de la Plata (1995), and today is a member of the Department of Paleontology, Museo de Historia Natural, San Rafael, Mendoza Province. He co-wrote "A New Pterosaur from the Jurassic of Cuba" (2004).

Delalande

Delalande's Sandveld Lizard *Nucras lalandii* **Milne-Edwards,** 1829

Delalande's Gecko *Tarentola delalandii* **Duméril** and **Bibron,** 1836 [Alt. Tenerife Wall Gecko]

Skink sp. *Mabuya delalandii* Duméril and Bibron, 1839

Pierre Antoine Delalande (1787–1823) worked for Muséum National d'Histoire Naturelle, Paris. He collected in the region around Rio de Janeiro (1816) with Auguste de Saint-Hilaire, and in the African Cape with his nephew, Jules Verreaux, and Andrew Smith (1818). Later Geoffroy Saint-Hilaire employed him as a taxidermist. Three birds are named after him. See also **Lalande.**

De la Sagra

Cuban Galliwasp *Diploglossus delasagra* **Cocteau,** 1838

Ramón de la Sagra (1801–1871) was a Spanish economist

and botanist who lived and worked in Cuba and died in Switzerland. He was Director, Havana Botanical Gardens (1822–1834). During this time he established a model farm and collected widely. After returning to Spain he devoted himself to the study of political economy before joining the French revolution (1854). He wrote *Historia física, política y natural de la isla de Cuba* (13 vols., 1839–1861). A bird is named after him. See also **Sagre.**

Del Campo

Del Campo's Leaf-toed Gecko *Phyllodactylus delcampoi* Mosauer, 1936

Professor Rafael Martín del Campo y Sanchez (1910–1987) was a herpetologist at the Institute of Biology, Universidad Nacional Autónoma de México, where he was Curator, National Collection of Amphibians and Reptiles (1940–1979). He collaborated closely with the American herpetologist Bogert, and they co-wrote *The Gila Monster and Its Allies* (1956). See also **Rafael** and **Martin del Campo.**

Delcourt

Delcourt's Sticky-toed Gecko *Hoplodactylus delcourti* **Bauer** and **Russell,** 1986 EXTINCT [Alt. Kawekaweau]

Alain Delcourt worked at Muséum d'Histoire Naturelle de Marseille (1979). He discovered that the collection included a preserved specimen of a large gecko. It was unlabeled but was believed to have been collected 150 or more years earlier. He sent a photograph of it to Bauer and Russell, who eventually examined it and identified it with the "Kawekaweau"—a giant forest lizard in the Maori oral tradition. The Marseilles specimen is the only known example of what was the world's largest gecko, last seen alive in 1870.

Delean

Pernatty Knob-tailed Gecko *Nephrurus deleani* Harvey, 1983

Dr. J. Steven "Steve" C. Delean is a statistician who works at the Faculty of Tropical Environmental Science and Geography, James Cook University, North Queensland. He is also associated with the Australian Institute of Marine Science.

De Lema

See **Lema.**

Dell

Darling Range Southwest Ctenotus *Ctenotus delli* **Storr,** 1974

Dr. John Dell collected the holotype of this skink. He was a herpetologist at the Western Australian Museum, Perth, and is now an honorary associate there. He has made a study of Rottnest Island reptiles.

Delisle

Delisle's Wedge-snouted Skink *Sphenops delislei* **Lataste,** 1876

Dr. Fernand Delisle (1836–1911) was a physician and anthropologist who worked at Muséum National d'Histoire Naturelle, Paris. He was an early user of photography for recording human facial and cranial types.

Del Solar

Leaf-toed Gecko sp. *Phyllodactylus delsolari* Venegas et al., 2008

Gustavo del Solar (1937–2008) was a conservationist who owned tamarind plantations in northern Peru. He rediscovered the White-winged Guan (1977). It is perhaps unusual for an ornithologist with no herpetological interests to have a reptile named after him, but it is so in this case and was clearly done to honor him in the year that he died.

Denison

Australian elapid snake genus *Denisonia* **Krefft,** 1869

Sir William Thomas Denison (1804–1871) became Lieutenant Governor of Van Diemen's Land (Tasmania) (1846) and Governor-General of New South Wales, Van Diemen's Land, Victoria, South Australia, and Western Australia (1854). He was Governor of Madras (1861–1866), returning to England via the newly opened Suez Canal. His last public appointment sounds contemporary: he was Chairman of an enquiry into the pollution of rivers in Britain. He was a conchologist with a collection of 8,000 species of Australian shells. Krefft was appointed Assistant Curator of the Australian Museum (1860) on Denison's recommendation. Port Denison, Western Australia, is named after him.

Denzer

Denzer's Forest Dragon *Gonocephalus denzeri* **Manthey,** 1991

Dr. Wolfgang Denzer is a German herpetologist who works closely with Manthey. They have co-written various articles, particularly on agamid lizards.

Deplanche

Skink sp. *Sigaloseps deplanchei* **Bavay,** 1869

Emile Deplanche (1824–1875) was a surgeon in the French navy. He co-wrote *Essais sur la Nouvelle Calédonie* (1863). A bird is named after him.

Deppe

Deppe's Arboreal Alligator Lizard *Abronia deppii* **Wiegmann,** 1828

Blackbelly Racerunner *Aspidoscelis deppei* Wiegmann, 1834

Mexican Bullsnake *Pituophis deppei* **Duméril,** 1853 [Alt. Mexican Pine Snake]

Deppe's Centipede Snake *Tantilla deppei* **Bocourt,** 1883

Ferdinand Deppe (1794–1861) was a horticulturalist, collector, and artist. He first arrived in Mexico (1824) with Count von Sack, who seems to have been an irresolute "expedition leader" who soon returned to Germany while Deppe stayed in Mexico until 1827. He made a brief visit home, returned to Mexico with botanist Wilhelm Schiede, and stayed until 1836. Many of the specimens Deppe collected went to Museum für Naturkunde Berlin. A squirrel and a fish are named after him.

De Queiroz

De Queiroz's Spinytail Iguana *Ctenosaura oedirhina* De Queiroz, 1987

Dr. Kevin De Queiroz (b. 1956) is an American zoologist and herpetologist who joined the Division of Reptiles and Amphibians at the Smithsonian (1990) as an Assistant Curator, becoming an Associate Curator (1994) and Curator (1999). His master's degree was awarded by San Diego State University (1985) and his doctorate by Berkeley (1989). He favors the PhyloCode as a way of naming taxa, as opposed to the Linnaean system, and co-wrote "Phylogenetic Taxonomy" (1992).

Deraniyagala

Deraniyagala's Gecko *Cnemaspis podihuna* Deraniyagala, 1944

Deraniyagala's Snake Skink *Nessia deraniyagalai* **Tayor,** 1950 [Alt. Deraniyagala's Nessia]

Deraniyagala's Earth Snake *Uropeltis ruhunae* Deraniyagala, 1954

Deraniyagala's Striped Lacerta *Ophisops minor* Deraniyagala, 1971

Sri Lanka Rough-sided Snake *Aspidura deraniyagalae* **Gans** and Fetcho, 1982

Deraniyagala's Tree Skink *Lankascincus deraniyagalae* **Greer,** 1991

Professor Paulus Edward Peiris Deraniyagala (1900–1976) was a scientist and zoologist who was Director of all National Museums, Ceylon (Sri Lanka) (1939–1963). He had a wide range of interests, including reptiles and elephants. In his youth he was a notable flyweight boxer who defeated the champion of the British Empire's Armed Forces (1923).

Derjugin

Derjugin's Lizard *Darevskia derjugini* **Nikolsky,** 1898 [Syn. *Lacerta derjugini*]

Professor Dr. Konstantin Michailovich Derjugin (1878–1938) was a hydrobiologist at Leningrad State University.

Dermal

Natricine snake sp. *Hologerrhum dermali* **Brown, Leviton,** Ferner, and **Sison,** 2001

"Dermal" is the nickname of Dr. Ronald Crombie (q.v.). He gave Brown and Ferner much guidance during their work on Philippine reptiles and amphibians.

De Rooij

De Rooij's Bow-fingered Gecko *Cyrtodactylus sermowaiensis* De Rooij, 1915

De Rooij's Forest Skink *Sphenomorphus derooyae* **de Jong,** 1927

Dr. Petronella Johanna "Nelly" De Rooij (1883–1964) was a Dutch zoologist who graduated from Universiteit van Amsterdam (1904). At that time discrimination against women in Dutch law meant that she could get no further education, so she went to Switzerland to study and was awarded her doctorate by Universität Zürich (1907). She started work as a Curator of Reptiles and Amphibians, Zoological Museum, Universiteit van Amsterdam, but following administrative reforms (1922) she was forced to resign, which ended her scientific career. She wrote *The Reptiles of the Indo-Australian Archipelago* (2 vols., 1915–1917). Some specimens were sent to her by other Dutch women who were living in the Dutch East Indies (Indonesia). See also **Petronella.**

De Saix

Desaix's Bush Viper *Atheris desaixi* **Ashe,** 1968 [Alt. Ashe's Bush Viper]

Frank De Saix is an American who was a Peace Corps volunteer in Kenya when he collected the viper holotype.

De Schauensee

Deschauensee's Keelback *Amphiesma deschauenseei* **Taylor,** 1934

De Schauensee's Anaconda *Eunectes deschauenseei* **Dunn** and **Conant,** 1936

Dr. Rodolphe Mayer Deschauensee (1901–1984) was Curator and Director, Academy of Natural Sciences, Philadelphia. He was an ornithologist but collected anything and everything when on an expedition—and he went on many, including at least three to Siam (Thailand), during the third of which (1933–1934) the holotype of the keelback was collected.

De Silva

Agamid lizard sp. *Calotes desilvai* Bahir and Maduwage, 2005

Professor Anslem De Silva is a Sri Lankan herpetologist and a patron of herpetology. He founded, and is President of, the Amphibian and Reptile Research Organisation (ARROS). He hosted the Fourth World Congress of Herpetology (2001).

Despax

Despax's Ground Snake *Atractus paucidens* Despax, 1910

Despax's Parrot Snake *Leptophis riveti* Despax, 1910

Despax's Slender Snake *Tachymenis elongata* Despax, 1910

Raymond Justin Marie Despax (1886–1950) was a French zoologist, entomologist, and herpetologist. He wrote "Sur trois collections de reptiles et de batraciens provenant de l'archipel Malais" (1912). An amphibian is named after him.

Deuve

Kukri Snake sp. *Oligodon deuvei* David, Vogel, and van Rooijen, 2008

Colonel Jean Deuve (1918–2008) fought in WW2, was wounded (1940), and eventually came under British command in India as a member of the Special Operations Executive. Although he was not there, his unit blew up the Bridge on the River Kwai. He was parachuted (1945) into Laos to organize anti-Japanese guerrillas and stayed until 1964, being successively chief of intelligence, director of Lao police and political adviser to the Prime Minister. Under cover of being a military attaché, he was head of French intelligence in Japan (1965–1968), returning to Paris (1969) responsible for Eastern Europe, Asia, and Oceania and (1974–1978) senior director of all intelligence gathering and infrastructure abroad. He retired (1979) to write on modern Laotian history and the history of the medieval Duchy of Normandy. When in Laos he became renowned as an expert on local snakes and became an associate of Muséum National d'Histoire Naturelle, Paris. He wrote *Serpentes du Laos* (1970).

De Vet

Moluccan Bow-fingered Gecko *Cyrtodactylus deveti* **Brongersma,** 1948

Dr. Arnold C. De Vet (1904–2001) was a neurosurgeon at St. Ursula Clinic, Wassenaar, Holland (1936–1970), and a founding member of the World Federation of Neurosurgical Societies. The clinic was named after his wife, who had died in an accident.

De Vis

De Vis' Four-fingered Skink *Lygisaurus foliorum* De Vis, 1884

De Vis' Whipsnake *Demansia vestigiata* De Vis, 1884

De Vis' Bloodsucker *Ctenophorus inermis* De Vis, 1888

De Vis' Emo Skink *Emoia pallidiceps* De Vis, 1890

De Vis' Banded Snake *Denisonia devisi* **Waite** and **Longman,** 1920

Charles Walter De Vis (1829–1915) was a zoologist. He gave up his work as a clergyman (1862) to become Curator, Queens Park Museum, Manchester. He emigrated to Australia (1870), where he became Librarian, School of Arts, Rockhampton, Queensland. He published many popular articles under the pen name of "Thickthorn," which brought him to the attention of the Trustees of the Queensland Museum, who recruited him to be its first Director (1882–1905); he remained a consultant until 1912. He was a founding member of the Royal Society of Queensland (1884) and its President (1888–1889), and a founder and first Vice President of the Australasian Ornithologists' Union (1901). He wrote around 50 papers on herpetological subjects (1881–1911). Two mammals and two birds are named after him.

De Waal

De Waal's Agama *Agama (hispida) makanikarika* **Fitzsimons,** 1932 [Alt. Makgadikgadi Spiny Agama]

S. W. P. De Waal is a zoologist who worked at the Department of Zoology, University of Cape Town, where he became head of the Department of Lower Vertebrates (1971). He wrote "The Testudines (Reptilia) of the Orange Free State, South Africa" (1980).

De Witte

Witte's Five-toed Skink *Leptosiaphos dewittei* **Loveridge,** 1934

Witte's Spider Gecko *Agamura misonnei* Witte, 1973 [Alt. Misonne's Spider Gecko]

Dr. Gaston-François de Witte (1897–1980) worked in the Belgian Congo (1933–1935 and 1946–1949). He was originally a colonial administrator but served as a naturalist and collector for Institut des Parcs Nationales Congo-Belge, Tervuren, from 1938. See also **Witte.**

De Zwaan

Dibamid lizard sp. *Dibamus dezwaani* **Das** and **Lim,** 2005

Gecko sp. *Cnemaspis dezwaani* Das, 2005

Professor Dr. Johannes Pieter Kleiweg de Zwaan (1875–1971) was an anthropologist at Universiteit van Amsterdam. He originally qualified as a physician and, after working for a short time as a ship's doctor, went as doctor on an expedition (1907) to central Sumatra. He then visited Java, Bali, Lombok, Japan, and India, to which he returned (1910) to conduct research on Nias Island. He collected both holotypes.

Diana

Diana's Coral Snake *Micrurus diana* **Roze,** 1983

In Roman mythology, Diana was goddess of the moon and of hunting. Roze's flamboyant etymology states, "This brilliantly coloured and beautiful coral snake is dedicated to Diana, the goddess of forests, animals and the moon who should be adored and invoked to protect the endangered nature, particularly animals."

Diard

Diard's Blind Snake *Typhlops diardi* **Schlegel,** 1839

Pierre-Medard Diard (1794–1863) was a French explorer who collected in Southeast Asia (1827–1848). He created Buitenzorg Botanical Gardens, Java. Two birds and a mammal are named after him.

Dice

Dice's Shortnose Skink *Eumeces brevirostris dicei* **Ruthven** and **Gaige,** 1933

Dr. Lee Raymond Dice (1887–1977) was an ecologist. He was Director, Laboratory of Vertebrate Biology, University of Michigan, and later Professor of Human Genetics and Professor of Internal Medicine, retiring as Emeritus. He kept a mouse colony for study there for 50 years (1925–1975). He was President of the American Society of Mammalogists (1947–1949). He wrote *Life Zones and Mammalian Distribution* (1923). A mammal is named after him.

Dickerson

Dickerson's Worm Lizard *Cadea palirostrata* Dickerson, 1916

Dickerson's Gecko *Cnemaspis dickersonae* **Schmidt,** 1919

Dickerson's Side-blotched Lizard *Uta concinna* Dickerson, 1919

Dickerson's Desert Whiptail *Aspidoscelis tigris dickersonae* **Van Denburgh** and **Slevin,** 1921

Dickerson's Earless Lizard *Holbrookia maculata dickersonae* Schmidt, 1921

Dickerson's Collared Lizard *Crotaphytus dickersonae* Schmidt, 1922 [Alt. Mexican Collared Lizard; Syn. *C. collaris dickersonae*]

Mary Cynthia Dickerson (1866–1923) was a herpetologist who founded one of the longest-running and most important museum-based herpetological programs in the world. She had to fund her own education, so she taught high school to earn enough to cover her studies, eventually graduating from the University of Chicago (1897). She taught in various institutions (1897–1905) before resigning to take up writing full time. She taught ecology at Stanford (1907) and was one of four curators appointed to the newly formed Department of Ichthyology and Herpetology, American Museum of Natural History (1909). The department was split (1919), and she became Curator of Herpetology (1920) but soon "retired" (1921). In fact she was asked to take a rest because of her erratic behavior, and when she did not do so, she was removed for psychiatric assessment. She returned to the museum in a deranged state and was committed to an institution, where, sadly, she died (1923). She wrote *The Frog Book* (1906).

Diehl

Diehl's Little Ground Snake *Stegonotus diehli* Lindholm, 1905

Wilhelm Diehl (1874–1940) was a Protestant missionary at Bogadjim, New Guinea. His first wife died of malaria (1904). He remarried (1907), and as their second child was born in Germany (1916) we assume that he and his wife returned home just before or on the outbreak of WW1. He made a herpetological collection in and around Astrolabe Bay, German New Guinea, and sent it to the Natural History Museum, Wiesbaden.

Dighton

Travancore Cat Snake *Boiga dightoni* **Boulenger,** 1894

S. M. Dighton. A typically brief Boulenger etymology noted only that the holotype "was obtained by Mr. S. Dighton at Pirmaad, at an altitude of 3,300 feet [1,000 meters], in January, 1893." We believe this is the "S. M. Dighton" who was known to be a tea planter at Travancore, Kerala (1888).

Diguet

Garter Snake sp. *Thamnophis digueti* **Mocquard,** 1899 [Junior syn. of *T. hammondii* Kennicott, 1860]

Spiny Lizard sp. *Sceloporus digueti* Mocquard, 1899

Leon Diguet (1859–1926) was a chemical engineer and geologist who was employed at a copper mine at Santa Rosalia, Baja California (1889–1892). He became interested in local natural history and collected specimens for Muséum National d'Histoire Naturelle, Paris, which was so impressed that they sent him back to Mexico and employed him as a full-time explorer and collector. He explored Baja California (1893–1894), identifying important early rock paintings and reporting on them in *L'Anthropologie* (1895). He revisited Mexico several times before WW1. He wrote *Les cactacées utiles du Mexique* (published posthumously, 1928). The Barrel Cactus *Ferocactus diguetii* is named after him.

Dinnik

Dinnik's Viper *Vipera dinniki* **Nikolsky,** 1913

Nikolai Yakovlevich Dinnik was a zoologist who specialized in Caucasus natural history. He taught mathematics and physics in the local high school in Stavropol Kavkazskii. He wrote *Animals of the Caucasus* (1914). His son, Aleksandr Nikolaevich Dinnik, was a distinguished Professor of Mechanics and a Russian academician.

Dione

Dione's Ratsnake *Elaphe dione* **Pallas,** 1773

According to Homer, Dione was the mother of the goddess Aphrodite.

Distant

Distant's Thread Snake *Leptotyphlops distanti* **Boulenger,** 1892

William Lucas Distant (1845–1922) was a British entomologist. After a trip on his father's whaling ship (1867) to the Malay Peninsula, visiting the Nicobar Islands (1868), he became completely fascinated by entomology. For much of his early life he worked at a tannery in London, but he was able to make two long visits to Transvaal, the second of which was to last four years. He worked for the British Museum of Natural History (1899–1920), which bought his collection of 50,000 specimens (1920). He edited *The Zoologist* magazine and wrote *Synonymic Catalogue of Homoptera* (1906).

Ditmars

Rock Horned Lizard *Phrynosoma ditmarsi* **Stejneger,** 1906

Dr. Raymond Lee Ditmars (1876–1942), a zoologist and herpetologist, was Curator of Reptiles at the Bronx Zoo, New York. He worked in the Department of Entomology at the American Museum of Natural History (1893–1897), resigning because he could make more money as a stenographer. He became a reporter for the *New York Times* (1898), and on one of his first assignments he discovered that the newly founded New York Zoological Society had opened a zoo in the Bronx. He quickly got himself a job there and spent the rest of his life at the zoo. He began as Assistant Curator of Reptiles (1899), then was put in charge of mammals (1926) and of insects (1940). He wrote *The Reptile Book* (1930). He was also an early proponent of filming animals and left an enormous archive of film material.

Dixon

Dixon's Leaf-toed gecko *Phyllodactylus dixoni* Rivero-Blanco and **Lancini,** 1968
Dixon's Bachia *Bachia huallagana* Dixon, 1973
Gray-checkered Whiptail *Aspidoscelis dixoni* Scudday, 1973
Dixon's Anotosaura *Anotosaura brachylepis* Dixon, 1974
Dixon's Milk Snake *Lampropelitis triangulum dixoni* Quinn, 1983
Dixon's Ground Snake *Liophis atraventer* Dixon and **Thomas,** 1985
Gecko genus *Dixonius* **Bauer** et al., 1997
Neotropical House Snake sp. *Thamnodynastes dixoni* Bailey and Thomas, 2007

Dr. James Ray Dixon (b. 1928) took his bachelor's (1950), his master's (1957), and his doctorate (1961) in zoology, all at Texas A&M University. He was Curator of Reptiles at the Ross Allen Reptile Institute (1954–1955) and Associate Professor, Veterinary Medicine, Texas A&M University (1951–1961). He was Assistant Professor, Wildlife Management, New Mexico State University (1961–1965), and became Curator of Herpetology, Los Angeles County Museum. He was Professor of Wildlife and Fishery Sciences, Texas A&M University, from 1971, and today he is Professor Emeritus and Curator Emeritus, Texas Cooperative Wildlife Collection. He wrote *Amphibians and Reptiles of Texas* (1987).

Doederlein

Dwarf/Reed Snake sp. *Calamaria doederleini* Gough, 1902

Dr. Ludwig Heinrich Philipp Döderlein (1855–1936) was a zoologist and paleontologist. He studied at Erlangen and Munich universities, and his doctorate was awarded by Strasbourg (1877). He taught in Japan (1879–1881), making a collection of Japanese fauna (1880–1881). He became Curator and Director, Zoological Collections, Strasbourg (1882), and was appointed as Assistant Professor (1883) at Université de Strasbourg, becoming Professor of Zoology (1891). He was sacked and expelled from the city (1919) because he was a German; Alsace and Lorraine had been restored to France by the Treaty of Versailles. He was head of the Zoological Collections, München Staatliches Museum für Naturkunde, with the title of Honorary Professor Emeritus of Taxonomy (1923–1927).

Doello-Jurado

Freiberg's Iguana *Stenocerus doellojuradoi* **Freiberg,** 1944

Professor Martín Doello-Jurado (1884–1948) was a zoologist, malacologist, and paleontologist. He was

Director, Museo Argentino de Ciencias Naturales "Bernardino Rivadavia," Buenos Aires (1923–1946).

Dolasi

Keelback (snake) sp. *Tropidonophis dolasii* Kraus and **Allison,** 2004

Dolasi Salepuna gave the describers help in their field research in Papua New Guinea.

Dollfus

Coffee Anole *Anolis dollfusianus* **Bocourt,** 1873

Auguste Dollfus (1840–1869) was a traveler, mining engineer, and geologist who took part in the French scientific expedition to Mexico (1864–1867). He investigated a number of volcanos during his expeditions.

Domergue

Domergue's Leaf Chameleon *Brookesia thieli* **Brygoo** and Domergue, 1969 [Alt. Thiel's Pygmy Chameleon]

Worm Snake sp. *Typhlops domerguei* **Roux-Estève,** 1980

Charles Antoine Domergue (1914–2008) was a naturalist, ornithologist, speleologist, and herpetologist. He worked closely with Brygoo and spent much of his life in Madagascar. He had a laboratory at L'Institut Pasteur de Madagascar, where he was still working when he died. He was a Professor at the University of Tuléar, Madagascar, and a member of the Academy of Madagascar. He previously worked for Muséum National d'Histoire Naturelle, Paris. He wrote *Notes sur les chamaleo de Madagascar* (1973).

Donnelly

Donnelly's Arthrosaura *Arthrosaura synaptolepis* Donnelly et al., 1992

Dr. Maureen Ann Donnelly (b. 1954) is a biologist and herpetologist at Florida International University, where she is Associate Professor of Biology. Before that she was a researcher at the American Museum of Natural History and spent time in Costa Rica investigating poison-dart frogs. Her bachelor's degree was awarded by California State University (1977) and her doctorate by the University of Miami (1987).

Donoso

Donoso's Lava Lizard *Tropidurus atacamensis* Donoso-Barros, 1966

Donoso's Steppe Iguana *Urostrophus valeriae* Donoso-Barros, 1966

Donoso's Forest Iguana *Liolaemus silvanae* Donoso-Barros, 1970 [Syn. *Vilcunia silvanae*]

See **Donoso-Barros.**

Donoso-Barros

Patagonian Tortoise *Geochelone (chilensis) donosobarrosi* **Freiberg,** 1973

Donoso-Barros' Snake *Incaspis cercostropha* Donoso-Barros, 1974

Donoso-Barros' Tree Iguana *Liolaemus donosobarrosi* **Cei,** 1974

Tree Iguana sp. *Liolaemus donosoi* **Ortiz,** 1975 [Syn. *L. constanzae* Donoso-Barros, 1961, according to some]

Dr. Roberto Donoso-Barros (1922–1975) was a Chilean naturalist and herpetologist. He originally qualified as a physician (1947). He became Professor of Biology at Universidad de Chile (1954), leaving to take up a similar post at Universidad de Concepción (1965). He wrote *Reptiles de Chile* (1961). He was killed in a car crash. See also **Donoso.**

D'Orbigny

Bolivian Burrowing Snake *Apostolepis dorbignyi* **Schlegel,** 1837

D'Orbigny's Bachia *Bachia dorbignyi* **Duméril** and **Bibron,** 1839

D'Orbigny's Banded Anole *Pristidactylus fasciatus* D'Orbigny and Bibron, 1847

South American Hognose Snake *Lystrophis dorbignyi* Duméril, Bibron, and **Duméril,** 1854

D'Orbigny's Tree Iguana *Liolaemus dorbignyi* **Koslowsky,** 1898

Alcide Dessalines d'Orbigny (1802–1857) was a traveler, collector, illustrator, and naturalist. His father, Charles-Marie Dessalines d'Orbigny (1770–1856), was a ship's surgeon. Muséum National d'Histoire Naturelle, Paris, sent him to South America (1826). There the Spanish briefly imprisoned him, mistaking his compass and barometer for "instruments of espionage." After prison, he lived for a year with the Guarani Indians, learning their language. He spent five years in Argentina and then traveled north along the Chilean and Peruvian coasts before moving into Bolivia; he returned to France in 1834. Once home he donated thousands of specimens of all kind to the Natural History Museum, Paris. His fossil collection led him to determine that there were many geological layers, revealing that they must have been laid down over millions of years. This was the first time such an idea had been put forward. He wrote *Dictionnaire universel d'histoire naturelle* (1839–1849). Five birds and two mammals are named after him.

Doria

Sarawak Water Snake *Enhydris doriae* **Peters,** 1871

Doria's Anglehead Lizard *Gonocephalus doriae* Peters, 1871 [Alt. Peters' Forest Dragon]

Middle Eastern Short-fingered Gecko *Stenodactylus doriae* **Blanford,** 1874
Lacertid lizard sp. *Latastia doriai* **Bedriaga,** 1884
Nigerian Agama *Agama doriae* **Boulenger,** 1885
Doria's Ground Skink *Scincella doriae* Boulenger, 1887
Doria's Green Snake *Cyclophiops doriae* Boulenger, 1888
Barred Keelback *Tropidonophis doriae* Boulenger, 1897 [Syn. *Amphiesma doriae*]

Marchese Giacomo Doria (1840–1913) was a zoologist who collected in Persia (Iran) with de Filippi (1862–1863) and in Borneo with Beccari (1865–1866). He founded Museo Civico di Storia Naturale, Turin (1867–1913), becoming its first Director. Six mammals and a bird are named after him.

Dorr

Dorr's Racer *Coluber dorri* **Lataste,** 1888

Commandant Emile Dorr (1857–1907) was an officer in the French Marine Infantry and a botanical collector in Africa. He was in Senegal (1887) and Madagascar (1897). He collected the snake holotype (1887).

Douglas, A. M.

Kimberley Crevice-skink *Egernia douglasi* **Glauert,** 1956
Douglas' Skink *Glaphyromorphus douglasi* **Storr,** 1967

Athol M. Douglas is a zoologist who worked at the Western Australian Museum, Perth (1950s–1980s). He wrote *Tigers in Western Australia?* (1986) about possible Thylacine survival on the Australian mainland. A mammal is named after him.

Douglas, D.

Short-horned Lizard *Phrynosoma douglasi* **Bell,** 1833

David Douglas (1799–1834) was a botanist and traveler who collected in North America (1823–1834) and Hawaii (1834) for the Royal Horticultural Society, London. He suffered from eye problems and consequently fell into a pit trap in Hawaii, where he was gored to death by a feral bull that had been similarly caught. The Douglas fir is named after him, and he also introduced the Sitka Spruce and the Lodge-pole Pine to the UK. A bird and a mammal are named after him.

Doumergue

Doumergue's Fringe-fingered Lizard *Acanthodactylus spinicauda* Doumergue, 1901

François Doumergue (1858–1938) was a French zoologist and naturalist who lived in Oran, Algeria. He was President of the local geographical and archeological society. He wrote *Essai sur la faune erpetologique de l'Oranie* (1901).

Downs

Downs' Earth Snake *Geophis immaculatus* Downs, 1967
Savage's Earth Snake *Geophis downsi* **Savage,** 1981

Dr. Floyd Leslie Downs (b. 1938) is a zoologist. He received his bachelor's degree (1958) from Cornell and both his master's (1960) and doctorate (1965) from the University of Michigan. He was a member of the faculty at the College of Wooster until retiring as Emeritus Professor (1998). He co-edited *Salamanders of Ohio* (1989).

Drapiez

White-spotted Cat Snake *Boiga drapiezii* **Boie,** 1827

Pierre Auguste Joseph Drapiez (1778–1856) was a refugee from France who became a Professor at Muséum des Sciences Naturelles, Brussels. He wrote *Dictionnaire classique des sciences naturelles* (1853).

Drewes

Drewes' Worm Snake *Leptotyphlops drewesi* **Wallach,** 1996

Dr. Robert Clifton Drewes (b. 1942) is Curator and Chairman, Department of Herpetology, California Academy of Sciences. He took his bachelor's degree at San Francisco State University (1969) and his doctorate at the University of California, Los Angeles (1981). He worked at Nairobi Snake Park, Kenya (1969–1970). He taught at a number of institutions, including Harvard, the University of Kansas, and, as Affiliate Professor, the University of Idaho, before joining the California Academy of Sciences (1970). His major area of interest is Africa, and he has visited 30 different countries there. He co-wrote *Pocket Guide to the Reptiles and Amphibians of East Africa* (2006).

Dring

Dring's Gecko *Cnemaspis argus* Dring, 1979 [Alt. Argus' Rock Gecko]
Dring's Borneo Rock Gecko *Cnemaspis dringi* **Das** and **Bauer,** 1998
Dring's False Bloodsucker *Pseudocalotes dringi* Hallermann and **Böhme,** 2000

Dr. Julian Christopher Mark Dring (b. 1951) of the Natural History Museum, London, is a herpetologist who collected in Sarawak (late 1970s). He was responsible for the herpetological component of the Gunong Lawit expedition. He wrote *Collection of Amphibians and Reptiles from Borneo*. Two amphibians are named after him.

Drucker-Colin

Fence Lizard sp. *Sceloporus druckercolini* Perez-Ramos and Saldanha de la Riva, 2008

Probably named after Dr. René Raúl Drucker-Colín (b. 1937), a Mexican physiologist and neurobiologist who studies sleep and sleep disorders. He is Director for

Promotion of Sciences at Universidad Nacional Autónoma de México. He has been President of the Mexican Academy of Sciences and a member of the President of the Republic's Scientific Advisory Council.

Drummond-Hay

Drummond-Hay's Rough-sided Snake *Aspidura drummondhayi* **Boulenger,** 1904

Drummond-Hay's Earth Snake *Rhinophis drummondhayi* **Wall,** 1921

Henry Maurice Drummond-Hay (1869–1932) was a planter who lived in Ceylon (Sri Lanka) and devoted his time to natural history in general and herpetology in particular. When Wall visited him (1920), Drummond-Hay had been collecting herpetological specimens for at least 24 years. Wall wrote in *The Snakes of Ceylon* (1921), "I found his bungalow a veritable museum, stocked with specimens of every kind. . . . [He] is one of those rare naturalists, who shuns rather than seeks the limelight." His father, Colonel Henry Maurice Drummond-Hay (1814–1896), was a noted illustrator, botanist, ichthyologist, and ornithologist.

Dubois

Dubois' Sea Snake *Aipysurus duboisi* **Bavay,** 1869

Charles Fréderic Dubois (1804–1867) was a Belgian naturalist, as was his son Alphonse Joseph Charles Dubois (1839–1921). The sea snake was named in memory of the father. They co-wrote *Les oiseaux de l'Europe* (1868–1872), completed by Alphonse and published after Charles Fréderic's death. Two birds are named after one or both of them.

Duellman

Duellman's Cat-eyed Snake *Leptodeira annulata cussiliris* Duellman, 1958

Duellman's Earth Snake *Geophis incomptus* Duellman, 1959

Duellman's Pygmy Leaf-toed Gecko *Phyllodactylus duellmani* **Dixon,** 1960

Sierra Juarez Earth Snake *Geophis duellmani* **H. M. Smith** and Holland, 1969

Duellman's Anole *Anolis duellmani* **Fitch** and **Henderson,** 1973 [Alt. Duellman's Pygmy Anole]

Duellman's Dwarf Iguana *Enyalioides cofanorum* Duellman, 1973

Duellman's Teiid *Alopoglossus atriventris* Duellman, 1973

Duellman's Tree Iguana *Liolaemus duellmani* **Cei,** 1978

Dr. William Edward Duellman (b. 1930) is a herpetologist regarded as the world authority on neotropic frogs. He became Curator of Herpetology, University of Kansas (1959), and retired (1997) as Curator Emeritus and Professor Emeritus, Department of Ecology and Biological Evolution. His wife, Linda Trueb, is also a herpetologist and Curator of Herpetology, University of Kansas. They co-wrote *Biology of Amphibians* (1986).

Duerden

Duerden's Burrowing Asp *Atractaspis duerdeni* Gough, 1907

Dr. James Edwin Duerden (1865–1937) was an Englishman who was Professor of Zoology, Rhodes University College, Grahamstown, South Africa (1905–1932). He worked in Dublin as a demonstrator at the Royal College of Science for Ireland (1893–1895), and he became Curator, Museum, Institute of Jamaica (1895), after which he had temporary jobs at the universities of North Carolina and Michigan before taking up his post in South Africa. He wrote "South African Tortoises of the Genus *Homopus*, with Description of a New Species" (1906).

Dugand

Dugand's Blind Snake *Leptotyphlops dugandi* **Dunn,** 1944

Armando Dugand (1906–1971) was a naturalist and Director, Institute of Natural Sciences, Universidad Nacional de Colombia, Bogotá (1940–1953). He co-founded the magazine *Caldasia* (1940). He led several bird-collecting expeditions into the Colombian interior (1940s). A bird is named after him.

Dugès, A. A.

Dugès' Horned Lizard *Phrynosoma orbiculare dugesi* **Duméril** and **Bocourt,** 1870

Dugès' Spiny Lizard *Sceloporus dugesii* Bocourt, 1873

Dugès' Ring-necked Snake *Diadophis punctatus dugesi* Villada, 1875

Dugès' Earth Snake *Geophis dugesii* Bocourt, 1883

Dugès' Skink *Eumeces dugesi* Thominot, 1883

Dugès' Brownsnake *Storeria dekayi anomala* Dugès, 1888

Professor Alfredo Augusto Dugès (1826–1910) was Professor of Natural History, Universidad de Guanajuato, Mexico. He is regarded as being the father of Mexican herpetology, as he was the first to define Mexican herpetofauna in Linnaean terms. His father was Antoine Louis Dugès (q.v.).

Dugès, A. L.

Dugès' Lizard *Lacerta dugesii* **Milne-Edwards,** 1829 [Alt. Madeiran Wall Lizard; Syn. *Teira dugesii*]

Dr. Antoine Louis Dugès (1797–1838) was a French physician. He was interested in comparing the anatomy of man with that of animals. He wrote *Recherches sur l'ostéologie et la myologiedes batrichiens à leurs différents ages* (1835).

Dumas

Skink sp. *Trachylepis dumasi* Nussbaum and **Raxworthy,** 1995

Dr. Philip Conrad Dumas (1923–1992) was a zoologist and herpetologist whose master's degree and doctorate were awarded by Oregon State University. He taught at the Department of Biological Sciences, University of Idaho, Moscow, Idaho (1953–1965), and was Professor of Biological Sciences, Central Washington University (1966–1989). An annual Philip C. Dumas Lecture in Biology is sponsored by Central Washington University, and Nussbaum delivered the inaugural lecture.

Duméril

Big-headed Amazon River Turtle *Peltocephalus dumeriliana* Schweigger, 1812
Duméril's Fringe-fingered Lizard *Acanthodactylus dumerilii* **Milne-Edwards,** 1829
Duméril's Monitor *Varanus dumerilii* **Schlegel,** 1839
Duméril's Short-legged Skink *Brachymeles talinis* Duméril and **Bibron,** 1839
Duméril's Skink *Gongylus brachypoda* Duméril and Bibron, 1839
Duméril's Graceful Brown Snake *Urotheca dumerilii* Bibron, 1840
Duméril's Dtella *Gehyra baliola* Duméril, 1851
Duméril's Lava Lizard *Ophryoessoides tricristatus* Duméril, 1851
Duméril's Madagascar Swift *Oplurus quadrimaculatus* Duméril, 1851
Duméril's Wedge-snouted Skink *Stenocercus trachycephalus* Duméril, 1851
Duméril's Worm Lizard *Leposternon octostegum* Duméril, 1851
Duméril's Diadem Snake *Elapomorphus lemniscatus* Duméril, Bibron, and Duméril, 1854
Duméril's Kukri Snake *Oligodon sublineatus* Duméril, Bibron, and Duméril, 1854
Duméril's Tropical Gecko *Perochirus ateles* Duméril, 1856
Duméril's Wedge-snouted Skink *Sphenops sphenopsiformis* Duméril, 1856
Duméril's Coral Snake *Micrurus dumerilii* **Jan,** 1858
Duméril's Boa *Acrantophis dumerili* Jan, 1860
Duméril's False Coral Snake *Oxyrhopus clathratus* Duméril, Bibron, and Duméril, 1864
Duméril's Frog-eating Snake *Stegonotus dumerilii* **Günther,** 1865
Whorltail Iguana sp. *Stenocercus dumerilii* **Steindachner,** 1867
Duméril's Wolf Snake *Lycodon dumerili* **Boulenger,** 1893

Dr. André Marie Constant Duméril (1774–1860) was a zoologist and herpetologist. He qualified as a physician in 1793. He was Professor of Anatomy, Muséum National d'Histoire Naturelle, Paris (1801–1812), changing to Professor of Herpetology and Ichthyology (1813–1857). He built up the largest herpetological collection of the time. Toward the end of his career he was assisted by his son, Auguste Henri André, also a distinguished zoologist, who later took over his father's professorship (1857). Some amphibians are named after him.

Dumnui

Dumnui's Bent-toed Gecko *Cyrtodactylus dumnuii* **Bauer** et al., 2010

Sophon Dumnui is Director of the Zoological Park Organization, Thailand.

Dunger

Dunger's File Snake *Mehelya egbensis* Dunger, 1966

Gerald T. Dunger is an American herpetologist who specializes in the herpetofauna of West Africa. He wrote *The Lizards of Nigeria* (published in sections from 1971).

Dunmall

Dunmall's Snake *Furina dunmalli* Worrell, 1955

William "Bill" Dunmall, who collected examples of this snake, lived near Bundaberg, Queensland.

Dunn

Dunn's Tree Snake *Sibynomorphus vagrans* Dunn, 1923
Dunn's Emo Skink *Emoia similis* Dunn, 1927
Dunn's Tropical Ground Snake *Trimetopon simile* Dunn, 1930
Dunn's Earth Snake *Geophis dunni* **Schmidt,** 1932
Dunn's Spiny-tailed Iguana *Morunasaurus groi* Dunn, 1933
Dunn's Mabuya *Mabuya guaporicola* Dunn, 1935
Dunn's Anole *Anolis dunni* **H. M. Smith,** 1936
Dunn's Roadguarder *Crisantophis nevermanni* Dunn, 1937
Dunn's Hognose Viper *Porthidium dunni* **Hartweg** and **Oliver,** 1938
Dunn's Least Gecko *Sphaerodactylus dunni* Schmidt, 1938
Dunn's Water Snake *Hydromorphus dunni* **Slevin,** 1942
Dunn's Ameiva *Ameiva niceforoi* Dunn, 1943
Dunn's Saphenophis Snake *Saphenophis antioquiensis* Dunn, 1943
Dunn's Mud Turtle *Kinosternon dunni* Schmidt, 1947
Dunn's Tinyfoot Teiid *Micrablepharus dunni* **Laurent,** 1949
Dunn's Ground Snake *Atractus dunni* **Savage,** 1955
Dunn's Snail-sucker *Sibon dunni* **Peters,** 1957
Natricine snake sp. *Sinonatrix dunni* Malnate, 1968
[Said to be a synonym of *Natrix tessellata*]

Dr. Emmett Reid Dunn (1894–1956) took both his bachelor's degree (1915) and his master's (1916) at

Haverford College, where he became Professor of Biology (1934); and his doctorate (1921) at Harvard, where he worked at the Museum of Comparative Zoology. He was Assistant Professor of Zoology, Smith College (1916–1928). He visited London, Paris, and Berlin to study the collections there (1928). He was Secretary of *Copeia*, the journal of the American Society of Ichthyologists and Herpetologists (1924–1929), and was the society's President (1930–1931). From 1937 he was closely associated with the Philadelphia Academy of Natural Sciences, becoming Curator of Herpetology (1944). He tried to become an army officer in WW1 but was rejected because it was thought that his weekend pursuit of salamanders and snakes was unbecoming in an officer and a gentleman. Instead he served as an Ensign in the U.S. Navy (1917–1918). He wrote *American Caecilians* (1942). See also **Gro.**

Duperrey

Duperrey's Window-eyed Skink *Pseudemoia duperreyi* **Gray,** 1838 [Syn. *Bassiana duperreyi*]

Captain Louis-Isadore Duperrey (1786–1865) was a French naval officer. He entered the service in 1802. He was second in command and hydrologist on board *L'Uranie* during its circumnavigation (1817–1820). He was appointed (1821) to command *La Coquille* for its circumnavigation (1822–1825).

Duquesney

Duquesney's Galliwasp *Celestus duquesneyi* **Grant,** 1940

Douglas DuQuesnay, "whose quick eye located the type," as Grant wrote in the original description, appears to have had thespian leanings, for a man with that name appeared in the chorus of a pantomime called *Soliday and the Wicked Bird* in Jamaica (1943–1944). We assume that Grant spelled the binomial with *-eyi* instead of *-ayi* in error.

Durheim

Durheim's Kukri Snake *Oligodon durheimi* Baumann, 1913

Herr Durheim sent the Berner Naturhistorischen Museum a collection of animal specimens from Sumatra (1907), which is all the etymology says.

Durrell

Durrell's Night Gecko *Nactus serpensinsula durrellorum* **Arnold** and **Jones,** 1994

Gerald "Gerry" Malcolm Durrell (1925–1995) is best known for the Durrell Wildlife Preservation Trust, Jersey, Channel Islands. He was born in India and first went to England (1928) upon his father's death. The family lived a Bohemian existence on Corfu (1935–1939). His first expedition (1947), to British Cameroons (Cameroon), was financed by his inheritance from his father. He sold the animals he brought back and so financed further expeditions to British Guiana (Guyana), but spent so much on feeding his collected animals that he went broke. He founded his zoo in Jersey (1958) with the help of his first wife, Jacqueline Sonia Wolfenden, from whom he was later divorced (1979). He married Lee McGeorge Wilson (1979), a naturalist, zookeeper, and author from Tennessee. She has carried on the work that Gerald started. He wrote *My Family and Other Animals* (1956), which was a financial success and provided funding for more expeditions. A mammal is named after him and an amphibian after both of them.

Dusen

Yellow Tegu *Tupinambis duseni* Lönnberg, 1910

Dr. Per Dusén (1855–1926) was a Swedish naturalist, botanist, cartographer, explorer, and bryologist. His first overseas collecting expedition was to Cameroon (1890). He was in Argentina on the Princeton expeditions to Patagonia (1896–1899). He was on board the *Antarctic*, responsible for the cartography, as a member of Nathorst's expedition to Spitzbergen (1899). Princeton awarded him an honorary doctorate (1904). He collected the holotype of this lizard (1900).

Dussumier

Aldabra Tortoise *Dipsochelys dussumieri* **Gray,** 1831

Round Island Keel-scaled Boa *Casarea dussumieri* **Schlegel,** 1837

Western Ghats Flying Lizard *Draco dussumieri* **Duméril** and **Bibron,** 1837

Dussumier's Forest Skink *Sphenomorphus dussumieri* Duméril and Bibron, 1839

Dussumier's Smooth Water Snake *Enhydris dussumierii* Duméril, Bibron, and **Duméril,** 1854

Jean-Jacques Dussumier (1792–1883) was a collector, traveler, and trader and a shipowner in the French merchant navy. He was interested in cetaceans and reported on sightings he had made while at sea, keeping up a correspondence on the subject with Cuvier, who wrote a number of the formal scientific descriptions. Otherwise he seems to have collected mainly molluscs and fishes. A bird and a mammal are named after him.

Dutton

Skink sp. *Scelotes duttoni* **Broadley,** 1990

Paul Dutton is an ecologist who has been pressing for the conservation of Bazaruto Archipelago, off Mozambique, since 1989, when he was employed by South Africa's Endangered Wildlife Trust to be warden of Marine

National Park. He lived on Bazaruto Island for many years during the civil war period, studied the unique habitats, and trained community "guardas da fauna" to protect marine turtles through their sustainable utilization. Dutton is now an independent consultant based in Durban, South Africa.

Duvaucel

Three-striped Roofed Turtle *Emys duvaucelii* **Duméril** and **Bibron,** 1835 [Junior syn. of *Kachuga dhongoka* Gray, 1834]

Duvaucel's Gecko *Hoplodactylus duvaucelii* Duméril and Bibron, 1836 [Alt. Northern Sticky-toed Gecko]

Alfred Duvaucel (1792–1824) was a French naturalist who explored India. He was the son, from her first marriage, of Mme. Cuvier. He became Naturalist to the King (1807). His stepfather sent him with Diard (q.v.) to India to collect for Muséum National d'Histoire Naturelle, Paris (1818), and together they established a botanical garden in Chandernagor (1818). Sir Thomas Raffles hired them to collect natural history objects in Sumatra (1819). However, when Raffles discovered that they had sent most of the material they had collected to Muséum National d'Histoire Naturelle, Paris, rather than to him, they were summarily dismissed. Duvaucel's Gecko was erroneously named after him when the museum specimens taken to Europe were credited to him. Only later were the animals found to have actually come from New Zealand. He died in Madras (Chennai), India. A bird and a deer are named after him.

Dwyer

Variable Black-naped Snake *Suta dwyeri* Worrell, 1956

John Dwyer was an Australian herpetologist and reptile collector who collected for Taronga Zoo, Sydney. He was involved in the collection of venom to produce antivenin. He went with Worrell to Queensland in search of taipans for Taronga (1952).

Dymond

Dymond's Japalure *Japalura dymondi* **Boulenger,** 1906

Rev. Francis "Frank" John Dymond (1866–1932) was a Methodist missionary in China. Boulenger based his description on "four specimens from Tongchuan fu [Yunnan], obtained by the Rev. F. J. Dymond." Dymond appears to have been a regular correspondent with, and collector for, Boulenger. Dymond supplied the holotype of the Yunnan Box Turtle *Cuara yunnanensis*, later declared to be extinct but recently rediscovered.

E

Eastwood

Eastwood's Whip Lizard *Tetradactylus eastwoodae* **Methuen** and **Hewitt,** 1913 [Alt. Eastwood's Longtailed Seps] EXTINCT

Miss A. Eastwood collected the holotype (1911) and presented it to the Transvaal Museum (1912). The lizard was last seen in 1928 and is assumed extinct due to habitat loss.

Ebenau

Ebenau's Leaf-tailed Gecko *Uroplatus ebenaui* **Boettger,** 1879

Ebenau's Leaf Chameleon *Brookesia ebenaui* Boettger, 1880

Karl Ebenau, a zoologist who was German Consul in Madagascar (1880–1890), accompanied Stumpff and Boettger, who split the specimens they had gathered between them, and wrote (1880) a list of new reptiles and amphibians that Ebenau had collected on Nossi-Bé Island.

Eberhardt

Eberhardt's Kukri Snake *Oligodon eberhardti* Pellegrin, 1910

Philippe Albert Eberhardt (1872–1942) was a Swiss botanist who was in Tonkin (Vietnam) (1906–1920). He collected for a number of institutions, including the California Academy of Sciences; the Herbarium, Museum Wiesbaden; and Harvard.

Ebner

Ebner's Cylindrical Skink *Chalcides ebneri* **Werner,** 1931

Ebner's Viper *Vipera ursinii ebneri* Knöpfler and **Sochurek,** 1955

Richard Ebner (1885–1961), a schoolteacher and entomologist, was Werner's traveling companion in Morocco. He traveled widely in Europe, North Africa, and Central Asia. He bequeathed his huge collection of Orthoptera to Naturhistorisches Museum Wien. This skink has not been sighted since 1970.

Echidna

Ground snake sp. *Atractus echidna* Passos et al., 2009

In Greek mythology Echidna was a female monster, half woman and half serpent, and mother of many other monsters including the Hydra, the Chimaera, and the Sphinx.

Echternacht

Echternacht's Ameiva *Ameiva anomala* Echternacht, 1977

Dr. Arthur Charles "Sandy" Echternacht (b. 1939) is a herpetologist at the Department of Zoology, University of Tennessee, Knoxville, where he is Professor of Ecology and Evolutionary Biology. He graduated from the University of Kansas (1970). He wrote "A New Species of Lizard of the Genus *Ameiva* (Teiidae) from the Pacific Lowlands of Colombia" (1977).

Edio

Edio's Ground Snake *Atractus edioi* Da Silva et al., 2005

Dr. Edio Laudelino da Luz is a Brazilian engineer who worked on the Cana Brava hydroelectric power project. He was the Director responsible for environmental matters in the managing consortium.

Edward Newton

Rodrigues Blue-dotted Day Gecko *Phelsuma edwardnewtoni* **Vinson** and Vinson, 1969 EXTINCT

Sir Edward Newton (1832–1897) was a colonial administrator in Mauritius (1859–1877) and an amateur ornithologist who visited Madagascar (ca. 1862). He sent the remains of two extinct birds, the Dodo *Raphus cucullatus* and the Solitaire *Pezophaps solitaria*, to his brother, Alfred, Professor of Zoology and Comparative Anatomy, Cambridge (1866–1907). The brothers jointly published "On the Osteology of the Solitaire" (1869). The gecko is thought to have become extinct.

Edwards

Edwards' Middle American Ameiva *Ameiva festiva edwardsii* **Bocourt,** 1873

Sir Alphonse Milne-Edwards (1835–1900) was a zoologist and paleontologist and was Director, Muséum National d'Histoire Naturelle, Paris, when Bocourt was a taxidermist there. He worked closely with Prince Albert I and may have encouraged the Prince to establish Musée Oceanographique, Monte Carlo. The Prix Alphonse Milne-Edwards was created (1903) in his memory. He named a crab after Bocourt, ostensibly honoring the collector, possibly as a quid pro quo. He wrote *Histoire naturelle de l'oiseaux.* Two birds and seven mammals are named after him. The holotype was collected during the French Scientific Mission to Mexico and Central America. Therefore, it is highly likely that *edwardsii* refers to Milne-Edwards.

Edwards, A.

Myall Slider *Lerista edwardsae* **Storr,** 1982

Adrienne Edwards is a herpetologist. She was working at the Department of Herpetology, South Australian Museum, Adelaide, when Storr described this skink. She co-wrote *Guidelines for Vertebrate Surveys in South Australia* (2000).

Edwards, G.

Northern Ringneck Snake *Diadophis punctatus edwardsi* **Merrem,** 1820

George Edwards (1694–1773) was an illustrator, naturalist, and ornithologist. He was Librarian, Royal College of Physicians, London (1733–1764), and corresponded regularly with Linnaeus. Four volumes of Edwards' *A Natural History of Birds* were published between 1743 and 1751.

Edwards, L. A.

Edwards' Rattlesnake *Sistrurus catenatus edwardsii* **Baird** and **Girard,** 1853 [Alt. Desert Massasauga]

Colonel Dr. Lewis A. Edwards (1824–1877) was a surgeon. He joined the army (1846) and took part in the Mexican War (1846–1848). He was in various military posts (1848–1854) and collected in Arkansas, in Mexico, and on the Pacific Railroad Survey for the Smithsonian. He was posted to the office of the Surgeon-General in Washington, DC (1854), and was Attending Surgeon there (1856–1862). He worked in various military hospitals and was Chief Medical Officer, Bureau of Freedmen, Refugees, and Abandoned Lands (1866–1869).

Eew

Anole sp. *Anolis eewi* **Roze,** 1958

E. E. W. (see **Williams, E. E.**). This is one of those playful binomials so beloved of zoologists.

Egeria

Blue-tailed Shinning-skink *Cryptoblepharus egeriae* **Boulenger,** 1888

The skink is named after HMS *Egeria*, which called at Christmas Island (1887).

Eggel

Usambara Five-toed Skink *Proscelotes eggeli* **Tornier,** 1902 [Syn. *Scelotes eggeli*]

Dr. Eggel was a German army physician in East Africa and German South-West Africa (now Namibia), where his regiment was stationed (1904). He collected reptiles, particularly chameleons, for the Berlin Museum in general and Tornier in particular.

Ehmann

Ehmann's Ctenotus *Ctenotus ehmanni* **Storr,** 1985

Harald "Harry" F. W. Ehmann is a herpetologist at the Department of Environment and Natural Resources, South Australia. He co-wrote *Australian Reptiles and Frogs* (1995).

Eigenmann

Eigenmann's Prionodactylus *Cercosaura eigenmanni* **Griffin,** 1917

Professor Dr. Carl Henry Eigenmann (1863–1927) was a German-born American ichthyologist. He graduated from Indiana University with a bachelor's degree (1886), being awarded a doctorate by the same university (1889). He and his wife, Rosa Smith, also a noted ichthyologist, spent time at Harvard studying Agassiz's collection (1887). He was Curator, San Diego Natural History Society (1888), then became Professor of Zoology, Indiana University (1891). He traveled in much of the Americas, his last expedition being to Peru, Bolivia, and Chile (1918–1919).

Eiselt

Colubrid snake sp. *Pseudorabdion eiselti* **Inger** and **Leviton,** 1961

Dwarf/Reed Snake sp. *Calamaria eiselti* Inger and **Marx,** 1965

Eiselt's Dwarf Racer *Eirenis eiselti* **Schmidtler** and Schmidtler, 1978

Eiselt's Pond Turtle *Emys orbicularis eiselti* Fritz, **Baran, Budak,** and Amthauer, 1998

Dr. Josef Eiselt (1912–2001) was a herpetologist. He joined Naturhistorisches Museum Wien as a volunteer (1939). He served in the German armed forces for the duration of WW2. He returned to Vienna to find that his job had been filled by someone else and worked as a laborer for the British occupation forces (1946–1949). He taught and worked as a scientific assistant (1950–1952), and rejoined the staff of Naturhistorisches Museum Wien (1952). After 1962 he started to make trips abroad to collect and research the herpetology of Turkey, Iraq, and Afghanistan. He was Director, Vertebrate Collections (1972–1977). In retirement he traveled even more.

Eisenman

Eisenman's Bent-toed Gecko *Cyrtodactylus eisenmani* Ngo Van Tri, 2008

Dr. Stephanie Eisenman is Director, World Wildlife Fund, USA. The etymology says that she "has greatly contributed to wildlife conservation around the world."

Eisentraut

Eisentraut's Chameleon *Chamaeleo eisentrauti* **Mertens,** 1968

Professor Dr. Martin Eisentraut (1902–1994) was a zoologist, poet, and collector. He was on the staff of the Berlin Zoological Museum when he went to West Africa (1938). He was Curator of Mammals, Staatlisches Museum für Naturkunde, Stuttgart (1950–1957), and was Director, Zoologisches Forschungsmuseum Alexander Koenig, Bonn (1958–1973). He made six trips to Cameroon and Bioko (1954–1973). He wrote *Notes on the Birds of Fernando Pó Island, Spanish Equatorial Africa* (1968) and a slim volume of poems. Four mammals and a bird are named after him.

Eladio

South American Gecko *Gonatodes eladioi* Nascimento, Avila-Pires, and **Cunha,** 1987

Dr. Eladio Da Cruz Lima (1900–1943) was a Justice, Supreme Court, State of Pará, Brazil. He was also an artist, literary critic, archeologist, and zoologist associated with Museu Paraense Emilio Goeldi, Belém, Brazil. He wrote *Mammals of Amazonia. Volume I: General Introduction and Primates* (1943).

Elder

Jewelled Gecko *Strophurus elderi* **Stirling** and **Zietz,** 1893

Sir Thomas Elder (1818–1897) was a businessman, pastoralist, and philanthropist. Born in Scotland, he arrived in Adelaide in 1854 and formed a business partnership. The company financed copper mines (1859), invested heavily in land, and became one of the world's largest sellers of wool. Elder personally owned 18,000 square kilometers (6,900 square miles) of Australia. He was the first to see that camels would be the answer to the problem of transportation in the dry center of the continent and imported 124 breeding animals, with Afghan drivers to manage them. He financed two major expeditions, Warburton's 6,500-kilometer (4,000-mile) journey from the center of Australia to the west coast (1872–1873) and Giles' expedition (1875). He bore the entire cost of financing the Elder exploring expedition in Western Australia (1891–1892). He visited Spain (1860) and noted, after visiting picture galleries, that "picture seeing is more fatiguing than people think." He was clearly broad-minded, as he left legacies to the Presbyterians, the Methodists, and the Anglicans.

Eleodor

Eleodor's Tree Iguana *Liolaemus eleodori* **Cei, Etheridge,** and Videla, 1985

Don Eleodoro Sánchez worked for the Fauna Division of the Agricultural Subsection of the Government of San Juan Province, Argentina. He was continuous and intelligent in his support during the investigation of an area of the provincial reserve of San Guillermo.

Elisa

Elisa's Leaf-toed Gecko *Asaccus elisae* Werner, 1895 [Alt. Werner's Gecko]

We surmise that Elisa may have been Werner's wife or daughter, but unfortunately he gives no clue in the original description.

Ellenberger

Amphisbaena sp. *Dalophia ellenbergeri* **Angel,** 1920

Ellenberger's Long-tailed Seps *Tetradactylus ellenbergeri* Angel, 1922 [Alt. Ellen's Whip Lizard]

Victor Ellenberger (1879–1972) came from a family of Swiss Protestant evangelical missionaries in Southern Africa. He also was a naturalist and anthropologist. He was born in Lesotho, sent to France for his secondary education, and returned to the mission in Africa. He served in Barotseland, Zambia (1903–1917), and Basutoland-Lesotho (1917–1934). He took French nationality (1929) and, when he left Africa, went to France and was in charge of a parish near Paris (1935–1947). He wrote *La fin tragique des bushmen* (1953).

Elliot, G. F. S.

Elliot's Chameleon *Chamaeleo ellioti* **Günther,** 1895

Captain George Francis Scott Elliot (1862–1934) was a botanist, traveler, and author. He took bachelors' degrees in mathematics at Cambridge (1882) and in science at Edinburgh (1885). He went to South Africa (1885) and from there to Mauritius via Madagascar. His next expeditions were to Tripoli and Egypt, and, as botanist, with the French/English Commission to define the Sierra Leone boundary. From West Africa he set out for Uganda, where he collected reptiles. He was Professor of Botany, Glasgow Veterinary College (1896–1904). As he had been a soldier, he reenlisted on the outbreak of WW1 (1914) at the age of 52. Posted to Egypt (1915), he fought at the Battle of Romani, during which nearly all the men in his command were killed or wounded, and later fought at Gaza. He was ordered to return (1917) to the UK, but his ship was torpedoed. He survived but arrived home wearing an Italian officer's uniform and a pair of white slippers.

Elliot, W.

Elliot's Shieldtail *Uropeltis ellioti* **Gray,** 1858

Elliot's Forest Lizard *Calotes ellioti* **Günther,** 1864

Sir Walter Elliot (1803–1887) was a career civil servant in the Indian Civil Service, Honourable East India Company, Madras (1821–1860). He was Commissioner for the administration of the Northern Circars (1845–1854) and a member of the Council of the Governor of Madras (1854–1860). He was a distinguished Orientalist, and his interests included botany, zoology, Indian languages, numismatics, and archeology. He was a regular correspondent of Charles Darwin's. His Indian herbarium was given to the Edinburgh Botanic Garden. He retired to Scotland and, despite blindness, worked on local natural history projects. Two mammals are named after him.

Elliott

Australian Brown Snake sp. *Pseudonaja elliotti* Hoser, 2003

Adam Elliott is a private keeper of reptiles and a wildlife conservationist based in Melbourne, Australia.

Elsey

Australian turtle genus *Elseya* **Gray,** 1867

Dr. Joseph Ravenscroft Elsey (1834–1858) was a surgeon, explorer, and physician who qualified in London (1855). He was Assistant Naturalist on the North Australian exploring expedition led by Augustus Gregory (1855–1856). He was then on St. Kitts, West Indies (1857–1858). The pastoral property, Elsey Station, was the setting for Mrs. Aeneas Gunn's book on tropical outback life in Australia, *We of the Never-Never.*

Emerson

Ground Snake sp. *Atractus emersoni* Silva Haad, 2004

See **Belluomini.**

Emigdio

Emigdio's Ground Snake *Atractus emigdioi* Gonzalez-Sponga, 1971

Emigdio González Sponga collected the holotype, and we assume is related to the describer.

Emin

Emin Pasha's Worm Snake *Leptotyphlops emini* **Boulenger,** 1890

Boulenger's Sandveld Lizard *Nucras emini* Boulenger, 1898

Emin Pasha (1840–1892) was the name by which Eduard Schnitzer became known. He was a German physician who worked in Albania (then under Turkish rule) and acquired the name *Emin*, meaning "faithful one." He was an explorer, naturalist, collector, and administrator in Africa, making important contributions to the knowledge of the Sudan and Central Africa. He became Medical Officer to the staff of General Charles G. Gordon, British Governor-General and Administrator, the Sudan (1876). Gordon, who was killed by the Mahdi at Khartoum (1885), appointed Emin, with the title of Bey, to be Pasha (Governor) of the southern Sudanese province of Equatoria (1878). A Sudanese uprising forced him to retreat (1885) into what is now Uganda. Henry Morton Stanley (on his last expedition) led a search party (1888) to rescue Emin, only to find that he didn't want to be rescued. Emin's claim to fame was that he abolished slavery in the territories he commanded—which is probably why he was beheaded by slave traders in the region of Lake Tanganyika. Three birds and two mammals are named after him, as is Mount Emin in the Ruwenzori range.

Emma Gray

Emma Gray's Forest Lizard *Calotes emma* **Gray,** 1845

Mary Emma Gray (1787–1876) was the wife of the describer. She was his amanuensis, after a severe stroke paralyzed his right side, also acting as an artist and, sometimes, co-author with him. She was a conchologist and algologist.

Emmel

Emmel's Ground Snake *Atractus emmeli* **Boettger,** 1888

Ferdinand Emmel sent two specimens of this snake to Naturmuseum Senckenberg, Frankfurt.

Emmott

Noonbah Robust Slider *Lerista emmotti* **Ingram, Couper,** and Donnellan, 1993

Emmott's Short-necked Turtle *Emydura macquarii emmotti* **Cann** et al., 2003 [Alt. Cooper Creek Turtle]

Angus Emmott (b. 1962) is a grazier and natural historian and a well-known "Friend of the Queensland Museum." He is Chairman of the Community Advisory Committee to support "whole-of-basin" management of the Lake Eyre basin.

Emoll

Nicaraguan Slider *Trachemys emolli* Legler, 1990

Dr. Edward Moll ("E. Moll.") was Chairman of the World Conservation Union's Freshwater Chelonian Specialist Group and on the staff of Eastern Illinois University, where he is now Professor Emeritus. He was Volunteer Naturalist, Mason Audubon Center; was an Adjunct Professor, School of Renewable Natural Resources, University of Arizona; and is currently a Board Director, Tuscon Herpetologists Society. He and his brother, Don, have both been involved with turtle conservation and have traveled the world to study and observe them; their collections are in the Field Museum. They co-wrote *The Ecology, Exploitation, and Conservation of River Turtles.*

Emory

Emory's Rat Snake *Pantherophis emoryi* **Baird** and **Girard,** 1853 [Alt. Great Plains Rat Snake; Syn. *Elaphe emoryi*]

Texas Spiny Softshell Turtle *Apalone spinifera emoryi* **Agassiz,** 1857

Brigadier-General William Hemsley Emory (1811–1887) graduated from West Point (1831), served as a second lieutenant, resigned to be a civil engineer (1836) then rejoined the army (1838). His specialty was mapping the U.S. borders. He was on an expedition that produced a new map of Texan claims west of the Rio Grande (1844) and was a regimental commander in California and Mexico during the Mexican-American War (1846–1848). He was Chief Astronomer for the California-Mexico Boundary Survey (1848–1853) during which he collected zoological specimens along the Rio Grande for the Smithsonian. He was stationed in Indian Territory at the start of the American Civil War (1861) and in danger of capture by the Confederate army, but he successfully attacked, then retreated to Fort Leavenworth. After the defeat of the Confederacy he held a number of commands, including his final posting as commander of federal troops in Louisiana, Arkansas and Mississippi (1871–1875). He was forcibly retired (1876). He wrote *Notes of a Military Reconnaissance from Fort Leavenworth in Missouri to San Diego, California* (1848).

Engdahl

Engdahl's Burrowing Viper *Atractaspis engdahli* Lonnberg and **Andersson,** 1913

Reverend Theodor Engdahl was a Swedish missionary who made a collection of reptiles from Kismayu and Mofi, on the Juba River, British Somaliland (Somalia). He was certainly in Kismayu (1906), as his reports on a school there are still extant. He collected the viper holotype.

Engle

Engel's Mabuya *Mabuya englei* **Taylor,** 1925

Captain Francis G. Engle (1888–1974) joined the U.S. Coast and Geodetic Survey (1907). He commanded a number of ships, including the steamship *Pathfinder* (q.v.). He served in the U.S. Navy in both world wars and retired twice (1937 and 1943). Taylor wrote, "Captain Engle extended me innumerable courtesies while I was on his ship and assisted me greatly in making collections on the Mindanao coast."

Entrecasteaux

Entrecasteaux's Window-eyed Skink *Pseudemoia entrecasteauxii* **Duméril** and **Bibron,** 1839

Admiral Antoine Raymond Joseph de Bruni d'Entrecasteaux (1739–1793) entered the French navy as an adolescent (1754). His career included being Governor for a time of the French colony of Mauritius. He sailed (1791) with the vessels *Ésperance* and *Récherche* to discover the fate of La Perouse's expedition, unheard of since it sailed from Botany Bay (1788). His search around mainland Australia and Tasmania for La Perouse (1792–1793) ended with his death from scurvy, and the whole expedition ground to a halt in the Dutch East Indies (Indonesia) (1794) on hearing of the declaration of a republic in France.

Eranga Viraj

Eranga Viraj's Shieldtail Snake *Rhinophis erangaviraji* Wickramasinghe et al. 2009

Eranga Viraj Dayarathne was an Instructor, Reptiles Group of the Young Zoologists' Association of Sri Lanka, Department of National Zoological Gardens.

Ercolini

Lanza's Racerunner *Mesalina ercolinii* **Lanza** and Poggesi, 1975 [Syn. *Eremias ercolinii*]

Professor Antonio Ercolini was primarily an ichthyologist who worked at Department of Animal Biology and Genetics, Natural History Museum, Università degli Studi di Firenze (1954–1996). He collected fishes at Bud-Bud, Somalia (1975). Lanza works at the same university.

Erdelen

Horned Agama sp. *Ceratophora erdeleni* Pethiyagoda and Manamendra-arachchi, 1998

Dr. Walter R. Erdelen (b. 1951) is a zoologist, botanist, geneticist, and chemist who is, since 2001, Assistant Director-General for Natural Sciences, UNESCO. His bachelor's degree (1973), master's (1977), and doctorate (1983) were awarded by Ludwig-Maximilians-Universität München. He worked in Sri Lanka, India, and the Maldives (1977–1981) and in London and Munich (1981–1988). He was Senior Lecturer, University of Saarland (1988–1993). He was at University of Würzburg (1993–1997), becoming Professor of Ecology and Biogeography (1995). He was Visiting Professor, Department of Biology, Indonesian Institute of Technology, Bandung (1997–2001).

Erdis

False Coral Snake sp. *Oxyrhopus erdisii* **Barbour,** 1913

Ellwood C. Erdis (1867–1944) was the chief engineer of the National Geographic Society and Yale University Peruvian expedition (1912), during which the snake holotype was collected. He was a topographer and archeologist who superintended the excavation of the Inca site at Machu Picchu.

Eremchenko

Skink sp. *Asymblepharus eremchenkoi* Panfilov, 1999

Dr. Valery Konstantinovich Eremchenko (b. 1949) is a herpetologist who is Professor of Ecology, University of Bishkek, Kyrghyzstan. He co-wrote "On the Ecology of the Gecko *Teratoscincus scincus*" (2007).

Erhard, A.

Erhard's Pond Turtle *Mauremys leprosa erhardi* **Schleich,** 1996

Andreas Erhard (b. 1960) is a naturalist long involved in environmental protection. Since the 1980s he has started many social and educational initiatives to improve the environment, including reorganizing vocational training of staff at German gasoline stations. He was publicly honored by Bavaria (1993), but his work stretches far beyond his native province.

Erhard, D.

Erhard's Wall Lizard *Podarcis erhardii* **Bedriaga,** 1882

Dr. D. Erhard was a German naturalist who spent many years in the Cyclades and wrote *Fauna der Cycladen* (1858).

Eric Smith

Guerrero Long-tailed Rattlesnake *Crotalus ericsmithi* **Campbell,** 2007

Dr. Eric N. Smith is Assistant Professor of Biology, University of Texas, Arlington, where he took his doctorate. His principal interests are systematics, biogeography, herpetology, and his black Labrador, Chester. He knows Latin America well, having been brought up in Guatemala and having collected in Venezuela and Ecuador. He wrote "Two New Species of *Eleutherodactylus* (Anura: Leptodactylidae), of the *alfredi* Group, from Mountains of the Caribbean Region of Guatemala" (2005).

Erik

Ground Snake sp. *Atractus eriki* Esqueda, La Marca, and Bazo, 2005

Erik La Marca (b. 1990) is the youngest son of Venezuelan herpetologist Enrique La Marca, the second author, who told us that Erik, who first discovered the snake, is not a herpetologist but a private pilot pursuing a career in commercial aviation.

Erlanger

Ethiopian House Snake *Lamprophis erlangeri* **Sternfeld,** 1908

Baron Carlo von Erlanger (1872–1904) was a German collector. He traveled in the Tunisian Sahara (1893 and 1897). He visited Abyssinia (Ethiopia) and Somaliland (1900–1901), accompanied for part of the time by O. R. Neumann. He died in a car accident in Salzburg, so must have been one of the first victims of a road traffic accident. Among other taxa named after him are two mammals and three birds.

Ernest

Ernest's Anole *Anolis ernestwilliamsi* **Lazell,** 1983

See **Williams, E. E.**

Ernst

Ernst's Map Turtle *Graptemys ernsti* Lovich and **McCoy,** 1992 [Alt. Escambia Map Turtle]

Dr. Carl Henry Ernst (b. 1938) is a herpetologist at Department of Zoology, Division of Amphibians and Reptiles, at the Smithsonian and Professor Emeritus of Biology, George Mason University, Virginia. He co-wrote *Snakes of the United States and Canada* (2003).

Ernst Keller

Gecko sp. *Gekko ernstkelleri* Rösler et al., 2006

Ernst Keller is a German with an interest in gecko preservation. The authors wrote, "The new species is named in honor of Ernst Keller, who takes a keen interest in the gecko fauna of the Philippines, a country that he has visited many times. By taking patronage of this unique gecko from Panay, Mr. Keller supports nature conservation in the area."

Escarchados

Tree Iguana sp. *Liolaemus escarchadosi* Scolaro, 1997

Los Escarchados is the name of an area in Patagonia.

Eschscholtz

Eschscholtz's Lizard *Dicamptodon ensatus* Eschscholtz, 1833 [Alt. California Giant Salamander]

"Eschscholtz's Lizard" is an archaic name for this salamander, which is why an amphibian appears in a book about reptiles. Dr. Johann Friedrich von Eschscholtz (1793–1831) was a Russian physician who took part in Kotzebue's *Predpriaetie* expedition as chief naturalist. He wrote the expedition report, *Zoologischer Atlas*, part 5 of which, containing all the salamanders, appeared after his death.

Espinal

Coffee Snake sp. *Ninia espinali* McCranie and **Wilson,** 1995

Mario R. Espinal is a Honduran biologist, whom the describers called a "good and longtime friend and occasional field companion." He co-wrote "The Herpetofauna of Parque Nacional La Muralla, Honduras" (2001).

Espinoza

Tree Iguana sp. *Liolaemus espinozai* Abdala, 2005

Dr. Robert Earl Espinoza (b. 1967) is a herpetologist and biologist who is Associate Professor, University of Nevada, Reno. He co-wrote "Two New Cryptic Species of *Liolaemus* (Iguania, Tropiduridae) from Northwestern Argentina—Resolution of the Purported Reproductive Bimodality of *Liolaemus-Alticolor*" (1999).

Essex

Essex's Leaf-toed Gecko *Goggia essexi* **Hewitt,** 1925

Essex's Mountain Lizard *Tropidosaura essexi* Hewitt, 1927

Robert Essex worked as a collector at Albany Museum when Hewitt was Director there. He wrote "Descriptions of Two New Species of the Genus *Acontias* and Notes on Some Other Lizards Found in the Cape Province" (1925).

Essington

Essington's Ctenotus *Ctenotus essingtonii* **Gray,** 1842

Named after Port Essington, Northern Territory, Australia.

Etheridge

Etheridge's Anole *Anolis etheridgei* **Williams,** 1962

Tan Racer *Coluber constrictor etheridgei* **Wilson,** 1970

Curlytail Lizard sp. *Leiocephalus etheridgei* Pregill, 1981 EXTINCT

Etheridge's Lava Lizard *Tropidurus etheridgei* **Cei,** 1982

Snake genus *Etheridgeum* **Wallach,** 1988

Etheridge's Long-nosed Snake *Rhinocheilus etheridgei* **Grismer,** 1990 [Syn. *R. lecontei etheridgei*]

Tree Iguana sp. *Liolaemus etheridgei* **Laurent,** 1998

Etheridge's Blind Snake *Typhlops etheridgei* Wallach, 2002

Dr. Richard Emmett Etheridge (b. 1929) is a biologist and herpetologist. He took both his bachelor's degree (1951) and his master's (1952) at Tulane University. He served in the U.S. Navy (1952–1956) as a sonar operator. He was a teaching assistant at the University of Michigan (1958–1959), which awarded his doctorate (1959). He worked at the University of Southern California and Los Angeles County Museum (1959–1960), and at San Diego State University (1961–1997) as Professor of Biology and Curator of Herpetology at the university's museum and is now Professor Emeritus. Wallach was taught by Etheridge and created the genus *Etheridgeum* in his honor.

Etienne

Chameleon sp. *Chamaeleo etiennei* **Schmidt,** 1919

Dr. Etienne was a Belgian physician and entomologist in the Congo early in the 20th century. He accompanied King Albert on his journey through the Congo basin. He was extremely helpful to American Museum of Natural History's Congo expedition when they were based in the Congo's Banana region.

Eugene

Eugene's Anole *Anolis eugenegrahami* **Schwartz,** 1978

Eugene D. Graham Jr. was the co-discoverer of the lizard with Thomas Thurmond and Schwartz. With William Sommer, they all collected extensively along the coast of Haiti (1978). He wrote "A New Species of Lizard (*Sphaerodactylus*) from Northwest Haiti" (1981).

Eurydice

Brown-backed Yellow-lined Ctenotus *Ctenotus eurydice* **Czechura** and **Wombey,** 1982

In Greek mythology Eurydice was the wife of Orpheus. The describers say, "The name was arbitrarily chosen."

Everett

Colubrid snake sp. *Calamaria everetti* **Boulenger,** 1893

Everett's Kukri Snake *Oligodon everetti* Boulenger, 1893

Alfred Hart Everett (1848–1898) was a British colonial administrator who worked in the East Indies. He collected widely, and it is believed that a jawbone of an orangutan *Pongo pygmaeus,* which he found in a cave, may have been used in the "Piltdown Man" hoax. He was interested in all aspects of natural history and anthropology. He collected in Borneo and also published reports on the island's caves and volcanic phenomena. His death made the front page of the *Sarawak Gazette.* Five mammals and ten birds plus other taxa are named after him. See **Alfred.**

Evermann

Evermann's Anole *Anolis evermanni* **Stejneger,** 1904

Dr. Barton Warren Evermann (1853–1932) was a schoolteacher (1876–1886) and a student at Indiana University, where he was awarded his bachelor's degree (1886), master's (1888), and doctorate (1891). He worked for the Bureau of Fishes in Washington (1891–1914) in different capacities, combining his various roles with lecturing on zoology at Cornell (1900–1903), Yale (1903–1906), and, later, Stanford, after he became Director of the Museum, California Academy of Sciences (1914).

Eversmann

Comb-toed Gecko *Crossobamon eversmanni* **Wiegmann,** 1834

Dr. Alexander Eduard Friedrich Eversmann (1794–1860)—in the Russian style, Eduard Aleksandrovich Eversmann—was a pioneer Russian/German physician and entomologist. After education in Germany he worked for two years as a physician at an armaments factory. He became disenchanted with medicine and followed his fascination for zoology, eventually becoming Professor of Zoology and Botany, University of Kazan, Russia. He traveled in remote areas of Russia's Asian empire. He concentrated on Lepidoptera, but not to the exclusion of giving detailed scientific descriptions of other taxa, and became the greatest expert on the fauna of southern Russia. Two mammals and five birds are named after him.

Ewerbeck

Ewerbeck's Round-headed Worm Lizard *Chirindia ewerbecki* **Werner,** 1910

Karl Ewerbeck was District Officer and a customs official at Lindi, Tanganyika (Tanzania), when it was a German colony. He collected the lizard holotype (1903). He was also involved with a German university expedition to his area (1900).

Exocet

Christmas Island Blind Snake *Ramphotyphlops exocoeti* **Boulenger,** 1887

Nowadays we think of "Exocet" as a missile, but the etymology means "flying fish," and Boulenger so named the snake because officers from HMS *Flying Fish* collected the first specimen.

Eydoux

Spine-tailed Sea Snake *Aipysurus eydouxii* **Gray,** 1849

Joseph Fortuné Théodore Eydoux (1802–1841) was a French naturalist who became a naval surgeon (1821). He was aboard *La Favorite* in the East Indies (1830–1832). He also was a member of the crew of *La Bonite*, which circumnavigated the globe (1836–1837). He co-wrote *Voyage autour du monde exécuté pendant les années 1836 et 1837 sur la corvette La Bonite* (1841). He died in Martinique.

F

Fabian

Fabian's Lizard *Liolaemus fabiani* **Yáñez** and **Núñez,** 1983 [Alt. Yanez's Tree Iguana]

Professor Dr. M. Fabian Jaksic works for the Department of Ecology, Universidad Catolica de Chile, where he is Director, Center for Advanced Studies in Ecology and Biodiversity, and President, Chile Committee on the Environment. He was Assistant Curator of Herpetology, University of California (1979–1982), while studying for his doctorate there (awarded 1982). He co-wrote "Spatial Distribution of the Old World Rabbit *(Oryctolagus cuniculus)* in Central Chile" (1979).

Fairchild

Fairchild's Anole *Anolis fairchildi* **Barbour** and **Shreve,** 1935

Dr. David Grandison Fairchild (1869–1954) was a botanist and explorer after whom the botanical gardens in Coral Gables, Florida, were named. He took his bachelor's (1888) and master's (1889) degrees at Kansas State College of Agriculture. His doctorate was honorary, awarded by Oberlin College (1915). He became a plant explorer for U.S. Department of Agriculture, traveling the world before managing the Plant Introduction Program. Among his successes was bringing flowering cherry trees from Japan to Washington. Barbour, a lifelong friend, was a Director of the Fairfield Tropical Botanic Garden and often stayed with him when in Florida. He wrote *The World Was My Garden: Travels of a Plant Explorer* (1938).

Falla

Falla's Skink *Oligosoma fallai* **McCann,** 1955 [Alt. McCann's Ground Skink]

Sir Robert Alexander Falla (1901–1979) was an ornithologist and museum administrator in New Zealand. He was Assistant Zoologist with the Sir Douglas Mawson–led British, Australian, and New Zealand Antarctic research expedition (1929–1931). He was Director, Dominion Museum, Wellington (1947–1966). A bird is named after him.

Farr

Alligator lizard sp. *Gerrhonotus farri* Bryson and Graham, 2010

William L. Farr (b. 1958) is an artist and amateur herpetologist whose "day job" since 2000 has been as a zookeeper in the Herpetology Department at Houston Zoo. He collected the holotype.

Faust

Wolf Snake sp. *Lycodon fausti* Gaulke, 2002

Dr. Richard Faust (1927–2000) was President, Frankfurt Zoological Society (1987–2000). He studied paleontology, anthropology, and zoology at Johannes Gutenberg-Universität Mainz. He worked at Frankfurt Zoo (1952–1992), being Curator of Birds (1952), Assistant Director (1958), and Director (1974). Frankfurt Zoological Society is a major sponsor of PESCP (Philippine Endemic Species Conservation Project).

Fea

Fea's Viper *Azemiops feae* **Boulenger,** 1888

Bow-fingered Gecko *Cyrtodactylus feae* Boulenger, 1893

Fea's Chameleon *Chamaeleo feae* Boulenger, 1906

St. Thomas Beaked Snake *Rhinotyphlops feae* Boulenger, 1906

Ugly Worm Lizard *Cynisca feae* Boulenger, 1906

Leonardo Fea (1852–1903) was an explorer, zoologist, painter, and naturalist and an assistant at Museo Civico di Storia Naturale di Genova. He liked remote places and visited the Cape Verde Islands (1898), Burma (Myanmar), China, and Equatorial Guinea. Two mammals, two birds, and five amphibians are named after him.

Fedtschenko

Fedtschenko's Bow-fingered Gecko *Cyrtopodion fedtschenkoi* **Strauch,** 1887 [Alt. Fedtschenko's Grasping Gecko; Syn. *Tenuidactylus fedtschenkoi*]

Alexei Pavlovich Fedtschenko (or Fedchenko) (1844–1873), a graduate in zoology and geology from Moscow University, was a naturalist and explorer of Central Asia. The Fedchenko Glacier in the Pamirs is named after him, as is the asteroid 3195 Fedchenko. He died while climbing on Mont Blanc. After his death the Russian government published accounts of his discoveries and explorations.

Fehlmann

Fehlmann's Four-clawed Gecko *Gehyra fehlmanni* **Taylor,** 1962 [Alt. Fehlmann's Dtella]

Dr. Herman Adair Fehlmann (1917–2005) of George Vanderbilt Foundation, Bangkok, and the Smithsonian was primarily an ichthyologist. His herpetological specimens were deposited in the Stanford University Museum.

Feick

Feick's Dwarf Boa *Tropidophis feicki* **Schwartz,** 1957

John R. Feick was an Associate Professor of Biology, St. Anselm's College, Manchester, New Hampshire (1961–2005). His bachelor's degree in biology is from Albright College (1958), and his master's in zoology was awarded by the University of Pennsylvania (1962). He worked closely with Schwartz.

Ferguson, H. S.

Cardamom Hills Earth Snake *Rhinophis fergusonianus* **Boulenger,** 1896

Harold S. Ferguson (1852–1921) was a planter, zoologist, and herpetologist who spent most of his life in Travancore, southern India. He was associated with the museum at Trivandrum (1880–1904), becoming Director (1894). When he left India he wrote *Travancore Batrachians* (1904). An amphibian is named after him.

Ferguson, W.

Ferguson's Day Gecko *Cnemaspis scalpensis* Ferguson, 1877

William Ferguson (1820–1887) was a surveyor who lived in Ceylon (Sri Lanka) (1839–1887). He was a keen amateur naturalist, taking a great interest in Ceylon's botany and measured the native trees. He wrote *Reptile Fauna of Ceylon* (1877).

Fernand

Fernand's Skink *Lygosoma fernandi* **Burton,** 1836 [Alt. Fire Skink]

Named after the island of Fernando Po (Bioko, Equatorial Guinea).

Ferrara

Ferrara's Mabuya *Mabuya ferrarai* **Lanza,** 1978

Dr. Franco Ferrara of Consiglio Nazionale delle Ricerche, Centro di Studio per la Faunistica ed Ecologia Tropicali, Florence, is an isopod specialist.

Ferrari-Perez

Spiny Lizard sp. *Sceloporus ferrariperezi* **Cope,** 1885 [Junior syn. of *S. torquatus* Wiegmann, 1828]

Professor Dr. Fernando Ferrari-Perez (d. 1927) was a zoologist. The Geographical and Exploring Commission, Republic of Mexico, was established (1877) to catalogue the natural history and resources of Mexico, and he was made Naturalist there (1879). He gave Cope access to, and considerable help with, the herpetology collection (1885). He wrote *Catalogue of Animals Collected by the Geographical and Exploring Commission of the Republic of Mexico* (1886).

Ferreira

Gymnophthalmid lizard sp. *Leposoma ferreirai* **Rodrigues** and Avila-pires, 2005

Alexandre Rodrigues Ferreira (1756–1815) was the first naturalist to explore the Amazon and Pantanal biomes in the states of Pará and Mato Grosso, and was known as "the Brazilian Humboldt." He was educated at Universidade de Coimbra, Portugal, and taught natural history there until going to work at Museu da Ajuda, Lisbon (1778–1783). The Portuguese government sponsored him to explore in Brazil. He followed the course of the Amazon and its tributaries (1783–1792), studying the indigenous people, their languages, and the fauna and flora of the region. He returned to Lisbon (1793) and was Director, Natural History Museum and Botanical Gardens. He never returned to Brazil.

Ferron

Wolf Snake sp. *Lycodon ferroni* **Lanza,** 1999

Cédric Ferron is a French speleologist who collected the snake holotype while on a caving expedition to Samar, Philippines.

Festa

Peracca's Large-scaled Lizard *Ptychoglossus festae* **Peracca,** 1896

Amazonian Scaly-eyed Gecko *Lepidoblepharis festae* Peracca, 1897

Drab Ground Snake *Liophis festae* Peracca, 1897

Peracca's Whorl-tailed Iguana *Stenocercus festae* Peracca, 1897

Peracca's Teiid *Alopoglossus festae* Peracca, 1904

Veronica's Anole *Anolis festae* Peracca, 1904

Dr. Enrico Festa (1868–1939) graduated from Universita di Torino (1891). He visited Egypt, Palestine, Jordan, Lebanon, and Syria (1893). He collected in Panama and Ecuador (1895–1898). He worked for Museo e Instituto di Zoologia Sistematica dell'Universita di Torino (1899–1923) as Deputy Assistant Professor (1899), retiring as Honorary Vice Director. A bird is named after him.

Field

Field's Horned Viper *Pseudocerastes fieldi* **Schmidt,** 1930

Henry Field (1902–1986), an anthropologist, was a grandson of Marshall Field (1834–1906), who founded the famous Field Museum in Chicago. Henry worked at the museum as an Assistant Curator of Physical Anthropology, later becoming the Head Curator (1934–1941). During this period he participated in many expeditions, including the North Arabian Desert expedition (1927–1928).

Filippi

Filippi's Ground Snake *Atractus favae* Filippi, 1840

See **De Filippi.**

Finch

Agama sp. *Agama finchi* **Böhme** et al., 2005

Brian W. Finch is an Australian ornithologist who discovered this new species. He co-wrote *Species-Checklist of the Birds of New Guinea* (1985).

Finsch

Finsch's Monitor Lizard *Varanus finschi* **Böhme,** Horn, and **Ziegler,** 1994

Friedrich Hermann Otto Finsch (1839–1917) was an ethnographer, ornithologist, naturalist, and traveler. He visited many areas from Lapland to the South Seas. He was the Director of a number of museums, including Bremen (1884) and Brunswick. Bismarck appointed him Imperial Commissioner for the German Colony of Kaiser-Wilhelm-Land (New Guinea) (1884). He founded the town of Finschhafen (1885), which remained the seat of German administration until 1918. He co-wrote *Die Vogel Ost Afrika*. Many birds and a mammal are named after him.

Fionn

Peninsula Crevice-dragon *Ctenophorus fionni* **Procter,** 1923

In letters to her sister, Joan Beauchamp Procter refers to a "Fionn," whom she was "missing terribly." We know no more than that but think she had this same Fionn in mind when she named the lizard. The original description throws no light on this, so maybe Proctor wanted to keep the identity of Fionn private.

Fischer

Fischer's Cat Snake *Boiga pulverulenta* Fischer, 1856

Fischer's Tree Boa *Epicrates striatus* Fischer, 1856 [Alt. Haitian Boa]

Fischer's Longtail Snake *Enulius flavitorques unicolor* Fischer, 1881

Fischer's Chameleon *Kinyongia fischeri* **Reichenow,** 1887 [Alt. Nguru Two-horned Chameleon]

Fischer's Thick-toed Gecko *Pachydactylus laevigatus* Fischer, 1888

Fischer's Dwarf Gecko *Lygodactylus fischeri* **Boulenger,** 1890

Fischer's Snail Sucker *Sibon fischeri* Boulenger, 1894

Dr. Johann Gustav Fischer (1819–1889) was a vertebrate zoologist. He earned his first degree in Leipzig and his doctorate in Berlin (1843), having written his thesis on the cranial nerves in amphibians and reptiles. He went on to teach in various secondary schools, later establishing one himself. In his latter years he was a volunteer Warden in charge of the fish, amphibian, and reptile collection at Museum für Naturkunde Berlin.

Fisk

Tent Tortoise *Testudo fiski* **Boulenger,** 1886 [Junior syn. of *Psammobates tentorius verroxii* Smith, 1839]

Fisk's House Snake *Lamprophis fiskii* Boulenger, 1887

Rev. George H. R. Fisk was Chaplain at the Breakwater Convict Station, South Africa (1870–1876). He may be the same Rev. George Fisk who, when vicar of Walsall (1842), produced illustrated diaries of a grand tour from England through the Continent and Near East to Palestine and back. A number of fish are named after him.

Fitch, H. S.

Fitch's Gartersnake *Thamnophis sirtalis fitchi* **Fox,** 1951 [Alt. Valley Gartersnake]

Fitch's Anole *Anolis fitchi* **Williams** and **Duellman,** 1984

Dr. Henry Sheldon Fitch (1909–2009) was a herpetologist. He grew up in Oregon and as a child loved snakes, for "the real bonus was in seeing horrified adults scatter." He did his graduate work at Berkeley, gaining a master's (1933) and a doctorate (1937) in zoology. His employment as a field biologist with the U.S. Fish and Wildlife Service (1938–1947) was interrupted by a stint as a pharmacist in the U.S. Army (1941–1945). He was at University of Kansas (1948–1980), first as an Instructor, then Assistant Professor (1949) and Professor (1958). From 1965 onward he did extensive field work in Central America. He retired officially (1980) but continued actively collecting snakes and writing scientific papers (2007). He died three months before his 100th birthday.

Fitch, S. P.

Chameleon sp. *Chamaeleo harennae fitchi* **Necas,** 2004

Steve Fitch Paul (b. 1957) is a Free Methodist minister, currently serving as the Southern California Conference Superintendent, which is the equivalent of being an Anglican Bishop. By 2001 the conference had initiated medical, educational, and church work throughout much of southern Ethiopia, a country Fitch visited often. He developed an interest in African flora and fauna and started the Eden Reforestation Projects, which has planted millions of seedlings in Ethiopia, Sudan, and Madagascar. He collected the holotype in 2002.

Fittkau

Tree Iguana sp. *Liolaemus fittkaui* **Laurent,** 1986

Professor Dr. Ernst Josef Fittkau (b. 1927) is a scientist, entomologist (specializing in Diptera), and herpetologist who was Director, Zoologische Staatssammlung München (1976–1992), and still actively researching in retirement. He first went to South America (1960) and is proud that he has collected on every continent except Antarctica. He is also interested in the history of natural science and the people who shaped it, writing an appreciation of Spix (1983), as well as "Crocodiles and the Nutrient Metabolism of Amazonian Waters" (1973).

Fitzgerald

Fitzgerald's Tree Iguana *Liolaemus fitzgeraldi* **Boulenger,** 1899

Major Edward Arthur Fitzgerald (1871–1930) was an

officer in the British army and a mountaineer, traveler, and explorer. He led a number of expeditions to New Zealand and to South America. He made the first ascent of Aconcagua (1897), the highest mountain in South America.

Fitzinger

Pygmy Keeled Lizard *Algyroides fitzingeri* **Wiegmann,** 1834

Fitzinger's Tree Iguana *Liolaemus fitzingerii* **Duméril** and **Bibron,** 1837

Fitzinger's False Coral Snake *Oxyrhopus fitzingeri* **Tschudi,** 1845

Fitzinger's Ground Skink *Leiolopisma eulepis* **Fitzinger,** 1853

Fitzinger's Coral Snake *Micrurus tener fitzingeri* **Jan,** 1858

Fitzinger's Blind Snake *Leptotyphlops fitzingeri* Jan, 1861

Leopold Joseph Franz Johann Fitzinger (1802–1884) was a zoologist who had a considerable influence on herpetology; he created 70 of the genus names in use today. He read botany at Universität Wien and worked at Naturhistorisches Museum Wien (1817–1861). He became Director of the Munich and Budapest zoos (1861). He wrote *Neue Classification der Reptilien* (1826) and *Systema reptilium* (1843).

FitzSimons, F. W.

FitzSimons' Whip Lizard *Tetradactylus africanus fitzsimonsi* **Hewitt,** 1915

Frederick William FitzSimons (1871–1951) was Director, Port Elizabeth Museum and Snake Park, and father of Dr. Vivian Frederick Maynard FitzSimons (q.v.). He was a dynamic personality, appointed (1906) to run a "sleepy" museum, and quickly energized it and the local inhabitants. He wrote *Snakes* (1932).

FitzSimons, V.

FitzSimons' Burrowing Skink *Typhlacontias brevipes* FitzSimons, 1938

FitzSimons' Leaf-toed Gecko *Goggia microlepidota* FitzSimons, 1939

FitzSimons' Blind Legless Skink *Typhlosaurus gariepensis* FitzSimons, 1941

FitzSimons' Flat Lizard *Platysaurus orientalis fitzsimonsi* **Loveridge,** 1944

Kalahari Garter Snake *Elapsoidea sundevallii fitzsimonsi* Loveridge, 1944

FitzSimons' Thick-toed Gecko *Pachydactylus fitzsimonsi* Loveridge, 1947 [Alt. Button-scaled Gecko; Syn. *Chondrodactylus fitzsimonsi*]

FitzSimons' Dwarf Gecko *Lygodactylus bernardi* FitzSimons, 1958 [Alt. Bernard's Dwarf Gecko]

FitzSimons' Dwarf Burrowing Skink *Scelotes fitzsimonsi* **Broadley,** 1994

Dr. Vivian Frederick Maynard FitzSimons (1901–1975) was a herpetologist whose father was Frederick William FitzSimons (q.v.). His bachelor's degree (1921) and master's (1923) were awarded by Rhodes University. He worked at the Transvaal Museum (1924–1966), first as Senior Assistant, Zoology, then as Curator, Department of Lower Vertebrates and Invertebrates, and Director (1946). His research was mainly on lizards and snakes. He spent 1930 on an expedition to Bechuanaland (Botswana).

Fletcher

Fletcher's Blind Snake *Typhlops fletcheri* **Wall,** 1919

F. W. F. Fletcher was a planter in the Wynaad area, near the Nilgiri Hills, southern India. He wrote *Sport on the Nilgiris and in Wynaad* (1911).

Flower

Flower's Worm Snake *Typhlops floweri* **Boulenger,** 1899

Thai False Bloodsucker *Pseudocalotes floweri* Boulenger, 1912

Flower's Tortoise *Testudo graeca floweri* Bodenheimer, 1935

Captain Stanley Smyth Flower (1871–1946) was Director, Cairo Zoological Gardens, Giza, Egypt (1898–1924). He had previously spent two years as Scientific Adviser to the Siamese government. Flower visited the zoo at Madras (Chennai) as an adviser (1913) and described many zoos of the time. He wrote *Zoological Gardens of the World* (2 vols., 1908–1914). Three mammals are named after him.

Fonseca

Fonseca's Lancehead *Bothrops fonsecai* **Hoge** and **Belluomini,** 1959

Dr. Flavio da Fonseca was Director, Parasitology Laboratory, Instituto Butantan, São Paulo, and Professor, School of Medicine, Universidade de São Paulo. He was an expert on the venom of *Bothrops* snakes. He wrote *Animais peconhentos* (1949).

Forbes, D. M.

Forbes' Graceful Brown Snake *Rhadinaea forbesi* **H. M. Smith,** 1942

See **Forbes, Mr. and Mrs.**

Forbes, H. O.

Forbes' Kukri Snake *Oligodon forbesi* **Boulenger,** 1883

Forbes' Forest Skink *Sphenomorphus forbesi* Boulenger, 1888

Socotra Leaf-toed Gecko *Hemidactylus forbesii* Boulenger, 1899

Henry Ogg Forbes (1851–1932) was an explorer and

collector, who on one expedition retraced Wallace's footsteps in the Moluccas. After a number of ill-fated expeditions in New Guinea, Forbes was appointed meteorological observer, Port Moresby. He was Director, Canterbury Museum, New Zealand (1890–1893). He also worked at Liverpool Museum and was editor of its *Bulletin*, in which Boulenger published the description of the gecko. Forbes wrote *Naturalist's Wanderings in the Eastern Archipelago* (1885).

Forbes, Mr. and Mrs.

Forbes' Forest Ground Skink *Scincella forbesorum* **Taylor,** 1937

Forbes' Anole *Anolis forbesi* Smith and van Gelder, 1955

Mr. and Mrs. Dyfrig McHattie Forbes lived near Vera Cruz, Mexico. Visiting herpetologists were always welcome at their house; Hobart Smith and his wife, Rozella, stayed with the Forbes (who were planters) while on honeymoon. Mr. Forbes was an amateur herpetologist who had discovered a new species of salamander that was adapted to life in saline water. See also **Forbes, D. M.**

Forcart

Dwarf/Reed Snake sp. *Calamaria forcarti* **Inger** and **Marx,** 1965

Dr. Lothar Forcart (1902–1990) was a malacologist at Naturhistorisches Museum Basel. He wrote *Nomenclature Remarks on Some Generic Names of the Snake Family Boidae* (1951).

Ford, G. H.

Ford's Boa *Epicrates fordi* **Günther,** 1861

George Henry Ford (1809–1876) was South African–born. Sir Andrew Smith employed him to make drawings and paintings of specimens he collected, and Ford was also employed by the Cape Town Museum (1825). Ford followed Smith when he returned to London (1837). He was employed as an artist at the British Museum, where he stayed for the rest of his life. Günther says, "I have named it after Mr. Ford, whose merits in herpetology are well known by his truly artistical drawings."

Ford, H. A.

West African Striped Lizard *Poromera fordii* **Hallowell,** 1857

Dr. Henry A. Ford was a medical missionary in Gabon, where he made a collection of natural history specimens that he presented to the Philadelphia Academy of Natural Sciences (1851). The collection included the first gorilla skeleton to be seen in the USA.

Ford, J. R.

Mallee Dragon *Ctenophorus fordi* **Storr,** 1965

Dr. Julian Ralph Ford (1932–1987), an Honorary Associate, Western Australian Museum, was an ornithologist and herpetologist. He received degrees in chemistry and zoology (1953) and a doctorate (1984) from the University of Western Australia. He worked for Shell as an industrial chemist (1954–1960), then switched to lecturing and was a Senior Lecturer in Chemistry, Curtin University, Western Australia (1987). He collected the lizard holotype and many of the paratypes. He co-wrote *Northern Extension of the Known Range of the Brush Bronzewing* (1959).

Fornasini

Fornasini's Blind Snake *Typhlops fornasinii* Bianconi, 1847

Cavaliere Carlo Antonio Fornasini (b. 1805) was a collector who operated in the area around Inhambane, Mozambique, from 1839. He left Italy for unknown reasons (probably political), traveling to Portugal and thence to Mozambique. He mainly collected spiders and botanical specimens, including the first example of the cycad that Bertolini named *Encephalartos ferox* (1851).

Forrer

Forrer's Parrot Snake *Leptophis diplotropis forreri* **H. M. Smith,** 1943

Alphonse Forrer (1836–1899) was born in England but emigrated to the USA. He joined the Union army at the outbreak of the American Civil War (1861). After 1865 he was employed by the British Museum to collect zoological specimens in western Mexico and the USA, and was still collecting in Mexico in the 1880s. He also supplied specimens to other museums and in that connection made four trips to Europe. Other taxa including an amphibian are named after him.

Forskal

Forskal's Sand Snake *Psammophis schokari* Forsskål, 1775

Peter Forsskål (1736–1763) was a Swedish traveler and naturalist who was at Uppsala Universitet (1751) and studied at Georg-August-Universität Göttingen (1756). He was Professor of Natural History, Københavns Universitet (1760). He was a member of the Danish expedition to Arabia (1761–1763) and during that expedition died of the plague in Yemen. His journals and notes were published posthumously (1775).

Forsten

Forsten's Tortoise *Indotestudo forstenii* **Schlegel** and **Müller,** 1844

Forsten's Cat Snake *Boiga forsteni* **Duméril, Bibron,** and **Duméril,** 1854

Forsten's Pointed Snake *Rabdion forsteni* Duméril, Bibron, and Duméril, 1854

Eltio Alegondas Forsten (1811–1843) collected in the East

Indies (1838–1843). He was primarily a botanist and was interested in the pharmaceutical properties of plants. Three birds are named after him. He wrote *Dissertatio botanico-pharmaceutico-medica inauguralis de cedrela febrifuga* (1836).

Forster

Forster's Tree Iguana *Liolaemus forsteri* **Laurent,** 1982

Dr. Walter Forster (1910–1986) was an entomologist who worked at Zoologische Staatssammlung München (1939–1975), retiring as Director. He led two extensive expeditions to Bolivia and elsewhere in South America (1949–1954).

Forsyth

Forsyth's Toadhead Agama *Phrynocephalus forsythii* **Anderson,** 1872

Sir Thomas Douglas Forsyth (1827–1886) originally went to India with the Honourable East India Company. He was sent to Yarkand (1870) to visit Yakub Beg, then ruler of independent Chinese Turkestan. He failed to meet him so he made another expedition, the famous Second Yarkand Expedition (1873) in which Dr. Ferdinand Stoliczka (1838–1874) played an important part, dying of spinal meningitis on the return journey. Forsyth was instrumental in preventing a war between British India and Burma (Myanmar) (1875). He wrote *Report of a Mission to Yarkund in 1873* (1875). This agama was collected on the first expedition.

Fouda

Elba Gecko *Hemidactylus foudaii* **Baha el Din,** 2003

Dr. Moustafa Mokhtar Fouda (b. 1950) is Director, Nature Conservation Sector, Egyptian Environmental Affairs Agency. He and Baha el Din work closely together on conservation issues; in the original description Baha el Din describes Fouda as a "colleague and friend."

Fowler

Fowler's Galliwasp *Celestus fowleri* **Schwartz,** 1971
Fowler's Anole *Anolis fowleri* Schwartz 1973
Andros Island Boa *Epicrates striatus fowleri* **Sheplan** and Schwartz, 1974

Danny C. Fowler often worked with Schwartz; they co-wrote "The Anura of Jamaica: A Progress Report," appearing in *Studies on the Fauna of Curacao and Other Caribbean Islands* (1973). An amphibian is named after him.

Fox, S.

Fox's Lizard *Liolaemus foxi* **Núñez, Navarro,** and **Veloso,** 2001

Dr. Stanley Forrest Fox (b. 1946) joined the faculty at Oklahoma State University (1977) and has been Regents Professor of Zoology and Curator of Herpetology since 1992. After a master's degree at University of Illinois, he was awarded a doctorate by Yale (1973). His major interest is lizards, especially those in Mexico and Chile. He co-wrote "Field Use of Sprint Speed by Collared Lizards (*Crotaphytus collaris*): Compensation and Sexual Selection" (2006).

Fox, W.

Fox's Mountain Meadow Snake *Adelophis foxi* **Rossman** and Blaney, 1968

Dr. Wade Fox Jr. (1920–1964) was a zoologist and herpetologist at the University of California, Berkeley. He was on the editorial board of *Copeia* and President of the Herpetologists' League when he had a fatal heart attack. The etymology states, "In recognition of the major contributions of the late Wade Fox to the systematics of the genus *Thamnophis*, we are pleased to name this garter snake ally in his honor." He wrote "Relationships of the Garter Snake, *Thamnophis ordinoides*" (1948).

Fraas

Fraas' Lizard *Lacerta fraasii* Lehrs, 1910 [Syn. *Parvilacerta fraasii*]

Dr. Eberhard Fraas (1862–1915) was a geologist and paleontologist. He studied at the universities of Leipzig and Munich. He worked at Staatlisches Museum für Naturkunde, Stuttgart (1891–1915), starting as an Assistant at the Mineral Collection, being promoted to Curator of Geology, Paleontology, and Mineralogy (1894). He visited Spain, Sardinia, Italy, the Balkans, and the western part of North America. He went to German East Africa (1907), where he discovered fossils of Jurassic dinosaurs. He was regarded as being at the height of his powers when he died of "extreme debilitation caused by dysentery that he had caught in East Africa."

Francisco Paiva

Ground Snake sp. *Atractus franciscopaivai* Silva Haad, 2004

Dr. Francisco Paiva do Nascimento is a herpetologist at the Museu Paraense Emilio Goeldi, Belém, Brazil.

Franco

Ground snake sp. *Atractus francoi* Passos et al., 2010

Dr. Francisco Luis Franco is Curator of Herpetology, Instituto Butantan, Brazil, where a disastrous fire (2010) destroyed most of the collection. Universidade Estadual Paulista Júlio de Mesquita Filho awarded his bachelor's degree (1987), Pontifícia Universidade Católica do Rio Grande do Sul his master's (1994), and Universidade de São Paulo his doctorate (1999). He drew the authors' attention to this previously undescribed species.

Fraser, C.

Fraser's Scalyfoot *Delma fraseri* **Gray,** 1831

Charles Fraser (1788–1831) was a botanist and gardener. He joined the British army (1815) and was sent to Australia as part of a convict guard detachment, arriving in Sydney (1816), but his skills as a horticulturalist were quickly recognized and he was appointed Superintendent of Royal Botanic Gardens. He was discharged from the army (1821) and was appointed Colonial Botanist. He traveled widely as a field collector, being a member of three of Oxley's expeditions (1817–1819), and later visited Tasmania, New Zealand, and Norfolk Island. He accompanied Stirling on his expedition to the Swan River, Western Australia (1827), and was heavily blamed for writing a report that overstated the quality of the soil in the area. He was sent by the Governor to collect plants and establish a public garden in Brisbane (1828).

Fraser, L.

Fraser's Anole *Anolis fraseri* **Günther,** 1859

Fraser's Ground Snake *Liophis fraseri* **Boulenger,** 1894 [Syn. *L. epinephelus fraseri*]

Centipede Snake sp. *Tantilla fraseri* Günther, 1895

Louis Fraser (1810–1866) was a zoologist, collector, curator, explorer, zookeeper, consul, author, dealer, and taxidermist. He was employed at the museum of the Zoological Society, London (1832–1841 and 1842–1846), first as an office boy, then as Clerk, Assistant Curator, and Curator. He collected in West Africa (1841–1842), being official naturalist on the Niger expedition. He took charge of Lord Derby's zoological collections (1848–1850). He was Vice Consul at Whydah (Ouida), and collected in the Bights of Benin (Nigeria) (1851–1853), Ecuador (1857–1859), and Guatemala and California (1860). He tried to establish himself as a natural history dealer, opening a shop in London to sell exotic birds. This venture appears not to have been a success, as he returned to the USA and died there. He published *Zoologica Typica* (14 parts, 1845–1849). Seven birds and a mammal are named after him.

Fred Parker

Fred Parker's Blind Snake *Typhlops fredparkeri* **Wallach,** 1996

Fred Parker's White-lipped Python *Leiopython fredparkeri* Schleip, 2008

See also **Parker, F.**

Freiberg

Freiberg's Blind Snake *Leptotyphlops australis* Freiberg and Orejas-Miranda, 1968

Freiberg's Iguana *Stenocerus doellojuradoi* Freiberg, 1944

Dr. Marcos Abraham Freiberg (1911–1990) was a herpetologist and ichthyologist at Museo Argentino de Ciencias Naturales "Bernardino Rivadavia," Buenos Aires. He wrote *Turtles of South America* (1981).

Freitas

Burrowing Snake sp. *Apostolepis freitasi* **De Lema,** 2004

Marco Antonio de Freitas is an amateur herpetologist who collaborates with De Lema, who wrote, "The name is a homage to Marco Antonio de Freitas, devoted amateur herpetologist and efficient collaborator to my studies." Freitas wrote *Serpentes da Bahia e do Brasil* (1999).

Freminville

Freminville's Scorpion-eating Snake *Stenorrhina freminvillei* **Duméril, Bibron,** and **Duméril,** 1854

Christophe-Paulin de la Poix Chevalier de Fréminville (1787–1848) was a naval officer and naturalist. He was on board *La Syrène*, which in 1806 attempted to discover the Northwest Passage (1806). He was an expert of the history and archeology of the late Middle Ages and on the history of Brittany, and of the Templars in particular. Toward the end of his life an old episode affected him and he became deranged. He was in command of the French frigate *La Néréide* in the West Indies (1822). He fell from some rocks and was lucky enough to be rescued from drowning and nursed back to health by a beautiful local girl, Caroline. He had to sail to Martinique but when he returned, he found she had drowned herself, thinking that he had deserted her. He took away some of her dresses as keepsakes (1842) and spent the last six years of his life wearing her old clothes.

Frere

Major Skink *Egernia frerei* **Günther,** 1897

Stout Bar-sided Skink *Eulamprus frerei* **Greer,** 1992

Named after Mount Bartle Frere in Queensland.

Fritts

Fritts' Whorltail Iguana *Stenocercus apurimacus* Fritts, 1972

Dr. Thomas Harold Fritts (b. 1945) is a wildlife biologist whose bachelor's and master's degrees in zoology were awarded by the University of Illinois. He started his doctorate at the University of Kansas (1968). He taught at St. Edward's University and became Curator of Amphibians and Reptiles, San Diego Natural History Museum (1976). He joined the Fish and Wildlife Section of the Biological Survey, New Orleans (1982), and transferred to Washington, DC (1984), becoming Section Chief (1988). He moved to the Mid-continent Ecology Science Center (1988). He was President, Charles Darwin Foundation for the Galapagos Islands (2002). He has made a particular study of the problems caused by Brown Tree Snakes.

Frost, C.

Frost's Lerista *Lerista frosti* Zietz, 1920 [Alt. Centralian Slider]

Charles Frost (d. 1915) was an Australian naturalist. In 1895 he co-authored the description of a new species of skink, which he called *Rhodona tetradactyla*. This name had already been taken, so Zietz renamed it after Frost. An amphibian is named after him.

Frost, D.

Frost's Iguana *Microlophus koepckeorum* **Mertens,** 1956

Frost's Arboreal Alligator Lizard *Abronia frosti* **Campbell** et al., 1998

Dr. Darrell Richmond Frost (b. 1951) is a Curator and Associate Dean of Science at the American Museum of Natural History. He is a specialist on amphibian and lizard systematics. He was President of the Society for the Study of Amphibians and Reptiles (1998). We were told that Mertens decided to give the iguana its vernacular name "Frost's" as he had once met Frost at a reptile conference.

Fruhstorfer

Javan Bluebelly Snake *Tetralepis fruhstorferi* **Boettger,** 1892

Hans Fruhstorfer (1866–1922) was a German professional collector. He collected butterflies, but his descriptions and paperwork were hurried and sometimes inaccurate. His commercial success as a dealer in butterflies enabled him to take an early retirement and enjoy lifelong financial independence—but he died young after an unsuccessful cancer operation. His butterfly collection is scattered among a number of museums, including the Natural History Museums in Paris and London. He collected the snake holotype.

Fry

Fry's Lerista *Lerista picturata* Fry, 1914

Dene Barrett Fry (1893–1917) worked at New South Wales University as a taxonomist and collector (1912–1915). He died in France during WW1. He wrote "On a Collection of Reptiles and Batrachians from Western Australia" (1914).

Fuelleborn

Flapjack Chameleon *Chamaeleo fuelleborni* **Tornier,** 1900

Dr. Friederich Fülleborn (1866–1933) was a physician who worked in Tanganyika (Tanzania) (1896–1900). He became a Professor at Hamburg University and was an expert on tropical diseases. Among other taxa named after him are four birds.

Fugler

Pichincha Snake *Emmochliophis fugleri* **Fritts** and **H. M. Smith,** 1969 [Alt. Pinchinda Snake, apparently a misspelling of Pichincha]

Professor Dr. Charles M. Fugler (1929–1999) taught biology at the University of North Carolina (1990). Louisiana State University, Museum of Natural Science, has a Charles M. Fugler Fellowship in Tropical Vertebrate Biology. George Key collected the holotype, and Fugler presented it to the University of Illinois Museum of Natural History. He wrote "Biological Notes on *Rana tigrina* [*sic*] in Bangladesh and Preliminary Bibliography" (1984).

Fuhn

Fuhn's Five-toed Skink *Leptosiaphos fuhni* **Perret,** 1973

Fuhn's Snake-eyed Skink *Cryptoblepharus fuhni* **Covacevich** and **Ingram,** 1978

Dr. Ion Eduard Fuhn (1916–1987) was a Romanian herpetologist. Originally interested in birds, he developed an interest in herpetology under the influence of the German herpetologist Wolterstorff. He first studied philosophy, completing a doctorate (1946). He then worked at Ministry for Foreign Affairs but rather lost interest, particularly when his family property was nationalized by the Socialist regime. Fuhn joined the Academy of Science, directing the sections of Herpetology and Arachnology (1954–1976).

Fuller

Lake Disappointment Ground Gecko *Diplodactylus fulleri* **Storr,** 1978

Phillip John Fuller is an ornithologist who has researched and published on Lake Disappointment. He works for the Western Australian Department of Environment and Conservation. He wrote "Breeding of Aquatic Birds in Mid-western Australia" (1963). Like Storr, he was a member of the Western Australian Naturalists Club.

Funaioli

Archer's Post Gecko *Hemidactylus funaiolii* **Lanza,** 1978 [Alt. Kenya Leaf-toed Gecko]

Dr. Ugo Funaioli is a zoologist with an interest in taxidermy. He is associated with the museum attached to Instituto Statale della Ss. Annunziata, Florence, where he has been helping to restore the mounted specimens on display. He and Lanza made several expeditions together to Somalia (1959–1970). He wrote *Fauna and Caccia in Somalia* (1957).

G

Gabo

Burrowing Snake sp. *Apostolepis gaboi* **Rodrigues,** 1993

"Gabo" is Rodrigues' nickname for Dr. Gabriel Omar Skuk Sugliano, a zoologist and biologist, who received his bachelor's degree (1986) from the Faculty of Sciences, Universidad de la República, Uruguay, and his master's (1994) and doctorate (1999) from Universidade de São Paulo. He is Adjunct Professor, Universidade Federal de Alagoas, Brazil, where he specializes in herpetological taxonomy. He co-wrote "Description of the Tadpole of *Hylomantis granulosa* (Anura: Hylidae)" (2007). He collected the snake holotype.

Gabrielle

Gabrielle's Gecko *Paragehyra gabriellae* Nussbaum and **Raxworthy,** 1994

Gabriellà Raharimanana is a Malagasy herpetologist who assisted in the fieldwork and the preparation of specimens. She participated in the field surveys carried out by the authors over several years.

Gadow, C. M.

Gadow's Spiny Lizard *Sceloporus gadoviae* **Boulenger,** 1905

Clara Maud Gadow, née Paget, was the wife of Dr. Hans Friedrich Gadow (q.v.).

Gadow, H. F.

Gadow's Anole *Anolis gadovii* **Boulenger,** 1905

Gadow's Alligator Lizard *Mesaspis gadovii* Boulenger, 1913

Gadow's Tree Lizard *Urosaurus gadovi* **Schmidt,** 1921

Dr. Hans Friedrich Gadow (1855–1928) was a German zoologist. His main contribution to zoology was to devise a method of taxonomy based on comparisons of 40 characteristics in birds. Dr. Albert Carl Günther of the Natural History Museum, London, encouraged him. Gadow became Curator, Strickland Collection, Cambridge (1884), and also lectured on the morphology of vertebrates. He co-wrote *A Dictionary of Birds* (1893–1896). A bird is named after him.

Gaerdes

Mayer's Sand Lizard *Pedioplanis gaerdesi* **Mertens,** 1954

Jan Gaerdes (1889–1981) was a pharmacist who first went to German South-West Africa (Namibia) (1913). He served in the German army there during WW1. He leased a farm at Kalidona (1925), finally buying it outright (1940). He was a keen hunter and zoologist and wrote articles on the status of game in Namibia. In the 1950s he sent a number of specimens to Mertens at Naturmuseum Senckenberg, Frankfurt.

Gaige

Skyros Lizard *Podarcis gaigeae* **Werner,** 1930

Gaige's Dwarf Gecko *Sphaerodactylus gaigeae* **Grant,** 1932 [Alt. Gaige's Least Gecko, Chevronated Sphaero]

Many-lined Skink *Eumeces multivirgatus gaigeae* **Taylor,** 1935

Gaige's Tropical Night Lizard *Lepidophyma gaigeae* Mosauer, 1936

Gaige's Pine Forest Snake *Rhadinaea gaigeae* **Bailey,** 1937

Gaige's Thirst Snake *Dipsas gaigeae* **Oliver,** 1937

Peruvian Rainbow Boa *Epicrates cenchria gaigeae* Stull, 1938

Big Bend Slider *Trachemys gaigeae* **Hartweg,** 1939

Gaige's Spiny Lizard *Sceloporus lundelli gaigeae* **H. M. Smith,** 1939

Gaige's Ground Snake *Atractus gaigeae* **Savage,** 1955

Helen Beulah Thompson Gaige (1890–1976) was a herpetologist who specialized in neotropical frogs. She studied at the University of Michigan, where she was Assistant Curator, Museum of Zoology (1910), Curator, Amphibians (1919), and Curator, Herpetology (1944). She became editor-in-chief of *Copeia* (1937). She married Frederick McMahon Gaige, an entomologist and onetime Director of the museum. In their honor the American Society of Ichthyologists and Herpetologists makes an annual award to a graduate student of herpetology. She co-wrote *The Herpetology of Michigan* (1928).

Gaimard

Colubrid snake sp. *Stenophis gaimardi* **Schlegel,** 1837

Dr. Joseph (or Jean, according to some sources) Paul Gaimard (1796–1858) was a French naval surgeon, explorer, and naturalist. He was on board *Uranie* (1817–1819) during its circumnavigation, interrupted by shipwreck on the Falklands, and continued the journey on board *Physicienne,* the ship that had rescued the expedition and then been purchased by them as a replacement. He was with Dumont D'Urville aboard the *Astrolabe* when they visited New Zealand (1826) and led the *Récherche* expedition to northern Europe, visiting Iceland, the Faeroe Islands, northern Norway, Archangel, and Spitsbergen (1838–1840). He was something of a dandy and, when visiting Iceland, handed out sketches of himself. A mammal, a bird, and a fish are named after him.

Galan

Galan's Rock Lizard *Iberolacerta galani* Arribas, Carranza, and Odierna, 2006

Dr. Pedro Galan Regalado is a herpetologist who is Professor of Zoology, Universidade da Coruña, Spain. He wrote *Anfibios e réptiles de Galicia* (2003). An amphibian is named after him.

Gallagher, D. S.

Gallagher's Kentropyx *Kentropyx vanzoi* **Gallagher** and **Dixon,** 1980

Daniel Stephen Gallagher Jr. was head of the Biology Department and Associate Professor of Biology, Howard Payne University, Texas (1987). He is now at the Department of Animal Science and Veterinary Pathobiology, Texas A&M University. He wrote "A Systematic Revision of the South American Lizard Genus *Kentropyx*" (1979).

Gallagher, M. D.

Gallagher's Gecko *Asaccus gallagheri* **Arnold,** 1972

Major Michael Desmond Gallagher (b. 1921) is a zoologist who lived in Oman (1977–1998). He was Curator, Oman Natural History Museum, from 1985. He was a member of the Zaire River expedition (1974–1975). He is a member of Royal Geographical Society, London. He wrote *Snakes of the Arabian Gulf and Oman* (1993). A bat is named after him.

Gallardo

Tree Iguana sp. *Liolaemus gallardoi* **Cei** and Scolaro, 1982

Para-anole sp. *Urostrophus gallardoi* **Etheridge** and **Williams,** 1991

José María Alfonso Felix Gallardo (1925–1994) was Director, Museo Argentino de Ciencias Naturales "Bernardino Rivadavia," Buenos Aires—like his grandfather Angel Gallardo (1867–1934) before him. He wrote *Anfibios y reptiles* (1994). A bird is named after him.

Gallot

Gallot's Lizard *Gallotia galloti* Oudart, 1839

D. Gallot was an amateur naturalist in the Canary Islands. He sent a lizard specimen to Muséum National d'Histoire Naturelle, Paris (1839), which was named in his honor as *Lacerta galloti*. Edwin Arnold (q.v.) later created a new genus, *Gallotia*, for this species and its relatives.

Gambel

Leopard Lizard genus *Gambelia* **Baird,** 1859

Dr. William Gambel (1821–1849) was a naturalist and collector, and the first ornithologist to spend any time in California. He started traveling after qualifying as a physician. He broke the first rule of natural history nomenclature etiquette—that is, Thou shall not name a species you discover after yourself. While riding the Santa Fe Trail (1842), Gambel collected a specimen and sent it to a museum, labeled "Gambel's Quail," apparently believing that Thomas Nuttall had already so named it. Gambel was mistaken, but the name became official nonetheless. He was Assistant Curator, Natural (now National) Academy of Sciences. He died of typhoid while attempting to cross the Sierra Nevada Mountains in midwinter, with a party on its way to the goldfields of California. Four birds are named after him.

Gammie

Gammie's Wolf Snake *Dinodon gammiei* **Blanford,** 1878

James Alexander Gammie (1839–1924) was a plant collector, naturalist, and plantation manager in Sikkim. He was in charge of Mungpu cinchona plantations (1865–1897).

Gane

Blind Snake sp. *Ramphotyphlops ganei* Aplin, 1998 [Syn. *Austrotyphlops ganei*]

Lori Gane is a collector and amateur herpetologist who collected the snake holotype (1991). He was a teacher in Pannawonica, Western Australia.

Gans

Amphisbaena sp. *Cynisca gansi* **Dunger,** 1968

Gans' Mabuya *Eutropis gansi* **Das,** 1991

Gans' Tree Skink *Lankascincus gansi* **Greer,** 1991

Worm Lizard sp. *Amphisbaena carlgansi* **Thomas** and Hedges, 1998

Gans' Gecko *Cyrtodactylus gansi* **Bauer,** 2003

Gans' Egg-eating Snake *Dasypeltis gansi* **Trape** and Mané, 2006

Dr. Carl Gans (1923–2009) was a zoologist who was born in Germany and emigrated to the USA (1939). He worked as an engineer (1947–1955). He was a Fellow in Biology (1957–1958), University of Florida. He taught biology at the State University of New York, Buffalo (1958–1971), then became Professor of Biology, University of Michigan. See also **Cege.**

Gardiner

Rotuman Forest Gecko *Lepidodactylus gardineri* **Boulenger,** 1897

Gardiner's Skink *Pamelaescincus gardineri* Boulenger, 1909

John Stanley Gardiner (1872–1946) was a British zoologist and oceanographer. He was Professor of Zoology and Comparative Anatomy, Cambridge (1909–1937). He traveled in the Pacific and Indian oceans, visiting the

Maldives in 1899 and Fiji sometime earlier. He wrote *The Natives of Rotuma* (1898). See **Stanley, G.**

Garman

Garman's Sea Snake *Hydrophis semperi* Garman, 1881 [Alt. Luzon Sea Snake]

Northern Prairie Lizard *Sceloporus undulatus garmani* **Boulenger,** 1882

Garman's Galliwasp *Celestus crusculus* Garman, 1887

Jamaican Giant Anole *Anolis garmani* **Stejneger,** 1899

Dr. Samuel Walton Garman (1843–1927) graduated in Illinois (1870), became a schoolteacher, and was Professor of Natural Science at a seminary in Illinois (1871–1872). He became Louis Agassiz's special student (1872) and from 1873 onward worked in the Herpetology and Ichthyology Section, Museum of Comparative Zoology, Harvard. He was in South America with Alexander Agassiz (1874) and surveyed Lake Titicaca. He wrote *Reptiles of Easter Island* (1908).

Garnier

Garnier's Skink *Phoboscincus garnieri* **Bavay,** 1869

Jules, sometimes "Jacques," Garnier (1839–1904) was a mining engineer, geologist, and explorer who discovered nickel ore in New Caledonia (1864). He visited England (1870) on a secret mission during which he bought three tons of gunpowder for French forces to use in the Franco-Prussian War. Later in his life he traveled in Canada.

Garnot

Garnot's House Gecko *Hemidactylus garnotii* **Duméril** and **Bibron,** 1836 [Alt. Indo-Pacific Gecko]

Prosper Garnot (1794–1838) was a French naval surgeon, naturalist, and collector working closely with Lesson. They were both on board *La Coquille* during its circumnavigation (1822–1825), and they co-authored the zoological section of *Voyage autour du monde exécuté par order du roi sur la corvette La Coquille pendant les années 1822–1825* (1828–1832). A bird is named after him.

Garrido

Garrido's Anole *Anolis pumilus* Garrido, 1988

Garrido's False Chameleon *Chamaeleolis barbatus* Garrido, 1982 [Alt. Garrido's Crested Anole; Syn. *Anolis barbatus*]

Escambray Twig Anole *Anolis garridoi* Diaz, Estrada and **Moreno,** 1996

Galliwasp sp. *Diploglossus garridoi* **Thomas** and Hedges, 1998

Orlando H. Garrido (b. 1931) is an ornithologist and Curator, Zoology Department, Museo Nacional de Historia Natural, Havana, Cuba. He has written books that are mainly aimed at young people to awaken their interest in natural history. He co-wrote *Field Guide to the Birds of Cuba* (2000). Two mammals are named after him. See **Lando.**

Garth

Pit-viper genus *Garthius* Malhotra and Thorpe, 2004

See **Underwood.**

Gasca

Colubrid snake sp. *Pseudoboodon gascae* **Peracca,** 1897

Captain A. Gasca was an Italian officer who collected reptiles in Eritrea.

Gascon

Clawed Gecko sp. *Pseudogonatodes gasconi* Avila-Pires and **Hoogmoed,** 2000

Dr. Claude Gascon is a Canadian ecologist who works closely with Avila-Pires and Hoogmoed in studies that concentrate on Amazon forests. He took his bachelor's degree at the University of Quebec, Montreal, and his doctorate at Florida State University. He is a senior member in the management of Conservation International. He is Visiting Professor, Department of Ecology, National Amazon Research Institute, and a Research Associate at the Smithsonian. He co-wrote *Lessons from Amazonia: The Ecology and Conservation of a Fragmented Forest* (2001).

Gasconi

African Helmeted Turtle *Pelomedusa gasconi* Rochebrune, 1884 [Junior syn. of *P. subrufa* Lacépède, 1788]

Alfred Gasconi (1842–1929) was born in the island of Saint-Louis and was the deputy for Senegal—where he had spent most of his life—in the parliament in Paris (1879–1889). He gave the turtle holotype to Rochebrune.

Gasperetti

Gasperetti's Horned Viper *Cerastes gasperettii* **Leviton** and **Anderson,** 1967

Leviton's Awl-headed Snake *Lytorhynchus gasperetti* Leviton, 1977 [Alt. Leviton's Leafnose Snake]

Wadi Kharrar Rock Gecko *Pristurus gasperetti* **Arnold,** 1986

John Gasperetti (1920–2001) had substandard eyesight and so was rejected for military service by the U.S. armed forces but instead was sent to Saudi Arabia as a surveyor and engineer for the oil company Aramco. He became a passionate devotee of Saudi Arabian wildlife in general and its reptiles in particular. He worked on a canal project in Afghanistan (1948–1949) but was soon back in the

Middle East with Getty Oil in the Saudi/Kuwaiti Neutral Zone. He was in Iraq (1955) but had to leave when a revolution broke out. He worked for the Ministry of Petroleum and Mineral Resources, Saudi Arabia, and for various engineering companies (1961–1977). He retired from surveying (1986) and joined the Saudi Meteorology and Environmental Protection Administration. He was disabled by a stroke (1996), which enforced his complete retirement (1997). He and his wife, Patricia, also a herpetologist, settled in Egypt. He never returned to the USA.

Gaudichaud

Gaudichaud's Ecpleopus *Ecpleopus gaudichaudii* **Duméril** and **Bibron,** 1839

Chilean Marked Gecko *Homonota gaudichaudii* Duméril and Bibron, 1836

Charles Gaudichaud-Beaupré (1789–1854) studied pharmacy and became a dispenser in the French navy (1810). Later he took part in several large expeditions as a naturalist: first around the world aboard the *Uranie* and *Physicienne* (1817–1820), to South America with the *Herminie* (1831–1833), and a circumnavigation on *Bonite* (1836–1837). Later Gaudichaud was appointed Professor in Pharmacy, attached to Muséum National d'Histoire Naturelle, Paris, and worked on the botanical collections from his expeditions. A bird is named after him.

Geay

Madagascan Speckled Hognose Snake *Leioheterodon geayi* **Mocquard,** 1905

Martin François Geay (1859–1910) was a pharmacist, natural history collector, and traveler. He led an expedition for the Ichthyology and Herpetology Laboratory, Muséum National d'Histoire Naturelle, Paris (1889–1891). He collected plant and animal specimens in Madagascar (1907) and during an expedition to French Guiana and Venezuela (1908). He collected the snake holotype.

Gemel

Leaf Turtle sp. *Cyclemys gemeli* Fritz et al., 2008

Richard Gemel (b. 1948) has been, since 1996, Technical Collections Manager, Herpetology Section, Naturhistorisches Museum Wien. He graduated in biology and physical education at Universität Wien and taught in Viennese schools (1973–1981 and 1991–1996). He wrote *Kurioses aus der Herpetologischen Sammlung des Naturhistorischen Museums Wien* (2005).

Gemminger

Cope's Forest Ground Skink *Scincella gemmingeri* **Cope,** 1864

Max Gemminger (1820–1887) was a zoologist, entomologist, and coleopterist who was Curator, Coleoptera Collection, Zoologische Staatssammlung München (1849–1887).

Geoffroy

Geoffroy's Side-necked Turtle *Phrynops geoffroanus* **Schweigger,** 1812

Étienne Geoffroy Saint-Hilaire (1772–1844) was a naturalist. He originally trained for the Church but abandoned theology to become Professor of Zoology (1793), when Jardin du Roi was renamed Musée National d'Histoire Naturelle. He expounded the theory that all animals conform to a single plan of structure. This was strongly opposed by Cuvier, who had been his friend, and a widely publicized debate between the two took place (1830). Despite their differences, the two men did not become enemies. Geoffroy gave one of the orations at Cuvier's funeral (1832). Modern developmental biologists have confirmed some of Geoffroy's ideas. He wrote *Philosophie anatomique* (1818–1822). Fifteen mammals and five birds are named after him.

Georges

Georges' Turtle *Elseya georgesi* **Cann,** 1997 [Alt. Bellinger River Turtle]

Professor Dr. Arthur Georges (b. 1953) is a freshwater turtle expert and Director, Applied Ecology Research Group, University of Canberra, Australia. His main interests are the conservation biology of native Australian species; the ecology, evolution, and systematics of Australian herpetofauna; and temperature-dependent sex determination. He wrote *The Australian Pig-nosed Turtle* (2000).

Gerard

Gerard's Water Snake *Gerarda prevostiana* **Eydoux** and **Gervais,** 1822 [Alt. Cat-eyed Water Snake]

Eydoux and Gervais originally described the species as *Coluber prevostianus*. Gray (q.v.) coined the name *Gerarda* (1849), and we have been unable to discover whom he had in mind. There are two strong candidates in Adam Gerard and Rev. Gerard E. Smith; Gray mentions both as having sent reptile specimens to him.

Gerard, P.

Gerard's Two-headed Snake *Chilorhinophis gerardi* **Boulenger,** 1913 [Alt. Gerard's Black and Yellow Burrowing Snake]

Dr. Pol Gérard (1886–1961) was a physician, histologist, anatomist, and naturalist. He was in the Belgian Congo early in the 20th century. He worked at Institut Royal des Sciences Naturelles de Belgique, becoming its Professor of Histology and "Administrateur" (1931–1961). He also was Professor of Pathological Anatomy, Faculty of

Medicine, Université Libre de Bruxelles, and was closely associated with Muséum des Sciences Naturelles, Brussels. He collected the snake holotype.

Gerrard

Pink-tongued Skink *Hemisphaeriodon gerrardii* **Gray,** 1845

Gerrard's Lerista *Lerista gerrardi* Gray, 1864

Edward Gerrard (1810–1910) worked as an attendant in Gray's department at the British Museum (1841–1896). He was Gray's "right-hand man" and looked after the galleries and storerooms. He also preserved and registered bottled animals and compiled a catalogue of osteological specimens at the British Museum.

Gervais

Philippine Dwarf Snake *Calamaria gervaisii* **Duméril, Bibron,** and **Duméril,** 1854

François Louis Paul Gervais (1816–1879) was a zoologist, paleontologist, and anatomist. He was a student of Blainville and succeeded him as Professor of Comparative Anatomy, Muséum National d'Histoire Naturelle, Paris (1868), where he had been Assistant (1835–1845). He was Professor of Zoology and Comparative Anatomy, Faculté des Sciences de Montpellier (1845–1868), and Head of the Faculty (1856). Four mammals are named after him.

Geyr

Sahara Mastigure *Uromastyx geyri* **Müller,** 1922

Professor Dr. Hans Freiherr Geyr von Schweppenburg (1884–1963) was an ornithologist who worked in the Sahara (1913–1914). He collected the lizard holotype.

Gezira

Tiger Snake sp. *Telescopus gezirae* **Broadley,** 1994

Al Jazirah or Gezira is one of the states that make up Sudan.

Gibbons, J. R. H.

Gibbons' Emo Skink *Emoia trossula* **Brown** and **Gibbons,** 1986

Dr. John Richard Hutchinson Gibbons (1946–1986) was Associate Professor of Biology at the School of Natural Resources, University of the South Pacific, Suva, Fiji, where a scholarship fund is now named after him. He and his entire family died in a boating accident. He wrote *A Brief Environmental History of Fiji* (1985).

Gibbons, J. W.

Pascagoula Map Turtle *Graptemys gibbonsi* Lovich and **McCoy,** 1992 [Alt. Pearl River Map Turtle]

James Whitfield "Whit" Gibbons (b. 1939) took his bachelor's (1961) and master's degrees (1963) at the University of Alabama and his doctorate (1967) at Michigan State University. He has been Professor of Ecology and Senior Research Ecologist, and is currently Director of Education Outreach at the Savannah River Ecology Laboratory, where he has worked for many years. He is also honorary Curator of Herpetology, Alabama Museum of Natural History (1994).

Gibson

Gibson's Gopher Snake *Pituophis lineaticollis gibsoni* **Stuart,** 1954

Dr. Colvin L. Gibson (b. 1918) was a physician who specialized in tropical medicine. He worked at Pan-American Sanitary Bureau's onchocerciasis station, Yepocapa, and gave Stuart a collection of amphibians and reptiles taken near it. He was publisher of *Tropical Medicine and Hygiene News* (1966–1986).

Gierra

Gierra's Blind Snake *Typhlops gierrai* **Mocquard,** 1897 [Alt. Usambara Spotted Worm Snake]

A. Gierra was an expert on north and east African languages. He presented a collection of reptiles from Tanga (in modern-day Tanzania) to Muséum National d'Histoire Naturelle, Paris, and specimens from Tanga to Entomological Society of France (1895). He collected the holotype.

Gigas

Giant Blue-tongued Skink *Tiliqua gigas* **Schneider,** 1801

False Water Cobra *Hydrodynastes gigas* **Duméril, Bibron,** and **Duméril,** 1854

Cape Verde Giant Gecko *Tarentola gigas* **Bocage,** 1875

Giant Garter Snake *Thamnophis gigas* **Fitch,** 1940

Perret's Nigerian Gecko *Cnemaspis gigas* **Perret,** 1986

Ground Snake sp. *Atractus gigas* **Myers** and Schargel, 2006

Gigas was a giant in Greek mythology, the child of Uranus and Gaea. The name is applied to taxa that are giants of their kind.

Gilbert, C.

Baja California Night Lizard *Xantusia vigilis gilberti* **Van Denburgh,** 1895

Gilbert's Skink *Eumeces gilberti* Van Denburgh, 1896 [Alt. Western Red-tailed Skink]

Gilbert's Leaf-toed Gecko *Phyllodactylus gilberti* **Heller,** 1903

Dr. Charles Henry Gilbert (1859–1928) was an ichthyologist and fishery biologist, whose main area of study was the Pacific Salmon. He received his bachelor's degree (1879) from Butler University, Indiana, but moved to take his master's (1882) and doctorate (1883) at Indiana University, the first-ever doctorate awarded by that university. Baird asked David Starr Jordan to do a survey of the U.S. West Coast Fisheries (1879), and Gilbert went as Jordan's assistant on an expedition from British Columbia to Southern California. This expedition lasted a year and was the start of a 50-year-long study of Pacific fishes by Gilbert and Jordan. Gilbert taught at Indiana University (1880–1884 and 1889), at the University of Cincinnati (1885–1888), and at the newly founded Stanford (1890–1925), retiring as Emeritus Professor. He served as naturalist-in-charge on cruises of the U.S. Fish Commission's vessel *Albatross* in Alaskan waters (1880s and 1890s), Hawaii (1902), and Japan (1906).

Gilbert, J.

Gilbert's Dragon *Lophognathus gilberti* **Gray,** 1842

John Gilbert (1812–1845) was a naturalist and explorer who was the principal collector of specimens for John Gould in southwestern Australia (1840–1842). Born in London, he was employed as a taxidermist by the Zoological Society, London. He became Curator, Shropshire and North Wales Natural History Society, Shrewsbury, but the society was short of funds so his contract was terminated (1837). He left for Australia (1838), where he was speared to death by Aborigines at the Gulf of Carpentaria while he was the naturalist on Leichhardt's expedition to Port Essington. Despite all his efforts he was poorly served by Gould, who left him with insufficient funds and equipment yet had high expectations of him. Gould barely acknowledged Gilbert's huge contribution of specimens, descriptions, and detailed observations, without which Gould's seminal work on Australia would be greatly diminished. Two mammals are named after him, one of which, Gilbert's Potoroo *Potorous gilbertii*, was rediscovered in 1994, having been officially proclaimed extinct.

Gillen

Gillen's Monitor *Varanus gilleni* Lucas and Frost 1895

Francis James Gillen (1855–1912) started work as a postal messenger (1867). He moved to Adelaide (1871) to work as a telegraph operator and attended evening classes at the School of Mines and Industries. He worked on the overland telegraph line (1875–1892) and was appointed telegraph and postmaster at Alice Springs, where he helped the Horn expedition (1894). He became very interested in ethnography and collected artifacts from the local tribes; but being a gambler, particularly on goldfields, he lost nearly everything. He was forced to sell his collection of ethnographic items to the National Museum of Victoria (1899). He became a special magistrate and protector of Aboriginals and was fearless in upholding the native peoples' rights, even charging a policeman with the murder of an Aboriginal. He met Sir Walter Baldwin Spencer in Alice Springs (1894 and 1896). They co-wrote *The Native Tribes of Central Australia* (1899) and set out to cross the continent; this endeavor, the Spencer-Gillen expedition (1901–1902), aroused considerable public interest. His last years were spent in an invalid's chair.

Girard, A.

Girard's Green Snake *Philothamnus girardi* **Bocage,** 1893

Alberto Arthur Alexandre Girard (1860–1914) was a French-Portuguese zoologist at Museu Bocage, Lisbon. Principally interested in marine zoology, he spent time aboard the trawler *Machado* observing species coming up from the depths.

Girard, C. F.

Girard's Whorltail Iguana *Stenocercus ornatus* **Gray,** 1845

Girard's Skink *Cyclodina aenea* **Girard,** 1857 [Alt. Copper Skink]

Central Texas Whip Snake *Masticophis taeniatus girardi* **Stejneger** and **Barbour,** 1917

Dr. Charles Frederic Girard (1822–1895) was a French herpetologist who was Louis Agassiz's pupil and assistant at Neuchâtel and moved with Agassiz to the USA. He was in Cambridge, Massachusetts (1847–1850), and worked with Baird (1850–1857), establishing the Smithsonian. He became an American citizen (1854) and while continuing his work at the Smithsonian, studied medicine and graduated from Georgetown College (1856). He briefly visited Europe (1860). During the American Civil War he sided with the Confederacy and supplied the Confederate army with medical and surgical supplies. He left the USA (1864), returned to France, and practiced medicine there, serving as a physician during the siege of Paris (1870).

Glauert

Glauert's Monitor *Varanus glauerti* **Mertens,** 1957 [Alt. Kakadu Sand Goanna, Kimberley Rock Monitor]

Ludwig Glauert (1879–1963) was born in England and trained as a geologist. He emigrated to Western Australia (1908) and joined the geological survey in Perth as a paleontologist. He worked for two years in Western Australian Museum as a volunteer, joining the permanent staff (1910) as Scientific Assistant and Keeper of Geology and Ethnology (1914). He worked on the Margaret River caves (1909–1915), studying remains from the Pleistocene. He served in the Australian army (1917–1919) and then studied Australian material in the British Museum before returning to Perth (1920) as Keeper of the Western Australian Museum's biological collections, becoming Curator (1927) and Director (1954). His interests were legion. He was the leading authority on Western Australian reptiles, used his own money to buy books for the museum, and helped with the taxidermy. He retired officially in 1956 but continued working on reptiles and scorpions.

Glaw

Chameleon sp. *Calumma glawi* **Böhme,** 1997

Dr. Frank Glaw (b. 1966) became Curator of Herpetology at Zoologische Staatssammlung München (1997). He is a specialist in systematics of Malagasy frogs and reptiles. He co-wrote *Field Guide to the Amphibians and Reptiles of Madagascar* (1992) with Miguel Vences, with whom he has described over 100 new species. He collected the holotype.

Gloyd

Eastern Fox Snake *Pantherophis gloydi* **Conant,** 1940
Oaxacan Small-headed Rattlesnake *Crotalus intermedius gloydi* **Taylor,** 1941
Gloyd's Hognose Snake *Heterodon nasicus gloydi* Edgren, 1952 [Alt. Dusty Hognose Snake]
Gloyd's Hump-nosed Viper *Hypnale walli* Gloyd, 1977 [Alt. Wall's Hump-nosed Viper]
Pit-viper genus *Gloydius* **Hoge** and Romano-Hoge, 1981

Dr. Howard Kay Gloyd (1902–1978) took his bachelor's degree in science at Ottawa University (1924) and a master's degree (1929) at Kansas State College, where he had been teaching since 1927. He was a graduate student at University of Michigan. He was Director, Chicago Academy of Sciences, for which he organized numerous reptile-collecting expeditions to Arizona (1936–1958). He joined the University of Arizona, Tucson, as Lecturer and Research Associate, Zoology Department (1958), retiring as Emeritus Professor (1974). He wrote "The Rattlesnakes, Genera *Sistrurus* and *Crotalus*" (1940). See also **Howard Gloyd.**

Gmelin

Gmelin's Bronzeback *Dendrelaphis pictus* Gmelin, 1789

Johann Friedrich Gmelin (1748–1804) belonged to a well-known family of German naturalists. He was Professor of Medicine, Georg-August-Universität Göttingen. Other members of his family were with Pallas (q.v.) on one of his expeditions. He published the 13th edition of Linnaeus' *Systema naturae* (3 vols., 1788–1796). Three mammals are named after him.

Godeffroy

Angle-headed Lizard *Hypsilurus godeffroyi* **Peters,** 1867 [Alt. Northern Forest Dragon; Syn. *Gonocephalus godeffroyi*]

Johann Caesar Godeffroy (1813–1885), a member of a German trading house that imported copra from the Pacific, was interested in ornithology. His family was of French extraction, having moved to Germany to avoid religious persecution. From 1845 his fleet of ships traded largely in the Pacific, and he established over 45 trading posts, buying land and property, laying the foundations of German colonial power. From 1857 he also collected natural history objects, with paid collectors on board his trading ships. He founded a museum in Hamburg and named it after himself (1860). He forgot to be sufficiently commercial, and the company was declared bankrupt (1879). Two birds are named after him.

Godman

Godman's Montane Pit-viper *Cerrophidion godmani* **Günther,** 1863
Godman's Graceful Brown Snake *Rhadinaea godmani* Günther, 1865
Godman's Anole *Anolis godmani* **Boulenger,** 1885
Godman's Earth Snake *Geophis godmani* Boulenger, 1894
Godman's Garter Snake *Thamnophis godmani* Günther, 1894

Dr. Frederick du Cane Godman (1834–1919) was a British naturalist who, with his friend Osbert Salvin, compiled the massive *Biologia Centrali Americana*. He was a lawyer but was wealthy enough that he had no need to work, so he devoted his life to natural history, particularly ornithology. He visited Norway, Russia, the Azores, Madeira, the Canary Islands, India, Egypt, South Africa, Guatemala, British Honduras (Belize), and Jamaica. He wrote *Natural History of the Azores, or Western Islands* (1870). His widow and daughters set up the "Godman Memorial Exploration Fund." Three mammals and two birds are named after him.

Goelet

Gymnophthalmid lizard sp. *Cercosaura goeleti* **Myers** and **Donnelly,** 1996

Robert Guestier Goelet (b. 1924) is a wealthy businessman, the head of an old family real-estate company. He has been involved in many cultural activities as well as being the eighth President of the American Museum of Natural History (1976–1989). He traveled widely, including at least 14 visits to Argentina, where he made a film on Argentine wildlife, *From the Pampas to Patagonia*. He sponsored the Terramar expedition to Venezuela (1994–1995).

Goetsch

Goetsch's Tree Iguana *Liolaemus goetschi* **Muller** and **Hellmich,** 1938

Dr. W. Goetsch was a German herpetologist and entomologist with a particular interest in ants and termites who worked at Zoologische Staatssammlung München. He made important collections in Southern Europe (1927–1934), Argentina (1927–1928), and Chile (1929–1931). He worked very closely with Hellmich.

Goetze

Goetze's Chameleon *Chamaeleo goetzei* **Tornier,** 1899

Walther Goetze (1872–ca. 1900) was a German botanist. He was a member of an expedition that explored the Tanzanian side of Lake Malawi and the Kinga Mountains (1897–1900). Many excellent photographs that he took during the expedition survived, but he did not.

Goldie

Goldie's Tree Cobra *Pseudohaje goldii* **Boulenger,** 1895 [Alt. Gold's Tree Cobra (in error)]

Sir George Dashwood Taubman Goldie (1846–1925) was a soldier and one of the great builders of the British Empire. He was the founder of what is now Nigeria. He played a role similar to that played in South Africa by Cecil Rhodes. Boulenger says that the tree cobra was "named in honour of Sir George Taubman Goldie, the Governor of the Royal Niger Company." The United Africa Company was formed (1879), renamed the National African Company (1881), and granted a charter as the Royal Niger Company (1886). This was achieved in the teeth of intense French and German opposition. It became impossible for the company to hold its position against the power of two such jealous states, and it handed over its territory to the British government (1900). Goldie spent time in Rhodesia (1903–1904) looking into local agitation for self-government—something not achieved until well into the 20th century.

Goldman

Goldman's Bunchgrass Lizard *Sceloporus goldmani* **H. M. Smith,** 1937

Major Edward Alphonso Goldman (1873–1946) was a field naturalist and mammalogist, born in Illinois. Edward Nelson (q.v.) hired him, first (1892) to assist his biological investigations of California and Mexico, then as Field Naturalist and eventually Senior Biologist with the U.S. Bureau of Biological Survey. Goldman collected in every region of Mexico (1892–1906), amassing an enormous fund of information on the natural history of the country. He was on the Biological Survey of Panama (1911–1912) during the construction of the canal. He assisted the U.S. government in negotiating with Mexico to protect migratory birds (1936). He had a honorary position with the Smithsonian, as an Associate in Zoology (1928–1946). He was President of the Biological Society of Washington (1927–1929) and American Society of Mammalogists (1946). He wrote *The Mammals of Panama* (1920). Many taxa including eight mammals and four birds are named after him, as is Goldman Peak in Baja California.

Goliath

Goliath Blind Snake *Typhlops golyathi* Dominguez and Moreno, 2009

Goliath of Gath (about 1,030 B.C.) was a Philistine warrior of giant size who was killed with a slingshot by David, later King of the Jews. (See Bible, 1 Sam. 17.4) His name is sometimes used in a binomial to denote the exceptional size of a species.

Golubev

Toadhead Agama sp. *Phrynocephalus golubewii* Shenbrot and Semyonov, 1990

Bow-fingered gecko sp. *Cyrtopodion golubevi* Nazarov, **Ananjeva,** and Rajabizadeh, 2010

Dr. Michael Leonidovich Golubev (1947–2005) was a Russian herpetologist. He emigrated from Russia in the 1990s to work at the University of California, Davis. He lived for a time in Seattle (2002). He co-wrote *Gecko Fauna of the USSR and Contiguous Regions* (1996).

Gomes

Gomes' Burrowing Snake *Apostolepis cearensis* Gomes, 1915

Gomes' Lizard-eating Snake *Phalotris nasutus* Gomes, 1915 [Syn. *Elapomorphus nasutus*]

Gomes' Pampas Snake *Phimophis iglesiasi* Gomes, 1915

Gomes' Green Racer *Philodryas oligolepis* Gomes, 1921

Sao Paulo Keelback *Helicops gomesi* **Amaral,** 1921

Dr. João Florêncio Gomes (1886–1919) was a physician, zoologist, and herpetologist. He graduated as a physician, Rio de Janeiro (1910), and joined Instituto Butantan, São Paulo, Brazil (1911). He wrote *Contribuição para o conhecimento dos ophidios do Brasil. 1. Descrição de quatro especies novas e um novo gênero de opisthóglyphos. 2. Ophidios do Museu Rocha (Ceará)* (1915). Gomes' description of the green racer was published posthumously.

Good

Good's Thick-toed Gecko *Pachydactylus goodi* **Bauer,** Lamb, and Branch, 2006

Dr. David Andrew Good (b. 1956) is a herpetologist who was a Curatorial Associate of Herpetology (1985–1989) and a Research Associate (1989–1990), Museum of

Vertebrate Zoology, University of California. He did a herpetofaunal survey of the Richtersveld (South Africa) with Bauer and Branch in the mid-1990s, when he was Curator of Herpetology, Louisiana State Museum. He is known for his work on salamanders and anguid lizards.

Goode

Goode's Desert Horned Lizard *Phrynosoma goodei* **Stejneger,** 1893

George Brown Goode (1851–1896) was an ichthyologist who was assistant to Baird, and Assistant Secretary of the Smithsonian from 1872. He went to Wesleyan University, Connecticut, to study natural sciences, then briefly to Harvard, returning to Wesleyan to assume charge of its Natural History Museum. He then worked for the U.S. Fish Commission and went on three expeditions. He met Baird and was persuaded to join the Smithsonian, overseeing much of the research undertaken by the U.S. Fish Commission. He wrote *Oceanic Ichthyology, a Treatise on the Deep-Sea and Pelagic Fishes of the World, Based Chiefly upon the Collections Made by the Steamers Blake, Albatross, and Fish Hawk in the Northwestern Atlantic* (1896). He died from pneumonia.

Gordon Gekko

Gordon Gekko's Gecko *Cnemaspis gordongekkoi* **Das,** 1993 [Syn. *Cyrtodactylus gordongekkoi*]

Gordon Gekko is the anti-hero of the 1987 film *Wall Street*, and we suppose it would have been just too much of a waste of a good joke not to exploit this name. In the film Gekko says, "Greed is good." Perhaps this species of gecko is greedier than most?

Gore

Gore's Bronzeback *Dendrelaphis gorei* **Wall,** 1910

Colonel St. George C. Gore (1849–1913) was Surveyor-General of India (1899–1904). He was in India for many years. While still a Lieutenant he was deputed (1878) by Colonel Everest (after whom Mount Everest is named) to explore and map the Pishin valley in what was the Northwest Province of India; it is now in Baluchistan, part of Pakistan.

Gorgon

Blue Anole *Anolis gorgonae* **Barbour,** 1905

Gymnophthalmid lizard sp. *Ptychoglossus gorgonae* **Harris,** 1994

Named after Gorgona Island, off the coast of Colombia, not after the monstrous Gorgons of Greek mythology.

Gorzug

Rio Grande Cooter *Pseudemys gorzugi* **Ward,** 1984

A contraction of George R. Zug (q.v.).

Goudot, J.-M.

Black Blind Snake *Leptotyphlops goudotii* **Duméril** and **Bibron,** 1844

Justin-Marie Goudot was a French zoologist who worked in Colombia (1822–1843). He was the first head of the Zoology Department, Museo Nacional de Colombia, having been asked (1822) by Cuvier and Humboldt, on behalf of the new Vice President, to create the department. He was a noted collector of invertebrates. He wrote *Nouvelles observations sur le tapir pinchaque* (1843). A bird is also named after him.

Goudot, J. P.

Forest Night Snake *Ithycyphus goudoti* **Schlegel,** 1837

Jules Prosper Goudot was a traveler, collector, and entomologist who worked in Madagascar in the first half of the 19th century. His nickname was "Bibikely," which is Malagasy for "insect," emphasizing his interest in bugs. He adopted local customs, married a local woman, and became so immersed in everything Malagasy that he resigned his position as collector for Muséum National d'Histoire Naturelle, Paris. He found the remains of a number of huge eggs and showed them to Professor Gervais of Muséum National d'Histoire Naturelle, Paris. These proved to be eggs of the extinct Elephant Bird (*Aepyornis*). Two mammals are named after him.

Gould

Gould's Monitor *Varanus gouldii* **Gray,** 1838

Gould's Black-headed Snake *Suta gouldii* Gray, 1841

John Gould (1804–1881) was the son of a gardener at Windsor Castle but became an ornithologist, artist, and taxidermist. The newly formed Zoological Society of London employed him as a taxidermist. He was certainly the most prolific publisher and original author of ornithological works in the world. In excess of 46 volumes of reference work were produced by him, in color (1830–1881). He published 41 works on birds, with 2,999 remarkably accurate illustrations by a team of artists, including his wife. He traveled to see birds in their natural habitats. Gould was a commercially minded man, and Victorian England was fascinated by the exotic, including those exquisite jewels the hummingbirds, a group with which his name is particularly associated. His superb paintings and prints of these and other birds were greatly sought after—so much so that he probably had trouble keeping up with the demand. At least 28 birds and 5 mammals are named after him.

Gow

Gow's Burrowing Skink *Anomalopus gowi* **Greer** and **Cogger,** 1985

Gow's Brown Snake *Pseudonaja gowi* Wells, 2002

Graeme Francis Gow (1940–2005) was an Australian herpetologist, who twice survived being bitten by an Inland Taipan. He was Curator of Reptiles, Taronga Zoo, Sydney, before establishing his Reptile World, Humpty Doo, Northern Territory. He wrote *Graeme Gow's Complete Guide to Australian Snakes* (1989) He died of cancer, but his illness was complicated by the effects of a bite from a death adder.

Grabowsky

Dwarf/Reed Snake sp. *Calamaria grabowskyi* **Fischer,** 1885

Indonesian Beauty Snake *Orthriophis taeniurus grabowskyi* Fischer, 1885

Friederich J. Grabowsky (1857–1929) was a zoologist and botanist. He worked in Borneo (1881–1884) and New Guinea (1885–1887), then he left the employment of the New Guinea Company and returned to Europe. He became Director, Zoological Gardens, Breslau, and Inspector, Brunswick Zoological Institute. He wrote *Vögel von Borneo: Im Südosten der Insel* (1883).

Graciela

Tree iguana sp. *Liolaemus gracielae* Abdala et al., 2009

Graciela Mirta Blanco is an Argentine herpetologist who graduated (1991) from the Faculty of Sciences, Universidad Nacional de Río Cuarto, Córdoba, Argentina, and now works at Universidad Nacional de San Juan, Rivadavia, Argentina. She co-wrote "*Phymaturus antofagastensis* Diet" (2008).

Graham, D. C.

Graham's Japalure *Japalura grahami* **Stejneger,** 1924

Rev. Dr. David Crockett Graham (1884–1961) was a Baptist missionary in Szechuan, China, from 1911 to 1918, then returned to the USA for postgraduate study. He started to correspond with the Smithsonian regarding collecting natural history specimens for them. He was back in Szechuan (1920–1926), during which period he encountered the Ch'uan Miao, a Chinese aboriginal tribe. The University of Chicago awarded his doctorate (1927), after which he returned to Szechuan (1928–1930). He became a Fellow of the Royal Geographical Society, London (1929), and during an expedition to Moupin acquired a Great Panda skin for the Smithsonian. He did advanced study in anthropology, ethnography, and archeology at Chicago and Harvard (1930–1932). He taught anthropology and archeology at West China Union University and was Curator, West China Union University Museum of Archeology, Art, and Ethnology, Chengtu, Szechuan Province (1932–1948). After retiring (1948) he lived in Colorado (1949–1961).

Graham, J. D.

Graham's Anole *Anolis grahami* **Gray,** 1846

Graham's Crayfish Snake *Regina grahamii* **Baird** and **Girard,** 1853

Texas Patchnose Snake *Salvadora grahamiae* Baird and Girard, 1853

Colonel James Duncan Graham (1799–1865) was a topographical engineer. He graduated from West Point (1817) and served as an artillery officer (1817–1829) until his particular skills were recognized and he transferred to the Corps of Topographical Engineers as a Captain, being promoted to Major (1838). He was the astronomer on the surveying party that established the boundary between the USA and Texas (then an independent nation) (1839–1840). He resurveyed the famous Mason-Dixon Line (1848–1850). He discovered that there are lunar tides on the Great Lakes (1854). He was promoted to Lieutenant-Colonel (1861) and Colonel (1863). He died from exposure.

Granchi

Tyrrhenian Lizard *Podarcis tiliguerta granchii* **Lanza** and Brizzi, 1974

Granchi's Leaf-toed Gecko *Hemidactylus granchii* Lanza, 1978

Edoardo Granchi is a herpetologist and taxidermist at Museo di Storia Naturale, Florence. This museum has a large collection of crabs—most appropriate, as *granchi* is Italian for crabs. He and Lanza have traveled together in Ethiopia.

Grandidier

Grandidier's Gecko *Geckolepis typica* Grandidier, 1867

Grandidier's Madagascar Ground Gecko *Paroedura androyensis* Grandidier, 1867

Grandidier's Velvet Gecko *Homopholis sakalava* Grandidier, 1867

Grandidier's Two-lined Skink *Euprepes bilineatus* Grandidier, 1869

Grandidier's Dwarf Gecko *Lygodactylus tolampyae* Grandidier, 1872

Grandidier's Madagascar Swift *Oplurus grandidieri* **Mocquard,** 1900

Grandidier's Water Snake *Liopholidophis grandidieri* Mocquard, 1904

Grandidier's Worm Snake *Xenotyphlops grandidieri* Mocquard, 1905

Alfred Grandidier (1836–1921) was a French explorer,

geographer, and ornithologist who collected in Madagascar (1865) and recovered bones of what turned out to be *Aepyornis maximus*, the Elephant Bird (1866). He wrote *Histoire naturelle des oiseaux de Madagascar* (1876). The Malagasy mineral grandidierite is named after him, and in the French part of the Antarctic there is a Mont Alfred Grandidier. Three mammals and a bird are named after him.

Grandison

Bunty's Dwarf Gecko *Lygodactylus grandisonae* **Pasteur,** 1962 [Alt. Kenyan Dwarf Gecko]

Grandison's Forest Skink *Sphenomorphus grandisonae* **Taylor,** 1962

Lanza's Writhing Skink *Lygosoma grandisonianum* **Lanza** and Carfi, 1966

Alice Georgie Cruikshank "Bunty" Grandison (b. 1927) is a herpetologist specializing in amphibians, and was Curator of Reptiles and Amphibians, Natural History Museum, London.

Grant, C.

Virgin Island Tree Boa *Epicrates monensis granti* Stull, 1933

Grant's Worm Snake *Typhlops granti* **Ruthven** and **Gaige,** 1935

Major Chapman Grant (1887–1983) was a grandson of Ulysses S. Grant, 18th President of the USA. He graduated from Williams College (1910) and became Assistant Curator, Entomology, Children's Museum, Brooklyn Institute of Arts and Sciences. He served in the U.S. Army (1913–1933). He visited the Caribbean, being stationed in Puerto Rico (1930s) and going on expeditions of the San Diego Museum of Natural History and University of Illinois (1950s). He started the journal *Herpetologica* (1932), which became the quarterly journal of the Herpetologists' League, which he co-founded (1936).

Grant, R. E.

Grant's African Ground Snake *Gonionotophis grantii* **Günther,** 1863

Dr. Robert Edmond Grant (1793–1874) was a physician and biologist. He was the first Professor of Comparative Anatomy at University College London, where the Grant Museum of Zoology and Comparative Anatomy is named after him. He held the Chair of Comparative Anatomy (1827–1874). We cannot be entirely sure that we have identified the right person, as Günther's description mentions only "Professor Grant," without making clear exactly who he had in mind. But in 1853 Grant became Swiney Lecturer in Geology at the British Museum, where Günther worked, and the two men would certainly have known each other.

Grant, W. R. O.

Grant's Leaf-toed Gecko *Hemidactylus granti* **Boulenger,** 1899

William Robert Ogilvie-Grant (1863–1924) was an ornithologist and Curator of Birds, Natural History Museum, London (1909–1918), where he had worked since 1882. He enlisted in the army at the beginning of WW1 and suffered a stroke while helping to build fortifications near London (1916). He collected in Somalia and Socotra Island. He wrote *A Hand-book to the Game Birds* (1895). Several birds are named after him.

Grauer

Grauer's Blind Snake *Rhinotyphlops graueri* **Sternfeld,** 1912 [Alt. Sternfeld's Beaked Snake]

Rwanda Five-toed Skink *Leptosiaphos graueri* Sternfeld, 1912

Rudolf Grauer (1870–1927) was an Austrian explorer and zoologist who collected in the then Belgian Congo (1909 and 1910–1911) on an expedition paid for by K. K. Naturhistorisches Hofmuseum. He suffered from actinomycosis contracted in Africa and eventually succumbed to this bacterial infection. Other taxa, including four birds and two mammals. are named after him.

Gravenhorst

Gravenhorst's Mabuya *Mabuya gravenhorstii* **Duméril** and **Bibron,** 1839

Gravenhorst's Tree Iguana *Liolaemus gravenhorstii* **Gray,** 1845

Johann Ludwig Christian (or Jean Louis Charles) Gravenhorst (1777–1857) was a zoologist and entomologist. He started studying law (1797) but changed course (1801) and studied zoology at Georg-August-Universität Göttingen. He went to Paris, where he met Cuvier (1802). He became Professor at Georg-August-Universität Göttingen (1805). He started teaching natural history at University in Frankfurt-am-Oder (1810), which university was transferred to Breslau (Wroclaw, Poland) (1811), and he became Director, Zoologischen Museum der Universität Breslau, donating his own collections. He started suffering from mental illness (1825), eventually ceasing all scientific work (1840). Sadly, by the time he died he had become completely withdrawn from reality. He wrote *Monographie coleopterorum* (1806).

Gray

Gray's Agama *Agama spinosa* Gray, 1831 [Alt. Spiny Agama]

Gray's Desert Racer *Coluber ventromaculatus* Gray, 1834 [Alt. Gray's Rat Snake; Syn. *Platyceps ventromaculatus*]

Gray's Kukri Snake *Oligodon dorsalis* Gray and **Hardwicke** 1835 [Alt. Bengalese Kukri Snake]
Gray's Snake Skink *Nessia burtonii* Gray, 1839 [Alt. Burton's Nessia]
Gray's Chinese Gecko *Gekko chinensis* Gray, 1842
Gray's Pit-viper *Trimeresaurus strigatus* Gray, 1842
Gray's Sea Snake *Hydrophis ornatus ornatus* Gray, 1842 [Alt. Ornate Reef Sea Snake]
Gray's Wall Gecko *Tarentola clypeata* Gray, 1842
Gray's Lava Lizard *Microlophus grayi* **Bell,** 1843 [Alt. Gray's Pacific Iguana, Floreana Lava Lizard; Syn. *Tropidurus grayi*]
Gray's Ornate Skink *Cyclodina ornata* Gray, 1843
Gray's Tree Gecko *Naultinus grayii* Bell, 1843 [Alt. Northland Green Gecko]
Gray's Blind Snake *Ramphotyphlops nigrescens* Gray, 1845
Gray's Dwarf Chameleon *Bradypodion ventrale* Gray, 1845
Gray's Ornate Gecko *Diplodactylus ornatus* Gray, 1845
Gray's Sticky-toed Gecko *Hoplodactylus granulatus* Gray, 1845 [Alt. Forest Gecko]
Gray's Mangrove Snake *Myron richardsonii* Gray, 1849 [Alt. Richardson's Mangrove Snake]
Gray's Scalyfoot *Delma grayi* **Andrew Smith,** 1849
Gray's Monitor Lizard *Varanus olivaceus* **Hallowell,** 1857
Gray's Earth Snake *Uropeltis melanogaster* Gray, 1858
Gray's Water Skink *Tropidophorus grayi* **Günther,** 1861
African Water Snake genus *Grayia* **Bocage,** 1866
Gray's Slider *Trachemys venusta grayi* **Bocourt,** 1868
Minor Snake-eyed Skink *Ablepharus grayanus* **Stoliczka,** 1872

John Edward Gray (1800–1875) was a zoologist and entomologist. He started work at the British Museum (1824) with a temporary appointment at 15 shillings (£0.75) per day, but became Keeper of Zoology. Gray was regarded as the leading authority on many reptiles, including turtles. He was also an ardent philatelist and claimed to be the world's first stamp collector. He worked at the museum with his brother George Robert Gray (1808–1872). J. E. Gray suffered a severe stroke that paralyzed his right side, including his writing hand (1869). Nevertheless, he continued to publish until the end of his life by dictating to his wife, Maria Emma, who had always worked with him as an artist and occasional co-author and after whom he named a lizard. (See **Emma Gray.**) The brothers co-wrote *Catalogue of the Mammalia and Birds of New Guinea in the Collection of the British Museum* (1859). They have a large number of taxa named after them.

Greef

Greef's Gecko *Hemidactylus greefii* **Bocage,** 1886

Dr. Richard Greef (1829–1892) was an entomologist, conchologist, and malacologist who collected on Sao Tome in the early 1880s. He became Professor of Zoology, Philipps-Universität Marburg. He wrote *Ueber die Fauna der Guinea-Inseln S. Thome und Rolas* (1884).

Green

Alpine Cool-Skink *Niveoscincus greeni* **Rawlinson,** 1975 [Alt. Northern Snow Skink]

Dr. Robert "Bob" Green (b. 1925) was a Tasmanian farmer. He was very interested in ornithology and photography, becoming a professional wildlife photographer (1953) and honorary ornithologist at the Queen Victoria Museum and Art Gallery, Launceston, Tasmania (1959). He sold his farm (1960) and joined the museum staff, becoming Curator (1962). The University of Tasmania awarded him an honorary doctorate (1987).

Greenway

Plana Cay Curlytail Lizard *Leiocephalus greenwayi* **Barbour** and **Shreve,** 1935
Ambergris Cay Dwarf Boa *Tropidophis greenwayi* Barbour and Shreve, 1936

James Cowan Greenway Jr. (1903–1989) was Curator of Birds, Museum of Comparative Zoology, Harvard. Among ornithologists he is known as co-editor of *Checklist of Birds of the World* (first published 1931). Greenway and his wife collected in the Bahamas (1930s). He wrote *Extinct and Vanishing Birds of the World* (1958). A bird is named after him.

Greenwell

Blind Snake sp. *Leptotyphlops greenwelli* **Wallach** and Boundy, 2005

Dr. John Richard Greenwell (1942–2005) was co-founder and first Secretary of the International Society of Cryptozoology. He studied reports of unknown animals, such as the Ri (reported to be a mermaid but turning out to be a dugong) and the Yeren (Chinese "wild man"). He was born in England, spent six years in South America, and settled in Tucson, Arizona, where he was Research Coordinator, Office of Arid Lands Studies, University of Arizona. His doctorate was honorary, awarded by Universidad Autónoma de Guadalajara, Mexico (1991). He was in pursuit of the Sasquatch (Bigfoot) when he died of cancer.

Greer

Greer's Island Skink *Geomyersia glabra* Greer and **Parker,** 1968
Greer's Ctenotus *Ctenotus greeri* **Storr,** 1979
Greer's Tree Skink *Lioscincus greeri* **Böhme,** 1979
Greer's Lerista *Lerista greeri* Storr, 1982
Haacke-Greer's Skink *Haackgreerius miopus* Greer and **Haacke,** 1982

Greer's Writhing Skink *Mochlus brevicaudis* Greer, **Grandison,** and Barbault, 1985
Greer's Elf Skink *Nannoscincus greeri* **Sadlier,** 1987
Greer's Blind Skink *Dibamus greeri* **Darevsky,** 1992
Greer's Water Skink *Cophoscincopus greeri* Böhme, **Schmitz,** and **Ziegler,** 2000
Litter Skink sp. *Lankascincus greeri* Batuwita and Pethiyagoda, 2007

Dr. Allen Eddy Greer Jr. is Principal Research Scientist, Herpetological Department, Australian Museum, Sydney. He worked at the Museum of Comparative Zoology, Harvard, where he was awarded his doctorate (1973), and was Curatorial Assistant, Herpetology, Museum of Vertebrate Zoology, University of California, Berkeley (1974–1975), before joining the Australian Museum. He wrote *Encyclopedia of Australian Reptiles.*

Greigert

Greigert's Shovelsnout Snake *Prosymna greigerti* **Mocquard,** 1906

Lieutenant Greigert was an officer in the French Colonial Infantry. He was in the Ivory Coast before 1902. We know very little from Mocquard's etymology except that Lieutenant Greigert obtained the snake holotype and sent it to Natural History Museum, Paris. He wrote *Notice geographique et historique sur la circonscription de Bouna* (1902).

Grey, B.

Grey's Sea Snake *Ephalophis greyae* **L. A. Smith,** 1931

Beatrice Grey collected the holotype (1930).

Grey, G.

Grey's Menetia *Menetia greyii* **Gray,** 1845

Sir George Grey (1812–1898) was a soldier, explorer, colonial governor, premier, and scholar. He explored Western Australia on government-financed expeditions (1837–1839). On the first expedition a native Australian, whom he shot, speared Grey; nevertheless, he championed the cause of assimilation. He became Governor of New Zealand (1845) and faced even greater difficulties than in Australia. Grey's greatest success was his management of Maori affairs, in which he scrupulously observed the terms of the Treaty of Waitangi, especially regarding land rights. He became Governor, Cape Colony, and High Commissioner, South Africa (1853). He sought to convert frontier tribes to Christianity to "civilize" them, supported mission schools, and built a hospital for African patients. He returned to New Zealand, where he was elected to its Parliament. Though politics left him little time to devote to scholarship, he was a keen naturalist and botanist and established extensive collections and important libraries at Cape Town and Auckland. Two mammals and a bird are named after him.

Grey, R.

Grey's Anole *Anolis greyi* **Barbour,** 1914

Robert M. Grey was a plant breeder (sugar cane) who became the first Superintendent of the botanical collections near the former Central Soledad, Cuba, which had been assembled by Edwin Atkins (1899). An amphibian is named after him.

Griffin, L. E.

Griffin's Dasia *Dasia griffini* **Taylor,** 1915 [Alt. Griffin's Keel-scaled Tree Skink]

Professor Lawrence Edmonds Griffin (1874–1949) was a herpetologist at Missouri Valley University. He collected in the Philippines early in the 20th century. He was associated with the Carnegie Museum, Pittsburgh, and is mentioned in their *Annals.* He wrote "A Check-list and Key of Philippine Snakes" (1911).

Griffin, M.

Griffin's Thick-toed Gecko *Pachydactylus griffini* **Bauer,** Lamb, and Branch, 2006

Michael "Mike" Griffin is Senior Support Specialist, Ministry of the Environment and Tourism, Windhoek, Namibia. He has contributed greatly to the knowledge and conservation of the herpetofauna of Namibia and for many years has provided support and advice to research expeditions. He wrote "The Species Diversity, Distribution, and Conservation of Namibian Mammals" (1998).

Griffin, P.

Griffin's Lerista *Lerista griffini* **Storr,** 1982

Philip Griffin (b. 1964) is an English-born herpetologist, photographer, and professional musician who grew up in Western Australia and lives in Adelaide. He worked with Storr on herpetological surveys of the Lake Eyre region. He co-wrote *Birds of the Houtman Albrolhos, Western Australia* (1986).

Grillo

Boulenger's Tree Lizard *Anisolepis grilli* **Boulenger,** 1891

Dr. Giuseppe Franco Grillo was a physician who emigrated to Brazil and worked in Palmeira, a predominately Italian settlement in southern Brazil. He supplied natural history specimens from southern Brazil to Boulenger and Thomas in London and to Doria in Genoa.

Grismer

Grismer's Collared Lizard *Crotaphytus grismeri* McGuire, 1994 [Alt. Sierra de Cucupah Collared Lizard]
Grismer's Bent-toed Gecko *Cyrtodactylus grismeri* Ngo Van Tri, 2008

Gecko sp. *Cyrtodactylus leegrismeri* Onn and Ahmad, 2010

Dr. Larry Lee Grismer (b. 1955) is a herpetologist who took his bachelor's (1981) and master's degrees (1986) at San Diego State University and obtained his doctorate from Loma Linda University (1994). He joined the faculty of La Sierra University, California (1994), and is now Professor of Biology. He has spent much time studying the herpetofauna of Baja California but now works in southeast Asia. He wrote *Amphibians and Reptiles of Baja California, Its Pacific Islands, and the Islands of the Sea of Cortes* (2002).

Griswold, D.

Griswold's Ameiva *Ameiva griswoldi* **Barbour,** 1916

Dr. Don W. Griswold was Director of the Rockefeller West Indian Hookworm Commission (1909–1914).

Griswold, J.

Dwarf/Reed Snake sp. *Calamaria griswoldi* **Loveridge,** 1938

John Augustus Griswold Jr. (1912–1991) was an aviculturist and ornithologist who was on the Harvard primate expeditions to Borneo (1936), Thailand (1937), and Peru (1939). He became Curator of Birds, Philadelphia Zoological Gardens (1947). He collected the snake holotype. An amphibian is named after him.

Griveaud

Marojejy Leaf Chameleon *Brookesia griveaudi* **Brygoo, Blanc,** and **Domergue,** 1974

Dr. Paul Griveaud (1907–1980) was an entomologist who worked for many years at the Entomological Laboratory, ORSTOM, Tananarine, Madagascar. He co-wrote *La protection des richesses naturelles, archéologiques et artistiques à Madagascar* (1968).

Gro

Dunn's Spinytail Iguana *Morunasaurus groi* **Dunn,** 1933

Lord Gro is a fictional character in E. R. Eddison's 1922 fantasy novel *The Worm Ouroboros*. Presumably Dunn was a fan.

Gronovi

Gronovi's Dwarf Burrowing Skink *Scelotes gronovii* **Daudin,** 1802

Laurens Theodorus Gronovius (1730–1777)— sometimes Laurentius Theodorus Gronovius or Laurens Theodore Gronow, or simply "Laurenti"—was a Dutch botanist and ichthyologist. He had one of the most extensive zoological and botanical collections of his day and is noted for having developed a novel way of preserving fish skins. Linnaeus acknowledged that Gronovius inspired him. He was the first person to use the word *scincus* (1754) to describe the group of reptiles that we call skinks today—something Daudin was well aware of when he wrote *Histoire naturelle, générale et particulière des reptiles* (1802). His father and both sons were also notable scholars. He wrote *Museum ichthyologicum* (1754), in which he described over 200 species of fish. He should not be confused with Josephus Nicolaus Laurenti (q.v.).

Grosse

Tree Iguana sp. *Liolaemus grosseorum* **Etheridge,** 2001

Monique Grosse requested that this species be "named to acknowledge the invaluable assistance given her in the field by her husband and children: Constantino, Ana, and Paul Grosse." Mrs. Grosse, better known under her maiden name, is Professor Dr. Monique Halloy, Instituto de Herpetologia, Fundación Miguel Lillo, Tucumán, Argentina, and is the stepdaughter of the Belgian-Argentine herpetologist Raymond Laurent (q.v.).

Grossmann

Marbled Gecko *Gekko grossmanni* **Günther,** 1994

Wolfgang Grossmann is a German herpetologist. He collected in Malaysia (2003). He co-wrote *Amphibians and Reptiles of Peninsular Malaysia and Thailand: An Illustrated Checklist* (1999).

Groundwater

Groundwater's Keelback *Amphiesma groundwateri* **M. A. Smith,** 1922

C. L. Groundwater lived in Bangkok (1905–1935) before retiring to Penang. He was a member of the St. Andrews Society (Bangkok Scots) and of the Bangkok St. Andrews Pipeband (a 1919 photograph shows him in Highland dress with bagpipes). In an article dated 1920 M. A. Smith says, "Mr. C. L. Groundwater I have to thank for his careful drawings of the heads of snakes" that illustrated the papers Smith was presenting with the type specimens to the British Museum of Natural History.

Grum-Grzimailo

Grum-Grzimailo's Toadhead Agama *Phrynocephalus grumgrzimailoi* **Bedriaga,** 1909

G. Y. Grum-Grzimailo (1860–1936) was a Russian entomologist who traveled widely in Central Asia (1884–1890). On some of his journeys his younger brother, Lieutenant M. Y. Grum-Grzimailo (b. 1862), accompanied him. We cannot be sure which brother is being honored, but the elder one is the more likely as he appears to have been the scientifically minded one.

Guarani

Lava Lizard sp. *Tropidurus guarani* **Alvarez, Cei,** and Scolaro, 1994

Named after the Guarani language and culture of Paraguay.

Guenther, A. C. L. G.

Günther's Writhing Skink *Lygosoma guentheri* **Peters,** 1854

Günther's Forest Racer *Dendrophidion brunneus* Günther, 1858

Günther's Forest Snake *Taeniophallus affinis* Günther, 1858

Günther's Green Tree Snake *Dipsadoboa unicolor* Günther, 1858

Günther's Island Racer *Arrhyton taeniatum* Günther, 1858

Günther's Keelback *Rhabdophis chrysargoides* Günther, 1858

Günther's Malaysian Coral Snake *Calliophis maculiceps* Günther, 1858

Günther's Mountain Snake *Xylophis stenorhynchus* Günther, 1858

Günther's Philippine Shrub Snake *Oxyrhabdium leporinum* Günther, 1858

Günther's Reed Snake *Liopeltis frenatus* Günther, 1858

Günther's Two-spotted Snake *Coniophanes bipunctatus* Günther, 1858

Günther's Whipsnake *Ahaetulla fronticincta* Günther, 1858

Günther's Mole Viper *Macrelaps microlepidotus* Günther, 1860 [Alt. Natal Black Snake]

Günther's Ground Snake *Atractus guentheri* **Wucherer,** 1861

Günther's Black-headed Snake *Suta nigriceps* Günther, 1863

Günther's Burrowing Snake *Plectrurus guentheri* **Beddome,** 1863

Günther's Dwarf Reed Snake *Pseudorabdion oxycephalum* Günther, 1863

Neon Blue-tailed Tree Lizard *Holaspis guentheri* **Gray,** 1863

Günther's Bloodsucker *Bronchocela smaragdina* Günther, 1864

Günther's Indian Gecko *Cyrtodactylus deccanensis* Günther, 1864

Günther's Kukri Snake *Oligodon cinereus* Günther, 1864

Gunther's Sea Snake *Microcephalophis cantoris* Günther, 1864

Günther's Vine Snake *Ahaetulla dispar* Günther, 1864

Günther's Worm Snake *Typhlops pammeces* Günther, 1864

Günther's Blind Snake *Ramphotyplops guentheri* Peters, 1865

Günther's Frog-eating Snake *Stegonotus batjanensis* Günther, 1865

Günther's Garter Snake *Elapsoidea guentheri* **Bocage,** 1866

Günther's Skink *Cyclodomorphus branchialis* Günther, 1867 [Alt. Common Slender Bluetongue]

Günther's Bronzeback *Dendrelaphis salmonis* Günther, 1872

Günther's Green Racer *Philodryas psammophideus* Günther, 1872

Günther's New Caledonian Gecko *Bavayia cyclura* Günther, 1872

Günther's Tropical Ground Snake *Trimetopon gracile* Günther, 1872

Günther's Black Snake *Bothrolycus ater* Günther, 1874

Günther's Emo Skink *Emoia lawesi* Günther, 1874

Günther's Earth Snake *Uropeltis liura* Günther, 1875

Günther's False Wolf Snake *Dinodon septentrionalis* Günther, 1875

Günther's Keelback *Amphiesma modesta* Günther, 1875

Günther's Rough-sided Snake *Aspidura guentheri* **Ferguson,** 1876

Günther's Tuatara *Sphenodon guntheri* **Buller,** 1877 [Alt. Brother's Island Tuatara]

Günther's Racer *Ditypophis vivax* Günther, 1881

Günther's Sun Tegu *Euspondylus guentheri* **O'Shaughnessy,** 1881

Günther's Worm Lizard *Leposternon guentheri* **Strauch,** 1881 [Junior syn. of *L. (Amphisbaena) microcephalum* Wagler, 1824]

Günther's False Coral Snake *Erythrolamprus guentheri* **Garman,** 1883

Günther's Day Gecko *Phelsuma guentheri* **Boulenger,** 1885 [Alt. Round Island Day Gecko]

Günther's Flying Lizard *Draco guentheri* Boulenger, 1885

Günther's Island Gecko *Christinus guentheri* Boulenger, 1885 [Alt. Lord Howe Island Southern Gecko]

Günther's Tropical Gecko *Perochirus guentheri* Boulenger, 1885

Günther's Whorltail Iguana *Stenocercus guentheri* Boulenger, 1885

Western Congo Worm Lizard *Monopeltis guentheri* Boulenger, 1885

Günther's Burrowing Skink *Scelotes guentheri* Boulenger, 1887

Günther's Cylindrical Skink *Chalcides guentheri* Boulenger, 1887

Günther's Ristella *Ristella guentheri* Boulenger, 1887

Günther's Many-tooth Snake *Sibynophis bistrigatus* Günther, 1889

Günther's Oriental Slender Snake *Trachischium guentheri* Boulenger, 1890
Günther's Dwarf Gecko *Lygodactylus miops* Günther, 1891
Günther's Lightbulb Lizard *Proctoporus guentheri* **Boettger,** 1891
Günther's False Fer-de-Lance *Xenodon guentheri* Boulenger, 1894
Black Centipede Eater *Aparallactus guentheri* Boulenger, 1895
Günther's Splendid Cat-eyed Snake *Leptodeira splendida* Günther, 1895
Milne Bay Ground Snake *Stegonotus guentheri* Boulenger, 1895
Günther's Tree Snake *Stenophis guentheri* Boulenger, 1896 [Syn. *Lycodryas guentheri*]
Ground Snake sp. *Liophis guentheri* **Peracca,** 1897
Günther's Toadhead Agama *Phrynocephalus guentheri* **Bedriaga,** 1907
Günther's Leaf-tail Gecko *Uroplatus guentheri* **Mocquard,** 1908
Spiny Lizard sp. *Sceloporus guentheri* **Stejneger,** 1918
Günther's Graceful Brown Snake *Rhadinaea guentheri* **Dunn,** 1938 [Syn. *Urotheca guentheri*]

Dr. Albert Carl Ludwig Gotthilf Günther (1830–1914) recognized (1867) that the tuatara was not a lizard but belonged to an entirely separate order among living reptiles. He was educated as a physician in Germany, then joined the British Museum (1856). He was appointed Keeper, Zoological Department (1857). He became a naturalized British subject (1862) and changed his second two Christian names to Charles Lewis. He became President, Biological Section, British Association for the Advancement of Science (1880), and was President of the Linnean Society (1881–1901). He wrote *The Reptiles of British India*. Three mammals are named after him.

Guenther, R.

Günther's Mangrove Monitor *Varanus rainerguentheri* **Ziegler, Böhme,** and **Schmitz,** 2007

Dr. Rainer Günther (b. 1941) was Curator of Herpetology, Natural History Museum, Humboldt-Universität, Berlin.

Guerin

Argentine Pampas Snake *Phimophis guerini* **Duméril, Bibron,** and **Duméril,** 1854

Félix Edouard Guérin-Méneville (1799–1874) was an entomologist who introduced silkworm breeding to France. There is a Felix-Edouard Guérin-Méneville Collection of Crustacea at the Academy of Natural Sciences, Philadelphia. He wrote *Iconographie du règne animal de G. Cuvier* 1829–44, a work that complements Cuvier's and Latreille's work, which lacked illustrations. A bird is named after him.

Guibé

Westem Dwarf Gecko *Lygodactylus guibei* **Pasteur,** 1964
Guibé's Chameleon *Calumma guibei* **Hillenius,** 1959 [Alt. North-western Chameleon]
False Coral Snake sp. *Oxyrhopus guibei* **Hoge** and Romano, 1977

Dr. Jean Guibé (1910–1999) was a French zoologist and herpetologist at Muséum National d'Histoire Naturelle, Paris, where he was Professor of Zoology (Reptiles and Fish) (1957–1975).

Guichard

Guichard's Rock Gecko *Pristurus guichardi* **Arnold,** 1986

Kenneth Mackinnon Guichard (1914–2002) was a British entomologist and art connoisseur. He worked on the Desert Locust Survey, Oman (1949–1950), and in the Sahara (1952). He was on Socotra Island, Yemen, as entomologist attached to the Middle East Command's 1967 expedition. Many of his specimens were stored in Natural History Museum, London, for nearly 20 years before being examined. He made his living through his flair for spotting good paintings and etchings, buying cheap and selling expensive.

Guichenot

Guichenot's Skink *Lampropholis guichenoti* **Duméril** and **Bibron,** 1839
Guichenot's Dwarf Iguana *Enyalioides laticeps* Guichenot, 1855
Guichenot's Giant Gecko *Rhacodactylus ciliatus* Guichenot, 1866

Antoine Alphone Guichenot (1809–1876) was a zoologist, herpetologist, and ichthyologist at Muséum National d'Histoire Naturelle, Paris. He taught there and also took part in expeditions for the museum. He did an extensive biological survey of Algeria. He stepped down to become Assistant Naturalist (1856–1872), a post he held until he retired.

Guillaumet

Chameleon sp. *Calumma guillaumeti* **Brygoo, Blanc,** and **Domergue,** 1974

Professor Dr. Jean-Louis Guillaumet is a French botanist attached to ORSTOM. He co-wrote *Flore et vegetation de Madagascar* (1974).

Guimbeau

Guimbeau's Day Gecko *Phelsuma guimbeaui* **Mertens,** 1963 [Alt. Orange-spotted Day Gecko]

Mr. B Guimbeau collected the holotype.

Guinea

Guinea's Sea Krait *Laticauda guineai* **Heatwole, Busack,** and **Cogger,** 2005

Dr. Michael Leonard Guinea works at the Faculty of Education, Health, and Science, Charles Darwin University. He is also a qualified diving officer and is involved in the university's research into sea turtles.

Guiral

Mocquard's File Snake *Mehelya guirali* **Mocquard,** 1887

Léon Guiral (1858–1885) was a former French naval quartermaster and a member of de Brazza's expeditions in the French Congo (arrived 1882). He supplied Mocquard with the second of two specimens on which the description of this snake was based. He died on Christmas Day 1885 at Libreville, and his *Le Congo français, du Gabon à Brazzaville,* was published posthumously (1889).

Gumprecht

Gumprecht's Green Pit-viper *Trimeresurus gumprechti* David, **Vogel, Pauwels,** and Vidal 2002

Andreas Gumprecht is a German herpetologist and taxonomist. He co-wrote *Amphibians and Reptiles of Peninsular Malaysia and Thailand: An Illustrated Checklist* (2002). Vogel says on his website that "this species was named, upon his own request, after Mr. Andreas Gumprecht."

Gunalen

Gunalen's Long-necked Turtle *Chelodina gunaleni* **McCord** and Joseph-Ouni, 2007

Danny Gunalen of Jakarta, Indonesia, is a collector, breeder, and dealer in reptiles. He was honored for his field expertise and his involvement in the discovery of this turtle. He is also an excellent photographer.

Gundlach

Gundlach's Anole *Anolis gundlachi* **Peters,** 1877

Johannes Christoph (Juan Cristóbal) Gundlach (1810–1896) was a German zoologist and ornithologist. He was a Curator at Philipps-Universität Marburg and later at Naturmuseum Senckenberg, Frankfurt, then lived in Cuba (1839–1896), except during the Civil War (1868–1878), when he stayed in Puerto Rico. He was zealous and single-minded, and tended to keep what he collected and describe it for science himself. An event that nearly killed him allowed him to follow his chosen profession. During a hunting accident he discharged a small gun so close to his nose that he lost his sense of smell. After that he could calmly dissect, macerate, and clean skeletons without difficulty. He wrote *Ornitología cubana.* Three birds and two mammals are named after him. See also **Juan Gundlach.**

Gununakuna

Tree Iguana sp. *Liolaemus gununakuna* Avila, 2004

Named after an aboriginal people living in Argentina.

Guppy

Solomons Scaly-toed Gecko *Lepidodactylus guppyi* **Boulenger,** 1884

Dr. Henry Brougham Guppy (1854–1926) was a traveler and collector who started as a navy surgeon in the China and Japan station on HMS *Hornet* (1877–1880), visiting Korea (1878), and on HMS *Lark* (1881–1884)—a ship involved in survey work in the West Pacific, particularly the Solomon Islands. He spent time in both Fiji (1896–1900) and Hawaii (1903–1904). He worked on botany in the West Indies (1907–1911) and Azores (1913–1914). He died in Martinique. He wrote the two-volume *Observations of a Naturalist in the Pacific between 1896 and 1899* (1903–1906). See also **Brougham.**

Gyi

Kapuas Mud Snake *Enhydris gyii* **Murphy, Voris,** and Auliya, 2005

Dr. Ko Ko Gyi was a Burmese herpetologist. He was Professor of Zoology, Rangoon Arts and Sciences University. He wrote "A Revision of Colubrid Snakes of the Subfamily Homalopsinae" (1970).

Gyldenstolpe

Gyldenstolpe's Isopachys *Isopachys gyldenstolpei* Lonnberg, 1916

Count Nils Gyldenstolpe (1886–1961) was a zoologist and ornithologist who was attached to Naturhistoriska Riksmuseet, Stockholm (1914–1961). He traveled in Siam (Thailand) (1911), Central Africa (1921), and New Guinea (1951). A bird is named after him.

H

Haacke

Haacke's Legless Skink *Typhlosaurus braini* Haacke, 1964 [Alt. Brain's Blind Legless Skink]
Haacke-Greer's Skink *Haackgreerius miopus* **Greer** and Haacke, 1982
Haacke's Thick-toed Gecko *Pachydactylus haackei* Branch, **Bauer,** and **Good,** 1996

Dr. Wulf Dietrich Haacke (b. 1936) was, until his retirement, Chief Herpetologist at the Transvaal Museum. He is a founder and honorary life member of the Transvaal Herpetological Association.

Haad

Blind Snake sp. *Liotyphlops haadi* Silva-Haad, Franco, and Maldonado, 2008

José Haad was the senior author's grandfather.

Haas, A.

Paraná False Boa *Pseudoboa haasi* **Boettger,** 1905

Albrecht Haas collected the holotype.

Haas, C. P. J. de

Haas' Bronzeback *Dendrelaphis haasi* de Rooijen and **Vogel** 2008

C. P. J. de Haas was a collector and herpetologist active in the Dutch East Indies (Indonesia) (1940s and 1950s). He wrote "Checklist of Snakes of the Indo-Australian Archipelago (Reptiles, Ophidia)" (1950).

Haas, G.

Haas' Toadhead Agama *Phrynocephalus nejdenses* Haas, 1957
Forest Skink sp. *Sphenomorphus haasi* **Inger** and **Hosmer,** 1965
Haas' Fringe-fingered Lizard *Acanthodactylus haasi* **Leviton** and **Anderson,** 1967

Georg Haas (1905–1981) was an Austrian-born Israeli zoologist, malacologist, and herpetologist who was Professor of Zoology, Hebrew University of Jerusalem, having joined the staff in 1932. He was educated at Akademisches Gymnasium before graduating from Universität Wien (1928). He did postdoctoral work in Berlin before leaving for Palestine. The Zoological Society of Israel offers an annual prize in his honor for the best Ph.D. research.

Habel

Marchena Lava Lizard *Microlophus habelii* **Steindachner,** 1876

Dr. Simeon Habel was a German-American traveler and naturalist who explored in Central and South America and spent six months (1868) on the Galapagos Islands. He wrote *The Sculptures of Santa Lucia Cosumalwhuapa in Guatemala* (1878).

Hackars

Hackars' Five-toed Skink *Leptosiaphos hackarsi* **Witte,** 1941

Henri-Martin Hackars (1881–1940) held the honorary rank of Lieutenant-Colonel. He was a District Commissioner in the colonial administration, Belgian Congo (Zaire), and was involved in the administration of Institut des Parcs Nationaux du Congo Belge.

Hades

Hades Blind Snake *Typhlops hades* Kraus, 2005

Although normally thought of as a place where departed souls end up, Hades was also the name of the Greek god of the Underworld. Kraus' etymology makes it clear that the snake is named after the god.

Haeckel

Haeckel's Toadhead Agama *Phrynocephalus haeckeli* **Bedriaga,** 1907

Dr. Ernst Haeckel (1834–1919) was an evolutionary biologist, zoologist, philosopher, and artist. He qualified as a physician in Berlin (1857). He studied zoology at Friedrich-Schiller-Universität Jena (1859–1862) and was Professor of Comparative Anatomy (1862–1909). He traveled extensively in the Canary Islands (1866–1867) and in Dalmatia, Egypt, Turkey, and Greece (1869–1873). He met Thomas Huxley and Charles Darwin, whose theories he embraced and promoted. He wrote *The Riddle of the Universe* (1901). An asteroid, 12323 Häckel, is named after him, as are two mountains, one in the USA and the other in New Zealand.

Haensch

Haensch's Whorltail Iguana *Stenocercus haenschi* **Werner,** 1901

Richard Haensch was a German collector and dealer, mainly of Lepidoptera, working in South America. He traveled over most of the countries of that continent (1884–1900).

Hagen

Hagen's Pit-viper *Trimeresurus hageni* Lidth De Jeude, 1886 [Syn. *Parias hageni*]

Dr. Bernhard Hagen (1853–1919) was a physician and amateur natural historian. After studying medicine at Munich he was employed by a Sumatran planting company. He undertook some collecting expeditions, accumulating mostly zoological specimens. The Astrolabe Company, New Guinea, employed him (1893). He returned to Germany (1895) but, with his wife, visited New Guinea again (1905). He was a section head, Naturmuseum Senckenberg, Frankfurt (1897–1904), where he founded the Ethnology Department. Two mammals are named after him.

Hagmann

Hagmann's Keelback *Helicops hagmanni* **Roux,** 1910

Dr. Gottfried A. Hagmann (1874–1946) was a Swiss zoologist who spent many years as Chief Zoologist, Museu Paraense Emilio Goeldi, Belém, Brazil, becoming a naturalized citizen. He collected the snake holotype. He wrote *Die eier von Caiman niger* (1902).

Hague

Hague's Anole *Anolis haguei* **Stuart,** 1942

Henry Hague collected in Guatemala. He was manager of the San Geronimo Estate near Vera Paz. He led a collecting expedition in Guatemala for the Smithsonian (1867).

Hajek

Hajek's Lizard *Liolaemus hajeki* **Núñez,** Pincheira-Donoso, and Garin, 2004

Professor Ernst Hajek is head of the School of Biological Sciences, Pontificia Universidad Católica de Chile. He is a member of the Environment Committee, Biological Society of Chile.

Hall, C.

Hall's Flat Gecko *Afroedura halli* **Hewitt,** 1935 [Formerly *Afroedura karroica halli*]

Charles Hall. The original text says the gecko holotype was "collected by Mr. Chas. Hall at Telle Junction near Palmietfontein, Herschel district, C.P."

Hall, W. P.

Hall's Spiny Lizard *Sceloporus megalepidurus halli* Dasmann and **H. M. Smith,** 1974

Dr. William Purington Hall III (b. 1939) is an evolutionary biologist whose doctorate was awarded by Harvard (1973). He worked as a demonstrator at Harvard (1968–1972); in the Department of Biology, University of Puerto Rico (1973–1976); at the University of Colorado (1976–1977); and in the Department of Genetics, University of Melbourne (1977–1979 and 1981–1982). After this varied academic career he worked in computer data-processing applications for a number of companies, including Bank of Melbourne. Since 1990 he has worked for a defense contractor involved in, among other activities, designing and constructing warships. He is now an Australian citizen.

Hallberg

Hallberg's Cloud Forest Snake *Cryophis hallbergi* **Bogert** and **Duellman,** 1963

Professor Dr. Thomas Boone Hallberg (b. 1923) is an American botanist who has lived for over 50 years in Mexico and is an expert on strains of maize. He works at Universidad de la Sierra Juárez, Oaxaca.

Hallier

Forest Skink sp. *Sphenomorphus hallieri* Lidth de Jeude, 1905

Dr. Johann Gottfried "Hans" Hallier (1868–1932) was a botanist who was educated at the Ludwig-Maximilians-Universität München and Friedrich-Schiller-Universität Jena. After a year as Assistant, University of Georg-August-Universität Göttingen, he was Botanical Assistant, Buitenzorg (Bogor) Herbarium, Java (1893–1896). He was Assistant Director, Botanical Laboratory, Ludwig-Maximilians-Universität München (1897–1898), and Assistant to the Director, Museum and Laboratory of Commercial Botany, Hamburg (1898–1908). He was Assistant, State Herbarium, Leiden (1908–1922). He was on the first Dutch scientific expedition to Borneo (1893–1894) and traveled and collected in the Far East (particularly the Philippines) (1903–1904). He visited Canada and the USA (1908–1910). He was a firm believer in the superiority of the German race.

Hallmann

Hallmann's Day Gecko *Phelsuma pusilla hallmanni* **Meier,** 1989

Dr. Gerhard Hallmann is a German herpetologist at Museum für Naturkunde Dortmund.

Hallowell

Hallowell's Coffee Sake *Ninia atrata* Hallowell, 1845

Hallowell's Ground Snake *Atractus fuliginosus* Hallowell, 1845

Hallowell's House Snake *Lamprophis virgatus* Hallowell, 1854

Hallowell's Centipede Snake *Tantilla vermiformis* Hallowell, 1860

Dr. Edward Hallowell (1808–1860) was an American herpetologist and physician in Philadelphia. He described over 50 new species of reptiles.

Hamilton

Hamilton's Terrapin *Geoclemys hamiltonii* **Gray,** 1831 [Alt. Black/Spotted Pond Turtle]

Dr. Francis Hamilton-Buchanan (1762–1829) was an ichthyologist and physician who qualified at Glasgow (1783) and was to have been a ship's surgeon, but his health was bad and remained poor until 1794, when he became well enough to work. He joined the Honourable East India Company's Bengal service as an Assistant Surgeon. He collected botanical specimens as he traveled. His botanical drawings were so admired that a number were presented to Joseph Banks, to whom he regularly sent specimens. He studied the fish of the Ganges and was often employed on survey work on all sorts of subjects, including fisheries. He became Superintendent of the Calcutta Botanical Gardens (1814). His family name at birth was Buchanan, but he dropped it and took the name Hamilton, his mother's maiden name (1815). He signed his name as "Francis Hamilton" or "Francis Hamilton (formerly Buchanan)." He wrote *Account of the Fishes of the Ganges* (1822). A bird is named after him.

Hammond

Hammond's Garter Snake *Thamnophis hammondii* **Kennicott,** 1860 [Alt. Two-striped Garter Snake]

Dr. William Alexander Hammond (1828–1900) was a physician, naturalist, and soldier. He started collecting on the Pacific Railroad Survey for the Smithsonian (ca. 1847). He qualified as a physician (1848) at New York University and joined the army (1849) as an Assistant Surgeon. He served (and collected birds for Baird) at various frontier stations (1849–1859). He became Professor of Anatomy and Physiology, University of Maryland (1859), but rejoined the army (1861) and served in the American Civil War as Surgeon-General of the Union army. He clashed with Edward M. Stanton, Secretary of War, and was court-martialed (1864, verdict reversed 1878). He lectured on nervous and mental diseases at the College of Physicians and Surgeons, New York (1865–1867), and was Professor of Nervous and Mental Diseases, Bellevue Hospital Medical College, New York (1867–1873), and then at New York University from 1874. He was a co-founder of New York Medical School (1882). From 1888 he was in private practice. A bird is named after him.

Hampton

Kukri Snake sp. *Oligodon hamptoni* **Boulenger,** 1900

Hampton's Slug Snake *Pareas hamptoni* Boulenger, 1905

Hampton's Japalure *Japalura hamptoni* **M. A. Smith,** 1935

Herbert Hampton was a friend of Boulenger, who inferred in his description that the holotypes of both snakes were collected by Hampton. The holotype of the japalure was collected in 1908 but not described for 27 years.

Han

Vietnamese Green Grass Lizard *Takydromus hani* Chou, Truong, and **Pauwels,** 2001

Professor Pao-the Han is a leading architect and designer in Taiwan. After early university education in Taiwan he took a master's degree in architecture at Harvard (1965). He has taught in a number of institutions, such as Taiwan National College of Arts (1995–2000). He was the first Director of the Taiwanese National Museum of Natural Science (1981–1995).

Hanitsch

Hanitsch's Reed Snake *Oreocalamus hanitschi* **Boulenger,** 1899 [Alt. Kalimantan Burrowing Snake]

Dr. Karl Richard Hanitsch (1860–1940) was a German biologist and museum curator. He was Demonstrator, Zoology, University College, Liverpool, and then Director, Raffles Library and Museum, Singapore (1895–1919). He wrote *An Expedition to Mt. Kinabalu, British North Borneo* (1900). A 50-cent postage stamp, issued by Christmas Island (1977), bears his portrait.

Hanlon

Hanlon's Ctenotus *Ctenotus hanloni* **Storr,** 1980 [Alt. Nimble Ctenotus]

Timothy Marcus Stephen "Mark" Hanlon is an Australian herpetologist who worked closely with Storr. They co-wrote *Herpetofauna of the Exmouth Region, Western Australia* (1980).

Hannah

King Cobra *Ophiophagus hannah* Cantor, 1836

Hannah was not a specific person. The name is probably derived from Greek mythology. Hamadryas was the nymph of the oak tree and mother of all the dryad nymphs of the lesser trees of the forest. The alternative name for this snake is Hamadryad.

Hannah C.

Skink sp. *Saproscincus hannahae* **Couper** and Keim, 1998

Hannah Couper is Patrick Couper's daughter.

Hannstein

Hannstein's Spot-lipped Snake *Rhadinaea hannsteini* **Stuart,** 1949

Walter Bernhard Hannstein (b. 1902) was a German whose father emigrated to Guatemala (1892). In the 1940s he bought a coffee plantation, Finca La Paz, which is still owned and managed by his descendants. He wrote a personal narrative of early-20th-century life in western Guatemala.

Harald Meier

Green Girdled Lizard *Zonosaurus haraldmeieri* **Brygoo** and **Böhme,** 1985

See **Meier.**

Hardegger

Hardegger's Orangetail Lizard *Philochortus hardeggeri* **Steindachner,** 1891

Dr. Dominik Kammel von Hardegger (1844–1915) was an Austrian physician who explored Somalia and the Harar area, Ethiopia (1875), and Egypt and Nubia (1884–1885). He was on expeditions organized by the Italian Geographical Society.

Hardwicke

Hardwicke's Bloodsucker *Brachysaura minor* Hardwicke and **Gray,** 1827

Hardwicke's Gecko *Eublepharis hardwickii* Gray, 1827

Hardwicke's Spiny-tailed Lizard *Uromastyx hardwickii* Gray, 1827

Hardwicke's Rat Snake *Coluber ventromaculatus* Gray, 1834 [Syn. *Platyceps ventromaculatus*]

Hardwicke's Sea Snake *Lapemis hardwickii* Gray, 1834

Major-General Thomas Hardwicke (1756–1835) served in the Bengal army of the Honourable East India Company. He was an amateur naturalist and collector who was the first to make the Red Panda *Ailurus fulgens* widely known, through a paper that he wrote (1821), "Description of a New Genus . . . from the Himalaya Chain of Hills between Nepaul and the Snowy Mountains." Cuvier stole a march on Hardwicke in formally naming the Red Panda because Hardwicke's return to England was delayed. He collected reptiles in India and published on them (1827) with Gray. A mammal and a bird are named after him.

Hardy, D. L.

Ground Snake sp. *Atractus davidhardi* Silva Haad, 2004

Dr. David L. Hardy (b. 1935) is a retired physician and anaesthetist who became a herpetologist and an expert on pit-vipers. He founded the Tucson Herpetology Society (1988).

Hardy, G. S.

Hardy's Skink *Cyclodina hardyi* Chapple et al., 2008

Dr. Graham S. Hardy is a herpetologist who worked at the National Museum of New Zealand, Te Papa, Wellington. He completed his doctorate and lectured in vertebrate zoology at Victoria University, Wellington (1975). He wrote "The New Zealand Scincidae (Reptilia: Lacertilia): A Taxonomic and Zoogeographic Study" (1977).

Hardy, J. D.

Escambray Small-headed Trope *Tropidophis hardyi* **Schwartz** and **Garrido,** 1975 [Syn. *T. nigriventris hardyi*]

Jerry David Hardy Jr. (b. 1929) is a zoologist and herpetologist at the Department of Zoology, University of Maryland. He has worked at the National Oceanographic Data Center's Laboratory Guantanamo Bay, where research on Cuban iguanas is carried out. He wrote "Notes on the Cuban Iguana" (2004).

Hardy, J. E.

Nidua Fringe-fingered Lizard *Acanthodactylus (scutellatus) hardyi* **G. Haas,** 1957

Dr. J. E. Hardy was head of the Department of Entomology, British Mandatory Government, Palestine (1942). He collected the lizard holotype.

Hardy, L. M.

Hardy's Hook-nosed Snake *Ficimia hardyi* Mendoza-Quijano and **H. M. Smith,** 1993 [Alt. Hidalgo Hook-nosed Snake]

Dr. Laurence McNeil Hardy (b. 1939) is a herpetologist who has been Director and Professor Emeritus, Museum of Life Sciences, Louisiana State University, since 1968; and is Research Associate, American Museum of Natural History, New York. His master's degree was from the University of Kansas (1964) and his doctorate from the University of New Mexico (1969). He wrote "The Genus *Syrrhophus* (Anura: Leptodactylidae) in Louisiana" (2004).

Harenna

Chameleon sp. *Chamaeleo harennae* **Largen,** 1995

Harenna is the name of an escarpment in the Bale Mountains, Ethiopia.

Harold

Harold's Legless Skink *Aprasia haroldi* **Storr,** 1978

Harold's Lerista *Lerista haroldi* Storr, 1978

Harold's Scalyfoot *Delma haroldi* Storr, 1987

Gregory Harold was a biologist who worked with Storr. They co-authored "Herpetofauna of the Zuytdorp Coast and Hinterland, Western Australia" (1983).

Harold Young

Harold Young's Supple Skink, *Lygosoma haroldyoungi* **Taylor,** 1962 [Alt. Harold's Writhing Skink, Banded Supple Skink]

Rev. Harold Young was a second-generation American Baptist missionary in Burma (Myanmar) who took over his father's mission (1932). He continued the mission until after WW2, when the newly independent Burmese government became suspicious of him and he moved the mission to Chiang Mai (Thailand), also becoming curator of the local zoo to which he donated his own collection. He trained radio operators for the CIA and translated intercepted messages for them, as did his son Oliver Gordon Young, who also provided training in maintenance and repair of radio sets in the 1950s and 1960s. Another of his sons, William "Bill" Young, is believed to have been a CIA agent, operating under consular cover, in Chiang Mai.

Harriett

White-crowned Snake *Cacophis harriettae* **Krefft** 1869

Harriett Scott (1830–1907) was an artist. However, even though Krefft was himself a very competent artist, Harriett Scott and her sister, Helena Forde, illustrated his *Mammals of Australia* (1871) with very fine black and white lithographs. Harriett had also supplied some of the illustrations for his *Snakes of Australia* (1869).

Harris

Harris' Anadia *Anadia altaserrania* Harris and **Ayala,** 1987

Dennis Martin Harris (b. 1950) worked in the Division of Amphibians and Reptiles, University of Michigan's Museum of Zoology (1980s). He and Ayala co-wrote "A New *Anadia* (Sauria: Teiidae) from Colombia and Restoration of *Anadia pamplonensis* Dunn to Species Status" (1987).

Hart

Hart's Glass Lizard *Ophisaurus harti* **Boulenger,** 1899 [Alt. Chinese Glass Lizard]

Sir Robert Hart, First Baronet (1835–1911), was a British consular and customs official in China. He was sent to Queen's University, Belfast, graduating in 1853. He was nominated as a student translator in the consular service and went to Hong Kong (1853). He became a supernumerary interpreter in Ningpo (1854), where he had to manage the consulate for several months. Cool-headed, he found favor and became Secretary to the allied commissioners governing the canton. He became an interpreter for the British Consulate (1858). He resigned (1859), becoming local Inspector of Customs. He rose to Inspector-General of Foreign Customs (1863–1907), retiring to become Vice Chancellor of his old university. Hart is the central figure in Lloyd Lofthouse's historical novel *My Splendid Concubine.*

Harter

Harter's Water Snake *Nerodia harteri* **Trapido,** 1941 [Alt. Brazos Water Snake]

Philip Harter (d. 1971) was an American amateur herpetologist. He was described in the etymology as "a most enthusiastic and energetic collector." He collected the snake holotype (1936).

Hartert

Werner's Gypsy Gecko *Hemiphyllodactylus harterti* **Werner,** 1900

Ernst Johann Otto Hartert (1859–1933) was a German ornithologist and oologist. He traveled extensively, often on behalf of his employer, Walter Rothschild (later Lord Rothschild). He was Ornithological Curator of Rothschild's private museum, Tring, which later became an annex to the British Museum (Natural History), London. He wrote *Die Vögel der paläarktischen Fauna*. Twelve birds are named after him.

Hartmann

Hartmann's Agama *Agama hartmanni* **Peters,** 1869

Dr. Karl Eduard Robert Hartmann (1832–1893) was a physician, anatomist, and anthropologist. He qualified in medicine at Humboldt-Universität, Berlin, where he became Professor of Anatomy (1873). He accompanied Baron Adalbert von Barnim on an expedition to Egypt and Nubia, based around the course of the Blue Nile. Adalbert died on the trip. Hartmann wrote *Reise in Nordost Africa* (1863).

Hartweg

Western Spiny Softshell Turtle *Apalone spinifera hartwegi* **Conant** and Goin, 1948

Hartweg's Emerald Lizard *Sceloporus taeniocnemis hartwegi* **Stuart** 1971

Dr. Norman Edouard "Kibe" Hartweg (1904–1964) was a herpetologist whose specialty was the distribution and taxonomy of turtles. He worked at the University of Michigan (1927–1964), where he had taken his doctorate (1934), and was Assistant Curator, Herpetology (1934), and Curator (1947).

Haseman

Haseman's Gecko *Gonatodes hasemani* **Griffin,** 1917

John Diederich Haseman (fl. 1890–1969) was a zoologist and ichthyologist. He graduated from Indiana University (1905) and taught there (1905–1906). As a graduate student, he was sent to Brazil as a last-minute substitute to represent the Carnegie Museum and to collect fishes on a museum-sponsored expedition. Upon arriving (1907), he found the main expedition about to set out. It was decided that he should run his own solo expedition, which lasted two and a half years and covered large areas of Argentina, Bolivia, Brazil, Paraguay, and Uruguay. He never went back to the university but continued to study ichthyology in the field. He wrote *Some Factors of Geographical Distribution in South America* (1912).

Hasselquist

Fan-footed Gecko *Ptyodactylus hasselquistii* Donndorff, 1798 [Alt. Yellow Fan-fingered Gecko]

Frederik Hasselquist (1722–1752) was a traveler and naturalist. He was one of Linnaeus' students at Uppsala. Linnaeus often lamented the lack of knowledge of the natural history of Palestine, and this inspired Hasselquist. He reached Smyrna (1749), visited Asia Minor, Egypt, and Cyprus in addition to Palestine, and made extensive collections. These eventually reached Sweden after he had died near Smyrna on his way home.

Hassler

Hassler's Anole *Anolis altavelensis* **Noble** and Hassler, 1933 [Alt. Noble's Anole]

William Grey "Bill" Hassler (1906–1979) was a herpetologist at the American Museum of Natural History, where he was an Assistant (1924–1937). He went on a number of expeditions to the West Indies (1929–1935), concentrating on the Dominican Republic. He was Director of the Fort Worth Museum of Science and History (1953–1962). He worked with Noble, and they co-wrote "Two New Species of Frogs, Five New Species and a New Race of Lizards from the Dominican Republic" (1933). The alternative name, Noble's Anole, may cause confusion, as two lizards have at times been given that vernacular name. See **Noble.**

Hatcher

Tree Iguana sp. *Liolaemus hatcheri* **Stejneger,** 1909

Professor John Bell Hatcher (1861–1904) was a geologist and paleontologist. He started working at Yale and was sent to investigate strange fossil horns that originated from Wyoming (1888). His work there (1888–1892) uncovered the remains of ceratopsid dinosaurs such as Torosaurus and Triceratops. His senior at Yale, Marsh, refused to allow his assistants to write and publish papers, prompting Hatcher to leave for Princeton as Curator of Vertebrate Paleontology (1893–1900). He joined the Carnegie Museum as Curator of Paleontology and Osteology (1900). He died of typhoid fever.

Hatori

Oriental Coral Snake sp. *Sinomicrurus hatori* Takahashi, 1930

S. Hatori was a Japanese herpetologist in Formosa (Taiwan), a Japanese possession until well into the 20th century. He published *A Survey Report of Venomous Snakes of Formosa* (1905).

Haug

Haug's Worm Lizard *Cynisca haugi* **Mocquard,** 1904

Pastor Ernest Haug (d. 1915) was a Protestant missionary and a correspondent of Muséum National d'Histoire Naturelle, Paris. He established his permanent residence at Ngomo, Ogowe River, Gabon. He made a major collection of the fishes of that area. Moreover, he took note of all their local names, in three different languages. Pellegrin made a study of the collection (1906) and published on them (1914). Because of his outstanding contributions to science Haug was posthumously awarded the Prix Secques of the Paris Zoological Society (1916). After his death his brother continued (1918) to send specimens that Ernest had collected.

Haughton

Goalpora Grass Lizard *Takydromus haughtonianus* **Jerdon,** 1870

Major-General Henry Lawrence Haughton (1883–1955) was in the Indian army (1904–1943). He left a number of photographs that he took in Kashmir (1910). He delivered a lecture (1953) on "Cyprus since the War," from which we infer he may have served there. He wrote *Sport and Folklore in the Himalaya* (1913).

Hayden

Hayden's Garter Snake *Thamnophis radix haydenii* **Kennicott,** 1860 [Alt. Western Plains Garter Snake]

Dr. Ferdinand Vandeveer Hayden (1829–1887) was a physician, geologist, and explorer. He took his degree in medicine (1853). He was part of a geological expedition to the Yellowstone and Missouri rivers (1854–1855) and was the geologist on the Warren expedition to Nebraska and Dakota (1856–1857). He went on another expedition to explore the Rocky Mountains and the Yellowstone (1859–1862) and collected vertebrates for the museum in Philadelphia. During the American Civil War (1862–1865) he served as a surgeon in the Union army. He was Professor of Geology, University of Pennsylvania (1865–1872), and served on the U.S. Geological Survey (1867–1886). His work led to the establishing of Yellowstone National Park. He has a shrew named after him.

Hayek

Sumatra Bloodsucker *Bronchocela hayeki* **Müller,** 1928

Hans von Hayek (1869–1940) was an Austrian artist who moved to Munich (1891). He became a painter of animals and outdoor scenes and traveled widely within Europe. He was a WW1 war artist and after 1918, using Munich as his base, made a number of long trips to Asia, visiting Ceylon (Sri Lanka) and Indonesia.

Heath, E. R.

Heath's Tropical Racer *Mastigodryas heathii* **Cope,** 1876

Dr. Edwin Ruthven Heath (1839–1932) was a traveler, pharmacist, physician, railway administrator, and diplomat. His father took Edwin to the goldfields of California (1849). Armed with two large six-shooters and a Bowie knife, he made his way (1853) alone from

California back to Wisconsin, via Nicaragua and New Orleans. He went to college and eventually decided on medicine as a career, graduating in 1863. He owned and ran a pharmacy in Kansas (1866–1868). He was appointed Physician to the American Legation, Chile (1869). He was in charge of the Pacasmayo Railway, Peru, until 1878, when he returned to the USA. He went on a new expedition (1879) to explore Bolivia. After covering 2,400 kilometers (1,500 miles) he came upon a railway construction camp where nearly all the workers were sick or dying. He stayed there for seven months treating the workers. He then continued the expedition, reaching Reyes, Bolivia, where he stayed until the end of 1880 and sent Cope a number of herpetological specimens. He returned to Kansas (1881). His final trip (1883) was to help a friend harvest his coffee crop in Guatemala. While there he practiced medicine and acted as Consul for Nicaragua, Bolivia, and Guatemala. He became an honorary member of the Royal Geographical Society, London (1883). A river in Bolivia is named after him.

Heath, H.

Heath's Worm Lizard *Amphisbaena heathi* **Schmidt,** 1936

Brazilian Mabuya *Mabuya heathi* Schmidt and **Inger** 1951

Dr. Harold Heath (1868–1951) received his doctorate (1898) from the University of Pennsylvania and was a marine invertebrate zoologist at Stanford (Palo Alto, California) and their Hopkins Marine Station. He eventually became Professor Emeritus of Embryology. Heath was on the Hopkins-Branner Brazilian expedition (1911) during which the holotype of the lizard was collected.

Heatwole

Heatwole's Anole *Anolis desechensis* Heatwole, 1976

Dr. Harold Franklin Heatwole (b. 1934) is a zoologist, herpetologist, and botanist. He took his bachelor's degree in botany (1955) in Indiana and his master's in zoology at the University of Michigan (1958). He has three doctorates: in zoology, University of Michigan (1960); in science, University of New England, New South Wales (1981); and in botany, University of Queensland (1987). He taught at the University of Michigan (1959–1960) and then at the University of Puerto Rico (1960–1966), leaving as Associate Professor to join the University of New England, New South Wales (1966–1991), rising to Associate Professor. He was Professor and head of the Zoology Department, North Carolina State University (1991–1996) and is now Emeritus. He was President of the Australian Society of Herpetologists (1977–1978) and is President of the Herpetologists' League. Among his seven books he co-wrote *Ecology of Reptiles* (1987) and is series editor for *Amphibian Biology*. He has suffered snakebite only once.

Hecht

Hecht's Caribbean Gecko *Aristelliger hechti* **Schwartz** and **Crombie,** 1975

Dr. Max Knobler Hecht (1925–2002) was an evolutionary biologist, paleontologist, and zoogeographer at the American Museum of Natural History, New York. He worked at Cornell University Museum (1950). He became Professor Emeritus in Biology, the Graduate Center, and a Research Associate in Vertebrate Evolution at the American Museum of Natural History. He was an expert on Crocodylia. The fossil crocodilian *Stratiotosuchus maxhechti* was named in his honor.

Hediger

Hediger's Snake *Parapistocalamus hedigeri* **Roux,** 1934

Dr. Heini Hediger (1908–1992) was a zoologist, biologist, and expert on animal behavior. He studied zoology, botany, and ethnology at Universität Basel (1928–1933) and was a Professor there (1942–1953) and at Universität Zürich (1953–1979). He was Director, Bern Zoo (1938–1943), Basel Zoo (1944–1953), and Zurich Zoo (1954–1973).

Heim

Heim's Forest Snake *Geodipsas heimi* **Angel,** 1936

Professor Roger Heim (1900–1979) was a mycologist and cryptogamist who founded *Revue de mycologie* (1936). He has been described as an adventurer, humane naturalist, and great administrator-leader. He led an expedition to Madagascar, which Angel wrote up (1936). He became Director, Muséum National d'Histoire Naturelle, Paris (1951–1965). He was also President of the "Foundation Singer Polignac," which sponsored many expeditions, notably to New Caledonia (1960). He published 560 articles, all of which he illustrated himself. He was a noted gastronome and liked to eat the nonpoisonous fungi. He also studied—presumably empirically—the hallucinogenic properties of others. He wrote *État actuel des dévastations forestières à Madagascar.*

Helen G.

Notaburi Forest Skink *Sphenomorphus helenae* **Cochran,** 1927

Helen Beulah Thompson Gaige (1890–1976) (q.v.).

Helen N.

Helen's Tiny Gecko *Tropiocolotes helenae* **Nikolsky,** 1907 [Alt. Khuristan Dwarf Gecko]

Helena Nikolsky was obviously connected to the describer, who gives her name but no other detail.

Helen T.

Helen's Worm Snake *Carphonis amoenus helenae* **Kennicott,** 1859 [Alt. Midwest Worm Snake]

Helen Tennison lived in Mississippi and collected specimens in that state for Kennicott. He wrote to Baird (1855), "My Mississippi cousin (Miss Helen Tennison) says she has got me snakes, lizards, salamanders, frogs, etc." and (1856), "On the whole I think my cousin (Miss Helen Tennisson of Monticello, Miss.) has done pretty well for a beginning but I must get her to do still more this summer." Many of her specimens are now in the collection of the Chicago Academy of Sciences.

Helena

Helena's Rat Snake *Elaphe helena* **Daudin,** 1803 [Alt. Trinket Snake; Syn. *Coelognathus helena*]

Daudin does not specify who Helena was, but in his text he states that the holotype was given to Dr. Patrick Russell who reported that it was an extremely pretty snake when alive. We think that Daudin had Helen of Troy, the classical embodiment of female beauty, in mind when he chose the binomial.

Helena P.

Helena's Ctenotus *Ctenotus helenae* **Storr,** 1969 [Alt. Clay-soil Ctenotus]

Helen Louise Pianka, née Dunlap, (b. 1940). The etymology says, "Named after Helen Louise Pianka, who accompanied her husband on all his desert trips and assisted substantially in collecting the material that made this revision possible" (i.e. Storr's revision of the genus *Ctenotus* in Western Australia). Helen was once married to Dr. Eric R. Pianka (q.v.).

Helena R. L.

Lava Lizard sp. *Tropidurus helenae* Manzani and Abe, 1990

Helena Ribas Lopes was a Brazilian herpetologist who graduated from the Institute of Biology, Universidade do Estado do Rio de Janeiro (1972). The etymology notes that the specific name *helenae* is "an expression of our deep friendship and honors the memory of Helena Ribas Lopes, a distinguished adviser and teacher." In 1987 she published her master's thesis as *Study of the Reproductive Biology of Teiu (Tupinambis Teguixin)*.

Heller

Southern Pacific Rattlesnake *Crotalus viridis helleri* Meek, 1905 [Syn. *C. oreganus helleri*]

Heller's Coral Snake *Micrurus lemniscatus helleri* Schmidt and Schmidt, 1925

Skink sp. *Panaspis helleri* **Loveridge,** 1932

Edmund Heller (1875–1939) was a zoologist and ornithologist. He collected in Colorado and the Mojave desert (1896–1897) while still a student at Stanford, from where he graduated (1901). He was in the Galapagos (1899) and in Alaska (1900). He was on the Field Museum's African expedition (1907). He became Curator, Mammals, Museum of Vertebrate Zoology, University of California (1907), and was on the Alexander Alaskan expedition (1908). He returned to Africa (1909–1912) with the Smithsonian-Roosevelt and Rainey expeditions and was on Lincoln Ellsworth's expedition to British Columbia and Alberta (1914). He served as Naturalist on the expedition to explore the rediscovered Inca city Machu Picchu, Peru (1915). He joined Roy Chapman Andrews on the American Museum of Natural History expedition to China (1916), and then, as Rainey had become official photographer for the Czech army in Siberia, Heller joined him in Russia. Heller led the Smithsonian Cape-to-Cairo expedition (1919), then worked briefly for the Roosevelt Wild Life Experiment Station, Yellowstone National Park, studying large game animals. He was Assistant Curator, Mammals, Field Museum, under Osgood (1919–1926), during which period he collected in Peru (1922–1923) and Africa (1923–1926), which was his last expedition. He was Director, Milwaukee Zoological Gardens (1928–1935), and Director, Fleishhacker Zoo, San Francisco (1936–1939). Three mammals and a bird are named after him.

Hellmich

Hellmich's Wolf Snake *Lycophidion hellmichi* **Laurent,** 1964

Hellmich's Tree Iguana *Liolaemus hellmichi* **Donoso-Barros,** 1975

Professor Dr. Walter Hellmich (1906–1974) was Director, Zoologische Staatssammlung München. He was interested in Himalayan fauna and flora and was Director, Nepal Research Centre, Katmandu (1972–1974).

Hemming

Somali Mabuya *Trachylepis hemmingi* **Gans, Laurent,** and Pandit, 1965

C. F. Hemming was an entomologist who was an expert on locusts and worked for the Desert Locust Survey, Somali Republic. He co-wrote *A Note on Wind-Stable Stone-Mantles in the Southern Sahara* (1968). He collected the skink holotype (1961).

Hemprich

Hemprich's Skink *Scincus hemprichii* **Wiegmann,** 1837

Hemprich's Coral Snake *Micrurus hemprichii* **Jan,** 1858

Wilhelm Friedrich Hemprich (1796–1825) was a physician, traveler, and collector. He studied medicine in Berlin and met Ehrenberg there. They were invited (1820) to serve as naturalists on an expedition to Egypt, after which they continued to journey and collect in Lebanon, Sinai, and Ethiopia. He co-wrote *Natural Historical Journeys in Egypt and Arabia* (1828), which Ehrenberg completed and published after Hemprich's death from fever in Massawa, Eritrea. Two birds and a mammal are named after him.

Hempstead

Hempstead's Pine Woods Snake *Rhadinaea hempsteadae* **Stuart** and **Bailey,** 1941

Maria Luisa Hempstead, née Dieseldorff. Hempstead is the family name of a dynasty of coffee planters in Guatemala. The family was certainly well established as in 1910 Robert Weir Hempstead married Maria Luisa Dieseldorff, whose father was also a coffee planter in Guatemala. The authors say in the etymology that the snake "is dedicated to Mrs. R. W. Hempstead of Cobán, Alta Verapaz, Guatemala, to whom the senior author is greatly indebted for the many courtesies she extended to him during his investigations in the Alta Verapaz."

Henderson, J. B.

Henderson's Anole *Anolis hendersoni* **Cochran,** 1923

Dr. John Brooks Henderson Jr. (1870–1923) was a physician, naturalist, and amateur malacologist, specializing in West Indian shells; his collection is in the Smithsonian. He graduated from Harvard (1891) and from Columbia Law School (now George Washington University) (1893), entering government service as a private secretary and later (1897) traveling to Europe and Turkey as a civilian observer of the armies of the great European powers. He collected in Cuba on the *Tomas Barrera* expedition (1914) and in Haiti (1917). He was a citizen member of the Smithsonian's Board of Regents (1911–1923). He wrote *The Cruise of the Tomas Barrera* (1916). Other taxa, principally shellfish, are named after him.

Henderson, R. W.

Henderson's Trope *Tropidophis hendersoni* Hedges and **Garrido,** 2002

Peten Centipede Snake *Tantilla hendersoni* Stafford, 2004

Robert William Henderson (b. 1945) is Senior Curator, Herpetology, Milwaukee Public Museum. He wrote *Neotropical Treeboas: Natural History of the* Corallus hortulanus *Complex* (2002).

Henkel

Henkel's Flat-tailed Gecko *Uroplatus henkeli* **Böhme** and **Ibisch,** 1990 [Alt. Frilled Leaf-tail Gecko]

Friedrich-Wilhelm Henkel (b. 1949) is a German herpetologist who specializes in keeping and breeding geckos. He co-wrote *Amphibians and Reptiles of Madagascar and the Mascarene, Seychelles, and Comoro Islands* (2000).

Hensel

Hensel's Snake *Ditaxodon taeniatus* Hensel, 1868

Reinhold Friedrich Hensel (1826–1881) was a zoologist, ichthyologist, paleontologist, and naturalist. He taught natural history in Berlin (1850–1860). He traveled to the southern Brazilian province Rio Grande do Sul (1863–1866) to make a study of fishes on behalf of Berliner Akademie, where he subsequently became Professor of Zoology (1867). He wrote *Beiträge zur Kenntnis der Wirbelthiere Süd-Brasiliens* (1868).

Henshaw

Henshaw's Night Lizard *Xantusia henshawi* **Stejneger,** 1893 [Alt. Granite Night Lizard]

Henry Wetherbee Henshaw (1850–1930) was a naturalist and ethnologist. He was the naturalist on the Wheeler survey of the American West (1872–1879). He worked for the U.S. Bureau of Ethnology (1879–1893) and edited *American Anthropologist* (1888–1893). He visited Hawaii several times (1894–1904). He joined the U.S. Department of Agriculture (1905), working in the Biological Survey, and from 1910 he was the official in charge.

Herbert

Herbert's Writhing Skink *Riopa herberti* **M. A. Smith,** 1916 [Alt. Herbert's Supple Skink; Syn. *Lygosoma herberti*]

E. G. Herbert (1870–1951) was an English collector and naturalist who worked in Siam (Thailand) for the Bombay-Burma Trading Corporation (1895–1920). He was a member of the Natural History Society of Siam, to which he presented a paper (1919). He wrote "Nests and Eggs of Birds in Central Siam" (1926). Two birds are named after him.

Hermann

Hermann's Tortoise *Testudo hermanni* **Gmelin,** 1789

Herrmann's Water Snake *Hydrodynastes bicinctus* Herrmann, 1804

Johannis Herrmann (sometimes "Hermann") (1738–1800) was Professor of Medicine and Natural History, Université de Strasbourg. He wrote *Observationes zoologicae quibus novae complures aliaeque animalium species, describuntur et illustrantur edidit Fridericus Ludovicus Hammer* (published posthumously, 1804).

Herman Núñez

′ Tree Iguana sp. *Liolaemus hermannunezi* Pincheira-Donoso, Scolaro, and **Schulte,** 2007

See **Núñez.**

Hermina

Sakishima Green Snake *Cyclophiops herminae* **Boettger,** 1895 [Syn. *Opheodrys herminae*]

Hermine Boettger was the describer's wife. An amphibian is also named after her.

Herminier

Martinique Curlytail Lizard *Leiocephalus herminieri* **Duméril** and **Bibron,** 1837 EXTINCT

Félix Louis L'Herminier (1779–1833) was a pharmacist and naturalist who left France (1794) for America and settled in Guadeloupe, West Indies. He was at one time the Royal Pharmacist and Naturalist. He sent specimens to the Academy of Natural Sciences. Philadelphia. He was exiled from Guadeloupe (1815) as a result of disturbances on the island. He went to Charleston, South Carolina, eventually settling on the island of Saint-Barthélemy, U.S. Virgin Islands (then a Danish colony). Two birds are named after him. The lizard is thought to have become extinct about 1840.

Hernan

Tree Iguana sp. *Liolaemus hernani* Sallaberry, **Núñez,** and **Yanez,** 1982

The etymology states, "Esta nueva especie se denominó *L. hernani*, en homenaje al padre de uno de los autores, el que falleció mientras se realizaba la excursion" which, roughly translated, means "This new species was named *L. hernani*, in tribute to the father of one of the authors, who passed away while the excursion was taking place." We have yet to identify the author referred to.

Hernandez

Hernandez's Helmeted Basilisk *Corytophanes hernandesii* **Wiegmann,** 1831

Hernandez's Short-horned Lizard *Phrynosoma (douglasii) hernandesi* **Girard,** 1858

Dr. Francisco Hernandez (1514–1587) was a Spanish physician and botanist who trained at Universidad de Salamanca. He became personal physician to King Philip II of Spain, who ordered Hernandez to go to Mexico, where he traveled for several years (1570–1577), visiting many famous Aztec sites, particularly the renowned gardens, which the Spaniards had not yet destroyed. He consulted Aztec physicians on the medicinal qualities of plants. After he returned to Spain he handed in a huge written report, which Philip II put in the Royal Library, Monastery of Escorial, and never had published. The monastery and library were destroyed by fire (1671), but some of his work survives through copies others had made.

Herre

Negros Scaly-toed Gecko *Lepidodactylus herrei* **Taylor,** 1923

Dr. Albert William Christian Theodore Herre (1868–1962) was an ichthyologist, ecologist, botanist, and lichenologist. He gained his bachelor's degree in botany and later took both his master's (1905) and doctorate (1908), both in ichthyology, at Stanford. He became acting head, Biology Department, University of Nevada (1909–1910), was vice principal of a high school, Oakland (1910–1912), taught at a local school, Washington State (1912–1915), and was then head of the Science Department, Western Washington College of Education (1915–1919). He left to go to the Philippines, where he was Chief of Fisheries, Bureau of Science, Manila (1919–1928). He was Curator of Zoology, Natural History Museum, Stanford (1928–1946). After retiring he returned to the Philippines (1947) as a member of the Fishery Program, U.S. Fish and Wildlife Bureau, and then worked in the School of Fisheries, University of Washington (1948–1957). After his second retirement he researched and collected lichens (1957–1962).

Herrera

Todos Santos Island Kingsnake *Lampropeltis herrerae* **Van Denburgh** and **Slevin,** 1923

Herrera's Mud Turtle *Kinosternon herrerai* **Stejneger,** 1925

Herrera's Alligator Lizard *Barisia herrerae* Zaldivar-Riverón and Nieto-Montes de Oca, 2002

Professor Dr. Alfonso Luis Herrera (1868–1942) was a naturalist and biologist. He was Director, Mexican Biological Institute (1897–1918). The Museum of Zoology, Universidad Nacional Autonoma de Mexico, is named after him.

Hervey

Hervey's Forest Dragon *Gonocephalus herveyi* **Boulenger,** 1887 [Junior syn. of *G. liogaster* Günther, 1872]

Dudley Francis Amelius Hervey (1849–1911) was a colonial administrator (1870s and 1880s) in Malaya and Resident Councillor, Malacca (1883–1885), before retiring (1893). He traveled extensively in the country and collected wherever he went, sending some specimens to Boulenger. He was a fellow of the Royal Geographical Society. His interest in natural history was mostly focused on botany, and he sent several collections to Kew and others to Calcutta.

Heward

Heward's Galliwasp *Celestus hewardi* **Gray,** 1838

Robert Heward (1791–1877) was a botanist who presented a considerable quantity of plants and other material to

Kew (1862). His main interest was in Australian plants, but he also left a small collection of Jamaican plants to the library of Chelsea Physic Garden, London. He was a member of the Linnean Society, London.

Hewitt

Hewitt's Spiny-tailed Lizard *Cordylus peersi* Hewitt, 1932 [Alt. Peers' Girdled Lizard]

Hewitt's Leaf-toed Gecko *Goggia hewitti* Branch, **Bauer,** and **Good,** 1995

Dr. John Hewitt (1880–1961) was the British-born Director of the Albany Museum, Grahamstown, South Africa (1910–1958). He had been Curator of the Sarawak Museum (1905–1908), during which time he collected entomological specimens in Borneo. Thereafter he was Assistant, Lower Vertebrates, Transvaal Museum (1909–1910). There he commenced his systematic work on the South African arachnids. He studied vertebrate zoology and archeology. A mammal is named after him.

Heyer

Vanzolini's Scaly-eyed Gecko *Lepidoblepharis heyerorum* **Vanzolini,** 1978

Dr. William Ronald Heyer (b. 1941) and his wife, Miriam, are both honored in this name. He was at the Field Museum (1969) and is now Curator of Herpetology at the Smithsonian's Department of Vertebrate Zoology. He has written many papers and studies jointly with Vanzolini. See also **Miriam.**

Hielscher

Hielscher's Day Gecko *Phelsuma hielscheri* Rösler, **Obst,** and **Seipp,** 2001

Michael Hielscher, a co-discoverer of this species, was a German biochemist who used chemical methods to identify gecko species.

Hikida

Hikida's Short-legged Skink *Brachymeles apus* Hikida, 1982

Hikida's Skink *Eumeces liui* Hikida and **Zhao,** 1989

Hikida's Bow-Fingered Gecko *Cyrtodactylus matsuii* Hikida, 1990

Hikida's Forest Dragon *Hypsilurus hikidanus* **Manthey** and **Denzer,** 2006

Dr. Tsutomu Hikida (b. 1951) is a zoologist and herpetologist at the Department of Zoology, Graduate School of Science, Kyoto University, Japan.

Hilaire

Hilaire's Side-necked Turtle *Phrynops hilarii* **Duméril** and **Bibron,** 1835

Isidore Geoffroy Saint-Hilaire (1805–1861) was a French zoologist. Having studied medicine and natural history, he became an assistant to his father Étienne (1824) at Muséum National d'Histoire Naturelle, Paris. He lectured on ornithology and taught zoology (1829–1832), becoming his father's deputy at the Faculty of Science (1837). He became Inspector, Académie de Paris (1840); Professor of the museum on the retirement of his father (1841); Inspector-General of the university (1844); member of the Royal Council for Public Instruction (1845); and Professor of Zoology at the Faculty of Sciences (1850). He founded and was President of the Acclimatization Society of Paris (1854). He was particularly interested in teratology—the study of what makes organisms deviate from normal.

Hilda

Skink sp. *Mabuya hildae* **Loveridge,** 1953 [Syn. *Trachylepis hildae*]

Miss Hilda L. Sloan was Loveridge's sister-in-law. She took part in his reptile-collecting expedition to Nyasaland (Malawi) (1948–1949).

Hildebrandt

Hildebrandt's Mabuya *Mabuya hildebrandtii* **Peters,** 1874

Northern Bark Snake *Hemirhagerrhis hildebrandtii* Peters, 1878

Hildebrandt's Skink *Paracontias hildebrandti* Peters, 1880

Johann Maria Hildebrandt (1847–1881) was a German botanist and explorer who collected and traveled in Arabia, East Africa, Madagascar, and the Comoro Islands (1872–1881). He was also interested in languages and wrote *Zeitschrift für Ethiopia* (1876), which deals with the vocabularies of dialects in the Johanna Islands. He died of yellow fever in Madagascar. Two birds and two mammals are named after him.

Hill

Hill's Ctenotus *Ctenotus hilli* **Storr,** 1969 [Alt. Top-end Lowlands Ctenotus]

Gerald Freer Hill (1880–1954) was a zoologist whose major research work was a taxonomic study of termites. He was naturalist and photographer of the Commonwealth Government Exploration Party, led by Henry Barclay, which explored Central Australia and the Northern Territory (1911). Alfred Ewart, Government Botanist, Victoria, told Hill that he would be paid 5 shillings (37 U.S. cents) for every new species he collected. Hill collected more than 600 specimens during the expedition. He stayed in the Northern Territory until 1917. He worked at the Australian Institute of Tropical Medicine (1919–1922) and then for CSIR (1928–1941). He collected most of the *Ctenotus* type series.

Hillenius

Dwarf/Reed Snake sp. *Calamaria hilleniusi* **Inger** and **Marx,** 1965

Hillenius' Short-nosed Chameleon *Calumma hilleniusi* **Brygoo, Blanc,** and **Domergue,** 1973

Dick Hillenius (1927–1987) was a biologist, writer, and teacher whose speciality was the taxonomy of toads, frogs, and chameleons. He was the first Curator of Reptiles and Amphibians, Zoological Museum, Artis Amsterdam, and made a collection of chameleons. His writing is regarded as more light-hearted (and therefore readable) than the heavy scientific prose of many academics. He wrote *De vreemde eilandbewoner* (1967).

Hind

Kenya Montane Viper *Montatheris hindii* **Boulenger,** 1910

Dr. Sidney Langford Hinde (1863–1931) was Medical Officer of the Interior, British East Africa; a Captain in the Congo Free State Forces; and a Provincial Commissioner, Kenya, as well as a naturalist and collector. He wrote *The Fall of the Congo Arabs* (1897). A bird and three mammals are named after him.

Hinds

Hinds' Kaieteur Lizard *Kaieteurosaurus hindsi* Kok, 2005

Samuel Archibald Anthony Hinds (b. 1943) is a chemical engineer who is Prime Minister of Guyana. He is a noted supporter of research in Kaieteur National Park, Guyana.

Hinkel

Skink sp. *Lepidothyris hinkeli* **Wagner, Böhme, Pauwels,** and **Schmitz,** 2009

Dr. Harald Hinkel is a German naturalist whose doctorate (1994) in herpetology was awarded by Johannes Gutenberg-Universität Mainz. He was drawn into the war in Rwanda (1992–1994) and became involved in disaster relief work (1996–2000), moving to do similar work in Somalia (2000), where he was lucky to survive being shot through the throat. He co-edited *Natur und Umwelt Ruandas—Einführung in die Flora und Fauna Ruandas* (1992).

Hobart

Hobart's Anadia *Anadia hobarti* LaMarca and Garcia-Perez, 1990

North Chiapas Anole *Anolis hobartsmithi* Nieto-Montes De Oca, 2001

See **Smith, H. M.**

Hodgson

Himalayan Trinket Snake *Orthriophis hodgsonii* **Günther,** 1860 [Syn. *Elaphe hodgsonii*]

Brian Houghton Hodgson (1800–1894) worked for the East India Company and was Assistant Resident, Nepal (1825–1843). His main interest was ornithology, and his collection of 9,512 specimens included 124 species that had never been described previously. He was interested in Buddhism and was among those who introduced that philosophy to Britain in the 19th century. He was also interested in the languages of Nepal and northern India. His patriotism verged on jingoism; he once said that Cuvier (who had stolen a march on Hardwicke by naming the Red Panda because Hardwicke's return to England was delayed) would "prevent England reaping the zoological harvest of her own domains." Five mammals and 19 birds are named after him.

Hoehnel

Von Höhnel's Chameleon *Chamaeleo hoehnelii* **Steindachner,** 1891 [Alt. High-casqued Chameleon, Helmeted Chameleon]

Rear Admiral Ludwig Ritter von Höhnel (1857–1942) was an Austrian officer who spent a lot of time ashore in Africa, exploring and acting as an ambassador. He was on Count Samuel Teleki Von Szek's expedition (1886) in eastern Equatorial Africa, as cartographer and keeper of the expedition log, during which lakes Rudolph and Stefanie were discovered and named. He accompanied William Aster Chanler in East Africa (1890s). He was sent by Emperor Franz-Josef II on a diplomatic mission to the court of Menelik II of Abyssinia (Ethiopia) to sign a treaty relating to commerce. He may have been short of funds after WW1 as he wrote his memoirs (1920s) *Over Land and Sea: Memoir of an Austrian Rear Admiral's Life in Europe and Africa, 1857–1909.*

Hoesch, U.

Day Gecko sp. *Phelsuma hoeschi* **Berghof** and Trautmann, 2009

Udo Hoesch discovered this species.

Hoesch, W.

Hoesch's Mabuya *Mabuya hoeschi* **Mertens,** 1954

Dr. Walter Hoesch (1896–1961) was a German zoologist who was badly wounded in WW1, invalided out of the army, and trained as a lawyer. He emigrated to South-West Africa (Namibia) (1930) and bought a farm but was wiped out by drought (1932). He started collecting specimens for Professor Karl Jordan, an entomologist, and accompanied Jordan on his South-West African tour (1933). He made a private collection of small animals and birds. He wrote *Die vögelwelt Deutsches-Südwestafrikas namentlich des Damara- und Namalandes* (1940). A bird is named after him.

Hoffmann

Hoffmann's Earth Snake *Geophis hoffmanni* **Peters,** 1859

Dr. Carl (or Karl) Hoffmann (1823–1859) was a German physician and naturalist. He went to Costa Rica (1853) with von Frantzius and began to explore the country, collecting mainly botanical specimens. He was later a physician in Costa Rica's army. Three birds and a mammal are named after him.

Hoge

Hoge's Mabuya *Mabuya macrorhyncha* Hoge, 1946
Hoge's Worm Lizard *Amphisbaena hogei* **Vanzolini,** 1950
Hoge's Ground Snake *Liophis paucidens* Hoge, 1953
Hoge's Keelback *Helicops hogei* **Lancini,** 1964
Hoge's Side-necked Turtle *Ranacephala hogei* **Mertens,** 1967 [Syn. *Phrynops hogei, Mesoclemmys hogei*]

Dr. Alphonse Richard Hoge (1912–1982) was a Brazilian-born Belgian herpetologist. Soon after his birth, the family returned to Belgium (1913). He qualified in medicine and natural sciences at Universiteit Gent (1934). He worked in medical research until 1939, when he left for Brazil to study uses of snake venom. He joined Instituto Butantan, São Paulo, Brazil (1946), and was Director of the Biology Department (1969–1982). See also **Alphonse.**

Hohenacker

Transcaucasian Rat Snake *Zamenis hohenackeri* **Strauch,** 1873

Dr. Rudolph Friedrich Hohenacker (1798–1874) was a Swiss physician and missionary. He went to a Swabian colony, Kirovabad (Gäncä, Azerbaijan), in South Caucasus (1821). He also collected plants. He returned to Switzerland (1841) and then lived in Germany (1842–1874). Hohenacker sent a specimen of this snake to Strauch in St. Petersburg.

Holbrook

Earless Lizard genus *Holbrookia* **Girard,** 1851
Holbrook's Spotted Lizard *Holbrookia maculata maculata* Girard, 1851
Speckled Kingsnake *Lampropeltis getula holbrooki* **Stejneger,** 1902 [Alt. Salt-and-pepper Kingsnake]

Dr. John Edwards Holbrook (1794–1871) was a zoologist and herpetologist who has been described as "the Father of North American Herpetology." He qualified as a physician (1818) and went to Edinburgh for postgraduate studies (1819). He visited Paris and became friendly with the great French naturalists of the day, including Cuvier and Duméril. He returned to the USA (1822) and practiced medicine in Charleston, South Carolina. He was Professor of Anatomy, Medical School of South Carolina (1824–1854). During the American Civil War he was a surgeon in the Confederate army, despite his age. After Union troops captured Charleston and his manuscripts and collections were looted, he gave up all scientific research. He was elected to the National Academy of Sciences (1868). He wrote *North American Herpetology* (1836–1842).

Hollinrake

Hollinrake's Bronzeback *Dendrelaphis hollinrakei* **Lazell,** 2002

Dr. James Barrie Hollinrake was administrator, Shek Kwu Chau Island (part of Hong Kong) (1971–1984) as Medical Superintendent, Government Drug Rehabilitation and Treatment Centre. He collected the snake holotype.

Holmberg

Holmberg's Desert Tegu *Dicrodon holmbergi* **Schmidt,** 1957

Professor Dr. Allan R. Holmberg (1909–1966) was an anthropologist at Cornell. He was a fellow at the Center for Advanced Study in Behavioral Science, Stanford (1954). His main interest in lizards was in their use as food by the local "Indians," with whom he lived (1940–1942) while a doctoral student. In the last 20 years of his life his main focus was the local peoples of Peru. He wrote *Nomads of the Long Bow: The Siriono of Eastern Bolivia* (1950). He collected the tegu holotype (1947).

Home

Home's Hinge-back Tortoise *Kinixys homeana* **Bell,** 1827

Sir Everard Home (1756–1832), an English naturalist and physician, published on human and animal anatomy. He is remembered for having kept a wombat at home at a time when it was fashionable to own exotic pets from Australia. The wombat lived in a domesticated state at Home's house in London for two years, became attached to people it knew, and was particularly good with children. Home authored 107 papers to the Royal Society, more than anyone else.

Hoogmoed

Hoogmoed's Scaly-eyed Gecko *Lepidoblepharis hoogmoedi* Avila-Pires, 1995
Gymnophthalmid lizard sp. *Arthrosaura hoogmoedi* Kok, 2008

Dr. Marinus Steven Hoogmoed (b. 1942) is a Dutch herpetologist who was at one time at Rijksmuseum van Natuurlijke Histoire, Leiden, but now lives in Belém, Brazil, and works at Museu Paraense Emílio Goeldi.

Hoogstraal

Hoogstraal's Cat Snake *Telescopus hoogstraali* **Schmidt** and **Marx,** 1956

Harold "Harry" Hoogstraal (1917–1986) was an expert in the field of medical zoology, parasitology, entomology, and ecology. He gained two degrees from the University of Illinois (1938 and 1942), then served in the U.S. Army (1942–1946). He made a number of field trips, to Mexico (1940), the Solomon Islands, New Guinea, and New Hebrides (1945) while serving at a military medical research establishment. He went to Mindanao and Palawan (1946–1947), then Africa and Madagascar (1948–1949). He took two further degrees after WW2 at the London School of Hygiene and Tropical Medicine. He was head of the Department of Medical Zoology, United States Medical Research Unit No. 3, Cairo (1950s and 1960s). He made a small collection of reptiles in Sinai and presented it to the Field Museum. He amassed the biggest collection of ticks outside of the British Museum and gave them to the Smithsonian. He died in Cairo of lung cancer on his 69th birthday. Three mammals are named after him.

Horsfield

Horsfield's Flying Gecko *Ptychozoon horsfieldi* **Gray,** 1827

Horsfield's Tortoise *Testudo horsfieldii* Gray, 1844 [Alt. Afghan Tortoise; Syn. *Agrionemys horsfieldii*]

Horsfield's Spiny Lizard *Salea horsfieldii* Gray, 1845

Dr. Thomas Horsfield (1773–1859) was an American naturalist. He qualified as a physician and became an explorer and collector of plants and animals. He began his career in Java (1796) while it was under Dutch rule, which the Honourable East India Company took over (1811) after Napoleon Bonaparte annexed Holland. He then worked for the East India Company and became a friend of Sir Thomas Raffles. Horsfield became unwell, so the company moved him to London (1819) to continue his research as Curator, and then Keeper, of the East India House Museum. He wrote *Zoological Researches in Java and the Neighbouring Islands* (1824). Ten birds and five mammals are named after him.

Horton

Horton's Mabuya *Mabuya croizati* Horton, 1973

Dr. David Robert Horton (b. 1945) is a herpetologist at the University of New England, New South Wales. He wrote "Evolution in the Genus *Mabuya*" (1973).

Horvath

Horvath's Rock Lizard *Iberolacerta horvathi* **Mehely,** 1904

Géza Horváth (1847–1937) was an entomologist. He was Director, Zoology Department, Magyar Természettudományi Múzeum, Budapest (1895–1923), and held a similar post at the University of Budapest from 1913. He was general editor of *General Catalogue of the Hemiptera.*

Hoser

Black White-lipped Python *Leiopython hoserae* Hoser, 1989 [Schleip, 2008]

Shireen Hoser is the describer's wife. As the original description was considered to be unsatisfactory, Schleip redescribed the python without changing the scientific name.

Hosmer

Hosmer's Egernia *Egernia hosmeri* **Kinghorn,** 1955

William Hosmer (1925–2002) was an Australian herpetologist who worked at the Field Museum. He was on the Spalding-Peterson expedition to the Northern Territory, Australia (1959–1960), during which he collected a new species of frog. He was a Field Associate, Department of Herpetology, American Museum of Natural History, New York (1962). He co-wrote "A New Skink from Australia" (1959).

Houston

Houston's Dragon *Ctenophorus vadnappa* Houston, 1974 [Alt. Red-barred Crevice-Dragon]

Dr. Terry Francis Houston is an entomologist at the Western Australian Museum, Perth. He is Curator of Insects and Director of the Department of Invertebrates.

How

Kimberley Deep-soil Blind Snake *Ramphotyphlops howi* **Storr,** 1983

Dr. Richard Alfred How (b. 1944) is a zoologist at the Western Australian Museum, Perth. He has published in regard to taxonomy and geographic morphological variation. He co-wrote "Reappraisal of the Reptiles on the Islands of the Houtman Abrolhos, Western Australia" (1998). A mammal is named after him.

Howard Gloyd

Castellana *Agkistrodon bilineatus howardgloydi* Conant, 1984

See **Gloyd.**

Howell

Zanzibar Dwarf Gecko *Lygodactylus howelli* **Pasteur** and **Broadley,** 1988

Tanzanian Dwarf Gecko *Lygodactylus kimhowelli* Pasteur, 1995

Blind Snake sp. *Leptotyphlops howelli* Broadley and **Wallach,** 2007

Kim Monroe Howell (b. 1945) is Professor of Zoology and

Marine Biology, University of Dar Es Salaam. He co-wrote *A Field Guide to the Reptiles of East Africa* (2006). A mammal is named after him.

Htun Win

Agamid lizard sp. *Calotes htunwini* **Zug** and **Vindum,** 2006

Htun Win (d. 2004) joined the Chatthin Wildlife Sanctuary, Myanmar (1993), becoming a forester (1995), then started work on herpetology (1997) and became a team leader (1999). He fell ill and died while surveying the herpetofauna of Kachin State.

Hubrecht

Hubrecht's Eyebrow Lizard *Phoxophrys tuberculata* Hubrecht, 1881

Dr. Ambrosius Arnold Willem Hubrecht (1853–1915) was a biologist. He studied at Universiteit Utrecht (1870–1873) and was awarded his doctorate (1874). He moved to Leiden (1873) and was Curator, Ichthyology and Herpetology, Rijksmuseum van Natuurlijke Histoire, Leiden (1875–1882), then returned to Utrecht as Professor of Zoology. The Huprecht Laboratory in Holland is named after him. He wrote "On a New Genus and Species of Agamidae from Sumatra" (1881).

Hudson

Hudson's Coffee Snake *Ninia hudsoni* **Parker,** 1940

C. A. Hudson was a collector, mainly of entomological specimens, for the Natural History Museum, London. He collected the snake holotype in the late 1930s.

Huebner

Inirida Worm Lizard *Mesobaena huebneri* **Mertens,** 1925

George Huebner (1862–1935) was a German pioneer photographer and collector who was also interested in ethnology. He traveled in Brazil and Peru (1885–1888). He had his own photographic studios, in Manaus, Brazil (1898–1920), and Belém (1906). He collected botanical specimens, especially orchids, in the Amazon basin (1921–1935). He collected the lizard holotype (1895).

Hughes

Hughes' Green Snake *Philothamnus hughesi* Trass and **Roux-Esteve,** 1990

Hughes' Saw-scaled Viper *Echis hughesi* **Cherlin,** 1990

Barry Hughes is a British herpetologist who works at the Department of Zoology, University of Ghana.

Hulse

Hulse's Tree Iguana *Liolaemus capillitas* Hulse, 1979

Dr. Arthur Charles Hulse (b. 1945) is Professor of Biology, Indiana University, Pennsylvania. He received his Ph.D. from the Department of Zoology, Arizona State University (1974). He co-wrote *Amphibians and Reptiles of Pennsylvania and the Northeast* (2001).

Hummelinck

Hummelinck's Anole *Anolis blanquillanus* Hummelinck, 1940

Dr. Pieter Wagenaar Hummelinck (1907–2003) was a naturalist. He took his first degree at Universiteit Utrecht (1935), starting work there in the Zoological Laboratory (1940). His speciality was the study of the fauna of the Netherlands Antilles, which he frequently visited, as well as other West Indian Islands. He retired in 1972, leaving behind his significant collections of land and marine animals at the university. When the lab closed down (1988) the collection was transferred to the Zoological Museum, Artis Amsterdam. A mammal is named after him.

Humphries

Humphries' Lerista *Lerista humphriesi* **Storr,** 1971 [Alt. Humphrey's Lerista (in error)]

Dr. Robert B. Humphries is an ecologist who studied zoology and botany and took his bachelor's degree (1972) at the University of Western Australia. Australian National University, Canberra, awarded his doctorate (1980). He works for the Western Australian Water Corporation, first as Environment Manager (1996–2004) and now as Manager of Sustainability. He has an interest in the conservation of Shark Bay, from whence the lizard comes, as did Storr. Storr said in his etymology that Humphries "has contributed many reptiles to the collections of the Western Australian Museum."

Hunsaker

Hunsaker's Spiny Lizard *Sceloporus hunsakeri* **Hall** and **H. M. Smith,** 1979

Dr. Don Hunsaker II (b. 1930), whose current major interest is avian ecology and behavior, is Professor Emeritus and Adjunct Professor, San Diego State University. He took his bachelor's and master's degrees at Texas Tech University and his doctorate at the University of Texas. He is Director, Environmental Trust Regional Conservation Management, Southern California. He conducts research at Hubbs Ocean Research Institute, San Diego, investigating sea turtles in the Atlantic and Pacific oceans.

Hussam

Hussam's Mussurana *Clelia hussami* Morato, Franco, and Sanches, 2003

Associate Professor Dr. Hussam El Dime Zaher is an

evolutionary biologist who is Curator, Herpetology, Department of Zoology, Universidade de São Paulo.

Hutchinson

Hutchinson's Ground Skink *Niveoscincus orocryptus* Hutchinson et al., 1988 [Alt. Heath Cool-Skink]

Dr. Mark Norman Hutchinson (b. 1954) is a herpetologist who is a researcher and Curator, Herpetology, South Australian Museum, Adelaide, where he has been since 1990. He was awarded his bachelor's degree (1977) and his doctorate (1984), both by LaTrobe University, Melbourne. He captured a previously unknown species of taipan in Australia's Central Ranges (2006).

Hutton

Hutton's Pit-viper *Trimeresurus huttoni* **M. A. Smith,** 1949 [Syn. *Tropidolaemus huttoni*]

Angus Hutton, who discovered the snake, was a planter and naturalist in the Madurai district of Tamil Nadu, southern India. He also collected (1948) the holotype of Salim Ali's Fruit Bat *Latidens salimalii*, described as a new species only in 1972.

Huynh

Bow-fingered Gecko sp. *Cyrtodactylus huynhi* Ngo and **Bauer,** 2008

Professor Dr. Dang Huy Huynh is Director, Institute of Ecology and Biological Resources, Hanoi, Vietnam. He wrote *The Endangered Primate Rescue Center at Cuc Phuong National Park: A Refuge for Confiscated Primates* (2004).

Hygom

Reinhardt's Lava Lizard *Tropidurus hygomi* **Reinhardt** and Lutken, 1861

Captain Vilhelm Johannes Willaius Hygom (b. 1818) was a Danish merchant seaman who took his master's ticket (1839). He was also a collector of marine and land specimens (1853–1861) for Steenstrup. He found a Giant Squid *Architeuthis dux* off the Bahamas (1855).

I

Ibanez

Anole sp. *Anolis ibanezi* Poe et al., 2009

Dr. Roberto Ibáñez Diaz is a Panamanian herpetologist at the Smithsonian Tropical Research Institute, Panama City, whose doctorate was awarded by the University of Connecticut. He has taught at a number of universities, including Panama, McGill University in Canada, and La Selva Biological Station, Costa Rica. He now holds the post of Regional Director of Draft Amphibian Conservation and Rescue.

Ibarra

Ibarra's Burrowing Snake *Adelphicos ibarrorum* **Campbell** and Brodie, 1988

Professor Jorge Alfonso Ibarra (1921–2000) was Director of Museo Nacional de Historia Natural, Guatemala, and Germán A. Ibarra is a naturalist and collector. Jorge headed the Zoological Museum, Faculty of Natural Science (1936–1947). He founded the museum, which now bears his name (1948), later becoming Director (1952). He also founded the magazine *Pro Natura* (1985). The etymology reads, "The name *ibarrorum* is in honor of Guatemala's first family of conservation, especially Jorge A. Ibarra, Director of the Museo Nacional de Historia Natural, and Germán A. Ibarra, naturalist and avid collector."

Iglesias

Gomes' Pampas Snake *Phimophis iglesiasi* **Gomes,** 1915

Sertao Lancehead *Bothrops iglesiasi* **Amaral,** 1923

Francisco Iglesias was a Brazilian agronomist and zoologist. He wrote *Sobre um mammifero ophiaphago do Brazil* (1917).

Ihering, H. F. I. and R. T. G. W.

Ihering's Fathead Anole *Enyalius iheringii* **Boulenger,** 1885

Tree Lizard sp. *Anisolepis iheringii* Boulenger, 1885 [Junior syn. of *A. undulatus* Wiegmann, 1834]

Ihering's Snake *Lioheterophis iheringi* **Amaral,** 1935

Dr. Hermann Friedrich Ibrecht von Ihering (sometimes Jhering) (1850–1930) was a German-Brazilian zoologist, malacologist, and geologist. He was trained as a physician and served in the German army. He went to Rio Grande do Sul, Brazil (1880), and founded the Museu de Zoologia, São Paulo (1894), spending 22 years as its first Director (1894–1916). He returned to Germany (1924) and died there. He co-wrote *Catálogos da fauna brasileira: As aves do Brazil* (1907) with his son, Rudolpho Teodoro Gaspar Wilhelm von Ihering (1883–1939), after whom Ihering's Snake is named. Three mammals and three birds are named after Hermann.

Ijima

Turtlehead Sea Snake *Emydocephalus ijimae* **Stejneger,** 1898

Isao Ijima (1861–1921) was Professor of Zoology, Tokyo University, and first President of the Ornithological Society of Japan. Two birds are named after him.

Illingworth

Illingworth's Gecko *Calodactylodes illingworthorum* **Deraniyagala,** 1953 [Alt. Sri Lanka Golden Gecko]

Margaret and Percy Illingworth. Originally named *illingworthi* in the binomial, Bauer and Das changed it to *illingworthorum* (2000) when they realized it honored two people.

Imke

Rooiberg Girdled Lizard *Cordylus imkeae* Mouton and Van Wyk, 1994

Imke Cordes is a German biologist who is a Research Associate, Department of Biology and Environmental Sciences, Carl von Ossietzky Universität, Oldenburg, Germany. She studied for her master's degree at the University of Stellenbosch, South Africa, and collected the lizard holotype on a field trip to Namaqualand.

Indraneildas

Gecko sp. *Cnemaspis indraneildasii* **Bauer** et al., 2002

See **Das.**

Inger

Inger's Bow-fingered Gecko *Cyrtodactylus pubisulcus* Inger, 1958

Inger's Ground Snake *Liophis ingeri* **Roze,** 1958

Inger's Mabuya *Mabuya clivicola* Inger et al., 1984

Inger's Tuberculate Gecko *Phyllodactylus tuberculosus ingeri* **Dixon,** 1964

Inger's Bow-fingered Gecko *Cyrtodactylus ingeri* **Hikida,** 1990

Dibamid lizard sp. *Dibamus ingeri* **Das** and **Lim,** 2003

Reed Snake sp. *Calamaria ingeri* **Grismer,** Kaiser, and Yaakob, 2004

Dr. Robert "Bob" Frederick Inger (b. 1920) is a herpetologist and ichthyologist who is Curator Emeritus of Amphibians and Reptiles, Field Museum, having started as a University of Chicago student volunteer. Since the 1950s his special subject has been the herpetology of Southeast Asia (particularly Borneo). He co-wrote *Living Reptiles of the World* (1957).

Ingoldby

Ingoldby's Stone Gecko *Cyrtopodion kachhensis ingoldbyi* **Procter,** 1923

Captain C. M. Ingoldby was a collector of many different zoological specimens. He collected birds in Zinjan (now Zanjan) and Bandar-e Gaz (Golestan Province), Iran (1919). He also worked in Waziristan, Pakistan (1922 and 1925). He collected mammal and botanical specimens in the Gold Coast (now Ghana) (1927). He wrote "A New Stone Gecko from the Himalaya" (1922).

Ingram, C. I.

Ingram's Brown Snake *Pseudonaja ingrami* **Boulenger,** 1908

Captain Collingwood Ingram (1880–1981) was an ornithologist and botanist whose father, Sir William Ingram (1847–1924), owned the *Illustrated London News.* Collingwood Ingram traveled the world collecting specimens, including visits to Japan and Australia (1907). He was an expert on cherry trees and found a species growing in England that had become extinct in its native Japan but was successfully reintroduced a few years later. In WW1 he was a fighter pilot in the Royal Flying Corps, and his war diaries have recently been discovered. He wrote *Isles of the Seven Seas* (1937).

Ingram, G. J.

Ingram's Ctenotus *Ctenotus ingrami* **Czechura** and **Wombey,** 1982

Ingram's Litter Skink *Lampropholis adonis* Ingram 1991

McIvor River Slider *Lerista ingrami* **Storr,** 1991

Dr. Glen Joseph Ingram (b. 1951) is a herpetologist who was Senior Curator, Amphibia and Birds, Queensland Museum, from which position he is now retired. He co-wrote *Atlas of Queensland's Frogs, Reptiles, Birds, and Mammals* (1991).

Ingrid

Campbell's Galliwasp *Diploglossus ingridae* Werler and **Campbell,** 2004

Ingrid Longstron Werler (1923–2003) was the wife of the senior author, John E. Werler (1923–2004). The John and Ingrid Werler Society at the Houston Zoo, Texas, was named in their honor.

Innes

Innes' Cobra *Walterinnesia aegyptia* **Lataste,** 1887 [Alt. Desert Cobra, Walter Innes' Snake, Egyptian Black Snake]

Dr. Walter Francis Innes Bey (1858–1937) was a physician and zoologist who was Librarian and Curator, Zoological Museum, at the School of Medicine, Cairo, Egypt (1885–1918). He traveled in Egypt, Eritrea, Sinai, and Sudan. He wrote *Voyage au Nil Blanc pour des recherches zoologiques* (1902).

Ionides

Liwale Round-snouted Worm Lizard *Loveridgea ionidesii* **Battersby,** 1950

Chameleon sp. *Brookesia brachyura ionidesi* **Loveridge,** 1951 [Syn. *Rieppeleon brachyurus*]

Ionides' Purple-glossed Snake *Amblyodipsas katangensis ionidesi* Loveridge, 1951

Ionides' Sharp-snouted Worm Lizard *Ancylocranium ionidesi* Loveridge, 1955

Ionides' Monitor *Varanus albigularis ionidesi* **Laurent,** 1964

Ionides' Blind Snake *Leptotyphlops ionidesi* **Broadley** and **Wallach,** 2007

Constantine John Philip Ionides (1901–1968) was known as the "Snake Man of British East Africa." His family, of Greek origin, had been settled in England for generations. He could be described as "colorful," though "iconoclastic" is better. He was expelled from school (1917), suspected of theft, and though innocent of that was known to be a poacher of pheasants and was found to have two loaded revolvers in his desk. He went to Sandhurst and was commissioned in the British army. He wanted to go to Africa, but his regiment was posted to India. He finally got transferred to Africa (1926), where he became, successively, an ivory poacher, big game hunter, game warden (he was responsible for expanding the Selous Game Reserve), and herpetologist (1926–1968). His varied experiences include losing his hearing in one ear from being trampled by a charging elephant; being bitten by several snakes, recording the sensations, and surviving despite never using antivenin; and having the population of an entire village flogged for disobeying him. He stated that after he died, he wanted his body to be thrown out for the hyenas to eat.

Irwin

Steve Irwin's Turtle *Elseya irwini* **Cann,** 1997

Stephen Robert "Steve" Irwin (1962–2006) was the well-known owner-manager of the Australia Zoo, Queensland, which his parents founded as the Queensland Reptile and Fauna Park. Steve worked as a crocodile trapper and became famous through his *Crocodile Hunter* TV footage, which screened in 130 countries. He appeared in the film *The Crocodile Hunter: Collision Course* (2002). He put a massive amount of time and money into conservation projects in Australia and around the world. Ironically, after years dicing with death with crocodiles, he died after being struck by a stingray's barb that pierced his heart while he was filming a documentary on Queensland's Great Barrier Reef.

Isabel

Tree Iguana sp. *Liolaemus isabelae* **Navarro,** 1993

Isabel Yermany is the wife of describer José Navarro.

Ishak

Forest Skink sp. *Sphenomorphus ishaki* **Grismer,** 2006

Muhamad Ishak Mat Sohor of Kampung Juara, Pulau Tioman, Malaysia, acted as Grismer's "guide and companion on many treks up Gunung Kajang and . . . taught him a wealth of information."

Isis

Mount Jukes Broad-tailed Gecko *Phyllurus isis* **Couper, Covacevich,** and **Moritz,** 1993

Isis was a goddess in Egyptian mythology, the wife and sister of Osiris and mother of Horus. She was worshipped as the archetypal wife and mother.

Iskandar

Iskandar's Wolf Gecko *Luperosaurus iskandari* **Brown, Supriatna,** and **Ota,** 2000

Flying Dragon sp. *Draco iskandari* Mcguire et al., 2007

Dr. Djoko Tjahono Iskandar (b. 1950) is an Indonesian zoologist and herpetologist. Since 1978 he has worked at Institut Teknologi Bandung, Java, where he is Professor of Biometrics and Ecology of Small Vertebrates. After obtaining his master's degree he was a Curator at the Museum Zoologicum Bogoriense. He spent time in France (1980–1984). He is the author of *Turtles and Crocodiles of Insular Southeast Asia and New Guinea* (2000).

Issel

Issel's Toadhead Agama *Phrynocephalus isseli* **Bedriaga,** 1907

Arturo Issel (1842–1922) was an Italian geologist, malacologist, and paleontologist. He conducted marine research along the Eritrean coast in the 1870s. A mammal is named after him.

Iturra

Tree Iguana sp. *Liolaemus patriciaiturrae* **Navarro** and **Núñez,** 1993

Dr. [Iris] Patricia Iturra-Constant (b. 1947) is a geneticist and biochemist at the Faculty of Medicine, Universidad de Chile, Santiago de Chile, where she is Professor of Biology and Natural Sciences. She co-wrote *Contribución sistemática al conocimiento de la herpetofauna del extremo norte de Chile* (1982).

Ivens

Ivens' Skink *Trachylepis ivensii* **Bocage,** 1879 [Syn. *Lubuya ivensii, Mabuya ivensii*]

Roberto Breakspeare Ivens (1850–1898) was a Portuguese naval officer, explorer, and colonial administrator. He joined the Portuguese navy at age 11 and was posted overseas (1871–1874). His ship visited São Tome and various South American ports (1875–1876). He was in Philadelphia at the Universal Exhibition (1876), then went on an expedition to Angola and Mozambique (1877). He revisited Mozambique (1885) and made the overland journey between there and Angola, one-third of which was through previously uncharted territory.

Iverson

Fujian Pond Turtle *Mauremys iversoni* **Pritchard** and **McCord,** 1991

Tamaulipas Blind Snake *Leptotyphlops dulcis iversoni* **H. M. Smith** et al., 1998

Professor Dr. John Burton Iverson III (b. 1949) is a herpetologist at the Department of Biology, Earlham College, Indiana, and also Director of the Joseph Moore Museum there. His primary interest is turtles. He is involved in programs in the Bahamas, Ecuador, Costa Rica, Mexico, and Nebraska. He wrote *Checklist with Distribution Maps of the Turtles of the World* (1986, revised 1992).

Iwasaki

Iwasaki's Snail-eater *Pareas iwasakii* **Maki,** 1937

Takuji Iwasaki (1869–1937) was the chief meteorologist at Ishigaki-jima, one of the main islands of the Sakishima Group, Ryukyu Archipelago (1905). He reported that local fishermen had brought ashore a marine animal that was thought to be a Pygmy Sperm Whale *Kogia breviceps* (1935).

J

Jackson, F. J.

Jackson's Centipede Eater *Aparallactus jacksonii* **Günther,** 1888

Jackson's Black Tree Snake *Thrasops jacksoni* Günther, 1895

Jackson's Chameleon *Chamaeleo jacksonii* **Boulenger,** 1896

Jackson's Forest Lizard *Adolfus jacksoni* Boulenger, 1899

Sir Frederick John Jackson (1859–1929) was an English administrator, diplomat, explorer, naturalist, and ornithologist. He led the British East Africa Company expedition to explore the new Kenya colony (1889), becoming its first Governor. He was also Governor of Uganda (1911–1918). He wrote *The Birds of Kenya Colony and the Uganda Protectorate* (published posthumously, 1938). Five mammals and nine birds are named after him.

Jackson, G.

Jackson's Lerista *Lerista jacksoni* **L. A. Smith** and Adams, 2007

Gregory Jackson is a layout artist who since 1976 has been responsible for the design of many of the Western Australian Museum's publications.

Jackson, J. F.

Jackson's Fathead Anole *Enyalius perditus* Jackson, 1978

James Frederick Jackson (b. 1943) is a zoologist and paleontologist. He wrote "Differentiation in the Genera *Enyalius* and *Strobilurus* (Iguanidae): Implications for Pleistocene Climatic Changes in Eastern Brazil" (1978).

Jacobi

Chapa Mountain Keelback *Opisthotropis jacobi* **Angel** and **Bourret,** 1933

Dr. Arnold F. V. Jacobi (1870–1948) was an entomologist who was Professor at Technische Hochschule and Director, Staatliches Museum für Tierkunde Dresden (1906–1937). He was in Tonkin (Vietnam) around 1900. He wrote *Mimikry und Verwandte Erscheinungen* (1913).

Jacobsen

Jacobsen's Worm Snake *Leptotyphlops jacobseni* **Broadley** and Broadley, 1999

Dr. Niels H. G. Jacobsen is a South African who worked for the Transvaal Chief Directorate, Nature and Environmental Conservation (1975–1995). Since 1995 he has been a freelance ecological consultant and herpetologist. His doctoral thesis (1989) so impressed Broadley that he named the snake after him. He wrote *Remarkable Reptiles of South Africa* (2005).

Jacobson

Jacobson's Bamboo Snake *Pseudoxenodon inornatus jacobsoni* van Lidth de Jeude, 1922

Jacobson's Gecko *Cnemaspis jacobsoni* **Das,** 2005

Edward Richard Jacobson (1870–1944) was a Dutch businessman and skilled amateur naturalist. He was manager of a trading company in Java, but he also lived for some years in Sumatra. He made extensive collections for Dutch museums, leaving his business (1910) to devote himself to natural history. His main interest was entomology, but he collected other taxa types too. He died in an internment camp during the Japanese occupation.

Jacova

Velvet Gecko sp. *Oedura jacovae* **Couper,** Keim and Hoskin 2007

This is one of those made-up binomials so beloved of zoologists: it stands for Jeanette Adelaide *Covacevich* (q.v.).

Jaeger

Jaeger's Ground Snake *Liophis jaegeri* **Günther,** 1858

Dr. Georg Friedrich Jäger (1785–1866) was a paleontologist who studied in Paris under Cuvier. He practiced as a physician in Stuttgart, where he and Kaup collaborated over excavations in Baden-Württemburg and he started cataloguing the contents of the King of Baden-Württemburg's "cabinet" (1835). He wrote *Über die fossilen Reptilien welche in Württemberg aufgefunden worden* (1828).

Jagor

Jagor's Water Snake *Enhydris jagorii* **Peters,** 1863

Jagor's Sphenomorphus *Sphenomorphus jagori* Peters, 1864

Professor Dr. Fedor Jagor (1817–1900) was a German ethnographer and naturalist who traveled in Asia, including the Philippines, collecting for Museum für Naturkunde Berlin in the second half of the 19th century. He wrote *Reisen in den Philippinen* (1873). Two mammals are named after him.

Jalla

Jalla's Sand Snake *Psammophis jallae* **Peracca,** 1896

Worm Lizard sp. *Tomuropeltis jallae* Peracca, 1910 [Junior syn. of *Dalophia pistillum* **Boettger,** 1895]

Rev. Luigi Jalla was an Italian missionary in Southern Rhodesia (now Zimbabwe). Peracca wrote (1886) that Jalla collected reptiles and amphibians "along the road from Kazangula to Bulawayo."

James

James' Tree Iguana *Liolaemus jamesi* **Boulenger,** 1891

Henry Berkeley James (1846–1892) was a British businessman who spent nearly 20 years in Chile. He began work as a clerk in Valparaíso and became Manager of a saltpeter mine near the Peruvian border (1871). He narrowly escaped death when an earthquake and tidal wave destroyed his home. He first collected Lepidoptera after a mule journey to the remote Chanchamayo in central Peru, where he went to see its endemic birds. Following the war between Chile and Peru and Bolivia over saltpeter mining (1879), he left for England, returning to Chile once more (1881) before retiring to England (1885). A bird is named after him.

Jameson

Jameson's Mamba *Dendroaspis jamesoni* Traill, 1843

Professor Robert Jameson (1774–1854) was a mineralogist, geologist, and naturalist who studied mining at Freiburg (1800) and was Regius Professor of Natural History, Edinburgh University (1804–1854). He was apprenticed to a surgeon but never qualified. Among his more celebrated pupils at Edinburgh was Charles Darwin, who said he found Jameson's lectures boring. Jameson was editor of the *Edinburgh New Philosophical Review* in which Traill published his description of the mamba. He wrote *Manual of Mineralogy* (1821).

Jan

Jan's Banded Snake *Simoselaps bertholdi* Jan, 1859
Jan's Blind Snake *Typhlops mirus* Jan, 1860
Jan's Pine Snake *Pituophis deppei jani* **Cope,** 1860 [Alt. Jan's Mexican Bullsnake]
Jan's Diadem Snake *Apostolepis dimidiatus* Jan, 1862
Jan's Earth Snake *Adelphicos quadrivirgatus* Jan, 1862
Jan's Kukri Snake *Rhynchocalamus melanocephalus* Jan, 1862
Jan's Shovelsnout Snake *Prosymna janii* Bianconi, 1862
Jan's Snake *Elapotinus picteti* Jan, 1862
Jan's False Coral Snake *Erythrolamprus bizona* Jan, 1863
Jan's Thread Coral Snake *Leptomicrurus narduccii* Jan, 1863
Jan's Forest Snake *Taeniophallus occipitalis* Jan, 1863
Jan's Green Racer *Philodryas varius* Jan, 1863
Jan's Ground Snake *Liophis albiventris* Jan, 1863
Jan's Hognose Snake *Lystrophis histricus* Jan, 1863
Jan's Snail-eater *Dipsas incerta* Jan, 1863
Jan's Tree Snake *Sibynomorphus vagus* Jan, 1863
Jan's Worm Snake *Typhlops exiguus* Jan, 1864
Jan's Cliff Racer *Coluber rhodorachis* Jan, 1865 [Alt. Braid Snake; Syn. *Platyceps rhodorachis*]
Texas Night Snake *Hypsiglena jani* **Dugès,** 1866
Jan's Snail-eater *Dipsas alternans* **Fischer,** 1885
Jan's Centipede Snake *Tantilla jani* **Günther,** 1895

Professor Georg (sometimes Giorgio or Georges) Jan (1791–1866) was an Austrian-born Italian taxonomist, zoologist, botanist, and writer. He started as an Assistant, Universität Wien, but moved to Italy to become Professor of Botany and Director, Botanical Gardens, Università degli Studi di Parma (then a duchy belonging to Austria). He was Director, Museo Civico di Storia Naturale, Milan, which was founded in 1838; a bequest stipulated that Jan, who contributed his own collections, was to be its Director. Both *Dipsas alternans* and *D. incerta* have at times been given the vernacular name "Jan's Snail-eater."

Janalee

Ground Snake sp. *Liophis janaleeae* **Dixon,** 2000

Professor Dr. Janalee Paige Caldwell (b. 1942) is a rainforest biologist and zoologist at Sam Noble Oklahoma Museum of Natural History, University of Oklahoma, where she is Curator of Amphibians. The University of Kansas awarded her doctorate (1974).

Janet

Skink genus *Janetaescincus* **Greer,** 1970

Janet Greer is the describer's younger sister.

Janeth

Ground Snake sp. *Atractus janethae* Silva Haad, 2004

Janeth Silva Collazos is the describer's sister.

Jansen

Jansen's Rat Snake *Elaphe jansenii* **Bleeker,** 1858 [Syn. *Gonyosoma jansenii*]

Albert Jacques Frédéric Jansen was an administrator in the Dutch East Indies (now Indonesia) and lived on Sulawesi. He was appointed Governor-General of Batavia (now Jakarta) (1848). He also collected botanical specimens, which he sent to the Dutch National herbarium in the 1850s. A fish is named after him.

Jarecki

Jarecki's Flying Dragon *Draco jareckii* **Lazell,** 1992

Dr. Henry G. Jarecki (b. 1933) started life in Stettin in Poland, qualified as a physician at Heidelberg, Germany, and practiced as a psychiatrist in the USA. He gave up medicine (1970) to become a commodities dealer and made a fortune trading in precious metals. He sold that business (1986) to run an investment bank. He bought the island of Guana, British Virgin Islands (1974), and had it run as a nature reserve. His niece, Dr. Lianna Louise Jarecki, is a herpetologist who works closely with Lazell.

Jarnold

Lined Rainbow Skink *Carlia jarnoldae* **Covacevich** and **Ingram,** 1975

"Jarnold" is Dr. Jennifer "Jenny" Mary Arnold (d. 1989), a zoologist who worked at the University of Western

Australia and the Environmental Protection Authority. After her death the Jennifer Arnold Memorial Research Grant was established at the University of Western Australia. She wrote "A Taxonomic Study of the Lygosomid Skinks of Queensland" and *Perth Wetlands Resource Book* (1966).

Jarujin

Jarujin's Forest Gecko *Cyrtodactylus jarujini* Ulber, 1993

Dr. Jarujin Nabhitabhata is a biologist who was Curator of Reptiles and Amphibians, Ecological Research Division, Thailand Institute of Scientific and Technological Research. He is currently Director, Natural History Museum, Bangkok. He co-wrote *A Photographic Guide to Snakes and Other Reptiles of Peninsular Malaysia, Singapore, and Thailand* (1998).

Jason

Jason's Mountain Reed Snake *Macrocalamus jasoni* **Grandison,** 1973

John Jason Gathorne-Hardy (b. 1968) originally trained as a zoologist and is now an artist. He succeeded to his father's courtesy title when his father became Fifth Earl of Cranbrook (1978). The father had been an Assistant, Sarawak Museum, and was involved with the Gunong Benom expedition (1967) during which the snake holotype was collected. Grandison obviously intended the honorific as a compliment to the young boy's father.

Jayakar

Jayakar's Lizard *Lacerta jayakari* **Boulenger,** 1887 [Syn. *Omanosaura jayakari*]

Arabian Sand Boa *Eryx jayakari* Boulenger, 1888

Jayakar's Agama *Trapelus jayakari* **Anderson,** 1896

Colonel Atmaram S. G. Jayakar (1844–1911) was an Indian surgeon. The Indian Medical Service sent him to Muscat (1878), and during his 30 years in the Oman area he studied the local wildlife and collected specimens, which he donated to Natural History Museum, London (1885–1899). He has a mammal named after him.

Jean

Jean's Spiny-tailed Gecko *Strophurus jeanae* **Storr,** 1988 [Alt. Southern Phasmid Gecko]

Miss Jean White is a member the Western Australian Museum's Department of Ornithology and Herpetology.

Jeanne

Lacertid lizard sp. *Psammodromus jeanneae* **Busack,** Salvador, and **Lawson,** 2006

Jeanne A. Visnaw (d. 2005) was the wife of Stephen D. Busack. She collected the holotype (1982).

Jellesma

Kabaena Bow-fingered Gecko *Cyrtodactylus jellesmae* **Boulenger,** 1897

Eeltje Jelles Jellesma (1851–1918) joined the Dutch East Indian Civil Service in 1870. He was in Celebes (now Sulawesi) in the period 1892–1903. He was the Resident, Manado, where he collected botanical specimens and facilitated the explorations of the Sarasins (q.v.).

Jensen

Jensen's Ground Snake *Xenopholis undulatus* Jensen, 1900

Professor Adolph Severin Jensen (1866–1953) was a zoologist and ichthyologist. He did much work on the fauna of Greenland and made several expeditions there. He was Malacological Curator, Zoological Museum, Københavns Universitet. He wrote *The Musk-Oxen in Greenland and Its Future* (1929).

Jerdon

Jerdon's Sea Snake *Kerilia jerdoni* **Gray,** 1849

Jerdon's Kukri Snake *Oligodon venustus* Jerdon, 1853

Jerdon's Snake-eye *Ophisops jerdonii* **Blyth,** 1853 [Alt. Jerdon's Cabrita, Syn. *Cabrita jerdonii*]

Jerdon's Many-tooth Snake *Sibynophis subpunctatus* **Duméril** and **Bibron,** 1854 [Alt. Jerdon's Polyodont]

Jerdon's Day Gecko *Cnemaspis jerdoni* **Theobald,** 1868

Jerdon's Forest Lizard *Calotes jerdoni* **Günther,** 1870

Jerdon's Pit-viper *Protobothrops jerdonii* Günther, 1875

Jerdon's Worm Snake *Typhlops jerdoni* **Boulenger,** 1890

Thomas Claverhill Jerdon (1811–1872) was a physician, zoologist, and botanist. He studied medicine at Edinburgh and became an Assistant Surgeon in the East India Company. He wrote *Birds of India* (1862–1864). Thirteen birds and a mammal are named after him.

Jicar

Jicar's Snake-Lizard *Lialis jicari* **Boulenger,** 1903 [Alt. Papua Snake-Lizard]

A. H. Jiear was Resident Magistrate, Daru, British New Guinea. He wrote regular reports, many of which have been the basis of later anthropological research. He presented the lizard holotype to the Natural History Museum, London, where Boulenger seems to have misread his name as "Jicar."

Jintakune

Jintakune's Kukri Snake *Oligodon jintakunei* **Pauwels, Wallach,** David, and **Chanhome,** 2002

Piboon Jintakune (b. 1960) is a herpetologist who collected the snake holotype in 1990. He has an M.S. degree and works at the Queen Saovabha Memorial

Institute, Thai Red Cross Society, Bangkok. He co-wrote "Venomous Snake Husbandry in Thailand" (2001).

Joan D.

Black-soil Ctenotus *Ctenotus joanae* **Storr,** 1969

Joan Maureen Dixon was Curator of Vertebrates, National Museum of Victoria; she is now Curator Emeritus. This skink was named after her "in appreciation of the loan of the splendid collection in her care."

Joan M.

Joan's Snake *Coniophanes joanae* **Myers,** 1966

Joan Wilson Myers is the describer's wife.

Joanna

Joanna's Leposoma *Leposoma ioanna* **Uzzell** and Barry, 1971

We do not know if this reptile is named after a person, as the etymology in the original description is completely unhelpful. We are told only that "the name *ioanna* is from the Greek Ιωαννα; it is feminine and used in apposition to the generic name."

Jobert

Jobert's Ground Snake *Psomophis joberti* **Sauvage,** 1884

Dr. C. Jobert was a French zoologist and collector of botanical and ichthyological specimens in Brazil. He observed the first examples of nematode disease in Brazilian coffee plantations (1878). He published a report (1878) in which he gave the ingredients and method of preparation of the poison curare, as practiced by the Tecuna Indians.

Joger

Joger's Saw-scaled Viper *Echis jogeri* **Cherlin,** 1990 [Alt. Joger's Carpet Viper, Mali Carpet Viper]

Professor Dr. Ulrich Joger is a German herpetologist who is Director, Staatlisches Naturhistorisches Museum, Braunschweig, which collaborates closely with the St. Petersburg Museum, where Cherlin is a staff member. He co-wrote "Evolution of Viperine Snakes" (1997).

Johan

Johan's Water Snake *Xenochrophis sanctijohannis* **Boulenger,** 1890

See **St. John, O. B. C.**

Johann

Johann's Keelback *Amphiesma johannis* **Boulenger,** 1908

Rev. John Graham ("Johannis" is just "John" in dog Latin). Graham sent three specimens of this snake to Boulenger. He also made a collection of fishes in Yunnan.

Johanna

Johanna's Skink *Amphiglossus johannae* **Günther,** 1880

Named after Johanna (Anjouan) Island in the Comoros.

John

John's Sand Boa *Eryx johnii* **Russell,** 1801 [Alt. Indian Sand Boa]

Rev. Dr. Christoph Samuel John (1747–1813) was a botanist and herpetologist and a medical missionary (1771–1813) at the Danish trading station of Tranquebar (now Tharangambadi), Tamil Nadu, not far from Madras (Chennai). It was a Danish colony from 1620 to 1845, when Denmark sold its possessions in India (including the Nicobar Islands) to Great Britain. Among John's friends was William Roxburgh, the botanist who lived in Madras in charge of the botanical gardens there. John was awarded an honorary doctorate (1795) for his studies in natural history. A mammal is also named after him.

Johnson

Centipede Snake sp. *Tantilla johnsoni* **Wilson,** Vaughan and **Dixon,** 1999

Dr. Jerry Douglas Johnson (b. 1947) is a zoologist and biologist. His bachelor's degree was awarded by Fort Hays State University (1972), his master's by the University of Texas, El Paso (1975), and his doctorate by Texas A&M University (1984). He was an Instructor in Biology at El Paso Community College (1975–2000) and became Professor of Biological Sciences, University of Texas, El Paso, and Director, Indio Mountains Research Station (2000).

Johnston

Johnston's Three-horned Chameleon *Chamaeleo (Trioceros) johnstoni*

Boulenger, 1901 [Alt. Johnston's Chameleon, Ruwenzori Three-horned Chameleon]

Johnston's Long-tailed Lizard *Latastia johnstonii* Boulenger, 1907

Sir Harry Hamilton Johnston (1858–1927) was a formidable explorer, colonial administrator, painter, photographer, cartographer, naturalist, and writer. He was a larger-than-life character and became known as the "Tiny Giant," being just 152 centimeters (5 feet) tall. He started exploring tropical Africa in 1882 and traveled over most of Africa, both south of the Sahara, meeting Henry Morton Stanley in the Congo (1883), and in East Africa (1884). He joined the colonial service (1885), serving in all parts of Africa, and established a British Protectorate in Nyasaland (now Malawi). Johnston was Queen Victoria's first Commissioner and Consul-General to British Central Africa. He spoke over 30 African languages as well as Arabic, Italian, Spanish, French, and Portuguese. He made the first Edison cylinder recordings in Africa, which have preserved his squeaky voice for posterity. He retired in 1904 to continue his pursuit of natural history. He discovered 100 new species, the Okapi *Okapia johnstoni* being the most memorable, and five other mammals and two birds are named after him.

Johnstone, R. A.

Johnston's Crocodile *Crocodylus johnstoni* **Krefft,** 1873 [Original spelling = *johnsoni* in error]

Robert Arthur Johnstone (1843–1905) was an explorer and policeman, sometimes known as "Snake," because he often teased children by producing snakes from his shirt. He managed a grazing property in Queensland (1865–1868) and a sugar plantation (1868–1871). He was appointed Sub-Inspector of Native Police, Cardwell District (1871). He was on George Dalrymple's northeast coast expedition. Johnstone discovered and named the Barron River when searching (1876) for a route over the ranges behind Trinity Bay to serve the new goldfields. His explorations led to the founding of Cairns and Innisfail. He was cleared of charges of having taken overextreme measures against the Aboriginals who killed the captain and crew of the *Maria*. When investigating the Green Island massacres (1873) he discovered the Johnstone River, now named after him, as is a type of possum.

Johnstone, R. E.

Rough Brown Rainbow-skink *Carlia johnstonei* **Storr,** 1974

Dr. Ronald "Ron" Eric Johnstone (b. 1949) is an ornithologist, West Australian Museum, Perth. He co-wrote *Lizards of Western Australia* (1981). A mammal is named after him.

Jonathan

Jonathan's Lancehead *Bothrops jonathani* **Harvey,** 1994

See **Campbell, J. A.**

Jones, C. R.

Jones' Girdled Lizard *Cordylus jonesii* **Boulenger,** 1891 [Alt. Limpopo Girdled Lizard, Jones' Armadillo Lizard]

The etymology says (and we cannot add to it) that the type specimen "was obtained by Mr. C. R. Jones . . . and presented by him to the British Museum."

Jones, R. E.

Jones' Imbricate Alligator Lizard *Barisia jonesi* Guillette and **H. M. Smith** 1982 [Syn. *B. imbricata jonesi*]

Dr. Richard Evan Jones (b. 1940) is an evolutionary biologist who was awarded his doctorate by the University of California (1968). He taught at the Department of Environmental, Population, and Organismic Biology, University of Colorado, where the two describers of this lizard were among his colleagues. He is now Professor Emeritus. He co-wrote "Further Observations on Arginine Vasotocin-Induced Oviposition and Parturition in Lizards" (1982).

Jordan

Jordan's Girdled Lizard *Cordylus jordani* **Parker,** 1936

Dr. Heinrich Ernst Karl Jordan (1861–1959) was an entomologist, botanist, and zoologist who after graduating at Georg-August-Universität Göttingen (1886) taught in a secondary school until going to work at Rothschild's Museum, Tring. He lived in England (1893–1959). He led an expedition to South-West Africa (Namibia) and Angola in the early 1930s. He was President of the Entomological Society, London (1929–1930). He described 2,575 new species himself, plus a further 851 as a collaborator with the Rothschilds.

José

Josés' Lizard *Liolaemus josephorum* **Núñez,** 2001

This lizard is named after two Chilean herpetologists, Professor José Navarro and José Yáñez. Navarro is a herpetologist and cytologist at Universidad de Chile. Yáñez, a zoologist and natural scientist, is Curator of Cetaceans, Museo Nacional de Historia Natural, Santiago de Chile. Navarro co-wrote "*Liolaemus patriciaiturrae* and *Liolaemus isabelae*, Two New Species of Lizards for Northern Chile: Biogeographic and Cytotaxonomic Aspects (Squamata, Tropiduridae)" (1993).

José, J. A.

Tree Iguana sp. *Liolaemus josei* Abdala, 2005

José Simón Abdala is Professor of Dentistry, Universidad Nacional de Cuyo, Mendoza, Argentina, and is the describer's father.

Joshua

Joshua's Blind Snake *Leptotyphlops joshuai* **Dunn,** 1944

The snake holotype, collected at a place called Jerico, Colombia, was unmarked and unremarked for years in a museum in Bogotá. In choosing the name Dunn may have been inspired by the biblical episode of Joshua's capture of the city of Jericho.

Josy

Sun Tegu sp. *Euspondylus josyi* Köhler, 2003

Köhler's description honored Franz-Josef "Josy" Hans for his support of taxonomic studies through the BIOPAT program.

Joynson

Joynson's Kukri Snake *Oligodon joynsoni* **M. A. Smith,** 1917

H. W. Joynson collected the holotype (1917). He played for the Singapore Cricket Club (1907) and was a member of the Natural History Society of Siam (1919). He wrote "Occurrence of the Rat Snake (*Zaocys carinatus*) in North Siam" (1927).

Juan Gundlach

Finca Ceres Anole *Anolis juangundlachi* **Garrido,** 1975

See **Gundlach.**

Juarez

Sierra Juarez Alligator Lizard *Mesaspis juarezi* Karges and **Wright,** 1987

Named after a mountain range in Oaxaca, Mexico.

Juarez, B.

Earth Snake sp. *Geophis juarezi* Nieto-Montes de Oca, 2003

Don Benito Pablo Juárez Garcia (1806–1872) was a Zapotec Indian who became both the first civilian and first indigenous person to be President of Mexico. He studied law at Universidad Autónoma Benito Juárez de Oaxaca, as the Institute of Arts and Sciences, Oaxaca, has been renamed after him. He received his degree (1831) and became a judge and Governor of his state (1841). He was Minister of Justice and Public Instruction (1855) and President of the Supreme Court (1857). During a conservative revolt (1858–1860) he acted as President but was forced to flee Mexico City. He was a ruler in exile during the period of the usurpation (1864–1867) by the French-backed Emperor Maximilian, whom he later had publicly executed. His great achievement was that during his years in government he succeeded in reducing the influence of the military and in curbing church power and wealthy landlords' privileges.

Judge

Barrier Skink *Oligosoma judgei* Patterson and Bell, 2009

Murray Judge and Bronwyn Judge of Oamaru, New Zealand, are doctors who are also keen rock climbers and members of the New Zealand Alpine Club. They discovered this skink in 2005 on Barrier Knob in the Darran Mountains, Fiordland. Because two persons are mentioned in the etymology, the scientific name should be the plural *judgeorum*.

Jukes, B.

Jukes' Turtle *Elseya jukesi* Wells, 2007

Brian Jukes was an Australian naturalist who was once resident in the area where the turtle is found (the holotype was taken at Pul Pul Creek, South Alligator River, Northern Territory). A noted herpetologist, Jukes collected in the Northern Territory, taking several holotypes of reptiles in the 1970s and 1980s.

Jukes, J.

Olive-brown Sea Snake *Hypotropis jukesii* **Gray,** 1846
[Junior syn. of *Aipysurus laevis* Lacépède, 1804]

Joseph Beete Jukes (1811–1869) was a geologist who took part in the geological survey of Newfoundland (1839–1840). He was on board HMS *Fly* (1842–1846), which was surveying the waters off Timor and northern Australia. He was Director of the Geological Survey of Ireland (1850–1869). He wrote *Narrative of the Surveying Voyage of HMS Fly* (1847). He died after a fall from his horse.

Julia C.

Julia's Ground Snake *Liophis juliae* **Cope,** 1879

Julia Collins (1866–1959), née Cope, was Edward D. Cope's only child.

Julia Z.

Tuxtlan Earth Snake *Geophis juliai* Pérez-Higareda, **H. M. Smith,** and López-Luna, 2001

Dr. Jordi Juliá Zertuche (1918–1985) was a Mexican entomologist and medical herpetologist who worked with Gonzalo Pérez-Higareda (one of the describers of this snake). He wrote "Mexican Reptiles of Significance for Public Health and Their Geographic Distribution" (1981)—in view of which it is ironic that he died of snakebite.

Julien

Aruba Leaf-toed Gecko *Phyllodactylus julieni* **Cope,** 1885

Dr. Alexis Anastay Julien (1840–1919) was an American geologist. He went to the guano island of Sombrero as the resident chemist (1860), staying in the Lesser Antilles until 1864. He sent his collections of birds and land shells to the Smithsonian. He surveyed the islets around Sankt Bartholomeus (1862, then a Swedish possession and now Saint Barthélemy, French West Indies) and visited Curacao, Bonaire, and Aruba (1881–1882). The University of New York awarded his doctorate (1882).

K

Kamdem Toham

Kamdem Toham's Gecko *Hemidactylus kamdemtohami* **Bauer** and **Pauwels,** 2002

Dr. André Kamdem Toham is a Cameroonese ichthyologist and landscape ecologist who is the Regional Representative, Gabon, for the World Wildlife Fund Central Africa.

Karanshah

Pit-viper sp. *Trimeresurus karanshahi* **Orlov** and Helfenberger, 1998 [Syn. *Himalayophis karanshahi*]

Professor Dr. Karan Bahadur Shah is a zoologist and researcher at the Nepalese National Natural History Museum, Tribhuvan University. He wrote "Checklist of the Herpetofauna of Nepal with English and Vernacular Names" (1998).

Karche

Naturelle Leaf Chameleon *Brookesia karchei* **Brygoo, Blanc,** and **Domergue,** 1970

Professor Dr. Jean-Paul Karche is a biogeologist and vulcanologist who was at the Laboratory of Geosciences, Université de Franche-Comté, in the 1980s. He worked at Université du Nord Madagascar (1960s and 1970s). He co-wrote *The Comores Archipelago in the Western Indian Ocean* (1986).

Karen

Burmese Leaf Gecko *Hemidactylus karenorum* **Theobald,** 1868

Named after the Karens, a hill tribe in Myanmar.

Karl Schmidt

Chinese Bamboo Snake *Pseudoxenodon karlschmidtii* **Pope,** 1928

Karl Schmidt's Lerista *Lerista karlschmidti* Marx and **Hosmer,** 1959 [Alt. Karl's Lerista, Lesser Robust Fine-lined Slider]

See **Schmidt, K. P.**

Karsten

Karsten's Girdled Lizard *Zonosaurus karsteni* **Grandidier,** 1869

Grandidier gives no explanation for his choice of name. It may be in honor of the botanist Hermann Karsten (1817–1908), or perhaps Grandidier made a spelling mistake for Kersten (q.v.).

Karunaratne

Karunaratne's Horned Lizard *Ceratophora karu* Pethiyagoda and Manamendra-Arachchi, 1998

G. Punchi Banda "Karu" Karunaratne (1930–1996) was a Sri Lankan zoologist, entomologist, and herpetologist who was Curator of Entomology at the Sri Lankan National Museum in Colombo. He wrote *Fauna of the Samanalawewa Area* (1992). An amphibian is named after him.

Kasner

Kasner's [Dwarf] Burrowing Skink *Scelotes kasneri* **FitzSimons,** 1939

J. H. Kasner collected the holotype in 1917, but we know nothing more about him. This is a good example of how much time can elapse between a species being collected and its being described.

Kästle

Kästle's Pond Turtle *Mauremys leprosa wernerkaestlei* **Schleich,** 1996

Werner Kästle is a German herpetologist. He works closely with Schleich, with whom he co-wrote *Amphibians and Reptiles of Nepal* (2002).

Kate

Leaf-tailed Gecko sp. *Saltuarius kateae* **Couper, Sadlier, Shea,** and Wilmer, 2008

Kate Couper is, we assume, the senior author's wife, as the description says that this gecko was named "for her ongoing support during the field component of this project."

Katrina

Water Python genus *Katrinus* Hoser, 1999

Katrina Hoser was the author's mother.

Kaulback

Smith's Japalure *Japalura kaulbacki* **M. A. Smith,** 1937

Kaulback's Lance-headed Pit-viper *Protobothrops kaulbacki* Smith, 1940

Lieutenant-Colonel Ronald John Henry Kaulback (1909–1995) was a botanist and explorer. He collected herpetofauna in Upper Burma (now Myanmar) (1930s). He went to Tibet and China with Frank Kingdon-Ward (1933) and was again in Tibet with J. Hanbury-Tracy (1935–1936). He wrote *Tibetan Trek* (1936).

Kaznakov

Kaznakov's Viper *Vipera kaznakovi* **Nikolsky,** 1909 [Alt. Caucasus Viper]

Aleksandr N. Kaznakov was a Russian naturalist who became Director, Caucasus Museum. He accompanied Kozlov on his expedition to Mongolia (1907–1909).

Kelaart

Kelaart's Gecko *Cnemaspsis kandiana* Kelaart, 1852
Kelaart's Slender Skink *Lankascincus taprobanensis* Kelaart, 1854

Lieutenant-Colonel Edward Frederick Kelaart (1819–1860) was a physician and zoologist born in Ceylon (now Sri Lanka). He qualified at Edinburgh and also studied in Paris. He was in the Ceylon medical service but also served in Gibraltar (1843–1845). He was appointed the Ceylon Government Naturalist, which paid £200 per annum (a lot of money in the mid-19th century) plus expenses on top of his army pay. One task was to investigate why the Ceylon Pearl Fisheries produced no profit. He investigated the life history of the pearl oyster and wrote four reports on the subject (published 1858–1863, the latter ones posthumously). In 1860 the Governor of Ceylon became ill and sailed for England. His health was of such concern that Kelaart, with his wife and five children, accompanied him as personal medical attendant. The Governor died two days before *Nubia* arrived at Southampton, and Kelaart died the following day. He wrote *Prodromus fauna Zeylanica* (1852). Four birds and three mammals are named after him.

Keller

Keller's Bark Snake *Hemirhagerrhis kelleri* **Boettger,** 1893

Professor Conrad Keller (1848–1930) was a German naturalist who went on an expedition to Somaliland. He wrote *Die Ostafrikanischen Inseln* (1898).

Kellogg

Kellogg's Coral Snake *Sinomicrurus kelloggi* **Pope,** 1928
[Syn. *Calliophis kelloggi, Hemibungarus kelloggi*]

Claude Rupert Kellogg (1886–1977), a zoologist and entomologist, worked and collected in Foochow District, China (1911–1941). Apart from teaching zoology at Anglo-American College, Foochow, as well as Fukien Christian University, he was a beekeeper and a missionary.

Kemp

Kemp's Ridley Turtle *Lepidochelys kempii* **Garman,** 1880

Richard Moore Kemp (1825–1908) is described as being a fisherman from Key West, Florida, and a man who was very interested in natural history. He was also a merchant and furniture dealer. He drew Garman's attention to the turtle (1877).

Kempton

Kempton's Anole *Anolis kemptoni* **Dunn,** 1940

Dr. Kempton Potter Aiken Taylor was a physician who qualified (1919) at the University of Pennsylvania. He was born Kempton Potter Aiken but took the surname, Taylor, of a relative who adopted him after his father, a doctor with severe financial problems, murdered his wife and committed suicide (1900). He worked in Guatemala (1928) and later in Panama, where Dunn, an old friend, visited him. His older brother, who was not adopted, was the poet Conrad Aiken (1889–1973).

Kendall

Kendall's Rock Gecko *Cnemaspis kendallii* **Gray,** 1845

Gray gave no information about Kendall when naming this gecko, and we have been unable to trace him.

Kendrick

Dark Broad-blazed Slider *Lerista kendricki* **Storr,** 1991

Dr. Peter G Kendrick is an Australian zoologist who was at the Western Australian Museum, where his doctorate was awarded (1991). Since 1989 he has been working for the Department of Conservation and Land Management, first on the Kimberley Rainforest Survey and in Pilbara (1992) as Regional Ecologist. He wrote "Two New Species of *Lerista* (Lacertilia: Scincidae) from the Cape Range and Kennedy Range of Western Australia" (1989).

Kenneally

Kenneally's Gecko *Diplodactylus kenneallyi* **Storr,** 1988

Kevin Francis Kenneally (b. 1945) is a botanist who is Scientific Coordinator of "Landscope," a program for the Western Australian Department of Conservation and Land Management. He first joined the herbarium that is part of CALM in 1972. When Storr was at the Western Australian Museum (1983) studying the herpetofauna of the Kimberley area, Kenneally was studying the flora of the area, on which he is considered a world authority. A number of plants are named after him.

Kennedy

Kennedy's Lerista *Lerista kennedyensis* **Kendrick,** 1989
[Alt. Kennedy Range Broad-blazed Slider]

Named after the Kennedy Range of mountains in Western Australia.

Kennedy, W. P.

Kennedy's Leafnose Snake *Lytorhynchus kennedyi* **Schmidt,** 1939

Dr. Walter P Kennedy was a physician who was Professor of Biology at the Royal College of Medicine, Baghdad, from where he sent specimens to the Field Museum. He worked at the Department of Physiology, Edinburgh University (1929), and at St. Mary's Hospital (1929–1932). During WW2 he was a Captain in the Royal Army Medical Corps, based in Baghdad. He wrote "Some Additions to the Fauna of Iraq" (1937).

Kennerly

Kennerly's Hog-nosed Snake *Heterodon nasicus kennerlyi* **Kennicott,** 1860 [Alt. Mexican Hog-nosed Snake]

Dr. Caleb Burwell Rowan Kennerly (1829–1861) graduated with a bachelor's degree from Dickinson College, Pennsylvania (1849), and then studied medicine, being awarded his doctorate by the University of Pennsylvania (1852). Baird helped him by getting him appointed as surgeon and naturalist to a number of expeditions sponsored by the government: the Pacific Railroad Survey (1853–1854), United States / Mexican Boundary Survey (1855–1857), and United States / United Kingdom joint Northwestern Boundary Survey (1857–1861). Kennerly kept up a correspondence with Baird, who credited him with many discoveries. He died of a sudden brain disorder while returning to Virginia from California and was buried at sea.

Kennicott

Kennicott's Water Snake *Thamnophis valida* Kennicott, 1860 [Alt. West Coast Garter Snake, Syn. *Nerodia valida*]

Robert Kennicott (1835–1866) was a naturalist who founded the Chicago Academy of Sciences and who explored the American Northwest (1857–1859). He worked for Baird at the Smithsonian, helping classify animals collected on the western frontier by army personnel involved in railroad surveys. He went to Canada and met Hudson Bay's chief trader, Bernard Ross, who became a close friend. After a period as a curator in Chicago he left to explore "Russian America" and spent the rest of his life in Alaska. He died of a heart attack. Two birds and several fish are named after him, as is the Alaskan town Kennicott.

Kersten

Kersten's Pygmy Chameleon *Rieppeleon kerstenii* **Peters,** 1898 [Alt. Kenya Stumptail Chameleon; Syn. *Rhampholeon kerstenii*]

Dr. Otto Kersten (1839–1900) was a German chemist and traveler. He was with Baron von der Decken in the unsuccessful attempt to climb Mount Kilimanjaro (1862). Kersten published six volumes of memoirs (1869–1879).

Key

Key's Teiid *Teuchocercus keyi* **Fritts** and **H. M. Smith,** 1969

Dr. George Key (1942–1999) was a physician and amateur herpetologist who collected amphibians and reptiles in Ecuador for the Harvard Museum of Comparative Zoology while working with the Peace Corps (1965–1966). His first degree was in zoology from the University of Iowa. His internship was in Panama, his residency in New Orleans, and he worked at a hospital in New Mexico (1975–1999). He was Associate Professor of Emergency Medicine, University of New Mexico (1994), where his knowledge of snakes and snakebite came in useful on more than one occasion. He kept turtles in his backyard and snakes in terraria at home. Whenever he went on walks with his family, he always carried his snake hook. In 1966 he collected the type series of this lizard, now in the University of Illinois Museum of Natural History. He died of a heart attack.

Keyserling

Giant Frog-eyed Gecko *Teratoscincus keyserlingii* **Strauch,** 1863

Alexander Friedrich Michael Lebrecht Nikolaus Arthur, Graf von Keyserling (1815–1891), was a Russian geologist, zoologist, botanist, and paleontologist of Baltic-German descent. He traveled in Estonia, northern Russia, and the Urals (1839–1846) at the behest of Tsar Nicholas I.

Khan

Khan's Bow-fingered Gecko *Cyrtopodion dattanensis* Khan, 1980

Professor Dr. Muhammad Sharif Khan (b. 1939) is now retired and residing in the USA. His bachelor's (1960) and master's degrees (1963) are from Punjab University, Lahore. He worked at a government college at Rabwah (1963–2001). He is Pakistan's leading herpetologist. He wrote *A Guide to the Snakes of Pakistan* (2002) and *Amphibians and Reptiles of Pakistan* (2006).

Kharin

Kharin's Sea Snake *Hydrophis vorisi* Kharin, 1984

Vladimir E. Kharin is a Russian herpetologist with the Institute of Marine Biology, Far East Branch, Russian Academy of Sciences, Vladivostok. He wrote "Sea Snakes of the Genus *Hydrophis sensu lato* (Serpentes, Hydrophiidae): On Taxonomic Status of the New Guinea *H. Obscurus*" (1984).

Kiester

Kiester's Emo Skink *Emoia ponapea* Kiester, 1982

Dr. Alan Ross Kiester is a herpetologist and biologist interested in ecology, evolution, and biogeography. The University of California in Berkeley awarded his bachelor's degree, and Harvard awarded his doctorate (1975). He taught at the University of Chicago and at Tulane University and is now a member of the Pacific Northwest Research Station, U.S. Department of Agriculture, Forest Service, where he is leader of the Global Biological Diversity Team.

Kikuchi

Botel Gecko *Gekko kikuchii* Oshima, 1912

Y. Kikuchi (1869–1921) was a collector for the Taipei Museum, Formosa (now Taiwan). A vole is named after him.

Kikuzato

Kikuzato's Stream Snake *Opisthotropis kikuzatoi* **Okada** and Takara, 1958 [Alt. Kikuzato's Brook Snake; Syn. *Liopeltis kikuzatoi*]

Kiyotasu Kikuzato collected the snake holotype (1956).

Kimberley

Kimberley Dtella *Gehyra kimberleyi* **Börner** and Schüttler, 1982

Named after the Kimberley area, northwest Australia.

King, F. W.

King's Nose-horned Lizard *Harpesaurus thescelorhinos* King, 1978

Dr. Frederick Wayne King (b. 1936) is a retired herpetologist and former Director of the Florida Museum of Natural History. He has undertaken herpetological work in Sarawak and is a specialist on crocodile conservation.

King, P. P.

Frilled Lizard *Chlamydosaurus kingi* **Gray,** 1825
King's Worm Lizard *Anops kingi* **Bell,** 1833
King's Skink *Egernia kingii* Gray, 1838
Madrean Alligator Lizard *Elgaria kingii* Gray, 1838
King's Tree Iguana *Liolaemus kingii* Bell, 1843
King's Sea Snake *Disteira kingii* **Boulenger,** 1896

Rear Admiral Philip Parker King (1791–1856) was an Australian-born British marine surveyor and collector. His father was Philip Gidley King, the third Governor of New South Wales. The family returned to England, and King entered the navy (1807). He commanded the cutter *Mermaid* (1818) and made a number of discoveries, including the Goulburn Islands. He carried out the first survey of the Great Barrier Reef (1819) and a second survey of the reef and the Torres Strait while commanding *Bathurst* (1821). He traveled in tropical America (1827–1832) commanding the British South American Survey.

King, R. D. and M.

King's Monitor *Varanus kingorum* **Storr,** 1980
Skink sp. *Lerista kingi* **L. A. Smith** and Adams, 2007

Dr. Richard Dennis King (1942–2002) was a Canadian-born Australian ecologist. His bachelor's and master's degrees (1968) were from the University of British Columbia and his doctorate (1978) from the University of Adelaide. He worked for the Western Australian Agricultural Protection Board (1978–1996). After retirement he became an honorary associate of the Western Australian Museum. He co-wrote *Monitors: The Biology of Varanid Lizards* (1999). The monitor is named for him and Dr. Max King (b. 1946), who secured the holotype. Max King is a geneticist and poet whose doctorate was awarded (1975) by the University of Adelaide. He worked at the Northern Territory Museum of Arts and Sciences, Darwin (1984–1993), and retired to breed cattle in New South Wales.

Kingdon-Ward

Kingdonward's Bloodsucker *Calotes kingdonwardi* **M. A. Smith,** 1935

Captain Francis "Frank" Kingdon-Ward (1885–1958) was an English botanist, collector, and explorer. He was one of those intrepid late Victorians who went everywhere, enduring the most enormous perils—in his case, storms, torrents, and an earthquake measuring over 9.5 on the Richter scale. He traveled extensively in Assam, Burma (now Myanmar), China, and Tibet. He served in the Indian army during WW1 and taught jungle survival techniques to Allied Forces during WW2, after previously avoiding capture by Japanese forces and having made his way alone through the Burmese jungle to India. He died of a stroke. He wrote *The Land of the Blue Poppy* (1913). A mammal and a bird are named after him.

Kinghorn

Kinghorn's Python *Morelia kinghorni* Stull, 1933
Kinghorn's Snake-eyed Skink *Proablepharus kinghorni* **Copland,** 1947

James Roy Kinghorn (1891–1983) was an ornithologist and herpetologist. He was on the staff of the Australian Museum, Sydney (1907–1956), becoming Assistant to the Director after service in the artillery of the Australian army (1915–1918). He became the resident radio and television naturalist for the Australian Broadcasting Corporation (1956). He wrote *The Snakes of Australia* (1929).

Kinkelin

Kinkelin's Graceful Brown Snake *Rhadinaea kinkelini* **Boettger,** 1898

Dr. Georg Friedrich Kinkelin (1836–1913) was a geologist and malacologist. After studying in Munich, he became a schoolmaster and lived in Frankfurt (1873). He joined the Senckenberg Natural History Society and was its Secretary (1874–1885). One of his closest friends was Boettger.

Kintore

Kintore's Egernia *Egernia kintorei* **Stirling** and **Zietz,** 1893 [Alt. Great Desert Skink]

Algernon Hawkins Thomond Keith-Falconer, Ninth Earl of Kintore (1852–1930), a British politician, was Governor of South Australia (1889–1895). In 1891 he visited the Northern Territory (1891), traveling overland on horseback to Alice Springs and from there, by train, back to Adelaide.

Kirby

Kirby's Least Gecko *Sphaerodactylus kirbyi* **Lazell,** 1994

Dr. Ian Earle Ayrton Kirby (1921–2006) was a veterinary surgeon as well as a vulcanologist, historian, archeologist, and Curator, St. Vincent and the Grenadines Museum. This museum contained a collection of pre-Columbian artifacts that he had unearthed. He studied at the Imperial College of Tropical Agriculture, Trinidad and Tobago (1942–1945). He received a scholarship that enabled him to go to Guelph University, Canada, to study veterinary medicine (1948).

Kirk

Kirk's Rock Agama *Agama kirkii* **Boulenger,** 1885

Dr. Sir John Kirk (1832–1922) was a diplomat, explorer, and naturalist. He was David Livingstone's chief assistant, physician, and naturalist during his second Zambesi expedition (1858–1863). He became Vice Consul, then Consul-General, in Zanzibar (1866–1886). He obtained the local sultan's agreement to a treaty abolishing the slave trade (1873). Three birds and three mammals are named after him.

Kirtland

Kirtland's Rattlesnake *Crotalophorus kirtlandi* **Holbrook,** 1842 [Junior syn. of *Sistrurus catenatus* Rafinesque, 1818]

Forest Vine Snake *Thelotornis kirtlandi* **Hallowell,** 1844

Kirtland's Water Snake *Clonophis kirtlandi* **Kennicott,** 1856

Dr. Jared Potter Kirtland (1793–1877) was a naturalist, botanist, doctor, legislator, teacher, and writer. He founded the Cleveland Museum of Natural History and Cleveland Medical College. He was a contemporary of Agassiz and Audubon. Kirtland made one important scientific discovery that brought him national attention: he asserted that the freshwater mussels *Unionacea* have distinct sexes, which formerly had been mistakenly classified as different species. Baird initially disputed this, but when Agassiz proved that Kirtland was right, Baird apologized by naming a bird after him. Kirtland was an accomplished horticulturalist and managed a large plantation of white mulberry trees for the rearing of silkworms. He is credited with originating 26 varieties of cherries and 6 of pears. Two birds are named after him.

Kisteumacher

Worm Lizard sp. *Leposternon kisteumacheri* Porto, Soares, and Caramaschi, 2000

Geraldo Kisteumacher is a Brazilian herpetologist who has worked with Caramaschi. They jointly described a frog (1989).

Kitaibel

Juniper Skink *Ablepharus kitaibelii* **Bibron** and Bory-St.-Vincent, 1833

Dr. Paul Kitaibel (1757–1817) was a Hungarian botanist. He turned to medicine and then botany after failing to become a lawyer or a priest. He became Professor of Botany at the University of Ofen (1809). Hungary has celebrated him, and the skink, on a postage stamp.

Kitchener

Skink sp. *Emoia kitcheneri* **How,** Durrant, **L. A. Smith,** and Saleh, 1998

Dr. Darrell J. Kitchener is a zoologist who was Curator of Mammals, Western Australian Museum, Perth. Since about 2000 he has been based in Indonesia.

Kitson

Skink sp. *Panaspis kitsoni* **Boulenger,** 1913

Sir Albert Ernest Kitson (1868–1937) was a geologist and naturalist. He was born in England but was taken by his parents to India (1869) and from there to Australia (1876). He worked as a geologist in Australia until 1907, when he became principal mineral surveyor for southern Nigeria, and he spent much of the rest of his life in Africa. He worked in other parts of West Africa, notably the Gold Coast (now Ghana) (1913–1930), where he discovered bauxite, manganese, and diamonds. He settled in England (1930). Kitson Avenue, Takoradi, Ghana, is named after him

Kizirian

Kizirian's Lightbulb Lizard *Proctoporus cashcaensis* Kizirian and Culoma, 1991 [Alt. Kizorian's Lightbulb Lizard; erroneous spelling]

Dr. David Alan Kizirian is a herpetologist and a Curatorial Assistant at the American Museum of Natural History. Before that he worked at the Los Angeles County Natural History Museum. His bachelor's degree was awarded by Texas A&M University (1984), his master's by the University of Texas, El Paso (1987), and his doctorate by the University of Kansas (1994). He has conducted extensive fieldwork in Ecuador, Peru, Mexico, and Vietnam. He co-wrote "A New Species of *Proctoporus* (Squamata Gymnophthalmidae) from Ecuador" (1991).

Klauber

Klauber's Dwarf Gecko *Sphaerodactylus klauberi* **Grant,** 1931 [Alt. Klauber's Least Gecko]
Banded Rock Rattlesnake *Crotalus lepidus klauberi* **Gloyd,** 1936
Klauber's Blind Snake *Leptotyphlops subcrotillus* Klauber, 1939
Spotted Chuckwalla *Sauromalus klauberi* **Shaw,** 1941
Tucson Shovel-nosed Snake *Chionactis occipitalis klauberi* **Stickel,** 1941
Klauber's (Spotted) Box Turtle, *Terrapene nelsoni klauberi* **Bogert,** 1943
San Diego Night Snake *Hypsigiena torquata klauberi* **Tanner,** 1944
Klauber's Leaf-toed Gecko *Hemidactylus klauberi* **Scortecci,** 1948
Night Lizard genus *Klauberina* **Savage,** 1957 [usually now synonymized with *Xantusia*]

Dr. Laurence Monroe Klauber (1883–1968) was an electrical engineer and inventor who became overall chief of a California utility. He was a keen herpetologist, becoming the world expert on rattlesnakes. Reptiles were still just his hobby when the San Diego Zoo asked him about some snakes that they had acquired (1923). He became the zoo's Curator of Reptiles (1923–1958). He gave around 36,000 specimens and all his notes and his library to the San Diego Natural History Museum. He wrote *Rattlesnakes: Their Habits, Life Histories, and Influence on Mankind* (1956).

Kleinmann

Kleinmann's Tortoise *Testudo kleinmanni* Lortet, 1883 [Alt. Egyptian Tortoise]

Edouard Kleinmann collected the holotype (1875). He was a stockbroker and one of the founders of the Crédit Lyonnais, and was in charge of that bank's overseas offices.

Klemmer

Kuala Lumpur Worm Snake *Typhlops klemmeri* **Taylor,** 1962
Klemmer's Dwarf Gecko *Lygodactylus klemmeri* **Pasteur,** 1964 [Alt. Malagasy Dwarf Gecko]
Klemmer's Day Gecko *Phelsuma klemmeri* **Seipp,** 1991

Dr. Konrad Klemmer (b. 1930) is a German herpetologist. He was Curator of Herpetology at Naturmuseum Senckenberg, Frankfurt, and Director (1990) when Seipp was sent on a collecting trip to Madagascar. Klemmer collected in Morocco and the western Sahara (1960s).

Kloss

Kloss' Sea Snake *Hydrophis klossi* **Boulenger,** 1912
Kloss' Emo Skink *Emoia klossi* Boulenger, 1914
Bearded Snake *Fimbrios klossi* **M. A. Smith,** 1920
Kloss' Forest Dragon *Gonocephalus klossi* Boulenger, 1920

Cecil Boden Kloss (1877–1949) was an ethnologist and zoologist. He was a member of the staff of the museum in Kuala Lumpur (1908), for which he traveled extensively as a collector. He started to work under Herbert Christopher Robinson (q.v.). He was the Director of the Raffles Museum in Singapore (1923–1932) and established its *Bulletin* (1928). Six birds and three mammals are named after him.

Kluge

Kluge's Helmeted Gecko *Diplodactylus galeatus* Kluge, 1963
Kluge's Dwarf Gecko *Lygodactylus klugei* **H. M. Smith, Martin,** and **Swain,** 1977
Kluge's Gecko *Diplodactylus klugei* Aplin and Adams, 1998
Bow-fingered Gecko sp. *Cyrtodactylus klugei* Kraus, 2008

Dr. Arnold Girard Kluge (b. 1935) has been at the University of Michigan since 1965. He is Professor of Zoology and Curator of Reptiles and Amphibians Emeritus in the university's Museum of Zoology. He is editor-in-chief of the journal *Cladistics*.

Knollman

Mount Isarog Forest Skink *Sphenomorphus knollmanae* Brown, Ferner, and Ruedas, 1995

Margy Knollman (d. 1989) was a teacher in a Montessori school in Loveland, Ohio. The etymology reads, "Named in honor of the late Margy Knollman, friend and teacher, who guided the senior author through his first scientific experiment at age seven and continued to encourage his herpetological pursuits until the time of her death." The Margy Knollman Nature Trail in Loveland is named after her.

Knox

Knox's Ocellated Sand Lizard *Meroles knoxii* **Milne-Edwards,** 1829

Dr. Robert Knox (1791–1862) was a physician, anatomist, natural scientist, and traveler. He graduated in medicine from the University of Edinburgh (1814), worked at St. Bartholomew's Hospital, London, and joined the army as an assistant surgeon (1815). He was stationed at the Cape of Good Hope (1817–1820). He went to France to study anatomy (1821) and met both Cuvier and Geoffroy Saint-Hilaire. He returned to Edinburgh (1822) and ran a private anatomy school (1826–1840). Among those who visited his dissecting theater was Audubon, who was shocked by the experience. Knox, like many anatomists of the period, bought corpses for dissection and did not ask too many questions about their provenance. He was a regular buyer of bodies from the notorious pair of "Resurrection Men," Burke and Hare. Although he escaped prosecution, his reputation was ruined, and his sources of income and influence dried up. He had suggested that there should be a Museum of Comparative Anatomy and he was its first Conservator, but he was forced to resign (1831). To have an income he turned to writing; his best-selling book was about fishing. He was a physician at the London Cancer Hospital (1856–1862).

Koch, C.

Koch's Gecko *Pachydactylus kochii* **FitzSimons,** 1959 [Alt. Cape Cross Thick-toed Gecko; Syn. *Colopus kochii*]

Koch's Chirping Gecko *Ptenopus kochi* **Haacke,** 1964 [Alt. Koch's Barking Gecko, Interdune Barking Gecko]

Dr. Charles Koch (1894–1970) was an Austrian-born entomologist. Following his research trip (1958) to the Namib Desert to study beetles, the Transvaal Museum decided to establish the Namib Desert Research Station (1962) with Koch as its first Director. He wrote "Some Aspects of Abundant Life in the Vegetationless Sand of the Namib Desert Dunes: Positive Psammotropism in Tenebrionid-Beetles" (1961).

Koch, K. L.

Koch's Day Gecko *Phelsuma (madagascariensis) kochi* **Mertens,** 1954

K. L. Koch was a German herpetologist and ornithologist. His bird collection is held at Naturmuseum Senckenberg, Frankfurt, where he worked.

Kock

Eastern Creek Lizard *Arthrosaura kockii* Lidth De Jeude, 1904

Dr. P. J. de Kock was a Dutch physician who was a collector of zoological specimens and Assistant to Lieutenant A. J. van Stockum, the leader of the expedition that found the source of the Saramacca River, Surinam (1902–1903). A mountain in Surinam, De Kock Berg, is named after him.

Koehler

Koehler's Gecko *Cnemaspis koehleri* **Mertens,** 1937

Max Köhler collected in West Africa with Dr. H. Graf during expeditions for Naturmuseum Senckenberg, Frankfurt (1930s). He collected the gecko holotype. An amphibian is named after him.

Koekkoek

Boenjoe Island Worm Snake *Typhlops koekkoeki* **Brongersma,** 1934

Marinus Adrianus Koekkoek (1873–1944) was a scientific illustrator and painter who worked at Rijksmuseum van Natuurlijke Histoire, Leiden (1918–1938). He did illustrations for van Oort's *Ornithologica neerlandica*, which were copied in *The Handbook of British Birds* by H. F. Witherby. He should not be confused with another artist who had exactly the same names, lived 1807–1870, and was a well-known landscape painter.

Koelliker

Koelliker's Glass Lizard *Ophisaurus koellikeri* **Günther,** 1873

Dr. Rudolph Albert von Koelliker, or just Albert Koelliker (1817–1906), was an anatomist, physiologist, and histologist. He became a student at Universität Zürich (1836), moving to Bonn (1838) and then to Berlin. He graduated in philosophy at Zurich (1841) and received his medical degree from Universität Heidelberg (1842). He was Professor of Physiology and Comparative Anatomy, Universität Zürich (1844–1847), and Universität Würzburg, Bavaria (1848–1906). He was an early proponent of the use of the microscope in biological research and a leading opponent of Darwinism.

Koepcke

Frost's Iguana *Microlophus koepckeorum* **Mertens,** 1956

Professor Hans Koepcke (d. 2000) and his wife, Maria (1924–1971) were both honored in the name of this lizard. She was killed in an air crash in the Andes on Christmas Eve of 1971; he was not on board the aircraft, but their daughter was and she, miraculously, survived. Hans was an ecologist at Universidad Nacional Mayor de San Marcos, while Maria—the more famous of the pair—was known as the "Mother of Peruvian Ornithology." She was born Maria Emilia Ana von Mikulicz-Radecki in Leipzig, Germany, and went to Peru in 1950. Her study of a coastal biome of Peru was a seminal work. Three birds and a mammal are named after her.

Koford

Coastal Leaf-toed Gecko *Phyllodactylus kofordi* **Dixon and Huey,** 1970

Dr. Carl B. Koford (1915–1979) was a naturalist, explorer, and conservationist, and an authority on the Californian Condor and on the flora and fauna of South America. He was Research Associate and Associate Research Ecologist, Museum of Vertebrate Zoology, University of California, Berkeley, where a Carl B. Koford Memorial Fund was established (1980). He served in the U.S. Navy during WW2, reaching the rank of Commander. He worked for eight years in Puerto Rico at Cayo Santiago Field Research Unit, National Institutes of Health. Two mammals are named after him.

Kohn

Kohn's Map Turtle *Graptemys kohnii* **Baur,** 1893 [Alt. Mississippi Map Turtle; Syn. *G. pseudogeographica kohnii*]

Joseph Gustave Kohn (1837–1906) was a wealthy citizen of New Orleans who donated his private collection in the late 1890s to the newly opened Tulane University Museum of Natural History. Sometime in the 1880s he purchased a living specimen of this turtle in the New Orleans Market. He also sent Baur the holotype of the Ringed Sawback Turtle *Graptemys oculifera*.

Koopman

Koopman's Anole *Anolis koopmani* **Rand,** 1961

Dr. Karl F. Koopman (1920–1997) was considered to be the world's leading authority on bat distribution and taxonomy and was a founding member of Bat Conservation International. He was on the staff of the American Museum of Natural History (1961–1990), latterly as Curator Emeritus. He was noted for his droll sense of humor, an example being his tongue-in-cheek support of global warming on the grounds that it would allow some of his favorite bat species to extend their ranges. He wrote about bats in works such as *Systematics of Indo-Australian Pipistrellus* (1973). Six mammals are named after him.

Koppes

Amaral's Blind Snake *Leptotyphlops koppesi* **Amaral,** 1954

S. J. Koppes collected the holotype in 1934.

Kopstein

Kopstein's Emo Skink *Emoia jakati* Kopstein, 1926

Kopstein's Bronzeback Snake *Dendrelaphis kopsteini* **Vogel** and de Rooijen, 2007

Skink sp. *Sphenomorphus capitolythos* **Shea** and Michels, 2008

Dr. P. Felix Kopstein (1893–1939) was an Austrian physician and naturalist. He collected in Albania (1914). He studied biology and medicine at the University of Vienna (1913–1920). He worked for the Dutch government in the Dutch East Indies (now Indonesia) from 1921, initially in Amboina. He made field trips to New Guinea and the Moluccas, and acted as local agent for Rijksmuseum van Natuurlijke Histoire, Leiden. He wrote *Een Zoölogische Reis Door de Tropen* (1930). The etymology of the skink species makes it clear that *capitolythos* is derived from *caput* (Latin for "head") and *lythos* (Greek for "stone"), in reference to the two German words in the name Kopstein, for "head" and "stone" respectively.

Kosciusko

Alpine Meadow-skink *Eulamprus kosciuskoi* **Kinghorn,** 1932

Named after Mount Kosciusko, Australia's highest mountain.

Koslow

Koslow's Toadhead Agama *Phrynocephalus koslowi* **Bedriaga,** 1907

General Petr Kuzmich Kozlov (1863–1935) was a researcher of Central Asia who was one of Prjevalksy's companions on his fourth expedition. He led the Mongolo-Tibetan expeditions (1899–1901) and (1923–1926) and the Mongolo-Sychuan expedition (1907–1909). Kozlov was sent to Tibet to improve relations there but stopped on the Silk Road (1908) when he discovered Khara-Khoto, the "Black City," which had been described by Marco Polo. He made excavations uncovering many scrolls, which he took back for study, and collected geographic and ethnographic materials. He wrote *Mongoliya I Kam*. He was the husband of a well-known Soviet (Russian) ornithologist, Dr. Elizaveta Vladimirovna Kozlova (1892–1975). Three birds (not described by his wife) and three mammals are named after him.

Koslowsky

Koslowsky's Tree Iguana *Liolaemus anomalus* Koslowsky, 1896

Tree Iguana sp. *Liolaemus koslowskyi* **Etheridge,** 1993

Julio Germán Koslowsky (1866–1923) was a Lithuanian Russian who became an Argentine citizen. He was a herpetologist and explorer who spent much time in Patagonia. In 1898 he tried to found a colony there for Lithuanians, Russians, and Poles but two years later only he and his family were still there. His grave was forgotten and was relocated only in 2003. Etheridge took Koslowsky's work as his starting point.

Kotschy

Kotschy's Gecko *Cyrtopodion kotschyi* **Steindachner,** 1870

Karl George Theodor Kotschy (1813–1866), an Austrian botanist, explorer, and collector, was the son of a botanically minded evangelical pedagogical theologian. He visited Cilicia, Syria, Egypt, and the Sudan (1836–1838), as well as Cyprus (1840) and Asia Minor (1842). He traveled in Persia (now Iran) (1842–1843) and returned to Vienna via Turkey. He next went to Egypt and Palestine (1855), Cyprus, Asia Minor, and Kurdistan (1859). A crustacean is named after him.

Koutou

Skink sp. *Leptosiaphos koutoui* Ineich, **Schmitz,** Chirio, and Lebreton, 2004

Denis Koulagna Koutou (b. 1958) was Director, Wildlife and Protected Areas, Ministry of Environment and Forest, Cameroon.

Kraal

Kei Island Worm Snake *Typhlops kraali* **Doria,** 1874

Captain P. F. Kraal was a Dutch military officer in the Moluccas when Beccari (q.v.) was visiting Amboina and the Aru Islands. He gave Beccari both advice and protection.

Kraepelin

Kelung Cat Snake *Boiga kraepelini* **Stejneger,** 1902

Kraepelin's Colobosaura *Colobosaura kraepelini* **Werner,** 1910

Professor Karl Mathias Friedrich Kraepelin (1848–1915), a zoologist and herpetologist, was a teacher who became Curator, Naturhistorisches Museum zu Hamburg.

Kramer

Kramer's Pit-viper *Trimeresurus macrops* Kramer, 1977 [Alt. Large-eyed Pit-viper; Syn. *Cryptelytrops macrops*]

Tana Delta Smooth Snake *Meizodon krameri* **Schätti,** 1985

Dr. Eugen Kramer (1921–2004) was a herpetologist at Naturhistorisches Museum Basel who worked very closely with Schätti of Museum d'Histoire Naturelle, Geneva. He originally trained as a mathematician and became Doctor of Mathematics at the Polytechnic School of Zurich (1955). During the 1950s he became interested in reptiles and began an intensive course of study of them. He amassed a very large private collection of over 11,000 specimens, which he presented to the Geneva museum (1973), and he became involved with the museum's joint project with Pontificia Universidad Católica del Ecuador to make a complete list of the reptiles and amphibians of Ecuador. He visited Ecuador three times (1985–1992). Apart from his interest in mathematics and herpetology, he was a gourmand and an excellent cook.

Krauss

Ghana Worm Lizard *Cynisca kraussi* **Peters,** 1878

Dr. Christian Ferdinand Friedrich von Krauss (1812–1890) was a collector, traveler, botanist, and scientist. He started out as an apothecary's apprentice and worked as a pharmacist but switched to study zoology, mineralogy, and chemistry at Eberhard Karls Universität Tübingen and Universität Heidelberg, where his doctorate was awarded (1836). He traveled to the Cape Province of South Africa (1838–1839), sailing to Natal with two other naturalists, Wahlberg and Delegorgue (1839). He returned to Europe (1840) after a further visit to Cape Town. He landed in England and sold 500 of his plant specimens to the British Museum before proceeding to Stuttgart, where he joined Staatlisches Museum für Naturkunde, becoming its Director (1856). A number of other taxa are named after him.

Krefft

Dwarf Crowned Snake *Cacophis krefftii* **Günther,** 1863

Krefft's Tiger Snake *Notechis ater* Krefft, 1866

Krefft's Turtle *Emydura macquarii krefftii* **Gray,** 1871

Johann Ludwig Gerhard Krefft (1830–1881) was a German adventurer, artist, and naturalist who settled in Australia. He went to the USA (1851) and worked as an artist in New York, but sailed for Australia (1852) to join the gold rush. He was a miner until 1857 when he joined the National Museum, Melbourne, as a collector and artist. He had a bad temper and feuded with the museum Trustees, which led to him being dismissed (1874). He refused to accept the decision and barricaded himself in his office, having to be carried bodily out of the building, still in his chair, and deposited in the street with the door locked behind him. He then set up a rival "Office of the Curator of the Australian Museum" and successfully sued the Trustees for a substantial sum of money. That was the end of his career, and he never worked seriously again, though he did write natural history articles for the Sydney press. His most famous discovery was of the Australian Lungfish *Neoceratodus forsteri*. He wrote *The Snakes of Australia* (1869). A bird and a mammal are named after him.

Kreutz

Anole sp. *Anolis kreutzi* McCranie, Köhler and **Wilson,** 2000

Jörg Kreutz was commemorated for having made hemipenial drawings of anoles. He worked with Köhler, with whom he co-wrote "*Norops macrophallus* (Werner, 1917), a Valid Species of Anole from Guatemala and El Salvador (Squamata: Sauria: Iguanidae)" (1999).

Krieg

Krieg's Tree Iguana *Liolaemus kriegi* **Muller** and **Hellmich,** 1939

Professor Dr. Hans Krieg (1888–1970) was a physician as well as an ethnographer, anthropologist, and zoologist. He was Director, Zoological Gardens, Munich (1927–1945). He was on the German Gran Chaco expedition (1925), made a second expedition to the same region (1932), and a third (1936) after the War of the Chaco. He led an expedition to Patagonia (1937–1938). He wrote *Zwischen Anden und Atlantik. Reisen einen Biologen in Südamerika* (1948).

Krisalys

Gecko sp. *Strophurus krisalys* **Sadlier,** O'Meally, and **Shea,** 2005

Kristin Alys Sadlier. We do not know what relationship Ms. Sadlier has to Ross Sadlier, the senior describer, but we think there must be one.

Krug

Krug's Anole *Anolis krugi* **Peters,** 1876

Carl (Karl) Wilhelm Leopold Krug (1833–1898) was a German businessman who went to work in Puerto Rico, where he became the German Vice Consul, acquired ownership of the firm he worked for, and married a wealthy landowner's daughter. His personal hobbies were zoology and botany. Krug and Gundlach (q.v.) collected the holotype of this anole when Gundlach was Krug's guest.

Kuchling

Kuchling's Long-necked Turtle *Chelodina kuchlingi* **Cann,** 1998

Dr. Gerald Kuchling is an Australian zoologist and herpetologist at the University of Western Australia. He authored *The Reproductive Biology of the Chelonia* (1999).

Kuehne

Van Denburgh's Rock Racer *Takydromus kuehnei* **Van Denburgh,** 1909 [Syn. *Platyplacopus kuehnei*]

Forest Skink sp. *Sphenomorphus kuehnei* **Roux,** 1910

Victor Kühne is an alias for Dr. John Cheesman "Snake" Thompson (1874–1943), an American naval surgeon and polymath with a vast knowledge of herpetology; he was also a psychoanalyst and a breeder of Siamese and Burmese cats, and he was fluent in Japanese. He graduated as a physician (1892) and joined the U.S. Navy (1897) and served in China during the Boxer Rebellion (1900), being part of the force that relieved Peking (Beijing). Using herpetology as cover, he traveled extensively in Asia, spying for the USA in Japan (1909–1911). He was recalled to the U.S. Navy (1917) for service in WW1, retiring in 1929. He was a co-founder of the Zoological Society of San Diego and worked for Van Denburgh at the California Academy of Sciences when Van Denburgh was Curator of Herpetology there; they co-wrote "Description of a New Species of Sea Snake from the Philippine Islands with a Note on the Palatine Teeth in the Proteroglypha" (1908).

Kuekenthal

Kuekenthal Emo Skink *Emoia kuekenthali* **Boettger,** 1895

Batjan Iridescent Snake *Calamorhabdium kuekenthali* Boettger, 1898

Professor Dr. Wilhelm "Willy" Georg Kükenthal (1861–1922) led the Bremen Geographical Society's expedition in the yacht *Berentine* to Kong Karls Land in the Arctic (1889). They ran aground, and the vessel was crushed by ice. Luckily the *Cecilie Maline,* a sealing vessel, saved everyone four days after they were stranded. He traveled in the East Indies (1894), including the Moluccas, where these reptiles originate.

Kugler

Kugler's Largescale Lizard *Ptychoglossus kugleri* **Roux,** 1927

Dr. Hans Gottfried Kugler (1893–1986) was a Swiss geologist and paleontologist who worked in the oil industry in Trinidad and Venezuela (1913–1959). He was responsible for the discovery of several oil fields in Venezuela. On his retirement from the oil business (1959) he returned to Switzerland and worked (1960–1986) as a volunteer at Naturhistorisches Museum Basel.

Kuhl

Kuhl's Galliwasp *Diploglossus monotropis* Kuhl, 1820

Forest Dragon sp. *Gonocephalus kuhli* **Schlegel,** 1848

Kuhl's Flying Gecko *Ptychozoon kuhli* **Stejneger,** 1902

Dr. Heinrich Kuhl (1797–1821) was a German ornithologist who was an assistant to Conrad Jacob Temminck. He also worked with Johan Conrad van Hasselt in Java for the Netherlands Committee for Natural Science. Kuhl died in Buitenzorg of a tropical disease. He wrote *Conspectus psittacorum* (1820). Other taxa, including six mammals and four birds, are named after him.

Kuhlmann

Kuhlmann's Tree Iguana *Liolaemus kuhlmanni* **Muller** and **Hellmich,** 1933

Dr. O. Kuhlmann collected in Chile for Zoologische Staatssammlung München, and obtained the holotype of this lizard (1928).

Kulzer

Kulzer's Lizard *Lacerta kulzeri* **Müller** and **Wettstein,** 1932 [Alt. Petra Lizard; Syn. *Phoenicolacerta kulzeri*]

Dr. Hans Kulzer (1889–1974) was a coleopterist who collected the lizard holotype (1931). He collected reptiles in the neighborhood of Van, Turkey (1912), for Zoologische Staatssammlung München, where he worked (1920–1948)—initially as a preparator but retiring as

Curator. His collection lay ignored for years, some animals from it being described only in the 1990s.

Kumarasinghe

Kumarasinghe's Day Gecko *Cnemaspis kumarasinghei* Wickramasinghe and **Munindradasa,** 2007

Siril Kumarasinghe (d. 2007), a wildlife ranger in Yala National Park, Sri Lanka, is said to have been killed by poachers. As the etymology puts it, "The species is an eponym . . . honouring Siril Kumarasinghe, for his sacrifices toward conserving the wildlife in the country, eventually giving away his own life to the very cause." (One of the authors of this book happened to arrive at Yala National Park on the very morning of this incident and was told by the wardens that there had been "an incident involving poachers." Shortly after this the army arrived in some strength, and we were told that the park was shut. Later the news media reported that the incident had not involved poachers but was rather an incursion by a unit of Tamil separatists; a number of government soldiers also lost their lives.)

Kumpol

Kumpol's Rock Gecko *Cnemaspis kumpoli* **Taylor,** 1963 [Alt. Trang Province Gecko]

Nai Kumpol Isarankura was Curator of Zoological Collections, Chulalongkorn University, Bangkok, Thailand. Taylor dedicated the species to Kumpol because he had "been very helpful to me in my study of the Thai faunas."

Kundu

Wolf Snake sp. *Lycodon kundui* **M. A. Smith,** 1943

Dr. Kundu, of the Harcourt Butler Institute of Public Health, Rangoon (now Yangon, Myanmar), collected the holotype of this snake.

Kuo

Kuo's Kukri Snake *Oligodon kunmingensis* Kou and **Wu,** 1993

Kou Zi-Tong co-wrote "A New Species of *Oligodon* from Yunnan (Serpentes: Colubridae)" (1993), in which this snake is described. His name is often spelled as "Kuo," through sheer carelessness.

Kur

Lacertid lizard sp. *Mesalina kuri* **Joger** and **Mayer,** 2002

Abd al-Kuri Island. See **Abd el Kuri.**

Kuroiwa

Kuroiwa's Ground Gecko *Goniurosaurus kuroiwae* Namiye, 1912

T. Kuroiwa collected the holotype (1909).

L

Labillardier

Labillardier's Ctenotus *Ctenotus labillardieri* **Duméril** and **Bibron,** 1839

Jacques J. H. de Labillardière (1755–1834) was a naval surgeon who served as botanist on the *Recherche* expedition (1791–1793). He was in Amboina (now Ambon), Indonesia (1792). Before the French Revolution he traveled widely in Europe and regularly corresponded with Sir Joseph Banks. He wrote *Novae Hollandiae plantarum specimen* (1804–1807).

Laborde

Labord's Chameleon *Furcifer labordi* **Grandidier,** 1872 ["Laborde's Chameleon" would be more correct]

Jean Gascon Laborde (1806–1878) was an adventurer who was shipwrecked on the coast of Madagascar (1831). He exercised considerable influence at the court of Queen Ranavalona I by making the first firearms in Madagascar. He was forced to leave after a failed coup d'état resulted in the expulsion of all foreigners, returning as French Consul (1861–1878). Grandidier was collecting in Madagascar in the years 1870–1872 and would have known Laborde in his capacity as consul.

Laboute

Laboute's Sea Snake *Hydrophis laboutei* **Rasmussen** and Ineich, 2000

Pierre Laboute (b. 1942) is a French marine biologist, professional diver, and underwater photographer. He worked for the IRD (L'Institut de Recherche pour le Développement) in New Caledonia (1996–2002), and set up his own diving and marine consultancy company there.

Lacépède

Lacépède's Ground Snake *Liophis cursor* Lacépède, 1789

Lacépède's Day Gecko *Phelsuma cepediana* **Merrem,** 1820 [Alt. Blue-tailed Day Gecko]

Bernard Germaine Etienne de la Ville, Comte de Lacépède (1756–1825), was a French naturalist. He came to the attention of Buffon, whose work on the classification of animals he was encouraged to continue. Buffon also got him a job at the Jardin du Roi (later Jardin des Plantes). Lacépède was active in politics and during "the Terror" lived in Normandy to avoid the guillotine. After his return to Paris he gave up scientific work for a political career and held several offices of state. Two mammals are named after him.

Lachesis

Bushmaster genus *Lachesis* **Daudin,** 1803

Lachesis was one of the three Fates in Greek mythology.

Lacroix

Lacroix Kukri Snake *Oligodon lacroixi* **Angel** and **Bourret,** 1933

Antoine François Alfred Lacroix (1863–1948) was a mineralogist and geologist. He took a doctorate in science (1889), but his supervisor Ferdinand André Fouqué consented to his graduation only on the condition that Lacroix marry his daughter. He was appointed Professor of Mineralogy at Muséum National d'Histoire Naturelle, Paris (1893), and Director of the Mineralogy Laboratory at École des Hautes Études (1896). He wrote *Contribution à la conaissance de la composition chimique et minéralogique des roches éruptives de l'Indochine* (1933).

Lalande

Delalande's Beaked Blind Snake *Rhinotyphlops lalandei* **Schlegel,** 1839

See **Delalande.**

Lally

Lally's Two-line Dragon *Diporiphora lalliae* **Storr,** 1974

Mrs. G. E. "Lally" Handley of Western Australian Museum is obviously a wizard on the typewriter, as Storr named this reptile after her "in appreciation of her excellence as a typist of scientific papers."

Lamar

Anole sp. *Anolis lamari* **Williams,** 1992

Snail-sucker (snake) sp. *Sibon lamari* Solorzano, 2001

William Wylly Lamar (b. 1950) is Adjunct Professor of Biology at the University of Texas, Tyler. He co-wrote *The Venomous Reptiles of Latin America* (1989).

Lambert

Lambert's Sea Snake *Hydrophis lamberti* **M. A. Smith,** 1917

Smith gives no indication in his original description as to the identity of the "Lambert" after whom he named this species. One possibility is Gustave Richard Lambert (b. 1846), a German-born photographer who was a member of the Natural History Society, Bangkok, at the same time as Smith.

Lamberton

Chameleon sp. *Chamaeleo lambertoni* **Angel,** 1921 [Junior syn. of *Furcifer lateralis* Gray, 1831]

Fito Leaf Chameleon *Brookesia lambertoni* **Brygoo** and **Domergue,** 1970

Charles Lamberton was a French paleontologist who

wrote about the subfossil fauna of Madagascar, having lived there (1927–1948) and having undertaken a number of paleontological expeditions to the southwest of the country (1930s). He was a Professor, Gallieni College, and Secretary, Malagasy Academy. He spent 50 years studying extinct lemurs. He wrote "On a New Kind of Fossil Lemur, the Malagasy Prohapalemur" (1936). A mammal is named after him.

Lampe

Irian Jaya Dtella *Gehyra lampei* **Andersson,** 1913

Eduard Lampe (1871–1919) collected for Museum Wiesbaden in the first two decades of the 20th century and worked at the museum cataloguing the collection and describing specimens, as did Andersson in Stockholm.

Lancelin

South-west Ctenotus *Ctenotus lancelini* **Ford,** 1969

Named after Lancelin Island, Western Australia.

Lancini

Lancini's Ground Snake *Atractus lancinii* **Roze,** 1961

Lancini's Sun Tegu *Euspondylus ampuedae* Lancini, 1968 [Syn. *Cercosaura ampuedae*]

Dr. Abdem Ramon Lancini (1934–2007) was a herpetologist who was Director, Museum de Ciencias Naturales, Caracas, Venezuela (1962–1991). He wrote *Serpientes de Venezuela* (1979).

Lando

Schwartz's Island Racer *Arrhyton landoi* **Schwartz,** 1965

Orlando H. Garrido (q.v.).

Lane

Lane's Leaf-toed Gecko *Phyllodactylus lanei* **H. M. Smith,** 1935

Lane's House Snake *Thamnodynastes lanei* Bailey, Thomas, and da Silva, 2005

Dr. Frederico Lane was an entomologist at Museu Paulista, São Paulo. According to Bailey, "He taught the senior author so much of what he knew of Brazil, its people and customs, and . . . with his charming wife, Aniuta, graciously opened their home to a sometimes lonely young colleague." Many taxa are named after him.

Lang, H.

Lang's Worm Lizard *Chirindia langi* **Fitzsimons,** 1939

Lang's Crag Lizard *Pseudocordylus langi* **Loveridge,** 1944 [Alt. Lang's Girdled Lizard; Syn. *Cordylus langi*]

Herbert Lang (1879–1957) was born in Germany and trained as a taxidermist. He later worked as such in both the Natural History Museum, Universität Zürich, and a commercial establishment in Paris. He moved to the USA (1903) and joined the American Museum of Natural History as a taxidermist. He led the museum's Congo expedition (1909–1915). On returning to New York he became Assistant in Mammalogy, then Assistant Curator (1919). He returned to Portuguese West Africa (now Angola) for the museum (1925) with Rudyerd Boulton. They covered 6,500 kilometers (4,000 miles) and collected 1,200 mammal specimens. He stayed on in Africa after the Angola expedition and took a job with Transvaal Museum, South Africa. He made a number of expeditions, including one for the American Museum of Natural History to the Kalahari Desert. He took over the management of a hotel in Pretoria (1935).

Lang, M.

Lang's Isopachys *Isopachys borealis* Lang and **Böhme,** 1990

Mathias Lang is a zoologist and herpetologist at Zoologisches Forschungsmuseum Alexander Koenig, Bonn, and a Research Associate at Koninklijk Belgisch Instituut voor Natuurwetenschappen, Brussels. He frequently works with Böhme, with whom he co-wrote "The Reptilian Fauna of the Late Oligocene Locality Rott near Bonn (Germany) with Special Reference to the Taxonomic Assignment of 'Lacerta' rottensis von Meyer, 1856" (1991).

Langer

Mussurana (snake) sp. *Clelia langeri* Reichle and Embert, 2005

Brother Andres Langer is a German Dominican monk who has been a missionary for over 30 years. He started collecting reptiles in the area of Pampagrande, Bolivia, in the last years of the 20th century and, with his helpers, has contributed over 1,400 specimens to the herpetological collection in Museo de Historia Natural, Noel Kempff Mercado, Santa Cruz de la Sierra, Bolivia.

Langsdorff

Langsdorff's Coral Snake *Micrurus langsdorffi* **Wagler,** 1824

Baron Georg Heinrich von Langsdorff (otherwise Grigoriy Ivanovich) (1774–1852) was a German physician, botanist, zoologist, traveler, naturalist, ethnographer, and diplomat. He graduated as a physician at Georg-August-Universität Göttingen (1797). He was elected as a "corresponding member" of the Academy of Science, St. Petersburg (1803), and was on Krusenstern's round-the-world expedition (1803–1806) on board *Nadezda*. He continued to travel widely, in Japan (1804–1805), the American Northwest (1805–1806), and Kamchatka, Siberia, and European Russia (1806–1808). He became an

Associate Professor in botany at the Academy of Science (1808), later moving to zoology (1809). He became Russian Consul General in Brazil (1813) and Chargé d'Affaires for Russia to Portugal in Rio de Janeiro (the Portuguese government was in exile during the Napoleonic Wars). He returned to Russia (1821) to organize an expedition in Brazil, which he led (1822–1828), but he caught a tropical fever that led to a psychological breakdown. He retired to Germany (1831). He wrote *Remarks and Observations on a Voyage around the World from 1803 to 1807* (1812). Two birds are named after him.

Lannom

Autlan Rattlesnake *Crotalus lannomi* **Tanner,** 1966

Joseph Robert Lannom Jr. collected the holotype and, until 2010, the only known specimen. He was involved in research into the effects of gamma radiation upon lizards. He co-wrote "Radiation Doses Sustained by Lizards in a Continuously Irradiated Natural Enclosure" (1968).

Lansberge

Lansberge's Hog-nosed Pit-viper *Porthidium lansbergii* **Schlegel,** 1841

Reinhart Frans van Lansberge (1804–1873) was a Dutch administrator in the West Indies. A bird is named after him.

Lanza

Lanza's Spiny Agama *Agama spinosa* **Gray,** 1831
Lanza's Writhing Skink *Lygosoma grandisonianum* Lanza and Carfi, 1966
Longtail Lizard sp. *Latastia lanzai* Arillo, Balletto and Spano, 1967
Skink sp. *Chalcides lanzai* **Pasteur,** 1967
Lanza's Racerunner *Eremias ercolinii* Lanza and Poggesi, 1975
Lanza's Leaf-toed Gecko *Hemidactylus ophiolepoides* Lanza, 1978

Benedetto Lanza (b. 1924) is an Italian herpetologist who was Professor of Biology and Director, Natural History Museum, Università degli Studi di Firenze. An agama that Lanza described was found to be a junior synonym of the Spiny Agama *Agama spinosa*, yet the eponymous common name has stuck.

Lar

Spotted Caribbean Gecko *Aristelliger lar* **Cope,** 1861 [Alt. Hispaniolan Giant Gecko]

Why Cope chose the name *lar* is a mystery to us. We think he may have been imitating Linnaeus, who had a penchant for whimsical names from the classics and who first used *lar* as a scientific name. In Roman mythology, a Lar (plural Lares) was a kind of guardian deity. Originally gods of cultivated fields, Lares were later regarded as minor gods of the household. The household Lar was often represented as a youthful figure holding a drinking horn and cup.

Largen

Racer (colubrid snake) sp. *Coluber largeni* **Schätti,** 2001
Largen's Gracile Blind Snake *Letheobia largeni* **Broadley** and **Wallach,** 2007

Dr. Malcolm John Largen is a herpetologist and a photographer of wildlife. He works at Liverpool Museum and was a member of the staff of the Biology Department, Haile Selassie University, Addis Ababa (1966–1977).

Lasalle

Lasalle's Ground Snake *Atractus lasallei* **Amaral,** 1931
Lasalle's Fishing Snake *Synophis lasallei* **Maria,** 1950

The ground snake is not named directly after a person, but after the Instituto de La Salle, Bogotá, Colombia. The holotype was part of a collection of snakes sent to Amaral by Brother Nicéforo Maria (see **Niceforo**), who was based at the Instituto de La Salle. As Brother Nicéforo described the fishing snake, it is reasonable to assume it also is named after the Instituto de La Salle.

Lataste

Lataste's Viper *Vipera latasti* **Bosca,** 1878
Lataste's Snake Skink *Ophiomorus latastii* **Boulenger,** 1887
Lataste's Lizard *Timon pater* Lataste, 1880 [Alt. North African Ocellated Lizard; Syn. *Lacerta pater*]

Professor Fernand Lataste (1847–1934) was a French zoologist. He made a collection of the reptiles and amphibians of Barbary (Morocco, Algeria, and Tunisia) (1880–1884). A few years later he turned his attention to South America, writing on the birds of Chile. He wrote *Étude de la faune des vertébrés de Barbarie* (1885). A mammal is named after him.

Latifi

Latifi's Viper *Vipera latifii* **Mertens, Darevsky,** and **Klemmer,** 1967
Zagros Tiny Gecko *Tropiocolotes latifi* **Leviton** and **Anderson,** 1972

Dr. Mahmoud Latifi (1930–2006) was an Iranian herpetologist who worked as a researcher at Institut d'État des Serums et Vaccins Razi, Teheran, Iran. He wrote *Snakes of Iran* (1991).

La Touche

Sichuan Mountain Keelback *Opisthotropis latouchii* **Boulenger,** 1899

John David Digues La Touche (1861–1935) was French

born and English educated. He was Inspector of Customs in China (1882–1921). Hoping to retire to Ireland, he died at sea on the way home from Majorca where he had spent the winter. He wrote *A Handbook of the Birds of Eastern China* (1925). Two birds and a mammal are named after him.

Laudahn

Gymnophthalmid lizard sp. *Riama laudahnae* Köhler and Lehr, 2004

Monika Laudahn is a technician at the Herpetological Department, Naturmuseum Senckenberg, Frankfurt, Germany.

Laurent

Laurent's Plated Lizard *Gerrhosaurus bulsi* Laurent, 1954

Laurent's Mountain Bush Viper *Atheris hispida* Laurent, 1955

File Snake sp. *Mehelya laurenti* **Witte,** 1959

Laurent's Five-toed Skink *Leptosiaphos hylophilus* Laurent, 1982

Tree Iguana sp. *Liolaemus laurenti* **Etheridge,** 1992

Sipo (colubrid snake) sp. *Chironius laurenti* **Dixon,** Wiest, and **Cei,** 1993

Dr. Raymond Ferdinand Louis-Philippe Laurent (1917–2005) was a Belgian herpetologist who worked for much of his life in Argentina and started the Herpetology Department, Fondación Miguel Lillo, Tucumán, Argentina (1975). Université Libre de Bruxelles awarded his doctorate (1940). He studied African herpetology. He wrote "Diagnoses préliminaires des quelques serpents venimeux" (1955).

Laurenti

Laurent's Whiptail *Cnemidophorus murinus* Laurenti, 1768

Josephus Nicolaus Laurenti (1735–1805) was an Austrian zoologist and anatomist. He has been described as the "Father of Herpetology," yet little seems to have been recorded about his life. He wrote *Specimen medicum, exhibens synopsin reptilium emendatam cum experimentis circa venena et antidota reptilium austracorum, quod authoritate et consensus* (1768), about the poison function of reptiles and amphibians. In it he defined 30 new genera of reptiles and amphibians (including *Bufo*, *Hyla*, *Gekko*, *Chamaeleo*, and *Iguana*).

Lavarack

Gulf Snapping Turtle *Elseya lavarackorum* **White** and Archer, 1994

Jim and Sue Lavarack are Australian paleontologists. They discovered the fossil of the turtle. It has subsequently been found to be still extant.

Lavilla

Neotropical Tree Snake sp. *Sibynomorphus lavillai* **Scrocchi,** Porto, and Rey, 1993

Tree Iguana sp. *Liolaemus lavillai* Abdala and **Lobo,** 2006

Esteban Orlando Lavilla is an Argentine herpetologist at Fondación Miguel Lillo, Tucumán, Argentina, where he is Director, Museum of Herpetology.

Lawder

Lawder's Bent-toed Gecko *Cyrtopodion lawderanum* **Stoliczka,** 1871

Stoliczka's laconic etymology refers to the gecko being named in honor of Mr. A. Lawder, who discovered it. We believe this to be A. W. Lawder who was, like Stoliczka, a member of the Geological Society, London, and based in India. It is possible that Lawder, like Stoliczka, was associated with the Geological Survey of India.

Lawes

Günther's Emo Skink *Emoia lawesi* **Günther,** 1874

Rev. William George Lawes (1839–1907) was a missionary. He sailed for the Pacific (1860) and was first sent to work on Savage Island (now Niue), where he served until 1872. After a visit to England he sailed to New Guinea, settling in Port Moresby, as one of the first permanent European residents (1874). He left for England on holiday (1878) having produced the first book ever written in a Papuan language. He was in great demand for his unrivaled knowledge of Papua and was liked and trusted by the local tribes. He returned to Port Moresby (1881) and acted as interpreter for the Protectorate proclamation (1884). He founded a new training college at Vatorata (1892) and served there for 10 years. He retired to Sydney (1906). A bird-of-paradise is named after him.

Lawrence

Lawrence's Dwarf Gecko *Lygodactylus lawrencei* **Hewitt,** 1926

Lawrence's Girdled Lizard *Cordylus lawrenci* **Fitz-Simons,** 1939

Dr. Reginald Frederick "Lawrie" Lawrence (1897–1987) was an entomologist who worked at the South African Museum from 1922. His university graduation was delayed until 1922 by two years' service in WW1, during which he was wounded (1918). His doctorate was from the University of Cape Town (1928). He was Director, Natal Museum, Pietermaritzburg (1935–1964). He left Natal (1966), moved to Grahamstown, and continued researching at the Albany Museum, returning (1984) to Pietermaritzburg to be near his two sons, his wife, Professor Ella Tratt Yule, having died (1978). There is an annual award in memory of him established by the Zoological Society, South Africa.

Lawton

Forest Skink sp. *Sphenomorphus lawtoni* **Brown** and **Alcala,** 1980

E. Lawton Alcala (d. 2007) was a biologist with Negros Oriental Environment and Natural Resources Division, Philippines, and a brother of the junior author. Silliman University awarded his bachelor's degree (1966). An amphibian is named after him.

Layard

Layard's Nessia *Nessia layardi* **Kelaart,** 1853

Edgar Leopold Layard (1824–1900) was born in Italy. He was in Ceylon (now Sri Lanka) (1844–1854) before going to Cape Colony, South Africa, as a civil servant on the Governor's staff. He worked as Curator, South African Museum (1855), in his spare time. Layard later worked in Brazil, Fiji, and New Caledonia. He wrote *The Birds of South Africa* (1867). Two mammals and nine birds are named after him.

Lazell

Cap-Haitien Least Gecko *Sphaerodactylus lazelli* **Shreve,** 1968

Lazell's Flying Dragon *Draco biaro* Lazell, 1987

Lazell's Blind Snake *Typhlops lazelli* **Wallach** and **Pauwels,** 2004

Dr. James Draper "Skip" Lazell (b. 1939) is a biologist and vertebrate zoologist. He received his Ph.D. from the University of Rhode Island (1970) and later worked as a herpetologist at Harvard Museum of Comparative Zoology. He is currently President of the Conservation Agency and is associated with the Peabody Museum, Yale. He wrote *This Broken Archipelago: Cape Cod and the Islands,* Amphibians, and Reptiles (1976).

Lea

Lea's Ctenotus *Ctenotus leae* **Boulenger,** 1887 [Alt. Orange-tailed Finesnout Ctenotus]

Rev. T. E. Lea sent the holotype to Boulenger.

Leach

Leach's Worm Snake *Typhlops punctatus* Leach, 1819 [Alt. Spotted Blind Snake]

Leach's Wolf Snake *Lycophidion irroratum* Leach, 1829

New Caledonia Giant Gecko *Rhacodactylus leachianus* **Cuvier,** 1829

Leach's Anole *Anolis leachii* **Gray,** 1837

Dr. William Elford Leach (1790–1836) was a zoologist. He originally studied medicine but he did not practice, instead being employed at the British Museum (1813–1821), where he became an expert on crustaceans. He was well known for idiosyncratic nomenclature: for example, he named (1818) nine genera after a "Caroline" (or various anagrams of that name), who may have been his mistress. He caught cholera in Italy and died from it. He wrote *The Zoological Miscellany* (1814). Four birds and a mammal are named after him.

Leache

Whiptail sp. *Cnemidophorus leachei* **Peracca,** 1897

Don Francisco Leache was the proprietor of a large property at San Lorenzo, Argentina. He was extremely hospitable and kind to Dr. Borelli (q.v.) during his stay at San Lorenzo.

Leal

Pestel Amphisbaena *Amphisbaena leali* **Thomas** and Hedges, 2006

Dr. Manuel Leal is an Assistant Professor, Biology Department, Duke University, Durham, North Carolina. Universidad de Puerto Rico awarded his bachelor's (1990) and master's degrees (1994). His doctorate was awarded (2000) by Washington University, St. Louis. He was a Postdoctoral Research Fellow at Union College, Schenectady, New York (2003). He co-wrote "Evidence for Habitat Partitioning Based on Adaptation to Environmental Light in a Pair of Sympatric Lizard Species." He and Thomas collected the holotype (1991).

Leber

Hispaniola Ameiva *Ameiva leberi* **Schwartz** and Klinikowski, 1966

Dr. David C. Leber was a herpetologist who worked closely with both describers; they were all on an expedition to Haiti in 1962. Leber collected in Hispaniola for some years (1960s). He illustrated Schwartz and Henderson's *A Guide to the Identification of the Amphibians and Reptiles of the West Indies, Exclusive of Hispaniola* (1985). He co-wrote "A Forest-dwelling Species of *Eleutherodactylus*" (1961). An amphibian is named after him.

LeBreton

Agama sp. *Agama lebretoni* Wagner, Barej, and Schmitz, 2009

Matthew LeBreton (b. 1973) is an Australian zoologist and herpetologist who became (2004) Ecology Research Coordinator, Global Viral Forecasting Initiative, Yaoundé, Cameroon. The University of New South Wales awarded his bachelor's degree in zoology (1996). He co-wrote *Atlas des reptiles de Cameroun* (2007).

Le Conte

Sharp-tailed Snake genus *Contia* **Baird** and **Girard,** 1853

Western Long-nosed Snake *Rhinocheilus lecontei* Baird and Girard, 1853

Dr. John Lawrence Le Conte (1825–1883) was an entomologist and biologist. He was also a physician during the American Civil War, reaching the rank of Lieutenant Colonel. His father, John Eatton Le Conte (1784–1860), was also a naturalist. While he was still a student Le Conte made a number of field trips to the Rocky Mountains and to Lake Superior. He made a second trip to Lake Superior accompanied by Louis Agassiz (1848). He went to California (1849) and explored the Colorado River. He moved (1852) to Philadelphia, which was his base for the rest of his life, though he made a number of overseas expeditions. He became Chief Clerk of the U.S. Mint in Philadelphia (1878), a complete career change. A bird and two mammals are named after him.

Lee

Chatham Leaf-toed Gecko *Phyllodactylus leei* **Cope,** 1889

Dr. Thomas Lee was a regular member of the group of naturalists used by the U.S. Fish Commission on the cruises of their research vessel, *Albatross*. He was aboard when they called at Cozumel Island off the Yucatan Peninsula (1885), and again in the Bahamas and the West Indies (1886). Chatham Island, in the Galapagos, is now called San Cristobal. Lee was on board when the *Albatross* visited Chatham Island (1888) en route from New York to San Francisco via Cape Horn.

Leech

Leech's Fathead Anole *Enyalioides leechii* **Boulenger,** 1885

John Henry Leech (1862–1900) was an explorer, collector, and entomologist, interested particularly in Coleoptera and Lepidoptera. He was educated at Cambridge. Leech collected in Morocco, the Canary Islands, Madeira, Japan, Korea, China, and the northwestern Himalayas, but from 1887 stayed in Britain to work on his collections, commissioning others to collect for him. He purchased (1889) the entomological periodical *The Entomologist*, partly as a means of publishing his own papers on his collections, but withdrew from it after objections were made about these inclusions. He wrote the three-volume *Butterflies from China, Japan, and Corea* (1892–1894).

Leeh

Leeh's Fathead Anole *Enyalioides leechii* **Boulenger,** 1885

This is a transcription error. See **Leech.**

Leeser

Urucum Worm Lizard *Amphisbaena leeseri* **Gans,** 1964

Leo Leeser (1871–1942) left his estate in such a way that, through the Leo Leeser Center for Tropical Biology, Gans was able to obtain a grant to "travel in pursuit of these studies." Leeser was murdered in the Theresienstadt concentration camp in Czechoslovakia (now Terezin, Czech Republic).

Lehmann, A.

Turkestan Rock Agama *Laudakia lehmanni* **Nikolsky,** 1896

Alexander Lehmann was a Russian biologist of German descent who traveled to Turkestan as a member of a Russian expedition (1839–1842).

Lehmann, F. C.

Lehmann's Ground Snake *Atractus lehmanni* **Boettger,** 1898

Friedrich Carl Lehmann (1850–1903) was a botanist who was German Consul in Colombia. He traveled extensively through Central and South America. He was in Costa Rica, Guatemala, and Panama in the early 1880s.

Lehmann(-Valencia)

Gymnophthalmid lizard sp. *Alopoglossus lehmanni* **Ayala** and **Harris,** 1984

Professor Frederico Carlos Lehmann-Valencia (1914–1974) was a biologist and conservationist who founded a natural history museum in Colombia, using his own vast collection as its basis (1936). He also founded Museo de Ciencias Naturales, Santa Teresita de Cali, Colombia (1963). The original etymology says the lizard was "named in memory of Carlos Lehmann Valencia, naturalist and leader in the preservation of wilderness areas in Colombia." Other taxa, including an amphibian, are named after him.

Lehner

Estado Falcón Worm Snake *Typhlops lehneri* **Roux,** 1926

Dr. Ernst Lehner was a geologist, paleontologist, and naturalist employed by the North Venezuela Petroleum Company in the state of Falcón and in Trinidad. He was in the Caribbean area for much of the 1920s and 1930s. He collected shells on Carriacou, a dependency of Grenada off the north coast of Venezuela. He wrote *Introduction to the Geology of Trinidad* (1935).

Leighton

Cape Sand Racer *Psammophis leightoni* **Boulenger,** 1902

Dr. Gerald Rowley Leighton (1868–1953) was a pathologist, zoologist, and herpetologist who qualified as a physician at Edinburgh (1895). He became Professor of Comparative Pathology and Bacteriology, Royal Veterinary

College, Edinburgh. He was in the Royal Army Medical Corps (1914–1918), reaching the rank of Lieutenant-Colonel. He founded and edited *Field Naturalist's Quarterly* (1902). He wrote *The Life-History of British Serpents and Their Local Distribution in the British Isles* (1901).

Leith

Leith's Sand Snake *Psammophis leithii* **Günther,** 1869
Indian Tent Turtle *Pangshura leithii* **Gray,** 1870 [Junior syn. of *Batagur tentoria* Gray, 1834]
Leith's Softshell Turtle *Nilssonia leithii* Gray, 1872
Leith's Tortoise *Testudo kleinmanni* Lortet, 1883 [Alt. Kleinmann's Tortoise, Egyptian Tortoise]

Dr. Andrew H. Leith was a physician employed in Bombay as Sanitary Commissioner. He started a system of registering deaths that recorded cause, age, gender, and the like, in an effort to identify the prevalence of smallpox (1848). He collected a tortoise specimen that Günther described as *Testudo leithii*, but this was later changed to *kleinmanni* when it was discovered that *T. leithii* had already been used for a fossil turtle found near Bombay.

Lema

Lema's Ground Snake *Liophis dilepis* **Cope,** 1862

Dr. Thales de Lema is a herpetologist at Instituto de Biociências da Pontifícia Universidade Católica do Rio Grande do Sul, Porto Alegre, Brazil, where he conducts research in the taxonomy and systematics of neotropical snakes. Lema's name has become associated with this snake when he helped clear up the uncertain taxonomy, which has led to changes in name and assignment since Cope first described it.

Lemos-Espinal

Chihuahuan Mezquite Lizard *Sceloporus lemosespinali* Guillermo, 2004

Julio Alberto Lemos-Espinal (b. 1959) is a Mexican herpetologist attached to the University of Colorado. He co-wrote "Ecological Observations of the Lizard *Xenosaurus grandis* in Cuautlapan, Veracruz" (1995). One of the co-authors of that article was Susy Sanoja Sarabia (see **Sanoja**), who is otherwise known as Mrs. Lemos-Espinal.

Leonard, G. R.

Leonard's Pipe Snake *Anomochilus leonardi* **M. A. Smith,** 1940

G. R. Leonard collected the holotype, but we know nothing more about him.

Leonard, P. M. R.

Burmese Rat Snake *Elaphe leonardi* **Wall,** 1921
Leonard's Keelback *Rhabdophis leonardi* Wall, 1923

P. M. R. Leonard, a member of the Burma Frontier Service, collected the holotypes. He wrote *Report of the Third Expedition to the "Triangle" for the Liberation of Slaves, Season 1928–29* (1929), which also dealt with the incidences of human sacrifice in Upper Burma at that time.

Leonhard

Kalahari Worm Lizard *Monopeltis leonhardi* **Werner,** 1910

See **Schultze.**

Leonhardi

Leonhard's Ctenotus *Ctenotus leonhardii* **Sternfeld,** 1919

Moritz Friherr von Leonhardi (1856–1910) was a German anthropologist. He studied law, but illness prevented his completion of a degree, so he studied privately, developing an interest in anthropology. His ill health also prevented travel, so he conducted his research through intermediaries. This led to ethnographic objects and zoological and botanical specimens being sent to him, which he passed to museums across Europe. Mostly he wrote as editor, publishing under the names of those who sent him material.

Leopold

Leopold Dtella *Gehyra leopoldi* **Brongersma,** 1930
Snake-eating Snake sp. *Polemon leopoldi* **Witte,** 1941

Prince Leopold III (1901–1983) became King of the Belgians (1934) and abdicated (1951) in favor of his son, Baudouin. Inspired by A. R. Wallace, Prince Leopold III and his first wife, Princess Astrid (1905–1935), visited the Dutch East Indian Archipelago (1928–1929). He also visited New Guinea and the Arfak Mountains, collecting zoological specimens.

Lepesme

Angel's Five-toed Skink *Lacertaspis lepesmei* **Angel,** 1940

Pierre Lepesme was an entomologist on the expedition to Cameroon during which the holotype of this reptile was collected. He wrote *Les insectes des palmiers* (1947). His daughters sold his collection to the Natural History Museum, Lyon (2002).

Leptien

Leptien's Mastigure *Uromastyx leptieni* Wilms and **Böhme,** 2000

Rolf Leptien is a German herpetologist who, according to Böhme, is an expert on Middle Eastern and African lizards.

Lerner

Lerner's Anole *Anolis smaragdinus lerneri* **Oliver,** 1948

Michael Lerner (1891–1978) founded the American

Museum of Natural History's Lerner Marine Laboratory, Bimini Island, Bahamas. He was a successful retailer of women's clothes and used his money to fund (and lead) a series of expeditions for the American Museum of Natural History, of which he was a trustee (1935–1941), to collect ethnographic and zoological specimens. His greatest interest was the collection and study of big game fish, and with his friend, the American novelist Ernest Hemingway, he and his wife, Helen (1902–1979), helped found the International Game Fish Association (1939).

Leschenault

Leschenault's Snake-eyed Lizard *Ophisops leschenaultii* **Milne-Edwards,** 1829 [Alt. Leschenault's Cabrita; Syn. *Cabrita leschenaulti*]

Skink sp. *Cryptoblepharus leschenault* **Cocteau,** 1832

Leschenault's Leaf-toed Gecko *Hemidactylus leschenaulti* **Duméril** and **Bibron,** 1836

Jean Baptiste Louis Claude Theodore Leschenault de la Tour (1773–1826) was a botanist who served as naturalist to two Kings of France, Louis XVIII (1814–1824) and Charles X (1824–1830). He was botanist on the voyage of *Casuarina*, *Géographe*, and *Naturaliste* (1801–1803), and he collected in Australia (1801–1802). He also collected in Java (1803–1806) and India (1816–1822) and visited the Cape Verde Islands, the Cape of Good Hope, Ceylon (now Sri Lanka), Brazil, and British Guiana (now Guyana). He wrote one of the first descriptions of coconuts and the extraction of their oil (1803). A bird and a mammal are named after him.

Lessona

Lessona's Agama *Trapelus lessonae* **De Filippi,** 1865

Brazilian Galliwasp *Diploglossus lessonae* **Peracca,** 1890

Professor Dr. Michele Lessona (1823–1894) was a zoologist. He graduated as a physician and practiced in Turin (1846) before going to Egypt as Director of the hospital at Karnak. While in Egypt he collected reptiles and presented them to Museo Regionale di Scienze Naturali di Torino upon his return home. He taught in secondary schools until becoming Professor of Mineralogy and Zoology, Università degli Studi di Genova (1954), then Professor of Zoology, Università di Bologna (1864). He returned to Turin (1865) to act as "locum tenens" for De Filippi, who was away on the *Magenta* expedition. During this expedition De Filippi died of cholera in Hong Kong (1867), and Lessona took his place as Professor of Zoology and Comparative Anatomy. He was an early proponent of Darwinism in Italy and was the first to translate *The Descent of Man* into Italian (1871). He entered politics and at the time of his death was a Senator of the Kingdom (1877).

Lesueur

Eastern Water Dragon *Physignathus lesueurii* **Gray,** 1831

Lesueur's Gecko *Oedura lesueurii* **Duméril** and **Bibron,** 1836 [Alt. Lesueur's Velvet Gecko]

Lesueur's Skink *Sphenomorphus lesueurii* Duméril and Bibron, 1839 [Junior syn. of *Ctenotus australis* Gray, 1838]

Charles Alexandre Lesueur (1778–1846) was a naturalist, artist, and explorer. He set sail for Australia aboard *Le Geographe* as an assistant gunner (1801). When the original artists jumped ship in Mauritius, Baudin appointed Lesueur as an official expedition artist. During the next four years he and Péron (q.v.) collected over 100,000 specimens representing 2,500 new species. Lesueur made 1,500 drawings, from which he produced a series of watercolors on vellum, which were published (1807–1816) in the expedition's official report. He went to live in the southern USA (1815–1837). He met Audubon (1824) and so admired his work that he urged Audubon to try again to get them published in France. He was appointed Curator, Muséum d'Histoire Naturelle du Havre, which was created to house his drawings and paintings (1845). A mammal and a bird are named after him.

Leuckart

Leuckart's Burrowing Skink *Anomalopus leuckartii* Weinland, 1862

Professor Dr. Karl Georg Friedrich Rudolf Leuckart (1822–1898) was a morphologist, parasitologist, and zoologist. He started studying medicine at Georg-August-Universität Göttingen, where he was appointed to lecture in zoology (1847). He became Professor, Universität Leipzig (1869). Leuckart was enormously influential in the study of parasites, including liver flukes, on which he worked with Weinland.

Levins

Isla Desecheko Least Gecko *Sphaerodactylus levinsi* Heatwole, 1968

Dr. Richard Levins (b. 1930), a former farmer in Puerto Rico, is now an ecologist and philosopher of science. He studied mathematics and plant breeding at Cornell and was awarded his doctorate by Columbia University. He taught at Universidad de Puerto Rico and University of Chicago and is currently Professor of Population Sciences at the Harvard School of Public Health. He has worked closely with Heatwole, with whom he co-wrote "Biogeography of the Puerto Rican Bank: Introduction of Species onto Palominitos Island" (1973).

Leviton

Leviton's Kukri Snake *Oligodon annamensis* Leviton, 1961

Leviton's Yellow-striped Slender Tree Skink *Lipinia pulchella levitoni* **Brown** and **Alcala,** 1963
Leviton's Rock Agama *Laudakia nuristanica* **Anderson** and Leviton, 1969
Santa Catalina Island Blind Snake *Leptotyphlops humilis levitoni* **Murphy,** 1975
Leviton's Awl-headed Snake *Lytorhynchus gasperetti* Leviton, 1977 [Alt. Leviton's Leafnose Snake]
Leviton's Cylindrical Skink *Chalcides levitoni* Pasteur, 1978
Leviton's Gecko *Asiocolotes levitoni* **Golubev** and Szczerbak, 1979
Mangrove Snake ssp. *Boiga dendrophila levitoni* Gaulke, Demegillo, and **Vogel** 2005

Dr. Alan Edward Leviton (b. 1930) took his bachelor's (1949), master's (1953), and doctorate (1960) at Stanford. He became Curator, Department of Herpetology, California Academy of Sciences (1957). He lectured in biology at Stanford (1962–1970). Since 1969 he has been Adjunct Professor of Biological Sciences, San Francisco State University. His specialty is the herpetofauna of Asia and, recently, of the Arabian Peninsula.

Lewis

Skink sp. *Saproscincus lewisi* **Couper** and Keim, 1998

Lewis Roberts collected the holotype of this small Australian skink.

Lewis, C. B.

Grand Cayman Blue Iguana *Cyclura lewisi* **Grant,** 1941

Charles Bernard Lewis Jr. (b. 1913) was Director, Institute of Jamaica, and President and Director, Jamaica Historical Society. His collections of plants (mostly 1940s-1950s) are held there. He was part of a collecting trip to Grand Cayman (1938) and collected the holotypes of a new butterfly and the eponymous iguana. He co-wrote "The Herpetology of the Cayman Islands, with an Appendix on the Cayman Islands and Marine Turtle" (1940).

Lhote

Lhote Orangetail Lizard *Philochortus lhotei* **Angel,** 1936

Henri Lhote (1903–1991) was a French explorer and ethnologist. He heard of unusual cave paintings in the central Sahara and set out to find and describe them. Lhote later referred to a curious painted figure as "Jabbaren," the "great Martian god." Although this picture and others probably represent ordinary humans in ritual masks and costumes, the popular press made much of this early "alien-contact" hypothesis. It was later borrowed by Erich von Däniken as part of his sensationalist claims of "ancient astronauts." Lhote co-wrote "Reptiles et amphibiens du Sahara central et du Soudan" (1938).

Lichtenfelder

Lichtenfelder's Gecko *Goniurosaurus lichtenfelderi* **Mocquard,** 1897

Lichtenfelder was an engineer who worked in Iles de Norway (Vietnam) and sent a collection of reptiles from there to Muséum National d'Histoire Naturelle, Paris. He also sent other taxa to the museum, including a crustacean, from Tonkin (1897). Mocquard gives no first name or initial, but the individual honored may be Charles Lichtenfelder, an engineer and architect who designed a number of buildings in Hanoi.

Lichtenstein

Lichtenstein's Green Racer *Philodryas olfersii* Lichtenstein, 1823
Lichtenstein's Short-fingered Gecko *Stenodactylus sthenodactylus* Lichtenstein, 1823
Lichtenstein's Water Snake *Lycodonomorphus rufulus* Lichtenstein, 1823
Lichtenstein's Toad-headed Agama *Phrynocephalus interscapularis* Lichtenstein, 1856
Lichtenstein's Night Adder *Causus lichtensteini* **Jan,** 1859

Martin Heinrich Carl Lichtenstein (1780–1857) was a physician, traveler, and zoologist who was head of Museum für Naturkunde Berlin from 1813 and founded the Berlin Zoo (1844). He traveled in South Africa (1802–1806) and while there became personal physician to the Dutch Governor of the Cape of Good Hope. Lichtenstein studied many species sent to the Berlin Museum by others, and "while he gave every species, or what he judged to be a species, a name, this was done without consulting the recent English and French literature. His only aim was to give the specimens in question a distinguishing mark for his personal needs. These names were used in Lichtenstein's registers and reappeared on the labels of the mounted specimens, but only exceptionally were they published by himself in connection with a scientific description." This caused much unnecessary confusion and trouble to others. He died at sea off Kiel—not of illness, as is sometimes reported, but rather when he fought a duel and came out second best. He wrote *Reisen in Sudlichen Africa* (1810). Seven birds and two mammals are named after him.

Lick

Cape Arboreal Spiny Lizard *Sceloporus licki* **Van Denburgh,** 1895

James Lick (1796–1876) was a cabinetmaker, piano builder, carpenter, and philanthropist. He went to Argentina (1821) and made enough money from building pianos to leave (1825) and spend a year in Europe. While sailing back to Argentina his ship was captured by the

Portuguese navy. He was taken to Montevideo as a prisoner-of-war but escaped and walked back to Buenos Aires. He moved his business to Valparaiso, Chile (1832), and then to Lima, Peru (1836). He thought that California would be lost by Mexico to the USA, so he moved himself, $30,000 in gold, 300 kilograms (660 pounds) of chocolate (which sold very quickly), and the piano-building business, and started investing in land in the small village of San Francisco (1848). Gold was discovered in the area a few days after he arrived, and the 1849 Gold Rush was on. He made a fortune out of land, fruit farming, hotels, and flour mills, supplying the needs of the prospectors, and became California's richest man. He left the greatest part of his fortune for the public good. He had an interest in astronomy, and part of his legacy was to build the Lick Observatory. He is buried under one of its pillars. A number of roads and buildings are named after him, but he would probably be most pleased by two eponymous astronomical phenomena: the Lick Crater on the moon and the asteroid 1951 Lick.

Lidski

Tibetan Toadhead Agama *Phrynocephalus lidskii* **Bedriaga,** 1909

S. A. Lidski was a Russian explorer, geographer, and naturalist who specialized in Central Asia. He traveled (1887) from Samarkand to Zarafshan via the Takhta-Karacha Pass and explored in the region of eastern Bokhara and Karategin (1888). He wrote *Materials for the Bibliography of Central Asia and the Neighbouring Countries. Russian Turkestan. Collections of Essays* (1899). We surmise that he may have been a member of Przewalski's fifth expedition to Central Asia, as the agama was described in Bedriaga's report on the scientific results of that expedition.

Liebmann

Liebmann's Earth Runner *Chersodromus liebmanni* **Reinhardt,** 1860

Professor Frederik Michael Liebmann (1813–1856) was a botanist. He undertook study tours in Norway and Germany and was a lecturer at the Danish Royal Veterinary and Agricultural College (1837–1840). He worked in Cuba and Mexico (1840–1845). He was Professor of Botany, København Universitet (1845–1856), and Director of the university's botanical gardens (1852–1856).

Lilford

Lilford's Wall Lizard *Podarcis lilfordi* **Günther,** 1874

Lord Lilford or Thomas Littleton Powys, Fourth Baron Lilford (1833–1896), kept a small menagerie when a schoolboy and a bigger one when at Oxford. Although he suffered ill health throughout his life, he traveled to study birds and other animals. He took up falconry, served in the militia in Dublin (1854–1855), and cruised the Mediterranean (1856–1858), making frequent return visits, especially to Spain (1864–1882). He was one of the founders of the British Ornithologists' Union (1858) and served as its President (1867). Two birds are named after him.

Lim

Tioman Round-eyed Gecko *Cnemaspis limi* **Das** and **Grismer,** 2003

Kok Peng "Kelvin" Lim (b. 1966) is a zoologist who is Curator of Vertebrates and a Collection Manager, Raffles Museum, Singapore. His first zoological interest was fish, but he now has an interest in a wide variety of taxa. He is known throughout the region for his herpetological expertise.

Lindheimer

Lindheimer's Rat Snake *Pantherophis obsoletus lindheimerii* **Baird** and **Girard,** 1853 [Alt. Texas Ratsnake; Syn. *Elaphe obsoleta lindheimerii*]

Ferdinand Jacob Lindheimer (1801–1879) was a lawyer and botanist. He studied at Bonn, Wiesbaden, and Jena universities. He had to leave Germany, having taken part in a failed insurrection attempt (1833), and went to the USA, arriving in Illinois (1834) and traveled by boat to New Orleans. He intended to go to Texas but was diverted to Mexico, where he worked (1833–1835). He left Mexico when the Texas Revolution began (1835) but was shipwrecked on the coast of Alabama. He then headed for Texas again but arrived one day too late for the decisive Battle of San Jacinto. He collected botanical specimens in Texas for Harvard, among other institutions (1836–1845). He met Prince Carl of Solms-Braunfels, who was setting up a German colony in Texas (1844). This was called New Braunfels, and Lindheimer spent the rest of his life there, setting up a German-language newspaper (1852) and supporting the Confederacy in the American Civil War. He is known as the "Father of Texas Botany," and his home in New Braunfels is now a public museum.

Lindner

Giant Cave Gecko *Pseudothecadactylus lindneri* **Cogger,** 1975

David A. Lindner is an Australian herpetologist. He works closely with Cogger, with whom he co-authored "Marine Turtles in Northern Australia" (1969).

Lindsay

Lindsay's Blind Snake *Leptotyphlops humilis lindsayi* **Murphy,** 1975

Dr. George Edmund Lindsay (1916–2002) was a botanist

who specialized in the cacti of Baja California, a place he first visited as soon as he got a driver's license (1930). He became the first Director, Desert Botanical Garden, Arizona (1938). During WW2 he was in the U.S. Air Force with the rank of Captain. He sold the family farm after he returned to California and resumed his studies, which he completed with a doctorate at Stanford (1956), and became Director, San Diego Natural History Society's Museum. He was Director, California Academy of Sciences, San Francisco (1963–1982). He was one of those who collected the snake holotype (1962).

Liner

Texas Scarlet Snake *Cemophora coccinea lineri* **Williams, Brown,** and **Wilson,** 1966

Liner's Tropical Night Lizard *Lepidophyma lineri* **H. M. Smith,** 1973

Liner's Garter Snake *Thamnophis lineri* **Rossman** and Burbrink 2005

Dr. Ernest Anthony Liner (1925–2010) spent 1955–1987 as a sales representative in the pharmaceuticals industry, but his passion was herpetology and, in recognition of his work in that field, the University of Colorado awarded him an honorary doctorate in science (1998). He was a Corporal in the U.S. Marine Corps in the Pacific from 1943, including the capture of Iwo Jima. He took a bachelor's degree (1951) at Southwestern Louisiana Institute and became a schoolteacher for a year. He worked in the Zoology Department, Tulane University, New Orleans (1952–1955), and then went into pharmaceuticals. He spent most of his free time in Mexico. His large collection is now at the American Museum of Natural History. In one way he was unique among herpetologists: he liked to eat reptiles. He was a master Cajun cook and enjoyed dining on reptiles for most of his life. He wrote *The Herpetological Cookbook* and followed it up with a revised and enlarged second edition called *The Culinary Herpetologist*. which contains over 950 recipes.

Linnaeus

Linnaeus' Lance Skink *Acontias meleagris* Linnaeus, 1758

Linnaeus' Sipo *Chironius carinatus* Linnaeus, 1758

Linnaeus' Writhing Skink *Lygosoma quadrupes* Linnaeus, 1766

See **Linné.**

Linné

Linné's Reed Snake *Calamaria linnaei* **Boie,** 1827 [Alt. Linné's Dwarf Snake]

Carl Linné (1707–1778) is much better known by the Latin form of his name, Carolus Linnaeus, or just Linnaeus. Late in life (1761) he was ennobled and so could call himself Carl von Linné. In the natural sciences he was undoubtedly one of the great heavyweights of all time, ranking with Darwin and Wallace. He is thought of primarily as a botanist, but he invented the system he published in *Systema naturae* that is still in use today, albeit with modifications, for naming, ranking, and classifying living organisms. He entered Lunds Universitet (1727) to study medicine and transferred to Uppsala Universitet (1728). At that time the study of botany was part of medical training. His first expedition was to Lapland (1732). He mounted an expedition to central Sweden (1734). He went to the Netherlands (1735) and finished his studies as a physician there before enrolling at Universiteit Leiden. He returned to Sweden (1738), lecturing and practicing medicine in Stockholm. He became Professor at Uppsala (1742) and restored the university's botanical garden. He bought the manor estate of Hammarby, outside Uppsala, where he built a small museum for his extensive personal collections (1758). This house and garden still exist and are now run by Uppsala University. His son, also named Carl, succeeded to his professorship at Uppsala but never was noteworthy as a botanist. When Carl the Younger died (1783) with no heirs, his mother and sisters sold the elder Linnaeus' library, manuscripts, and natural history collections to the English natural historian Sir James Edward Smith, who founded the Linnean Society of London to take care of them. Surprisingly few taxa are named after Linnaeus.

Linton

Linton's Dwarf Short-tailed Snake *Tantillita lintoni* **H. M. Smith,** 1940

Professor Linton Satterthwaite Jr. (1897–1978) was an archeologist who worked in Mexico (1930s) and published on the Mayan civilization. He was on the staff of the University of Pennsylvania Museum when he conducted excavations and a survey of Cahal Pech, a Mayan site in what is now Belize (1950–1951).

Lipetz

Lipetz's Tropical Night Lizard *Lepidophyma lipetzi* **H. M. Smith** and Alvarez del Toro, 1977

Dr. Milton L. Lipetz was a biochemist. He was a member of the faculty of the University of Colorado, of which he was Vice Chancellor, as well as Dean of the Graduate School, in the 1970s. He was Acting Chancellor of the Boulder campus (1981–1982).

Lister

Lister's Gecko *Lepidodactylus listeri* **Boulenger,** 1889 [Alt. Christmas Island Chained Gecko]

Joseph Jackson Lister (1857–1928) was a naturalist. He

joined HMS *Egeria* at Colombo as a volunteer naturalist and visited Christmas Island (1887). He made a "large biological and mineralogical collection." He wrote *On the Natural History of Christmas Island in the Indian Ocean* (1888).

Liu

Hikida's Skink *Plestiodon liui* **Hikida** and **Zhao,** 1989

Professor Ch'eng-chao Liu (1900–1976) was a Chinese herpetologist. He was Professor of Biology at West China Union University and later at the Chengdu Institute of Biology. He authored "Amphibians of Western China" (1950) and co-authored "Chinese Tailless Amphibians" (1961) with his wife, S.-Q. Hu.

Llanos

Llanos Side-necked Turtle *Podocnemis vogli* **Müller,** 1935 [Alt. Savannah Side-necked Turtle]

The Llanos is a tropical wet grassland plain in Colombia and Venezuela.

Llanos, F.

Leyte Sphenomorphus *Sphenomorphus llanosi* **Taylor,** 1919

Father Florencio Llanos was a Dominican priest. He was Rector of San Juan de Letran College, Manila (1914–1917), after which he became the Director, University of Santo Tomas, Philippines. Taylor studied the collections there.

Lobo

Tree Iguana sp. *Liolaemus loboi* Abdala 2003

Dr. Fernando José Lobo Gaviola (b. 1963) is an Argentine herpetologist and Adjunct Professor of Comparative Anatomy, School of Biology, Universidad Nacional de Salta, Argentina. He co-wrote "Two New Cryptic Species of *Liolaemus* (Iguania, Tropiduridae) from Northwestern Argentina—Resolution of the Purported Reproductive Bimodality of Liolaemus-Alticolor" (1999).

Löding

Löding's Pine Snake *Pituophis melanoleucus lodingi* **Blanchard,** 1924 [Alt. Black Pine Snake]

Peder Henry Löding (1869–1942) emigrated to the USA from Denmark. He lived in Mobile, Alabama, and was described by Blanchard as being a "pioneer student of Alabama reptiles and amphibians, through whose efforts have come to light the types upon which this species is based." He wrote the first comprehensive report of the herpetofauna of Alabama.

Lomi

Lomi's Blind Legless Skink *Typhlosaurus lomiae* **Haacke,** 1986

Ms. Lomi Wessels became Collection Manager, Lower Vertebrates and Invertebrates, Transvaal Museum, in 1976. Haacke is one of her colleagues.

Long

Forest Dragon sp. *Hypsilurus longii* **Macleay,** 1877

Mark H. Long donated the holotype.

Longman

Longman's Brown Snake *Pseudonaja carinata* Longman, 1915 [Junior syn. of *P. aspidorhyncha* McCoy, 1879]

Heber Albert Longman (1880–1954) was a paleontologist who was born in England and moved to Australia for medical reasons (1902). He spent many years in Queensland and published over 70 papers, mostly through the Queensland Museum. He described, among other fossils, *Kronosaurus queenslandicus* (1926), a huge marine reptile. A mammal is named after him.

Lopez-Jurado

Leaf-toed Gecko sp. *Hemidactylus lopezjuradoi* **Arnold** et al., 2008

Dr. Luis. Felipe López-Jurado is a Spanish herpetologist who is Professor of Biology at Universidad de Las Palmas de Gran Canaria. He co-edited *Marine Turtles: Recovery of Extinct Populations* (2007).

Lorenz

Lorenz's Tree Iguana *Liolaemus lorenzmuelleri* **Hellmich,** 1950

See **Müller, L.**

Lorenz, T. K.

Lorenz's Worm Snake *Ramphotyphlops lorenzi* **Werner,** 1909

Theodore K. Lorenz (1842–1909) was a German zoologist and ornithologist who did most of his work in Russia. He worked as a taxidermist at the Zoological Gardens, Imperial University, and explored in the Caucasus. He wrote *Die Vögel der Moskauer Gouvernements* (1894).

Loria

Boulenger's Bow-fingered Gecko *Cyrtodactylus loriae* **Boulenger,** 1897

Forest Skink sp. *Sphenomorphus loriae* Boulenger, 1897

Loria Forest Snake *Toxicocalamus loriae* Boulenger, 1898

Dr. Lamberto Loria (1855–1913) was an ethnologist who collected in New Guinea (1889–1890). He founded the first Italian Museum of Ethnography, Florence (1906). The museum was subsequently transferred to Rome, after he organized the first ethnography exhibition there (1911). Two mammals and a bird are named after him.

Lotiev

Lotiev's Viper *Vipera lotievi* **Nilson,** Tuniyev, **Orlov,** Hoggren, and Andren, 1995

K. Yu Lotiev is a Russian herpetologist who collected the viper holotype (1986). He co-wrote "Contribution to the Study of Intraspecific Variation of the Caucasian Lizard, *Lacerta caucasica*" (1992).

Loveridge

Loveridge's Forest Snake *Geodipsas procterae* Loveridge, 1922 [Alt. Uluguru Forest Snake; Syn. *Buhoma procterae*]

Loveridge's Large-scaled Lizard *Ptychoglossus nicefori* Loveridge, 1929

Loveridge's Ground Snake *Atractus loveridgei* **Amaral,** 1930

Kenya Sand Boa *Eryx colubrinus loveridgei* Stull, 1932

Loveridge's Writhing Skink *Lygosoma tanae* Loveridge, 1935

Loveridge's Anole *Anolis loveridgei* **Schmidt,** 1936

Loveridge's Gecko *Cnemaspis elgonensis* Loveridge, 1936

Loveridge's Garter Snake *Elapsoidea loveridgei* **Parker,** 1949 [Alt. East African Garter Snake]

Loveridge's Worm Snake *Typhlops loveridgei* Constable, 1949

Loveridge's Emo Skink *Emoia loveridgei* **Brown,** 1953

Loveridge's Green Snake *Philothamnus nitidus loveridgei* **Laurent,** 1960

Loveridge's Flat Gecko *Afroedura (transvaalica) loveridgei* **Broadley,** 1963

Loveridge's Limbless Skink *Melanoseps loveridgei* **Brygoo** and **Roux-Estève,** 1982

Arthur Loveridge (1891–1980) was a herpetologist and zoologist who is regarded as the father of East African herpetology. He worked in the Department of Reptiles, Harvard Museum of Comparative Zoology. He retired to St. Helena. He wrote "East African Reptiles and Amphibians in the United States National Museum" (1929) and a book with the enticing title *I Drank the Zambesi* (1954). A bird is named after him.

Lovi

Lovi's Reed Snake *Calamaria lovii* **Boulenger,** 1887

Named for Sir Hugh Brooke Low. Boulenger was a stickler for rules and spelled the scientific name with a *v* because there is no *w* in Latin. This has led people, over the years, to assume that the snake is named after someone called Lovi. The spelling of the scientific name is sometimes now unofficially "amended" to *lowi*. See **Low, H.**

Low, H.

Low's Dwarf Snake *Calamaria lowi* **Boulenger,** 1887 [Alt. Lovi's Reed Snake; Syn. *C. lovii*]

Sir Hugh Brooke Low (1824–1905) was a civil servant in the British Administration in Malaya and an amateur collector in the Malay Archipelago. He was the first successful British Administrator of Perak (1877–1889). His methods subsequently became models for British colonial operations in Malaya. Previously he had been an unremarkable Colonial Secretary of Labuan, an island off Borneo (1848–1877). There is a "Historical Trail" to the summit of Mount Kinabalu, Sabah, named after him, as he used to collect specimens from the summit, having been the first person to climb it (1851). He found many new orchids. Two mammals and a bird are named after him.

Low, T.

Low's Four-fingered Skink *Menetia timlowi* **Ingram,** 1977

Timothy "Tim" Low (b. 1956) is an Australian biologist, conservationist, author, and environmental consultant. He has been interested in reptiles since boyhood and has discovered several new species. He regularly contributes to Australian magazines and wrote *The New Nature: Winners and Losers in Wild Australia* (2002).

Lowe

Lowe's Garter Snake *Thamnophis sirtalis lowei* **Tanner,** 1988

El Muerto Side-blotched Lizard *Uta lowei* **Grismer,** 1994

Lowe's Tropical Night Lizard *Lepidophyma lowei* **Bezy** and Camarillo, 1997

Dr. Charles Herbert Lowe Jr. (1920–2002) was a herpetologist and biologist. He served in the U.S. Navy during WW2. He went to the University of California at Los Angeles (1946), where he was awarded a doctorate (1950). He then joined the University of Arizona, eventually becoming Professor of Ecology and Evolutionary Biology. He discovered about 20 new species and subspecies. He wrote *The Vertebrates of Arizona* (1964). After his death the "C. H. Lowe Herpetological Research Fund" was set up in his memory.

Lowery

Lowery's Alligator Lizard *Gerrhonotus liocephalus loweryi* Tihen, 1948

George Hines Lowery Jr. (1913–1978) was an ornithologist who was Founding Director, Museum of Zoology (1951–1961), and Professor of Zoology, Louisiana State University (from 1955). He played an important role in creating public awareness and support for a conservation ethic in Louisiana. He wrote *Louisiana Birds* (1955), winning the Louisiana Literary Award of that year. A bird is named after him.

Lucia

St. Lucia Anole *Anolis luciae* **Garman,** 1887

Named after the island of St. Lucia.

Lucila

Ground Snake sp. *Atractus lucilae* Silva Haad, 2004

Lucila Silva Collazos is the describer's sister.

Ludeking

Crested Lizard *Lophocalotes ludekingi* **Bleeker,** 1860

Dr. E. W. A. Ludeking was an army physician in the Dutch East Indies (now Indonesia). He was stationed in West Sumatra (1853–1861) and transferred to Ambon (1861). He went on leave in the Netherlands (1867–1869) and then returned to Ambon. He was a keen zoologist, making expeditions to the Moluccas and Aru Islands for Rijksmuseum van Natuurlijke Histoire, Leiden. He wrote *Schets van de residentie Amboina* (1968).

Ludovic

Glass Lizard sp. *Ophisaurus ludovici* **Mocquard,** 1905

Ludovici's Toadhead Agama *Phrynocephalus ludovici* Mocquard, 1910 [Junior syn. of *P. axillaris* Blanford, 1875]

See Léon Louis **Vaillant.** "Ludovic" is Latin for "Louis."

Lue

Japalure sp. *Japalura luei* **Ota, Chen,** and Shang, 1998

Professor Dr. Kuang-Yang Lue is a zoologist and herpetologist at the National Taiwan Normal University, Taipei. He received his Ph.D. from Mississippi State University (1976). He has worked with Ota (q.v.), with whom he co-wrote "Karyotypes of Two Lygosomine Skinks of the Genus *Sphenomorphus* from Taiwan" (1994).

Lugo

Lugo's Alligator Lizard *Gerrhonotus lugoi* **McCoy,** 1970

José "Pepe" Lugo Guajardo. The original description speaks of "his contributions to studies of the Cuatro Cienegas Basin fauna," and in the minutes of a meeting of the Desert Fishes Council, a tribute was paid to him for nearly 40 years of help and counsel and for acting as a guide for many scientists who had come to conduct research in the valley where he lived.

Lui

Chinese Leopard Gecko *Goniurosaurus luii* **Grismer,** Viets, and **Boyle,** 1999 [Alt. Chinese Cave Gecko]

Wai Lui is a Chinese herpetologist. According to Grismer the gecko was named in honor of Wai Lui "who spent six years tracking the existence of these populations." He collected the holotype (1995). Lui had long suspected that it represented a new species rather than being an example of *Goniurosaurus lichtenfelderi*, but Chinese researchers he sent it to disagreed. In the end he sent it to a U.S. university, which is how it came to be studied by the authors. He co-wrote "Ein Gecko von der alten Seidenstraße im Nordwesten Chinas, Teratoscincus roborowskii Von Bedriaga, 1906" (1997).

Lukban

Lukban's Loam-swimming Skink *Brachymeles lukbani* Siler et al., 2010

General Vicente R. Lukban (1860–1916) was a Filipino lawyer who became a revolutionary. He was imprisoned and tortured by the Spanish authorities (1896) but was released (1897) as the Philippine Revolution started. He went into exile in Hong Kong (1897), studied military theory there, returning to command troops (1898) against Spanish forces in Leyte, and he was instrumental in winning Philippine independence. He later fought against the Americans (1901–1902) until betrayed and captured. He was elected (1912) Governor of Tayabas (Quezon).

Lulu

Amphisbaena sp. *Dalophia luluae* **Witte** and **Laurent,** 1942

Named after the Lulu River, Belgian Congo.

Lumholtz

Dwarf/Reed Snake sp. *Calamaria lumholtzi* **Andersson,** 1923

Dr. Carl Sophus Lumholtz (1851–1922) was a Norwegian naturalist, ethnologist, humanist, and explorer. Having just graduated with a natural science degree (1880), he set off for northeastern Australia, where he spent time living with Aboriginal people until 1884. He organized a number of expeditions, including one to explore the Sierra Madre, Mexico (1890), for the American Museum of Natural History. He visited Borneo (1914), but a planned a trip to New Guinea was prevented by the outbreak of WW1. Lumholtz National Park in Queensland is named after him, as is a tree kangaroo.

Luna

Luna's Spiny Lizard *Sceloporus lunae* **Bocourt,** 1873

See **Rodriguez.**

Lund

Lund's Teiid *Heterodactylus lundii* **Reinhardt** and Lutken, 1862

Dr. Peter Wilhelm Lund (1801–1880) was a Danish

physician, botanist, zoologist, and paleontologist who lived and worked in Lagoa Santa, Minas Gerais. He first traveled to Brazil in 1833 and settled there for health reasons. His interest in fossils led him to explore many of the caves of the area. He assembled one of the most important mammal collections from a single locality in the Neotropics and made outstanding contributions toward describing the Pleistocene and recent mammal fauna of Brazil. He regularly corresponded with Charles Darwin. Two mammals are named after him.

Lundell

Lundell's Spiny Lizard *Sceloporus lundelli* **H. M. Smith** 1939

Cyrus Longworth Lundell (1907–1994) was primarily a botanist. The University of Texas holds his 6,000-volume collection of books on botany. He collected in the USA, Mexico, Guatemala, and Belize. He collected the lizard holotype.

Lutz, A.

Cerrado Lancehead *Bothrops lutzi* Mirando-Ribeiro, 1915

Adolpho Lutz (1855–1940) was a medical entomologist and parasitologist whose Swiss parents settled in Brazil (1849). He was taken to Switzerland (1857) to be educated and eventually qualified as a physician (1879) at Universität Bern. He did postgraduate work in Leipzig, Prague, Vienna, and London, studying under the great names of the day, like Lister and Pasteur. He finally returned to Brazil (1881) and set up a practice in the interior of the state of São Paulo, but returned to Europe (1888) to study dermatology in Hamburg. He spent time in Hawaii (1889–1991) directing work on leprosy. He returned to Brazil (1892) and traveled throughout the country before becoming head of the Bacteriological Institute, São Paulo (1893–1908). He was persuaded to move to Rio de Janeiro, where he served as Director, Medical Zoology, Instituto de Manguinhos (later renamed Instituto Oswaldo Cruz), from 1908 to 1938, although he retired officially in 1936. His daughter Bertha Maria Julia Lutz (q.v.) was a Brazilian zoologist.

Lutz, B. M. J.

Lutz's Tree Iguana *Liolaemus lutzae* **Mertens,** 1938

Bogert's Gecko *Bogertia lutzae* **Loveridge,** 1941

Bertha Maria Julia Lutz (1894–1976) was a zoologist and pioneering feminist who founded the Brazilian Federation for Feminine Progress (1922). Her father, Adolpho Lutz (q.v.), was Swiss and her mother was English. She was a member of the Brazilian Parliament for a short period (1936–1937) until Vargas' coup-d'état. Her main interest was amphibians.

Lynch

Lynch's Anole *Anolis lynchi* **Miyata,** 1985

Dr. John Douglas Lynch (b. 1942) is a herpetologist who moved from the USA to Colombia, where he is an Associate Professor, Instituto de Ciencias Naturales, Universidad Nacional de Colombia, Bogotá. He specializes in amphibians and is Curator of Amphibians at the university museum. He first visited Mexico (1964) while he was studying for his master's degree at the University of Illinois. He subsequently went to the University of Kansas for his doctorate and made a trip to Ecuador (1967), with the result that he spent the next 11 years researching Ecuadorian frogs. He visited Colombia (1979) and returned annually (1980–1996). He finally decided to move to Colombia and resigned his position at the University of Nebraska, where he had been Professor for 28 years. In 2000 he and a group of environmentalists were taken prisoner by the Colombian rebel National Liberation Army (2000); he had the good fortune to be released after only two days.

M

Maack

Amur Softshell Turtle *Pelodiscus maackii* **Brandt,** 1858

Richard Karlovich Maack (1825–1886) was a naturalist and ethnographer who was Professor of Natural History at Irkutsk (1852). He explored the Amur River (1855) and was Inspector of Schools in Eastern Siberia (1868–1879).

Macburnie

Python sp. *Morelia macburniei* Hoser, 2003

Cameron McBurnie is a member of the Victoria Association for Amateur Herpetologists.

MacClelland

MacClelland's Coral Snake *Sinomicrurus macclellandi* **Reinhardt,** 1844 [Syn. *Calliophis macclellandi, Hemibungarus macclellandi*]

Dr. John MacClelland (1805–1875) worked for the East India Company. He was one of a group of persons with botanical knowledge sent (1852) by the British government of India to Burma (now Myanmar) to conduct a timber survey. He pointed out that the hardwood industry should not depend upon teak alone.

MacCoy

See **McCoy, F.**

MacDougall

MacDougall's Tropical Night Lizard *Lepidophyma dontomasi* **H. M. Smith,** 1942

MacDougall's Snail Sucker *Sibon sartorii macdougalli* H. M. Smith, 1943

MacDougall's Graceful Brown Snake *Rhadinaea macdougalli* H. M. Smith and Langebartel, 1949

MacDougall's Spiny Lizard *Sceloporus macdougalli* H. M. Smith and Burnzahem, 1953

MacDougall's Variable Coral Snake *Micrurus diastema macdougalli* **Roze,** 1967

Thomas Baillie MacDougall (1896–1973) was a naturalist and collector who for many years spent his summers in Mexico. He wrote a number of papers, often as co-author, such as "New or Unusual Mexican Amphibians" (1949).

MacDowell

See **McDowell.**

Macfadyen

Macfadyen's Mastigure *Uromastyx macfadyeni* **Parker,** 1932

Dr. William Archibald Macfadyen was a geologist and paleontologist in British Somaliland (now part of Somalia). His publications include *The Geology of British Somaliland* (1933).

Macfarlan

MacFarlan's Skink *Lygisaurus macfarlani* **Günther,** 1877 [Alt. Translucent Litter-Skink]

Rev. Samuel Macfarlane (1837–1911) was a missionary. He was originally a railway mechanic. He moved to Manchester (1853) and decided to train as a missionary. He was ordained (1858) and sailed for Lifu in the Loyalty Islands (1859). The local French authorities disliked him, his mission being destroyed (1854) in a punitive raid. He toured eastern Australia (1867–1868). After the French succeeded in having him removed (1869), he planned to start a mission in New Guinea and sailed (1871) to reconnoiter. He arrived in the Torres Strait, landing on Darnley Island (1871). He went to London (1872) to get his plans for his new mission approved. By 1874 he was established at Somerset, Cape York Peninsula, moving to Murray Island (1877). He made 23 voyages, visited over 80 Torres Straits island villages, established 12 mission stations, learned something of six languages, and published translations in two of them (1874–1878). Among his journeys was one of 115 kilometers (70 miles) up the Fly River (1875) with D'Albertis. He returned to England (1886) and published *Among the Cannibals of New Guinea* (1888). He worked as an official at the London Missionary Society until retirement (1894). He sent a collection of reptiles taken in the Torres Straits islands to the British Museum.

MacGregor, A.

Tanimbar Death Adder *Acanthophis macgregori* Hoser, 2002

Andrew S. MacGregor is a retired policeman from Victoria. This reptile was named after him "for his efforts in trying to expose corruption in the Victoria Police and more recently in relation to the government version of events following the Port Arthur Massacre in Tasmania."

MacGregor, W.

MacGregor's Wolf Gecko *Luperosaurus macgregori* **Stejneger,** 1907

Sir William MacGregor (1846–1919) was a naturalist, collector, and diplomat. He was Chief Medical Officer in Fiji, Administrator in New Guinea, later Governor of Lagos and then of Newfoundland (1904–1909), and finally of Queensland (1909–1914). Two birds are named after him.

MacKay

MacKay's Burrowing Skink *Anomalopus mackayi* **Greer** and **Cogger,** 1985

Roy D. Mackay is a herpetologist and ornithologist. His

early career was as a taxidermist at the Australian Museum, Sydney. He worked at the Museum of Papua New Guinea, Port Moresby, and was appointed Preparator-in-Charge (1965). Most of his career was spent in PNG, where he survived being bitten by a death adder, and he has written on its fauna and flora. He published "Notes on a Collection of Reptiles and Amphibians from the Furneaux Islands, Bass Strait" (1955).

Mackinnon

Mackinnon's Wolf Snake *Lycodon mackinnoni* **Wall,** 1906

Mackinnon's Worm Snake *Typhlops mackinnoni* Wall, 1910

Philip W. Mackinnon, an all-round naturalist, lived in Mussoorie in Uttar Pradesh in the foothills of the Himalayas. He was probably a member of the Mackinnon family of local brewers. He is known to have collected the nest and eggs of the Forest Owlet *Athene (Heteroglaux) blewitti*, a species that was later thought to have become extinct but was rediscovered in 1997. He co-wrote "List of Butterflies from Mussoorie and Dun Valley" (1897).

Macklot

Macklot's Python *Liasis mackloti* **Duméril** and **Bibron,** 1844

Heinrich Christian Macklot (1799–1832) was a taxidermist who was appointed to assist members of the Dutch Natural Science Commission. He went on an expedition to New Guinea and Timor (1828–1830). Three birds and a bat are named after him.

Maclean

Carrot Rock Skink *Mabuya macleani* **Mayer** and **Lazell,** 2000

William P. Maclean III (1943–1991) graduated from Princeton and then undertook a doctorate in evolutionary biology at the University of Chicago. He became Assistant Professor (1969) and later full Professor and Chairman of the Department at the University of the Virgin Islands, St. Thomas, U.S. Virgin Islands. He was diagnosed with cancer (1988) but still went on an expedition to the South China Sea (1990). His publications include *Reptiles and Amphibians of the Virgin Islands* (1982).

Macleay

Macleay's Water Snake *Enhydris polylepis* **Fischer,** 1886

William John Macleay (1820–1891) was a politician and naturalist who wrote on entomology, ichthyology, and zoology. He took part in several expeditions, including one to New Guinea (1875), which expedition he financed, paying £3,000 to buy the barque *Chevert*. The whole of the Macleay family were avid naturalists and collectors, resulting in the Macleay Museum being built (1887) to house their natural history collection. They began collecting insects in the 18th century. Alexander Macleay (1767–1848), diplomat and entomologist, went to Sydney as Colonial Secretary (1826). He already had one of the largest private insect collections in the world. It was added to by his son, William Sharp Macleay, and expanded to include all aspects of natural history by William's cousin, William John Macleay, who donated the collections to Sydney University (1887).

MacMahon

See **McMahon.**

Macola

Macola's Tree Iguana *Liolaemus uspallatensis* Macola and Castro, 1982

Guido S. Macola is an Argentinian herpetologist. He has written a number of articles, often with Castro, such as their description of this iguanid: "Una nueva especies del género *Liolaemus* del área subandina Uspallata—Mendoza, Argentina" (1982).

Macquarie

Murray River Turtle *Emydura macquarii* **Gray,** 1831

Named after the Macquarie River, New South Wales, Australia. The river was named after Lachlan Macquarie (1762–1824), a former Governor of New South Wales.

Macrae

Blue Tree Monitor *Varanus macraei* **Böhme** and Jacobs, 2001

Duncan R. Macrae is a herpetologist who founded a Reptile Park on Bali, Indonesia, called Rimba. He has also supplied snake and amphibian venoms for research and production of antivenin. He is currently Director of Coastal Zone Management, a UK-based Conservation consultancy.

Macrinius

Macrinius' Anole *Anolis macrinii* **H. M. Smith,** 1968

Emil Macrinius collected many specimens in Oaxaca, Mexico, some described only several decades later. Smith had the specimens, which were in bad condition, on loan for 10 years before he described them.

Madge

Madge's Blind Snake *Typhlops madgemintonae* **Khan,** 1999

See **Minton, M.**

Magellan

Magellan's Tree Iguana *Liolaemus magellanicus* Hombron and Jacquinot, 1847

Named after the Straits of Magellan.

Mahendra

Leaf-toed Gecko sp. *Hemidactylus mahendrai* Shukla, 1983

Professor Dr. Beni Charan Mahendra (1904–1995) of St. John's College, Agra, was an Indian zoologist and herpetologist. He wrote *Handbook of the Snakes of India, Ceylon, Burma, Bangladesh, and Pakistan* (1984).

Main

Main's Ground Gecko *Diplodactylus maini* **Kluge,** 1962 [Syn. *Lucasium maini*]

Main's Menetia *Menetia maini* **Storr,** 1976

Dr. Albert "Bert" Russell Main (1919–2010), a well-known Australian ecologist, was married to arachnologist Barbara York Main. He served in the RAAF (1939–1942). After WW2 he studied at the University of Chicago. He was a researcher at the Department of Zoology, University of Western Australia (1952–1967), completing his doctorate there (1956) and becoming Professor of Zoology (1967), then Professor Emeritus after retiring. He served on a number of public bodies, including being the President of the National Parks Commission. An Australian rodent is named after him.

Maindron

Maindron's Skink *Sphenomorphus maindroni* **Sauvage,** 1878

Maurice Maindron (1857–1911) was a naturalist, collector, and entomologist. He joined the staff of Muséum National d'Histoire Naturelle, Paris (1875), and started on 25 years of almost continual travel. He traveled in New Guinea (1876–1877), Senegal (1879 and 1904), India (1880–1881, 1896, and 1900–1901), Indonesia (1884–1885), Djibouti and Somalia (1893), and Arabia (1896).

Mair

Mair's Keelback *Tropidonophis mairii* **Gray,** 1841

Dr. Mair was an army surgeon in the 39th Regiment of Foot. He collected the holotype of this snake during George Grey's expeditions to Western Australia (1837–1839). He sent specimens to John Edward Gray from the early 1830s.

Maki

Maki's Keelback *Amphiesma miyajimae* Maki, 1931

Ota's Japalure *Japalura makii* **Ota,** 1989

Professor Moichiro Maki (1886–1959) was a herpetologist at Kyoto Imperial University. He graduated from the University of Hiroshima (1909). After leaving university he worked for the colonial administration in Formosa (now Taiwan) until 1926, using his free time to study amphibians, reptiles, and insects. He worked at Kyoto University from 1927 to 1946, during which time he was awarded his Ph.D. (1932). He wrote the classic and superbly illustrated *Monograph of the Snakes of Japan* (3 vols., 1931). His wife was employed to catch the snakes, as he was deathly afraid of them.

Makolowode

Makolowode's Leaf-toed Gecko *Hemidactylus makolowodei* **Bauer** et al., 2006

Makolowode's Skink *Trachylepis makolowodei* Chirio et al., 2008

Paul Makolowode, a field herpetologist from Zimba, Central African Republic, has worked in his own country and in Cameroon.

Malcolm / Malcolm Smith

Malcolm's Pit-viper *Trimeresurus malcolmi* **Loveridge,** 1938 [Syn. *Parias malcolmi*]

Malcolm's Worm Snake *Typhlops malcolmi* **Taylor,** 1947

Smith's Bent-toed Gecko *Cyrtodactylus malcolmsmithi* Constable, 1949 [Alt. Malcolm's Bow-fingered Gecko]

See **Smith, M. A.**

Malleis

Malleis' Cat-eyed Snake *Leptodeira frenata malleisi* **Dunn** and **Stuart,** 1935

Harry Malleis (d. 1931) was probably an employee of the U.S. Fish and Wildlife Service (1919). He was in British Honduras (now Belize) (1923) and in Guatemala (1924–1925), involved in moving game birds to Georgia; he was primarily an ornithologist. According to the original description, he died at Flores, Guatemala, of blackwater fever contracted while exploring El Peten.

Mallimacci

Thorntail Mountain Lizard *Phymaturus mallimaccii* **Cei,** 1980

Dr. Hugo Salvador Mallimacci, a geologist with the Geological Mining Service of the Argentine army, was the first person to observe this species.

Manaute

Skink sp. *Nannoscincus manautei* **Sadlier, Bauer, Whitaker,** and Smith, 2004

Joseph Manauté, described as "a friend and colleague of the authors," is Assistant Minister for Agriculture in New

Caledonia, having previously worked in the Parks and Reserves Section of the Natural Resources Directorate.

Maness

Clawed Gecko sp. *Pseudogonatodes manessi* Avila-Pires and **Hoogmoed,** 2000

Scott Jay Maness (1948–1981) was a herpetologist with the U.S. Fish and Wildlife Service. He burned to death while helping fight a wildfire at Merritt Island National Wildlife Reserve, Florida. He took his bachelor's degree at California State University, after which he served for three years as a research zoologist with the U.S. Peace Corps.

Mann

Mann's Worm Lizard *Amphisbaena manni* **Barbour,** 1914

Fiji Scaly-toed Gecko *Lepidodactylus manni* **Schmidt,** 1923

Mann's Dwarf Gecko *Lygodactylus manni* **Loveridge,** 1928

Mann's Worm Snake *Typhlops manni* Loveridge 1941

William Montana Mann (1886–1960) ran away from home to join the circus. The circus owner, John Ringling, advised him to get an education instead. Mann followed that advice. He graduated from Staunton Military Academy, Virginia (1905), then worked as a ranch hand in Texas and New Mexico, where he also collected entomological specimens. He received his doctorate in entomology from Harvard (1915) and got a job at the Bureau of Entomology (U.S. Department of Agriculture) as a specialist in termites and ants. He was Director of the National Zoological Park (part of the Smithsonian) (1925–1956); on his retirement, he became Director Emeritus and an Honorary Research Assistant at the Smithsonian. During his career he took part in many expeditions, as well as making private trips, visiting over 35 countries. His wife, Lucille Quarry Mann (1897–1986), was a journalist, but after their marriage (1926) she traveled with him on his expeditions.

Manthey

Manthey's Forest Dragon *Gonocephalus lacunosus* Manthey and **Denzer,** 1991

Rock Agama genus *Mantheyus* Ananjeva and **Stuart,** 2001

Ulrich Manthey (b. 1946) is an engineer, but since the early 1980s he has been a freelance herpetologist. Since 1990 he has been a collaborator of Museum für Naturkunde Berlin. He works closely with Denzer; they co-wrote "A Revision of the Melanesian-Australian Angle Head Lizards of the Genus *Hypsilurus* (Sauria: Agamidae: Amphibolurinae), with Description of Four New Species and One New Subspecies" (2006). An amphibian is named after him.

Manuel, A.

Skink sp. *Chalcides manueli* **Hediger,** 1935

Albert Manuel of Rabat helped to organize Hediger's expedition to southern Morocco.

Manuel, J.

Tree Iguana sp. *Phrynosaura manueli* **Núñez** et al., 2003 [Syn. *Liolaemus manueli*]

Juan Manuel is the son of Isabel Yermany. See **Isabel.**

Manuela G.

Lacertid lizard sp. *Psammodromus manuelae* **Busack,** Salvador, and **Lawson,** 2006

Manuela González is the wife of one of the describers, Alfredo Salvador.

Manuela M.

Mountain Lizard sp. *Phymaturus manuelae* Scolaro and Ibargüengoytía, 2008

Manuela Martinez is the daughter of Nora Ibargüengoytía, the junior author, and her husband, Martin Martinez. Manuela has accompanied her mother to congresses and on fieldwork since birth.

Manzanares

Snail-eating Snake sp. *Sibon manzanaresi* McCranie, 2007

Tomás Manzanares Ruiz collected the holotype.

Mapuche

Tree Iguana sp. *Liolaemus mapuche* Abdala, 2002

Named after the Mapuche people, a Patagonian aboriginal tribe.

Maran

Gabon Mud Turtle *Pelusios marani* Bour, 2000

Jérôme Maran (b. 1973) is a herpetologist who has worked with Bour. They were co-authors of the description of another turtle, *Pelusios cupulatta* (2003). He has written a number of papers, including the curiously titled "The Turtles of Côte d'Ivoire and Inside Liberian Jails" (2006).

Marcano

Marcano's Anole *Anolis marcanoi* **Williams,** 1975

Marcano's Galliwasp *Celestus marcanoi* **Schwartz** and Inchaustegui, 1976

Eugenio de Jesus Marcano Fondeur (1923–2003) was a botanist who wrote 15 books, including one on poisonous and another on edible plants of the Dominican Republic. He was Professor, Universidad Autónoma de Santo Domingo (1975–2003), having previously been Professor, Instituto Politécnico Loyola de San Cristobal (1955–1975).

Marcella

Marcella's Graceful Brown Snake *Rhadinaea marcellae* **Taylor,** 1949

Marcella Newman. See **Newman, R. J. and M.**

March, D.

March's Palm Pit-viper *Bothriechis marchi* **Barbour** and **Loveridge,** 1929

Douglas H. March (d. 1939) was a well-known herpetologist who died from the bite of a fer-de-lance. He made a special collection and study of the herpetofauna of Honduras and had a serpentarium at Tela.

March, J.

Spanish Keeled Lizard *Algyroides marchi* **Valverde,** 1958 [Alt. Valverde's Lizard]

Juan March Ordinas (1880–1962) was a financier and adventurer. He was involved in smuggling tobacco between North Africa and Spain, and during WW1 he supplied goods to ships of both sides. He founded Bank March (1926) and became very influential under the monarchy but was imprisoned by the republican government. He escaped from prison and made it safely to Gibraltar, where he established himself as a British agent and became the main organizer and financier of Franco's rebellion, which led to the Spanish Civil War. After the nationalists won, he remained Franco's main backer. He created Fundación Juan March (1955). The etymology reads, "dedicandola a D. Juan March como reconocimiento por la inmensa ayuda que ha prestado a la investigación española con el establecimiento de la Fundación que lleva su nombre."

Marche Leon

Marche Leon Least Gecko *Sphaerodactylus elasmorhynchus* **Thomas,** 1966

Marche Leon is a place in Haiti.

Marcovan

Centipede Snake sp. *Tantilla marcovani* **De Lema,** 2004

Dr. Marcovan Porto is a biologist, comparative anatomist, and Assistant Professor at the Department of Zoology, Institute of Biosciences, Universidade de São Paulo, Brazil.

Marcy

Marcy's Checkered Garter Snake *Thamnophis marcianus* **Baird** and **Girard,** 1853

Major-General Randolph Barnes Marcy (1812–1887), who explored the American West, graduated from West Point (1832). He was an infantry officer (1833–1846), serving mostly in Michigan and Wisconsin. Promoted to Captain (1846), he fought in the Mexican War. He was in Texas and determined the route of the Marcy Trail from Fort Smith to Santa Fé (1847–1851). He commanded an expedition to find the source of the Red River (1852), which crossed over 1,600 kilometers (1,000 miles) of unexplored country in Oklahoma and Texas, discovering 25 new mammal and 10 new reptile species. Back home he found that the newspapers had reported him killed by Comanches. He published his report under the title *Exploration of the Red River of Louisiana in the Year 1852, with Reports on the Natural History of the Country* (1853). On the outbreak of the American Civil War, he was promoted to Colonel and became Inspector-General of the Army of the Potomac. He was inspector-general for various army departments (1863–1878) and eventually of the entire army, until he retired (1881).

Maren

Bronzeback (snake) sp. *Dendrelaphis marenae* Vogel and van Rooijen, 2008

Dr. Maren Gaulke is a German herpetologist. She works in the Philippines and wrote "Trimeresurus flavomaculatus (Gray, 1842), die Philippinen-Bambusotter" (2006). She collected some of the snake paratypes.

Margaret B.

Two-pored Dragon sp. *Diporiphora margaretae* **Storr,** 1974

Buff-snouted Blind Snake *Austrotyphlops margaretae* Storr, 1981

Skink sp. *Glaphyromorphus butlerorum* Aplin, **How** and Boeadi, 1993

Margaret Butler is the wife of Dr. W. H. Butler (q.v.). *Glaphyromorphus butlerorum* is named after both of them.

Margaret S.

Centralian Ranges Rock-Skink *Egernia margaretae* **Storr,** 1968 [Alt. Rock Egernia]

Margaret Anne Slater. Storr's etymology states, "Named after Margaret Anne, wife of K. R. Slater, in appreciation of her hospitality. Mr. Slater kindly donated the holotype to the Western Australian Museum." See **Slater, K.**

Maria

Khasi Hills Forest Lizard *Calotes maria* **Gray,** 1845

Possibly Gray named this after his wife, Mary Emma Gray, but he gives no etymology in his text. See **Emma Gray.**

Maria (del Rosario)

Maria's Worm Lizard *Blanus mariae* Albert and Fernandez, 2009

Maria del Rosario Aguilar Tortajada (1914–2002) was the

senior author's grandmother. She had a hard and tragic life. Her first husband, an army captain, was executed during the Spanish Civil War (1936–1939), and her first child died from rubella. Her second marriage was unhappy, but she inspired her granddaughter to become educated and take control of her own life. Eva Albert told us, "For this reason, I have made this small contribution to her memory. That is also dedicated to hundreds of women in my country that lived in worse times than I live now and that never had the freedom to choose." She died in a tragic accident.

Marias

Blemished Anole *Anolis mariarum* **Barbour,** 1932

Brother Nicéforo Maria (1888–1980) was a French monk and Brother Apolinar Maria (1877–1949) a missionary Colombian monk who was Director of Instituto de La Salle, Bogotá. The holotype was "collected by Brother Nicéforo Maria, and named for him and his distinguished colleague, Brother Apolinar Maria." See also **Niceforo.**

Marcella

Rattlesnake sp. *Crotalus marcellae* Perez-Garcia, 1995

Marcella Soda (1966–1992) was a biologist at the Department of Biology, Faculty of Science, Universidad de los Andes, Mérida, Venezuela. She published her degree thesis as *Ecological Relationships between Bat* Glossophaga longirostris *and Columnar Cacti in the Dry Pocket of Lagunillas, Merida, Venezuela* (1991).

Marie

Skink sp. *Nannoscincus mariei* **Bavay,** 1869

E. A. Marié (1835–1889) was a French collector. He traveled in New Caledonia (1869), Guadeloupe (1874), and Madagascar (1878). A bird is named after him.

Marisela

Marisela's Ground Snake *Atractus mariselae* **Lancini,** 1969

Marisela Urosa Zambrano. The etymology in the description of this snake is very short, merely saying that "this reptile . . . is named after Marisela Urosa Zambrano." We know nothing more about her.

Markus Comba

Gecko sp. *Gonydactylus markuscombaii* **Darevsky,** Helfenberger, **Orlov,** and Shah, 1998 [Syn. *Siwaligekko markuscombaii*]

Markus Comba (b. 1956) is a Swiss antiquarian bookseller and book-restorer who is also a naturalist. He has supported several expeditions to Central Asia and to Nepal.

Marshall

Marshall's Pygmy Chameleon *Rhampholeon marshalli* **Boulenger,** 1906 [Alt. Marshall's Stumptail Chameleon]

Sir Guy Anstruther Knox Marshall (1871–1959), an entomologist, was an expert on African and oriental weevils. He was born in India and sent to England to be educated. He was responsible for founding the *Bulletin of Entomological Research* (1909) and the *Review of Applied Entomology* (1913). He was the first Director of the Commonwealth Institute of Entomology. He was in East Africa for some time, as he is reported as having collected specimens near Salisbury, Southern Rhodesia (Harare, Zimbabwe) (1901–1902). He collected the chameleon holotype.

Marta

Colubrid snake sp. *Thamnosophis martae* **Glaw,** Franzen, and **Vences,** 2005 [Syn. *Bibilava martae*]

Marta Puente Molins gave "invaluable help in the field" to the authors.

Martens

Martens' Day Gecko *Phelsuma martensi* **Mertens,** 1961

Karel Martens was a German traveler, explorer, and herpetologist, being one of the earliest breeders of day geckos. He was a well-known visitor to Madagascar. Robert Mertens was one of his regular correspondents.

Martha

Galápagos Pink Land Iguana *Conolophus marthae* Gentile and Snell, 2009

Martha Rebecca Gentile was the second daughter of the first author. The etymology is perhaps the saddest we have seen: "Martha prematurely left this world. She was born dead, as consequence of a medical doctor's negligence, on August 20th 2003."

Martin Del Campo

Centipede Snake sp. *Tantilla martindelcampoi* **Taylor,** 1937

Martin del Campo's Alligator Lizard *Abronia martindelcampoi* Flores-Villela and Sánchez-H., 2003

See **Del Campo** and **Rafael.**

Martin, H.

Red Sea Lizard *Mesalina martini* **Boulenger,** 1897

Dr. Henri Martin had an extensive private collection of reptiles. It was dispersed after his death, but the holotype of this lizard, obtained in 1895, was kept by his son for his own private collection.

Martin, J. K. L.

Dutch Leaf-toed Gecko *Phyllodactylus martini* Lidth de Jeude, 1887

Dr. Johann Karl Ludwig Martin (1851–1942) was a German geologist. He was Professor of Geology at Universiteit Leiden (1877–1922). He was also Director of the Geological Museum there (1880–1922). He was best known for his paleontological and stratigraphical research on the fossil fauna of the Dutch East Indies. He studied at Georg-August-Universität Göttingen, which awarded his doctorate (1874). He then taught at Wismar, where he studied the glacial deposits. At Leiden his research was on the collections of fossils from the Dutch colonies at the newly created Geological Museum. He enlarged the collections by new purchases and expeditions to the Dutch colonies: the Dutch Antilles (1884), the Moluccas (1892), and Java (1910). After his retirement (1922) Martin continued his research on the stratigraphy of the Dutch East Indies.

Martinez-Rica

Pena de Francia Rock Lizard *Iberolacerta martinezricai* Arribas, 1996

Dr. Juan Pablo Martinez-Rica was Director of Instituto Pirenaico de Ecologia, Zaragoza, Spain, and is now Vice President, Royal Academy of Sciences. He studied the herpetology of the Balearic Islands (1950–1974). He wrote *Datos sobre la herpetologia de la provincia de Salamanca* (1979).

Martin Garcia

Martin Garcia Least Gecko *Sphaerodactylus ladae* **Thomas** and Hedges, 1988

This gecko gets its name from the Sierra Martin Garcia in the Dominican Republic. However, there is a story behind the scientific name *ladae*. The etymology reads, "In honor of a reliable companion who steered us into many otherwise inaccessible areas in Hispaniola." Obviously deliberately vague, it has the feel of a pun. We think it refers to a Lada car. These Russian cars were never considered pretty, but they were reliable and would start even in very heavy frosts and were rugged enough to drive along poorly maintained tracks.

Martins

False Boa sp. *Pseudoboa martinsi* Zaher, Oliveira, and Franco, 2008

Professor Marcio Roberto Martins Costa is a herpetologist at the Department of Ecology, Universidade de São Paulo, Brazil. The authors say, "The specific name . . . honors Dr. Marcio Martins for his invaluable contribution to knowledge of the natural history of Brazilian snakes."

Martin Stoll

Gecko sp. *Gonydactylus martinstolli* **Darevsky,** Helfenberger, **Orlov,** and Shah, 1998

Martin Stoll (b. 1956) is a Swiss photographer who is now an information technology specialist. He is a naturalist and gave support to the expedition to Nepal upon which the holotype was collected.

Martius

Amazon Water Snake *Hydrops martii* **Wagler,** 1824

Carl Friedrich Phillip von Martius (1794–1868) was a German botanist and ethnographer. He was a member, with Spix, of Wied-Neuwied's expedition in Brazil (1817–1820). He continued with Spix's work after his death. Martius founded *Flora Brasiliensis*. A bird is named after him.

Martori

Tree Iguana sp. *Liolaemus martorii* Abdala, 2003

Dr. Ricardo Armando Martori is an ecologist and herpetologist at the Faculty of Natural Sciences, Universidad de Río Cuarto, Argentina. He co-wrote "Temporal Variation and Size Class Distribution in a Herpetological Assemblage from Córdoba, Argentina" (2006).

Marx

Marx's Rough-scaled Lizard *Ichnotropis microlepidota* Marx, 1956

Marx's Worm Snake *Ramphotyphlops marxi* **Wallach,** 1993 [Syn. *Typhlops marxi*]

Hymen "Hy" Marx (1925–2007) was Emeritus Curator of Herpetology at the Field Museum, where he worked for over 40 years. He served in the U.S. Air Force (1943–1945). After WW2 he took a bachelor's degree in biology at Roosevelt University, Chicago (1949). He worked as an assistant for K. P. Schmidt and witnessed the occasion when Schmidt was fatally bitten by a boomslang.

Maryan

Maryan's Ctenotus *Ctenotus maryani* Aplin and Adams 1998

Brad Maryan is, as the description puts it, an "irrepressible enthusiast of the Australian herpetofauna." He is a Technical Officer at the Western Australian Museum, Perth, and has worked as a zookeeper. He co-wrote *Reptiles and Frogs in the Bush: South Western Australia* (2007). He does not appear to be on good terms with Raymond Hoser over the use of the term "Snakebuster," and according to one of Hoser's websites, Maryan pleaded guilty in 2000 to charges of wildlife trafficking, for which he was fined.

Maslin

Maslin's Racerunner *Aspidoscelis maslini* **Fritts,** 1969

Dr. Thomas Paul Maslin (1909–1984) was born to American missionary parents in China, where he lived until entering high school in Los Angeles (1927). He studied at the University of California, Berkeley (1928–1933), and acquired a lifelong interest in herpetology. He taught at an American school in China (1934–1936) and researched the local herpetofauna. He returned to the USA (1936) to do graduate study at Berkeley, taking his master's degree (1941). He worked at Stanford (1941–1945), first teaching and then as Curator at the Natural History Museum (1943). Stanford awarded his doctorate in zoology (1945). He taught in Colorado at the State University and at the University of Colorado, Boulder, combining the roles of Assistant Professor with Curator of Zoology (1945–1974). He became Emeritus Professor in 1975 and revisited China in 1981. His papers and his collection of over 60,000 specimens are at the University of Colorado.

Masters

Masters' Snake *Drysdalia mastersii* **Krefft,** 1866

George Masters (1837–1912) was an English entomologist who emigrated to Australia. He was Assistant Curator of the Australian Museum (1864–1874) and collected throughout Australia (1870s). He was Curator of the Macleay Collection in Sydney and Macleay's personal collector (1874–1912). Part of Macleay's donation to the museum was £6,000 to pay Masters' salary for life. He was in New Guinea in 1875 (see **Macleay** regarding the *Chevert* expedition). A bird is named after him.

Matschie

Matschie's Dwarf Gecko *Lygodactylus conradti* Matschie, 1892 [Alt. Conradt's Dwarf Gecko]

Togo Leaf-toed Gecko *Hemidactylus matschiei* **Tornier,** 1901

Dr. Paul Matschie (1861–1926) was a zoologist who was a pioneer in mammalian taxonomy. He was professor at the Zoological Museum, Humboldt-Universität, Berlin. He wrote many scientific papers and named many mammals (including the Mountain Gorilla), though he was overzealous in creating new "species" on the basis of minor differences. One bird and four mammals are named after him.

Matsui

Hikida's Bow-fingered Gecko *Cyrtodactylus matsuii* **Hikida,** 1990

Matsui's Water Skink *Tropidophorus matsuii* Hikida, **Orlov,** Nabhitabhata, and **Ota,** 2002

Professor Dr. Masafumi Matsui of Kyoto University (b. 1950) is a zoologist and herpetologist.

Matthew

Ground Snake sp. *Atractus matthewi* Markezich and Barrio-Amoros, 2004

Matthew Markezich (b. 1990) is the senior author's son.

Matuda

Matuda's Ratsnake *Pantherophis flavirufa matudai* **H. M. Smith,** 1941

Matuda's Arboreal Alligator Lizard *Abronia matudai* **Hartweg** and Tihen, 1946

Matuda's Anole *Anolis matudai* H. M. Smith, 1956

Dr. Eizi Matuda (1894–1978) was a Japanese botanist who moved to Mexico (1922) and became a Mexican citizen (1928). All his university education was in Formosa (now Taiwan), then controlled by Japan. The University of Tokyo awarded his doctorate (1962). He taught in Japan and traveled to study flora in mainland Asia (1914–1921). In Mexico he eventually became head of the Department of Botany, National Institute of Forestry.

Mawe

Mesoamerican River Turtle *Dermatemys mawii* **Gray,** 1847 [Alt. Central American River Turtle]

Lieutenant Mawe was a British naval officer who collected the holotype (1833). Unfortunately, Gray recorded nothing more about him.

Maximilian

Maximilian's Snake-necked Turtle *Hydromedusa maximiliani* **Mikan,** 1820

Gymnophthalmid lizard sp. *Micrablepharus maximiliani* **Reinhardt** and Lütken, 1862

Maximilian Alexander Philip, Prince zu Wied-Neuwied (1782–1867), was an aristocratic explorer. After studying natural history at Georg-August-Universität Göttingen, he entered the Prussian army (1802), ultimately becoming a Major-General. He collected in Brazil (1815–1817), Guyana (1821), and North America (1832–1834). He made a famous journey (1833) of some 8,000 kilometers (5,000 miles), principally up the Missouri River. He wrote *Reise nach Brasilien in den Jahren 1815 bis 1817* (1820). Two birds and four mammals are named after him. See also **Neuwied** and **Wied.**

Maxwell

Maxwell's Mountain Keelback *Opisthotropis maxwelli* **Boulenger,** 1914

Dr. John Preston Maxwell (1871–1961) was a Presbyterian missionary. He graduated from University College London and finished medical training at St. Bartholomew's. He left for Fujian, China, in 1898. He worked at Yungchun Hospital (1899–1919), then was Professor of Obstetrics and Gynecology at Peking Union Medical

College (1919–1937). He wrote *Osteomalacia in China* (1925). It is unclear when he left China, but it was probably after the Japanese capture of Peking (now Beijing) (1936), as he became a consultant at Newmarket General Hospital, England, in 1937. One of his obituaries states that he dedicated his life to China and epitomized Confucius' dictum, "Within the four seas all men are brothers." He presented the holotype of this snake to the British Museum.

Maya

Yucatán White-lipped Snake *Symphimus mayae* **Gaige,** 1936

Mayan Tropical Night Lizard *Lepidophyma mayae* **Bezy,** 1973

Named after the Maya, an indigenous people of the Yucatan Peninsula.

Mayer

Mayer's Sand Lizard *Pedioplanis gaerdesi* **Mertens,** 1954

Dr. Werner Mayer (b. 1943), a herpetologist at the Laboratory for Molecular Systematics at Naturhistorisches Museum Wien, is a specialist in African lizards. He co-wrote "The Parapatric Existence of Two Species of the *Pedioplanis* Complex (Reptilia: Sauria: Lacertidae) in Namibia" (1987), and as a result his name got attached to this species.

Maynard, C. J.

Great Inagua Ameiva *Ameiva maynardi* **Garman,** 1888

Maynard's Anole *Anolis maynardi* Garman, 1888

Professor Charles Johnson Maynard (1845–1929) was a zoologist, ornithologist, and lepidopterist. He was described as "Newtonville's [Massachusetts] enigmatic naturalist." He was a well-known observer of birds, particularly in Florida and the Bahamas. He was a contemporary of Outram Bangs in the Nuttall Ornithological Club. Among his publications is *Birds of Washington and Vicinity* (1898). A bird and a mammal are named after him.

Maynard, F. P.

Maynard's Longnose Sand Snake *Lytorhynchus maynardi* **Alcock** and Finn, 1897

Colonel F. P. Maynard (d. 1921) of the Indian Medical Service was an army physician and a naturalist. He was a member of the Afghan-Baluchi Boundary Commission (1896). He was Professor of Ophthalmic Surgery at the Medical College in Calcutta (1918).

McCall

Flat-tailed Horned Lizard *Phrynosoma mcallii* **Hallowell,** 1852

Brigadier General George Archibald McCall (1802–1868) was an amateur naturalist and collector. He graduated from West Point (1822) and fought in the Seminole and Mexican wars. When the American Civil War started he was appointed Brigadier General of Volunteers in the Union army, commanding the Pennsylvania Reserves division. He was captured and exchanged (1862). He remained on sick leave until his resignation (1863). He wrote a paper, "Some Remarks on the Habits &c., of Birds Met with in Western Texas, between San Antonio and the Rio Grande and in New Mexico" (1851), which he sent to Audubon. A bird is named after him.

McCord

McCord's Box Turtle *Cuora mccordi* **Ernst,** 1988

McCord's Snakeneck Turtle *Chelodina mccordi* Rhodin, 1994

Dr. William Patrick McCord (b. 1950) is a veterinary surgeon and a turtle specialist. He owns the East Fishkill Animal Hospital in New York State.

McCoy, C. J.

McCoy's Casquehead Iguana *Laemanctus serratus mccoyi* Perez-Higareda and **Vogt,** 1985

Dr. Clarence John "Jack" McCoy (1935–1993) took his bachelor's and master's degrees at Oklahoma State University and was awarded his doctorate by the University of Colorado while working as a Research Assistant at the university's museum. He joined the staff of the Carnegie Museum (1964), becoming Curator (1972). His publications include *Amphibians and Reptiles in Pennsylvania* (1982).

McCoy, F.

McCoy's Skink *Nannoscincus maccoyi* Lucas and Frost, 1894 [Alt. Maccoy's Elf Skink, Highlands Forest Skink]

Sir Frederick McCoy (1817–1899) was an Irish paleontologist and naturalist. He was educated at Cambridge and worked at the Woodwardian Museum on its fossil collection (1846–1850). He became Professor of Geology at Queen's College, Belfast (1850). When the Chair of Natural Science at the University of Melbourne was created (1854), McCoy was its first occupant. He taught many different subjects for about 30 years. He established and was the first Director of the National Museum of Natural History and Geology, Melbourne. His last publication was "Note on a New Australian Pterygotus" (1899).

McCrory

Bridle Snake sp. *Dryocalamus mccroryae* **Taylor,** 1922

Mrs. Ida M. McCrory collected the holotype (1920).

McDougall

McDougall's Kukri Snake *Oligodon mcdougalli* **Wall,** 1905

E. McDougall obtained the holotype. Other details of him are unknown.

McDowell

McDowell's Sea Snake *Hydrophis macdowelli* **Kharin,** 1983

Northern New Guinea Keelback *Tropidonophis mcdowelli* Malnate and **Underwood,** 1988

Worm Snake sp. *Typhlops mcdowelli* **Wallach,** 1996

McDowell's Bevelnosed Boa *Candoia paulsoni mcdowelli* **H. M. Smith** et al., 2001

Dr. Samuel Booker McDowell Jr. (b. 1928) is a herpetologist. He is Emeritus Professor of Zoology, Rutgers University; a Research Associate in the Herpetology Department, American Museum of Natural History, New York; and formerly of the Zoology Department, Columbia University. He has been described as "the most influential of all modern students of snake classification."

McGregor, R. C.

McGregor's Tree Viper *Trimeresurus mcgregori* **Taylor,** 1919 [Syn. *Parias flavomaculatus mcgregori*]

Richard Crittenden McGregor (1871–1936) was an Australian-born American ornithologist who collected the viper holotype (1907). At the time of his death he was Editor of the *Philippine Journal of Science* and Chief of the Publicity Division, Department of Agriculture and Commerce, Manila. He had been with the Bureau of Science since its inception (1902), for part of that time as Director. He started zoological collection on expeditions while an undergraduate. His first trip was to Panama collecting fish. He graduated from Stanford (1898). Although he was career scientist, a friend nevertheless said, "He gave me quite a shock one day when he said bluntly that he cared nothing whatever for science as such, that he liked birds better than anything in the world and that was all there was to it for him."

McGregor, W. R.

McGregor's Skink *Cyclodina macgregori* Robb, 1975

William R. "Barney" McGregor (1894–1977) was head of the Department of Zoology, University of Auckland, New Zealand. He was active in stopping the logging of kauri trees at Waipoua (1940s), and it was declared a sanctuary (1952) now named after him. Robb's etymology says the skink is named "in honour of Professor W. R. McGregor, for many years head of the Department of Zoology, University of Auckland, to whom I owe my interest in herpetology."

McGuire

McGuire's Rock Gecko *Cnemaspis mcguirei* **Grismer,** Grismer, Wood, and Chan, 2008

Dr. Jimmy A. McGuire took a bachelor's degree in business administration (1989) and a master's in biology (1994), both at San Diego State University, and a doctorate in zoology at the University of Texas, Austin (1998). He was a postdoctoral Fellow at the Smithsonian (1999) and Assistant Professor and Assistant Curator, Louisiana State University (2000–2002). Since 2003 he has been Assistant Professor and Curator of Herpetology, University of California. He wrote "The Effects of Desertification on Garter Snakes in Baja California" (1991).

McKenzie

McKenzie's Dragon *Ctenophorus mckenziei* **Storr,** 1981 [Alt. Dwarf Bicycle-Dragon]

Norman Leslie "Norm" McKenzie works for the Department of Conservation and Land Management, Western Australia (formerly Department of Fisheries and Wildlife). He graduated with a zoology degree from Monash and joined the Western Australian government as a Research Scientist (1970). Currently he is a Principal Research Scientist for the state's ecological survey program. He has published more than 190 scientific papers on a wide range of topics such as conservation and biogeography.

McMahon

McMahon's Desert Viper *Eristicophis macmahoni* **Alcock** and Finn, 1897

Dwarf Racer sp. *Eirenis mcmahoni* Wall, 1911

Colonel Sir Arthur Henry McMahon (sometimes MacMahon) (1862–1949) was an army officer. He was commissioned from Sandhurst (1882) and was posted to the India Staff Corps, entering the Punjab frontier force (1885). He transferred to the political department (1890) and acted as political agent for a number of small states. He was Commissioner for Baluchistan (1901–1903) and for Seistan (1903–1905), and returned to Baluchistan (1905) to act as arbitrator in the boundary dispute between Iran and Afghanistan (1906). He became Foreign Secretary to the Government of India (1911). He left India (1914) and became the first British High Commissioner for Egypt under the British Protectorate (1914–1916). He made a collection of reptiles in Baluchistan (1896) when he was arbitrating a boundary dispute between Baluchistan and Afghanistan.

McMillan

McMillan's Spiny-tailed Gecko *Strophurus mcmillani* **Storr,** 1978 [Alt. Short-tailed Striped Gecko]

Robert Peter McMillan (1921–2009) served in the Royal Air Force during WW2 as a Flying Officer. He gained a

B.Sc. from the University of Western Australia. He was an entomologist at the Western Australian Museum and later an Honorary Associate. Both McMillan and Storr had articles published about Rottnest Island in the *Western Australian Naturalist* (1962).

McNamara

McNamara's Burrowing Snake *Pseudorabdion mcnamarae* **Taylor,** 1917

Homer McNamara was a Superintendent of the La Carlota Agricultural Station in the Philippines. Taylor wrote, "I take pleasure in dedicating this species to Mr. Homer McNamara, superintendent of the La Carlota Agricultural Station, who rendered able assistance in making collections on the volcano [the Canlaon Volcano, Negros]." McNamara climbed the volcano with Taylor and another American who was a police lieutenant. Although McNamara assisted him in collecting specimens on the volcano, he was also involved in an incident that could have been Taylor's end, as Taylor recounted in his *Recollections of an Herpetologist*. As they rode through a mountain gorge, Homer's horse "reared . . . and its hooves struck me in the back and knocked me into the torrent, with my sacks and cans containing the precious collection. My horse, now free of me, managed to make the opposite bank, and while I was between the boulders I too managed to make the opposite bank. I emerged from the water perhaps a hundred feet farther downstream, but without a single specimen of serpent, lizard, or frog to show for the day's journey. The sacks containing two cans had been tucked in my belt but the water had pulled them away. Homer's horse had managed to plunge and swim through without serious mishap."

Mearns

Mearns' Rock Lizard *Petrosaurus mearnsi* **Stejneger,** 1894 [Alt. Banded Rock Lizard]

Lieutenant Colonel Edgar Alexander Mearns (1856–1916) was a surgeon in the U.S. Army. He was stationed in Mexico (1892–1894) and in the Philippines (1903–1907), and traveled in Africa (1909–1911). He published a great deal on natural history during the last decade of his life, including many descriptions of African birds. Among his finds in the Philippines was the rare Bagobo Babbler. He called it *Leonardia* (now *Trichastoma*) *woodi* after the commanding general of the American forces in Mindanao and Sulu. Mearns was a friend of Theodore Roosevelt and accompanied him to East Africa (1909). Childs Frick approached the Smithsonian (1911) for a scientist to accompany him on his trip to Africa, and Mearns was chosen. Frick agreed to pay Mearns' salary and expenses and to donate all collections to the Smithsonian. This was Mearns' last expedition. His life was beset with illness, including a "nervous breakdown complicated by malaria" and various parasitic conditions. He developed diabetes, and as this was before insulin treatment was available, nothing could be done for him. Five birds and seven mammals are named after him.

Mechel

Dwarf/Reed Snake sp. *Calamaria mecheli* Schenkel, 1901

Schenkel states that "A. v. Mechel" provided the holotype of this Sumatran snake. Although nothing more seems to be recorded about him, a "Hermann van Mechel" is on record as a planter, trader, and traveler who surveyed and mapped Lake Toba, Sumatra (1887). It seems likely that these two "v. Mechel"'s were related.

Mechow

Elongate Quill-snouted Snake *Xenocalamus mechowi* **Peters,** 1881

Major Friedrich Wilhelm Alexander von Mechow (1831–1890) was an explorer and naturalist who led several expeditions to Angola (1873–1882). He was born in Silesia and at a young age joined the Prussian army. He fought in the 1866 war and the Franco-Prussian War (1870–1871), being badly wounded in the battle of Wörth. He retired from the army in 1874. He explored the middle Kwango River in Angola (1880) and collected reptiles and amphibians. Of his life after his last expedition, little is known; even the year of his death is uncertain. Three birds and one mammal are named after him.

Medem

Medem's Coral Snake *Micrurus medemi* **Roze,** 1967

Medem's Worm Lizard *Amphisbaena medemi* **Gans** and Mathers, 1977

Medem's Neusticurus *Neusticurus medemi* **Dixon** and **Lamar,** 1981

Medem's Anole *Anolis medemi* **Ayala** and **Williams,** 1988

Professor Dr. Federico Medem (1912–1984) was born in Riga as Friedrich Johann Comte von Medem. He was of German origin but thought of himself as a Latvian. His family left Latvia after the Russian Revolution (1917) and moved to Germany. Medem studied at Humboldt-Universität Berlin and at Eberhard Karls Universität Tübingen. He worked for his doctorate at the marine biology station in Naples run by Gustav Kramer. He served in the Wehrmacht during WW2 and fought on the Russian front. After the war he worked in Germany and Switzerland. He moved to Colombia (1950), changed his name, and became a herpetologist and ardent conservationist at the research station at Villavicencio and at Universidad Nacional de Colombia, Bogotá. There is a herbarium named after him at Instituto Alexander von

Humboldt. He wrote numerous scientific papers (1950s–1980s), mostly on Colombian reptiles. He published two volumes that make up his *Los Crocodylia de Sur America* (1981–1983). A mammal is named after him.

Medici

East African Egg-eater *Dasypeltis medici* Bianconi, 1859

Dr. Michele Medici (1782–1859) was Emeritus Professor of Physiology at Università di Bologna, Italy. He qualified as a physician (1802) in Bologna, where he lived all his life. In addition to physiology he was interested in natural science, in pathology, and in the history of medicine and research. He published (1857) a history of the Bolognese school of anatomy. The original etymology of this snake, which states that Bianconi was once taught by Medici and regarded him as the best of teachers, is written in Latin: "Speciem hanc Michaeli Medici Physiologiae doctori emerito hujus Universitatis ornamento libens dico, optimo olim mihi Magistro."

Medusa

Venezuelan Pit-viper *Bothriopsis medusa* **Sternfeld,** 1920

Ground Snake sp. *Atractus medusa* Passos, Mueses-Cisneros, Lynch, and Fernandes, 2009

In Greek mythology, Medusa was one of the Gorgons, female monsters whose hair was made up of snakes. Her gaze could turn living things to stone. She was killed by Perseus, who avoided looking at her directly by using a highly polished shield in which to see her reflection.

Mehely

Mehely's Agama *Agama mehelyi* **Tornier,** 1902

Mehely's Lizard *Darevskia rudis bithynica* Mehely, 1909

Lajos Ludwig von Méhely (1862–1946/1952) was a biologist, herpetologist, and anthropologist. (He may have died before 1952, as he was a political prisoner after WW2.) In 1913 he became head of the Zoological Department of the museum, and Professor of Zoology and Anatomy, at the University Pázmány Péter, Budapest. He was extremely racist and while head of the Anthropology Department, Magyar Természettudományi Múzeum, Budapest, insisted that only his views and theories should be taught. He wrote a seminal, unpublished monograph called *Herpetologia hungarica*. The excellent herpetology collection was nearly entirely destroyed in the Budapest uprising (1956).

Meier

Skink sp. *Geoscincus haraldmeieri* **Böhme,** 1976

Harald Meier (b. 1922) is a herpetologist associated with the Senckenberg Museum, Frankfurt. He and Böhme published "Revision der madagassischen *Homopholis* (Blaesodactylus)-Arten (Sauria: Gekkonidae)" (1980). See also **Harald Meier.**

Meller

Meller's Chameleon *Chamaeleo melleri* **Gray,** 1865 [Alt. Giant One-horned Chameleon]

Charles James Meller (1836–1869) was a botanist who worked in Nyasaland (now Malawi) (1861) and on Mauritius (1865), where he was Superintendent of the Botanical Gardens. A bird and a mammal are named after him.

Mendes

Gecko sp. *Gonatodes alexandermendesi* **Cole** and Kok, 2006

Alexander Mendes is a Guyanese businessman, rancher, conservationist, adventurer, and explorer. He has provided logistical support and security for visiting naturalists.

Ménétries

Ménétries's Lizard *Ophisops elegans* Ménétries 1832 [Alt. Snake-eyed Lizard]

Édouard P. Ménétries (1802–1861) was a French zoologist who collected in Brazil and in Russia, where he settled. He studied at Muséum National d'Histoire Naturelle, Paris, under Cuvier, on whose recommendation he participated in Langsdorff's expedition to Brazil (1821–1825). He became Conservator of Collections of the Russian Academy of Sciences, St. Petersburg (1826). He explored and collected in the Caucasus (1829–1830) and wrote *Catalogue raisonée des objets de zoologie recueillis dans un voyage au Caucase et jusqu'aux frontières actuelles de la Perse* (1832). When the Zoological Museum of the Academy of Sciences was officially opened, Ménétries was first Curator of Entomology (1832–1861). He studied Siberian fauna and published one of the first works on Kazakh fauna. Three birds are named after him.

Merian

Argentine Giant Tegu *Tupinambis merianae* **Duméril** and **Bibron,** 1839

Anna Maria Sibylla Graff, née Merian (1647–1717), was born in Frankfurt into a Swiss family and lived most of her life in Germany, then in the Netherlands. She was a pioneering female naturalist, scientific illustrator, and trader in insect specimens. She was recognized as talented when she was 13, when she began studying insects and plants, painting what she saw. She married (1665) Johann Andreas Graff but continued to paint and give lessons. She was the first person to make clear the life cycle of a butterfly and used her illustrations for a book that she published: *Der Raupen wunderbare Verwandlung und sonderbare Blumennahrung* (The Caterpillar,

Marvellous Transformation and Strange Floral Food) (1679). She left her husband (1685) and took her mother and two daughters to a Labadist religious community in Friesland, the Netherlands, moving to Amsterdam in 1691; her husband divorced her in 1692. When her elder daughter married and moved to Surinam, Anna went too (1699). She stayed for two years, studying South American plants and animals. She was extremely critical of how the Dutch planters treated their black slaves and the local Amerindians. She caught malaria and had to return to Amsterdam (1701). She suffered a stroke (1715), became partially paralyzed, and died a pauper. Recently she has become something of a cult figure in Germany, where her portrait appeared on a bank note, schools are named after her, and a research ship, *Maria S. Merian*, was named in her honor, as is an amphibian.

Merrem

Merrem's Hump-nosed Viper *Hypnale hypnale* Merrem, 1820

Merrem's Madagascar Swift *Oplurus cyclurus* Merrem, 1820

Merrem's Ground Snake *Liophis miliaris merremi* **Wied,** 1821

Wagler's Snake *Waglerophis merremi* **Wagler,** 1824

Blasius Merrem (1761–1824) was a German zoologist who was the first person accurately to separate the reptiles and the amphibians, in *Versuch eines Systems der Amphibien* (1820). He combined the snakes and lizards into a single order, the Squamata, and also separated the crocodilians from the lizards. He was the first ornithologist to propose a division of birds into Ratitae (running birds) and Carinatae (flying birds).

Merriam

Merriam's Canyon Lizard *Sceloporus merriami* **Stejneger,** 1904

Clinton Hart Merriam (1855–1942) was a naturalist and physician. Through his father, a Congressman, he met Baird of the Smithsonian (1871), which led to his being invited to work as a naturalist in Yellowstone, Wyoming, as a member of the Hayden Geological Survey (1872). This experience guided his choice of further education; he studied biology and anatomy at Yale, graduating in medicine (1879). He soon forsook medicine for full-time scientific work (1883). Under his chairmanship, the Bird Migration Committee of the American Ornithologists' Union persuaded Congress to supply funds to study birds, as such work would benefit farmers. Merriam became the first Chief of the U.S. Biological Survey's Division of Economic Ornithology and Mammalogy. He undertook expeditions (1881) to both Death Valley and, as commander, the Bering Sea. He is most famed for his "life zone" theory, namely, that "temperature extremes were the principal desiderata in determining the geographic distribution of organisms." He has several mammals and a bird named after him.

Mertens

Mertens' Coral Snake *Micrurus mertensi* **Schmidt,** 1936

Mertens' Water Monitor *Varanus mertensi* **Glauert,** 1951

Mertens' Centipede Snake *Tantilla brevicauda* Mertens, 1952

Mertens' Lizard-eating Snake *Phalotris mertensi* **Hoge,** 1955

Mertens' Tropical Forest Snake *Umbrivaga mertensi* **Roze,** 1964

Mertens' Lined Snake *Tropidoclonion lineatum mertensi* **H. M. Smith,** 1965

Mertens' Day Gecko *Phelsuma robertmertensi* Meier, 1980

Robert Friedrich Wilhelm Mertens (1894–1975) was a German zoologist and herpetologist who was born in St. Petersburg. He left Russia (1912) to study medicine and natural history, obtaining his doctorate from Universität Leipzig (1915). After serving in the German army during WW1, he worked at the Senckenberg Museum in Frankfurt as an assistant (1919–1920). He was appointed (1920) to replace the dismissed Sternfeld (q.v.) in charge of herpetology, becoming Curator (1925) and Director (1947). He was a man of prodigious energy; his wife was his sole assistant (1920–1943). He was Chairman of the Zoology Section (1934–1955) and retired as Director Emeritus (1960). He lectured at Goethe-Universität Frankfurt am Main (1932–1939) and became Professor (1939). In spite of these responsibilities and huge workload, he still found time to publish about 800 scientific papers and 13 books. His first collecting trip was to Tunisia (1913), and during his time he visited 30 countries in search of specimens. During WW2 he had most of the Senckenberg collection evacuated to small towns, where the specimens were set up in locations like dance halls for use and study. He encouraged German soldiers fighting outside Germany to collect specimens for him, and a regular supply of reptiles and other creatures reached him courtesy of the German Field Post Office system. He died, aged 81, after a bite from a specimen of the Twig Snake *Thelotornis kirtlandi*, a South African snake that he had long kept as a pet. No antivenin existed then for this species. It took 18 very painful days for him to die, and he kept a diary of each day's events, remarking in it, with true gallows humor, "für einen Herpetologen einzig angemessene Ende" (a singularly appropriate end for a herpetologist). See also **Robert** and **Robert Mertens.**

Merton

Merton's Sea Snake *Parahydrophis mertoni* **Roux,** 1910 [Alt. Roux's Sea Snake]

Dr. Hugo Merton (1879–1939) was a zoologist. He visited the Kai and Aru islands in eastern Indonesia with Roux (1907–1908). He took a teaching post at Universität Heidelberg (1912), becoming Professor of Zoology (1920), but was dismissed (1935) when the Nazis "cleansed" the universities under the provisions of the Nuremberg Citizenship Laws. He moved to Britain (1937), taking up a position at the University of Edinburgh.

Messana

Messana's Racer *Coluber messanai* **Schätti** and **Lanza,** 1989 [Alt. Schatti's Racer]

Dr. Giuseppe Messana (b. 1944) is a biologist at the Institute for the Study of Ecosystems, Florence. His doctorate (1971) is from Università degli Studi di Firenze. He has written on the speleobiology of Somalia, has worked with Lanza, and published jointly with Chelazzi "Stenasellus costai sp. n., isopode freatobio gigante della Somalia" (1970). He is mainly interested in marine isopods and cave-dwelling invertebrates.

Mestre

Pinar del Rio Anole *Anolis mestrei* **Barbour** and **Ramsden,** 1916

Dr. Aristides Mestre y Hevia (1865–1952) was a physician, biologist, naturalist, and anthropologist. He was in joint charge (1903) of the Montané Anthropological Museum, Universidad de La Habana, Cuba. The original etymology reads, "For an old friend, Doctor Aristides Mestre, Adjunct Professor of Biology at the University of Havana."

Meszoely

Meszoely's Blind Snake *Typhlops meszoelyi* **Wallach,** 1999

Charles Aladar Maria Meszoely (b. 1933) is Professor of Biology at Boston University, where his interests include parasitology and vertebrate paleontology particularly focused on the fossil record of amphibians, lizards, and snakes. He is currently involved in two research projects, on the Isle of Wight, England, and in Wyoming, USA. He has worked for the Center for Vertebrate Studies at Northeastern University, where he was Wallach's graduate adviser. He has published a number of articles, such as, with R. E. Ford, "A New Eocene Frog (Palaeobatrachidae) from the British Islands" (1984).

Methuen

Methuen's Dwarf Gecko *Lygodactylus methueni* **FitzSimons,** 1937 [Alt. Woodbrush Dwarf Gecko]

Major Lord Paul Ayshford Methuen (1886–1974) was an artist and zoologist. He graduated in natural history at Oxford (1910). He went to Madagascar (1911) to collect subfossil lemurs for the Oxford University Natural History Museum. He made a very mixed but important collection that was not fully examined until 2000. He stayed in Africa for a time, working as an assistant at the Transvaal Natural History Museum. He was an administrator of two London landmarks, the Tate Gallery (1940–1945) and the National Gallery (1938–1945).

Metter

Metter's River Cooter *Pseudemys concinna metteri* **Ward,** 1984

Dean "Doc" Edward Metter (1932–2001) was for many years a member of the Faculty of Biology, University of Missouri, where he taught zoology, comparative anatomy, herpetology, and evolution. An annual award in his honor is given by the Society for the Study of Amphibians and Reptiles.

Mettetal

Mettetal's Amphisbaena *Blanus mettetali* **Bons,** 1963

M. Mettetal was head of the Laboratory of Animal Biology, Faculty of Sciences of Morocco.

Meyer, A. B.

Meyer's Emo Skink *Emoia kordoana* Meyer, 1874

Meyer's Legless Skink *Typhlosaurus meyeri* **Boettger,** 1894

Dr. Adolf Bernard Meyer (1840–1911) was an anthropologist and ornithologist who collected in the East Indies, New Guinea, and the Philippines. He was Professor at Ethnographische Museum, Dresden, becoming Director, Staatlisches Museum für Tierkunde, Dresden (1872). He wrote *The Birds of the Celebes and Neighbouring Islands* (1898). He was very interested in the evolution debate and corresponded with Wallace. Eight birds and one mammal are named after him.

Meyer, J. R.

Meyer's Anole *Anolis johnmeyeri* Wilson and McCranie, 1982

Dr. John Raymond Meyer is a research herpetologist and biogeographer at the Natural History Museum, London, who concentrates on Central American herpetology. He visited the Bay Islands, Honduras, with Wilson (1967). He co-wrote *A Guide to the Reptiles of Belize* (1998).

Meyerink

Meyerink's Kukri Snake *Oligodon meyerinkii* **Steindachner,** 1891

Steindachner gives no etymology in his description, but we think the snake is probably named after Hermann

Friedrick Meyerink (1850–1908), who was manager of the Borneo Company's German Hacienda Gomantong in the Philippines (1886). He was the German Consul to the Philippines (then a Spanish possession) and was based in Manila. There was political trouble in this period in the Philippines, and he appears to have arranged for a warship to be sent to look after German interests. He sent some ethnographic material to Berlin.

Michaelsen

Michaelsen's Spiny-tailed Gecko *Strophurus michaelseni* **Werner,** 1910 [Alt. Robust Striped Gecko]

Professor Wilhelm Michaelsen (1860–1937) was Curator of the Department of Invertebrates at Zoologischen Museum und Institut, Hamburg (1887–1923). He led the German South West Australian expedition.

Michele

Ground Snake sp. *Atractus micheleae* Esqueda and La Marca, 2005

Dr. Michele Ataroff Soler collected the holotype of this snake. She is a biologist and ecologist and a Professor at Centro de Investigaciones Ecologicas de Los Andes Tropicales at Universidad de los Andes, Mérida, Venezuela. This is also where she took her three degrees, culminating with a doctorate in tropical ecology (1990). She has published widely on ecological matters, especially soil erosion.

Mijares

Ground Snake sp. *Atractus mijaresi* Esqueda and La Marca, 2005

Abraham Mijares-Urrutia is a herpetologist at Universidad Francisco de Miranda, Venezuela.

Mikan

Neotropical Tree Snake sp. *Sibynomorphus mikanii* **Schlegel,** 1837

Johann Christian Mikan (1769–1844) was a German zoologist, botanist, and entomologist, and Professor of Botany at Prague (then controlled by Austria). He was one of the naturalists on the Austrian Brazil expedition (1817–1835).

Milius

Thick-tailed Gecko *Underwoodisaurus milii* Bory, 1823 [Alt. Barking Gecko]

Baron Pierre Bernard Milius (1773–1829) was a sailor, naturalist, and civil servant who took part in an exploratory voyage (1804) of the Mascarene Islands, Indian Ocean, under Nicolas Baudin, during which he became friends with Bory. He was Governor of Bourbon (now Réunion) (1818–1821), where he established a port and undertook agricultural projects. He was also despotic and despised by the locals. He was later appointed as Governor of French Guiana. He was present at the Battle of Navarino during the war of Greek independence.

Miller

Miller's Anole *Anolis milleri* **H. M. Smith,** 1950

Walter S. Miller was an ethnologist, anthropologist, linguist, and collector. He worked at the Summer Institute of Linguistics, Mexico City (1946), where he was engaged in making a philological study of the Mixes Indians. He made a herpetological collection, including this anole, in Oaxaca. Most of what he collected is either at the University of Illinois or at the Smithsonian. He wrote *Cuentos Mixes* (1956) about the indigenous people of Oaxaca.

Millot

Nosy Mamoko Skink *Paracontias milloti* **Angel,** 1949

Professor Dr. Jacques Millot (1897–1980) was a physician and naturalist at Muséum National d'Histoire Naturelle, Paris. He published on arachnology and on a fossil coelacanth that had been found in Madagascar, and later on live specimens. He spent many years in Madagascar in charge of the Scientific Institute of Madagascar and in 1953 started the periodical *Le naturaliste malagasy*.

Milne-Edwards

See **Edwards.**

Mindi

Qattara Gecko *Tarentola mindiae* **Baha El Din,** 1997

Mount Sinai Gecko *Hemidactylus mindiae* Baha El Din, 2005

Mindy Baha El Din (b. 1958) is the wife of Sherif Baha El Din (q.v.). She was born in Chicago, and her degree from Indiana University is in Arabic and economics. After graduating she took a course at Cornell in field ornithology and became an enthusiastic birder. She was employed by BirdLife International to establish an environmental education center at the Giza Zoological Gardens, Cairo (1988). Since 1992 she and her husband have been freelance wildlife consultants, organizing and guiding birding tours in Egypt.

Mingtao

Mingtao's Gecko *Gekko taibaiensis* Song, 1985

Mingtao Song (b. 1937) is a Chinese herpetologist.

Minton, M.

Madge's Blind Snake *Typhlops madgemintonae* **Khan,** 1999

Madge Alice Shortridge Rutherford Minton (1920–2004)

was the wife of Sherman Anthony Minton Jr. (q.v.). Like him, she was a herpetologist. In 1943 she joined the newly organized Women's Airforce Service Pilots (WASPs) and trained to ferry army airplanes to domestic military bases for delivery to combat flight crews. With her husband she co-wrote *Venomous Reptiles* (1989).

Minton, S. A.

Minton's Snake *Coluber karelini mintonorum* **Mertens,** 1969
Minton's Bent-toed Gecko *Cyrtopodion mintoni* **Golubev** and Szczerbak, 1981
Sherman's Blind Snake *Typhlops madgemintonae shermani*, **Khan,** 1999
Forest snake sp. *Toxicocalamus mintoni* Kraus, 2009

Sherman Anthony Minton Jr. (1919–1999) wanted to be a zoologist but his father insisted he do something "sensible," so he studied medicine but practiced zoology. He took his B.S. in zoology (1939), qualifying as a physician (1942). He was in the U.S. Navy (1943–1946), during which time his ship was hit by a kamikaze plane. After WW2 he enrolled at the University of Michigan and studied herpetology and microbiology. He was on the faculty of the Indiana University School of Medicine (1948–1984), retiring as Emeritus Professor (1984). He wrote *A Contribution to the Herpetology of West Pakistan* (1966). *Coluber karelini mintonorum* is named after Sherman and Madge Minton (see **Minton, M.**).

Minuth

Börner's Day Gecko *Phelsuma minuthi* **Börner,** 1980

Walter W. Minuth is a zoologist and herpetologist. He works closely with Börner, and they publish together, in such papers as "On the Taxonomy of the Indian Ocean Lizards of the *Phelsuma madagascariensis* Species Group (Reptilia, Gekkonidae)" (1984).

Mip Pugh

Python sp. *Morelia mippughae* Hoser, 2003

Mip Pugh is a herpetologist and breeder of reptiles who lives at Breakwater, near Geelong, Victoria, Australia. She and her husband, Mike, run a de facto hotel suite for countless other herpetologists who have enjoyed their hospitality.

Miriam

Miriam's Skink *Davewakeum miriamae* **Heyer,** 1972

Miriam Heyer discovered the holotype, and William Ronald Heyer named it after her. The scientific name of Vanzolini's Scaly-eyed Gecko *Lepidoblepharis heyerorum* is for them both.

Misonne

Misonne's Spider Gecko *Agamura misonnei* **Witte,** 1973 [Alt. Witte's Gecko]

Xavier Misonne is a Belgian zoologist and anthropologist. His long and varied career includes working for the World Health Organization, exploring mountains in Uganda, and staying with tribal peoples in Central Africa and Central Asia. He was Professor at Université Catholique de Louvain and organized scientific expeditions. He was Director of Institut Royal des Sciences Naturelles de Belgique (1978–1988). He has published a great many papers including, with Hayman and Verheyen, "The Bats of the Congo and of Rwanda and Burundi." One mammal is named after him.

Mitchell, B. L.

Mitchell's Flat Lizard *Platysaurus mitchelli* **Loveridge,** 1953

B. L. Mitchell was a naturalist who worked for the Department of Game, Fish, and Tsetse Control in Nyasaland (now Malawi) in the 1940s. He was the first person to recognize this genus (*Platysaurus*) in an area almost 240 kilometers (150 miles) northwest of its previously known range. He wrote *Some Reptiles and Amphibians of Nyasaland* (1950).

Mitchell, F. J.

Mitchell's Short-tailed Snake *Parasuta nigriceps* **Günther,** 1863 [Alt. Mallee Black-backed Snake; Syn. *Suta nigriceps*]
Mitchell's Water Monitor *Varanus mitchelli* **Mertens,** 1958
Pilbara Stone Gecko *Diplodactylus mitchelli* **Kluge,** 1963
Mitchell's Dtella *Gehyra fenestra* Mitchell, 1965
Mitchell's Bearded Dragon *Pogona (minor) mitchelli* Badham, 1976

Francis John Mitchell (1929–1970) was a herpetologist. He was a volunteer at the South Australian Museum, Adelaide, while still a student. Despite never graduating, he became Curator of Reptiles (1956), then Curator of Vertebrates (1966–1970). He expanded the herpetology collection and made a major contribution by initiating new research into reptile skeletons, which assisted the understanding of the evolution and relationships of reptiles and amphibians. He wrote *Adaptive Convergence in Australian Reptiles* (1958). His name became attached to the Short-tailed Snake long after it was described, probably because he redescribed it (1951) as *Denisonia nigrostriata brevicauda*—a description that has not stuck.

Mitchell, L. A.

Mitchell's Arboreal Alligator Lizard *Abronia mitchelli* **Campbell,** 1982

Lyndon A. Mitchell wrote a number of papers (1970s and 1980s) on the reproductive biology of reptiles, including one co-written with Campbell while at the University of Texas: "Miscellaneous Notes on the Reproductive Biology of Reptiles: The Uracoan Rattlesnake, *Crotalus vegrandis* Klauber (Serpentes, Viperidae)" (1979). He became Animal Care Manager at the Reptile Department, Dallas Zoo.

Mitchell, P. C.

Mitchell's Worm Lizard *Amphisbaena mitchelli* **Procter,** 1923

Sir Peter Chalmers Mitchell (1864–1945) was a journalist and zoologist. His early career was as a lecturer in Oxford and in London. As Secretary of the Zoological Society of London (1903–1935) he was responsible for many of the developments and improvements at the London Zoo. Joan Procter, who described this species, worked with him.

Mitchell, S. W.

Speckled Rattlesnake *Crotalus mitchelli* **Cope,** 1861

Dr. Silas Weir Mitchell (1829–1914) combined the professions of physician and novelist. He qualified as a physician in 1850. During the American Civil War he was in charge of a psychiatric hospital, which must have been one of the first in the world. After that war he became a specialist in neurology and is associated with inventing the idea of the "rest cure." Mitchell's Disease, Erythromelalgia, is named after him. He wrote short stories, historical novels, poetry, stories for children, and a monograph, *Researches on the Venom of the Rattlesnake* (1860), which is why a rattlesnake was named after him.

Mitra

Eastern Skink *Scincus mitranus* **Anderson,** 1871

Dr. Babu Rajendralala Mitra (1824–1891) was an archeologist and anthropologist. He read a paper to the Anthropological Society in London, "On the Gipsies of Bengal" (1867). Anderson described him as "my learned friend Babu Rajendralala Mitra" and explained that Mitra had obtained the skink holotype "from a Kashmir merchant, who stated that he brought the same from Arabia." He resigned as Vice President of the Science Association in 1882.

Mittleman

Mittleman's Tree Lizard *Urosaurus bicarinatus anonymorphus* Mittleman, 1940

Myron Budd Mittleman (b. 1918) is a herpetologist who was active in the Americas and Asia from the 1930s to the 1980s. He wrote a number of papers, such as "A Collection of Reptiles and Batrachians from Borneo and the Loo Choo Islands" (1952). He was also an inventor and lodged a number of patents.

Mivart

Boulenger's Emo Skink *Emoia mivarti* **Boulenger,** 1887

Dr. St. George Jackson Mivart (1827–1900) was best known as a naturalist but was also known for defending his Catholic faith from "scientific attacks." He was a physician and lawyer who practiced at the bar for a short while (1851) before following his natural scientific bent for research. He lectured in zoology at St. Mary's Hospital, London (1862). He was Professor of Biology at University College, Kensington (a short-lived Catholic University) (1874–1877). His publications included *Genesis of Species* (1871). By maintaining the creationist theory of the origin of the human soul, he attempted to reconcile his scientific evolutionism with the Catholic faith. However, Catholic authorities decided his orthodoxy to be questionable. In January 1900, after admonition and three formal notifications requiring him to sign a profession of faith, he was banned from receiving the sacraments by Cardinal Vaughan. He died of diabetes that same year, and a struggle ensued between his friends and the Roman Catholic Church as to who should bury him. Eventually he was buried in Kensal Green Catholic Cemetery (1904). His father was a wealthy man, owning the London hotel now known as Claridges.

Miyata

Miyata's Scaly-eyed Gecko *Lepidoblepharis miyatai* **Lamar,** 1985

Centipede Snake sp. *Tantilla miyatai* **Wilson** and Knight, 1987

Dr. Kenneth Ichiro Miyata (1951–1983) was a herpetologist who received his Ph.D. from the Museum of Comparative Zoology, Harvard (1980), and later worked for the Nature Conservancy. He traveled frequently to South America. He was also a famous fly-fisherman, but this interest led to his untimely death: in October 1983 he drowned while fishing alone on the Big Horn River in Montana.

Mjöberg

Atherton Tableland Skink *Glaphyromorphus mjobergi* Lönnberg and **Andersson,** 1915

Mjöberg's Forest Dragon *Gonocephalus mjobergi* **M. A. Smith,** 1925

Dr. Eric Georg Mjöberg (1882–1938) was a naturalist, entomologist, ethnographer, and explorer. He took his initial degree at Stockholms Universitet (1908) and his master's at Lunds Universitet (1912). He held various jobs in Sweden, including working at Naturhistoriska

Riksmuseet, Stockholm, and teaching in high schools (1903–1909). He led Swedish scientific expeditions in northwestern Australia (1910–1911) and Queensland (1912–1913). He worked in Sumatra at an experimental station (1919–1922), combining the duties with those of being Swedish Consul. He was Curator of the Sarawak Museum (1922–1924) and led a scientific expedition to Borneo (1925–1926).

Mocquard

Mocquard's File Snake *Mehelya guirali* Mocquard, 1887
Mocquard's Leaf-toed Gecko *Hemidactylus tropidolepis* Mocquard, 1888
Mocquard's African Ground Snake *Gonionotophis brussauxi* Mocquard, 1889
Mocquard's Eyebrow Lizard *Phoxophrys cephalum* Mocquard, 1890
Mocquard's House Gecko *Cosymbotus craspedotus* Mocquard, 1890 [Alt. Frilled Gecko; Syn. *Hemidactylus craspedotus*]
Mocquard's Small-eyed Snake *Hydrablabes praefrontalis* Mocquard, 1890
Mocquard's Swamp Snake *Tretanorhinus mocquardi* **Bocourt,** 1891
Mocquard's Litter Snake *Pseudorabdion collaris* Mocquard, 1892
Mocquard's Keeled Skink *Tropidophorus mocquardii* **Boulenger,** 1894
Mocquard's Dwarf Gecko *Lygodactylus verticillatus* Mocquard, 1895
Mocquard's Beauty Snake *Orthriophis taeniurus mocquardi* **Schultz,** 1896
Mocquard's Ground Gecko *Paroedura bastardi* Mocquard, 1900
Mocquard's Worm Snake *Typhlops decorsei* Mocquard, 1901
Mocquard's Agama *Agama sankaranica* Mocquard, 1905
Mocquard's Cylindrical Skink *Chalcides pulchellus* Mocquard, 1906
Mocquard's Writhing Skink *Lygosoma mocquardi* **Chabanaud,** 1917
Colubrid snake sp. *Alluaudina mocquardi* **Angel,** 1939
Mocquard's Blind Snake *Xenotyphlops mocquardi* **Wallach,** Mercurio, and Andreone, 2007

François Mocquard (1834–1917) was a herpetologist. He did much research on reptiles in Mexico and Central America (1870–1909). Among other things, he discovered (1905) a Malagasy blind snake, *Typhlops grandidieri*. With Bocourt and Duméril he published *Études sur les reptiles. Mission scientifique au Mexique et dans l'Amerique Centrale. Recherches zoologiques pour servir a l'histoire de la fauna de l'Amerique Centrale et du Mexique* (1883).

Modigliani

Modigliani's Flying Dragon *Draco modiglianii* Vinciguerra, 1892
Forest Skink sp. *Sphenomorphus modigliani* **Boulenger,** 1895
Modigliani's Nose-horned Lizard *Harpesaurus modigliani* **Vinciguerra,** 1933
Modigliani's Rock Gecko *Cnemaspis modiglianii* **Das,** 2005

Emilio Modigliani (1860–1932) was a zoologist and anthropologist who collected in Sumatra (1886–1894).

Moellendorff

Moellendorf's Rat Snake *Orthriophis moellendorffi* **Boettger,** 1886 [Syn. *Elaphe moellendorffi*]

Otto Franz von Möllendorff (1848–1903) was an expert on living and fossil molluscs. He started his career in China, where he had gone in 1873 to learn to be an interpreter. His elder brother, Paul Georg von Möllendorff, had been in China since 1869, and the brothers appear not to have left China until 1882. He wrote a number of articles, including "On the Supposed New Zealand Species of Leptopoma" (1893). Two mammals are named after him.

Mole

Mole's Gecko *Sphaerodactylus molei* **Boettger,** 1894 [Alt. Tobago Least Gecko]

Richard Richardson Mole (1860–1926) was a British resident of Port of Spain, Trinidad, who employed people to collect natural history specimen for him in Trinidad and Tobago. He was a founding member of the Trinidad Field Naturalists Club. He was a close collaborator with F. W. Urich, another club member. They wrote a number of papers together on Trinidadian fauna.

Molina

Molina's Lizard *Liolaemus molinai* Valladares et al., 2002

Abbot Juan Giovanni Ignazio (Ignacio) Molina (1740–1829) was a Chilean naturalist. He studied languages and natural history in a Jesuit college, and was appointed librarian at the college after becoming a member of the order. When Jesuits were banned, he left Chile for Italy (1768), where he was appointed Professor of Natural History at Università di Bologna (1774). All his natural history notes were lost en route to Italy, and he later rewrote what he could remember. He wrote *Saggio sulla storia naturale del Chili* (1782).

Moller

Gulfs Delma *Delma molleri* Lütken, 1863

Captain Möller from Ribe (Denmark) was a ship's captain who traded to Australia and brought the holotype to Denmark.

Molligoda

Molligoda's Day Gecko *Cnemaspis molligodai* Wickramasinghe and **Munindradasa,** 2007

Hayasinth Molligoda is a Sri Lankan honored for "his service and commitment to the conservation of Herpetofauna in the country."

Moloch

Moloch *Moloch horridus* **Gray,** 1841 [Alt. Thorny Devil]

Moloch was the sun-god of the Canaanites, to whom children were sacrificed. Christians later viewed him as a hideous demon.

Moltschanov

Toadhead Agama sp. *Phrynocephalus moltschanowi* **Nikolsky,** 1913

L. A. Molchanov (Moltschanov) was a Russian ornithologist and author. He worked in the Crimea (1903–1933). A bird is named after him.

Monard

Monard's Dwarf Skink *Afroablepharus duruarum* Monard, 1949

Professor Dr. Albert Monard (1886–1952) was a zoologist, naturalist, and explorer who made six expeditions to Africa. He taught at a high school in La Chaux-de-Fonds and was Curator of its Natural History Museum (1920–1952). His best-known work was *The Little Swiss Botanist* (1919), still used as a school textbook in the French-speaking cantons of Switzerland.

Monica

Monica's Thick-toed Gecko *Pachydactylus monicae* **Bauer,** Lamb, and Branch, 2006

Dr. Monica Frelow Bauer is the wife of the describer, Professor Aaron M. Bauer, who expresses his gratitude to her "for her support of systematic herpetology."

Monks

Monks' Rock Gecko *Cnemaspis monachorum* **Grismer** et al., 2009

Named after the monks at Wat Wanaram, Peninsular Malaysia, who allowed the describers of this gecko to hunt specimens in their caves.

Montañez

Tree Iguana sp. *Liolaemus montanezi* Cabrera and Monguillot, 2006

Alvaro Montañez is in charge of the guards at Parque Nacional San Guillermo, Argentina. He was recognized for his encouragement and support for all research activities in that reserve.

Montecristo

MonteCristo Arboreal Alligator Lizard *Abronia montecristoi* Hidalgo, 1983

This lizard is named after a mountain in El Salvador.

Montezuma

Sun-loving Litter-skink *Lygisaurus zuma* **Couper,** 1993

Montezuma II (1480–1520) was the last Aztec emperor and a sun worshipper. This skink is found in open forest habitats and loves basking in the sun.

Moojen

Brazilian Lancehead *Bothrops moojeni* **Hoge,** 1966

Dr. João Moojen de Oliveira (1904–1985) was a Brazilian zoologist who collected extensively from the 1930s to the 1950s. He collected a large proportion of the mammal specimens held by Universidade do Estado do Rio de Janeiro and Museu Nacional, Rio de Janeiro, where he was Curator of Mammals. He wrote what is regarded as the classic work on Brazilian rodents (1952). Three mammals are named after him.

Mora

Tuxtlan Road Guarder *Conophis morai* Perez-Higareda, Lopez-Luna, and **H. M. Smith,** 2002

Professor Dr. José Manuel Mora is a zoologist, herpetologist, and wildlife biologist whose major interest is the study of bats. He is Curator of Mammals at the museum, Universidad de Costa Rica. He published *Comparative Grouping Behavior of Juvenile Ctenosaurs and Iguanas* (1991), when he was a member of the Wildlife and Fisheries Sciences Department, Texas A&M University, where he was awarded his doctorate.

Morazan

Anole sp. *Anolis morazani* Townsend and **Wilson,** 2009

General José Francisco Morazán Quesada (1792–1842) was a statesman who was President of the now defunct Federal Republic of Central America (1830–1839), also serving as Head of State of Honduras (1827–1830), of Guatemala (1829), and of El Salvador (1839–1840). He invaded Costa Rica, became its dictator (1842), was accused of treason, and was executed by a firing squad.

Morelet

Morelet's Crocodile *Crocodylus moreletii* **Duméril** and **Bibron,** 1851 [Alt. Central American Crocodile]

Morelet's Alligator Lizard *Mesaspis moreleti* **Bocourt,** 1871

Pierre Marie Arthur Morelet (1809–1892) was a zoologist and malacologist who collected in the Canary Islands,

Guatemala, and Mexico, where he discovered the crocodile species (1850). He was a member of the Commission to Algeria at the start of the French occupation.

Moreno, F. J. P.

Tree Iguana sp. *Liolaemus morenoi* **Etheridge** and Christie, 2003

Dr. Francisco Josue Pascasio Moreno (1852–1919) was a zoologist and paleontologist. He traveled widely, exploring little-known parts of Argentina (1875–1879). He was Director of Museo de La Plata, which he founded, (1894–1905). He founded the Argentine Boy Scouts (1908) and a number of schools and children's homes. One mammal is named after him.

Moreno, L. V.

Zebra Dwarf Boa *Tropidophis morenoi* Hedges, **Garrido,** and Diaz, 2001

Luis V. Moreno is a herpetologist and Curator of the Department of Herpetology, Instituto de Ecología y Sistemática, Cuba. He has written numerous articles and scientific papers and was one of the editors of *The Iguanid Lizards of Cuba* (1999).

Morgan

Morgan's Blackhead Snake *Tantilla morgani* Hartweg, 1944

J. W. Morgan. Hartweg says that "the holotype was collected in 1939 near Necaxa, Mexico by J. W. Morgan" and that it was deposited in the University of Michigan's Museum of Zoology.

Morice

Kukri Snake sp. *Oligodon moricei* David, Vogel, and van Rooijen, 2008

Dr. Jean Claude Albert Morice (1848–1877) was a French naval surgeon, traveler, ethnographer, and naturalist. He was in Cochinchina (now Vietnam) (1872–1874 and 1875–1877) and wrote the first detailed account of the area's fauna. He made an important collection of sculptures that were lost in a shipwreck (1877) but recovered (1995).

Moritz

Leaf-tailed Gecko sp. *Saltuarius moritzi* **Couper, Sadlier, Shea,** and Wilmer, 2008

Professor Craig Charles Moritz is a population geneticist at the University of California, Berkeley, where he is Director of the Museum of Vertebrate Zoology. He has published widely, including, as co-author, *Tropical Rainforests: Past, Present, and Future* (2005).

Morne Dubois

Morne Dubois Least Gecko *Sphaerodactylus nycteropus* **Thomas** and **Schwartz,** 1977

Morne Dubois is a location in Haiti.

Morris

Arnhem Shovel-nosed Snake *Simoselaps morrisi* **Horner,** 1998

Ian James Morris (b. 1951) is an Australian naturalist and author who first collected and photographed this snake (1970). He is recognized for his contribution to research into and the understanding of the fauna of the Northern Territory and for his work in environmental education. He is also noted for his services to the indigenous peoples of Northern Territory, in recognition of which, and his educational activities, he was awarded the Order of Australia (2005).

Mostoufi

Lacertid lizard sp. *Lacerta mostoufii* Baloutch, 1976

Ahmad Mostoufi led an expedition (1972) to Iran's central desert. In his etymology Baloutch wrote, "Dédié au Pr. Ahmad Mostoufi, membre de la Faculté des Lettres, chef de l'expedition."

Mouhot

Mouhot's Kukri Snake *Oligodon mouhoti* **Boulenger,** 1914

Mouhot's Turtle *Cuora mouhotii* **Gray,** 1862 [Alt. Keeled Box Turtle; Syn. *Pyxidea mouhotii*]

Alexandre Henri Mouhot (1826–1861) was a French traveler who is best known for having rediscovered Ankhor Wat in Cambodia (ca. 1859). He was a philologer and taught languages in Russia (1844–1854). He started studying natural science in 1856 and married the daughter of Mungo Park, the English explorer—a connection that helped him when he decided to go to Indochina to collect botanical specimens (1857). The French authorities rejected his proposals, but the trip was supported by the Royal Geographical Society and the Zoological Society of London. He wrote *Travels in Indo-China* (published posthumously, 1864). He died in Laos.

Moyer

Moyer's Pygmy Chameleon *Rhampholeon moyeri* Menegon, Salvidio, and **Tilbury,** 2002

David C. Moyer is a Tanzanian researcher and conservationist with the Wildlife Conservation Society. He has published on herpetological subjects, co-writing "Within- and Between-Site Distribution of Frog Species on the Udzungwa Plateau, Tanzania" (2008). He continues to study Tanzanian wildlife and advocate its conservation.

Mueller, F.

Müller's Black-headed Snake *Micrelaps muelleri* **Boettger,** 1880
Müller's Snake *Rhinoplocephalus bicolor* Müller, 1885 [Alt. Square-nosed Snake]
Müller's Nessia *Nessia sarasinorum* Müller, 1889
Müller's Sand Boa *Gongylophis muelleri* **Boulenger,** 1892
Müller's Reed Snake *Calamaria muelleri* Boulenger, 1896

Dr. Friedrich "Fritz" Müller (1834–1895) was a physician and zoologist. He originally studied at Universität Basel and then at Würzburg and Prague, where he qualified as a physician (1857). He returned to Basel to practice medicine after further studies in Vienna, Paris, and Berlin. He gave public lectures in zoology at Universität Basel from 1868 onward. His main interests were arachnids, crustaceans, and reptiles.

Mueller, F. J. H.

Müller's Lerista *Lerista muelleri* **Fischer,** 1881

Baron Ferdinand Jacob Heinrich von Müller (1825–1896) was a German-born Australian botanist, geographer, explorer, physician, and naturalist. He was born at Rostock and, after education in Schleswig, was apprenticed to a chemist (1840). He studied botany at Christian-Albrechts-Universität zu Kiel, receiving his doctorate (1847). He had intended to practice medicine but was advised to go to a warmer climate for his health and left for Australia (1847). He first found employment in Adelaide as a chemist and contributed a few papers on botanical subjects to German periodicals. He moved to Melbourne (1851) and traveled within Victoria (1848–1852), describing a large number of plants. After he sent a paper to the Linnean Society at London on "The Flora of South Australia" (1852), he was appointed Government Botanist (1853). He was expedition naturalist for the exploration of the Victoria River and other parts of North Australia, and he was one of the four who reached Termination Lake (1856), continuing with Gregory's expedition overland to Moreton Bay. He was a member of the Victorian Institute for the Advancement of Science, later renamed the Royal Society of Victoria (1854–1872). He was a member of the society's "Exploration Committee," which established the Burke and Wills expedition (1860). He was Director of the Melbourne Botanic Gardens (1857–1873) and the benefactor of explorer Ernest Giles, the discoverer of Lake Amadeus and Kata Tjuta. Giles had originally wanted to name both after Müller, who found that embarrassing and prevailed upon Giles to desist. He wrote the 11-volume *Fragmenta phytographica Australiae* (1862–1881).

Mueller, L.

Müller's Leaf Chameleon *Kinyongia uthmoelleri* Müller, 1938 [Alt. Hanang Hornless Chameleon]

Lorenz Müller (1868–1953) was a herpetologist who trained as an artist in Paris and the Low Countries and worked as a scientific illustrator at Zoologische Staatssammlung München. He was particularly interested in herpetology, and as there was a vacancy at the Munich museum, he became de facto Curator (1903). He was a member of the museum's expedition to the Lower Amazon (1909–1910). He served in the German army during WW1 but was able to get posted to the Balkans and spent most of his time there collecting specimens. Afterward he returned to Munich and became Chief Curator of Zoology (1928). During WW2 both his private collections and the museum's were largely destroyed in air raids, but he set to rebuilding the collections. During his life he published more than 100 articles, monographs, and papers on herpetology. See also **Lorenz.**

Mueller, S.

Forest Skink sp. *Sphenomorphus muelleri* **Schlegel,** 1837
Müller's Crown Snake *Aspidomorphus muelleri* Schlegel, 1837
Müller's Blind Snake *Typhlops muelleri* Schlegel, 1839
Java Wolf Snake *Lycodon muelleri* **Duméril, Bibron,** and **Duméril,** 1854

Dr. Salomon Müller (1804–1864) was a naturalist who collected in Indonesia (1826), where he worked under Schlegel as a taxidermist assisting members of the Netherlands Natural Sciences Commission. He went to New Guinea and explored the interior of Timor. He collected in Java (1831) and explored western Sumatra (1833–1835). Three mammals and eight birds are named after him.

Muhlenberg

Muhlenberg's Turtle *Glyptemys muhlenbergii* Schoepf, 1801 [Alt. Bog Turtle]

Rev. Gotthilf Heinrich Ernst Muhlenberg (1753–1815) was a German Lutheran minister who emigrated to America (1770). His brother was a member of the Continental Congress (1776), so G. H. E. Muhlenberg had to leave Philadelphia for New Providence during the American War of Independence. He collected the holotype of the turtle, but his main interest was botany, and between 1778 and 1791 he listed more than 1,100 plants growing near New Providence. He was Pastor of the Lutheran Church at Lancaster, Pennsylvania (1779–1815), and became the first President of Franklin College (1787). He wrote a *Catalog of the Plants of North America* (1812).

Muller, D.

Muller's Velvet Gecko *Homopholis mulleri* **Visser,** 1987

Douglas Muller, a well-known amateur herpetologist, owns Command Farm near Huntleigh Siding, Transvaal, where he collected the holotype and paratypes for Visser.

Munindradasa

Munindradasa's Lanka Skink *Lankascincus munindradasai* Wickramasinghe et al., 2007

Dr. D. I. Amith Munindradasa (1966–2007) was a scientist, naturalist, and conservationist, but an electronic engineer by training. He graduated from the University of Moratuwa, Sri Lanka (1993), and was awarded a doctorate by Liverpool University. On his return to Sri Lanka he joined the faculty of the University of Moratuwa, becoming head of the Department of Electronics and Telecommunication Engineering. He made many field trips to study Sri Lankan fauna and was involved in the discovery of five *Cnemaspis* species. He died of pneumonia while in Israel on an official mission for the Sri Lankan Ministry of Defense.

Munoa

Munoa Worm Lizard *Amphisbaena munoai* Klappenbach, 1960

Rio Grande do Sul Blind Snake *Leptotyphlops munoai* Orejas-Miranda, 1961

Juan Ignacio Muñoa (1925–1960) studied medicine and vertebrate zoology but is best remembered as an anthropologist in the Museo Nacional de Historia Natural y Antropologia, Montevideo, Uruguay.

Murphy

Lichtenfelder's Gecko *Goniurosaurus murphyi* **Orlov** and **Darevsky,** 1999

Keeled Skink sp. *Tropidophorus murphyi* **Hikida, Orlov,** Nabhitabhata, and **Ota,** 2002

Dr. Robert "Dr. Bob" Ward Murphy (b. 1948) is Senior Curator of Herpetology at the Royal Ontario Museum (ROM), which he joined in 1984, and Professor in the Department of Ecology and Evolutionary Biology at the University of Toronto. He received a doctorate in biology from the University of California, Los Angeles (1982). His early work concentrated on the evolutionary genetics of reptiles from Baja California, Mexico. His was the first study to examine the genetic consequences of plate tectonics. He undertook postdoctoral studies at UCLA Medical School in flow cytometry and the diagnosis of forms of cancer, as well as the conservation genetics of fishes. He has collected amphibians and reptiles in many locations and has built up the world's largest collection of tissue samples for genetic research. He initiated the ROM's involvement in and commitment to working on the biodiversity crisis in Vietnam, which has resulted in significant exposure for the ROM, Ontario, and Canada. His involvement with the genetics of desert tortoises started in association with Dr. David Morafka (1998). He is committed to the ROM's display of living organisms discovered by ROM researchers, and to their conservation through research, education, and captive propagation. He has published innumerable scientific papers.

Murray, J.

Murray's Skink *Eulamprus murrayi* **Boulenger,** 1887 [Alt. Blue-speckled Forest Skink]

Sir John Murray (1841–1914) was a Canadian marine naturalist and oceanographer. He explored the Faroe Channel (1880–1882) and took part in and financed expeditions to Christmas Island. Boulenger doesn't specify which "Murray" he is naming this skink after. However, he notes that the holotype was acquired by the HMS *Challenger* expedition (1874–1876), which supports this candidate: Murray was in charge of collections on that expedition and edited *Report on the Scientific Results of the Voyage of "HMS Challenger"* (1880–1895). A bird is named after him.

Murray, L. T.

Murray's Mud Turtle *Kinosternon hirtipes murrayi* Glass and **Hartweg,** 1951 [Alt. Big Bend Mud Turtle]

Dr. Leo Tildon Murray (1902–1958) took both his master's degree (1931) and his doctorate (1935) at Cornell. He was Assistant Professor at a teachers' college in Pennsylvania (1935–1936), then Associate Professor and Director of the Museum at Baylor University, Waco, Texas (1936–1944), and Associate Professor of Zoology at Texas A&M University (1944–1946) before joining the U.S. Fish and Wildlife Service as an aquatic biologist (1946).

Mutahi

Bougainville's Scaly-toed Gecko *Lepidodactylus mutahi* **Brown** and **Parker,** 1977

Mutahi is a place in the northeast of Bougainville Island.

Myers, C. W.

Myers' Graceful Brown Snake *Rhadinaea myersi* **Rossman,** 1965

Myers' Anole *Anolis fungosus* Myers 1971

Worm Lizard sp. *Amphisbaena myersi* **Hoogmoed,** 1988

Colubrid snake sp. *Urotheca myersi* **Savage** and Lahanas, 1989

Myers' Large-scaled Lizard *Ptychoglossus myersi* **Harris,** 1994

Chocoan Toad-headed Viper *Bothrocophias myersi* Gutberlet and **Campbell,** 2001

Dr. Charles William Myers (b. 1936) was a Research Assistant at the University of Florida (1958–1960), where he took his master's degree. He has published many herpetological papers and articles, such as "An Enigmatic New Snake from the Peruvian Andes, with Notes on the Xenodontini (Colubridae: Xenodontinae)" (1986). He replaced Bogert (q.v.) at the American Museum of Natural History (1968). He retired in 1999 from active involvement, having been Chairman for two terms (1980–1987 and 1993–1998), but continues to do research there. See also **Charles Myers.**

Myers, G. S.

Myers' Snake *Myersophis alpestris* **Taylor,** 1963

Dr. George Sprague Myers (1905–1985), a biogeographer, herpetologist, and ichthyologist, was Professor Emeritus of Biological Sciences at Stanford. He was a keen natural historian with a lifelong interest in fish and amphibians. He published his first paper on ichthyology at age 15 and eventually wrote over 600 scientific papers and articles. He worked as a volunteer assistant at the American Museum of Natural History, New York (1922–1924). He enrolled at Indiana University part-time (1924), but when his sponsor fell ill he transferred to Stanford and graduated from there (1930), going on to complete his M.A. and doctorate (1933). He worked at the Smithsonian as Assistant Curator but was invited (1936) to return to Stanford as Assistant Professor in Biological Sciences and Curator of Zoological Collections. He developed courses in systematics for ichthyology and vertebrate paleontology and was appointed Full Professor (1938). During WW2 he spent over two years in Brazil on U.S. State Department funds to aid Museo Nacional and Divisão de Caça e Pesca—a program to maintain good relations with Latin America. He amassed an extensive library on ichthyology, herpetology, biogeography, the history of biology and exploration, and, as a sideline, the American Civil War. After retirement (1970) he became Visiting Professor of Ichthyology at the Museum of Comparative Zoology, Harvard.

Mys

Mys' Rainbow-Skink *Carlia mysi* **Zug,** 2004

Benoit Mys (d. 1990) died in a vehicle accident while doing fieldwork in northern Papua New Guinea. He published "The Zoogeography of the Scincid Lizards from North Papua New Guinea" (1988).

N

Nair

Ponmudi Rock Gecko *Cnemaspis nairi* **Inger, Marx** and Koshy, 1984

Dr. S. Madhavan Nair is a naturalist, museologist, and former Director of the National Museum of Natural History, New Delhi, where he served for 20 years. His early training included time at the Smithsonian (1969). In retirement he is Director of Education, New Delhi, for the World Wide Fund for Nature. He wrote *Endangered Animals of India and Their Conservation* (1992).

Nancy Coutu

Nancy Coutu's Mabuya *Mabuya nancycoutuae* Nussbaum and **Raxworthy,** 1998 [Syn. *Trachylepis nancycoutuae*]

Nancy Coutu (1967–1996) was an American Peace Corps volunteer who assisted the describers of this skink in their "Isalo expedition." Cattle thieves murdered her outside her village in Madagascar (1996). She died instantly from a hatchet blow, then was raped. The villagers guarded her until she could be identified. Nancy's journals, in a box with some clothes, arrived at her mother's home several weeks later. Her mother published the journals, with extracts from her letters home. The book and a school in Madagascar are both called *Souvenirs de Nancy*.

Nanuza

Rodrigues' Lava Lizard *Tropidurus nanuzae* **Rodrigues,** 1981

Dr. Nanuza Luiza de Menezes (b. 1934) is a botanist at Universidade de São Paulo, where Rodrigues is a colleague. She took all her degrees at São Paulo: bachelor's (1960), master's (1969), and doctorate (1971). She became a Professor (1984) and is now officially retired but still teaches as an Associate. She spent time at Kew (1979). She has spent much time investigating the trees of the Amazon basin.

Napoleon

Napoleon Skink *Egernia napoleonis* **Gray,** 1839

Named after Terre Napoléon, a former name given to parts of southern Australia.

Narducci

Jan's Thread Coral Snake *Micrurus narduccii* **Jan,** 1863 [Alt. Andean Black Coral Snake]

Dr. Louis Narducci was a Bolivian naturalist.

Nasrullah

Gecko sp. *Asaccus nasrullahi* **Werner,** 2006

Dr. Nasrullah Rastegar-Pouyani is a zoologist and herpetologist who is an Assistant Professor at the Department of Biology, Faculty of Science, Raza University, Iran. His bachelor's (1986) and master's degrees (1991) were both awarded by Iranian universities and his doctorate by Göteborgs Universitet (1999).

Natalia

Pricklenape (agamid lizard) sp. *Acanthosaura nataliae* **Orlov,** Truong, and Sang, 2006

Dr. Natalia Borisovna "Natasha" Ananjeva (b. 1946) is a Russian biologist and herpetologist who is head of the Herpetology Department, Zoological Institute, Russian Academy of Sciences, St. Petersburg. The describer, Orlov, is a member of her staff and has co-authored articles with her. See also **Anan.**

Natterer

Natterer's Hognose Snake *Lystrophis nattereri* **Steindachner,** 1867

Paraguayan Green Racer *Philodryas nattereri* Steindachner, 1870

Natterer's Gecko *Tropiocolotes nattereri* Steindachner, 1901

Dr. Johann Natterer (1787–1843) was a naturalist and collector. He studied botany, zoology, mineralogy, chemistry, and anatomy and was appointed as a taxidermist at the Naturhistorisches Museum Wien. He, Spix, and others went on an expedition to Brazil (1817), which started on the occasion of Archduchess Leopoldina's wedding to the Brazilian Crown Prince. Everyone traveled in two Austrian frigates. Natterer explored a potential river route to Paraguay (1818–1819) and went on five expeditions, exploring the Mato Grosso and the Amazon Basin (1821–1835). He returned to Vienna with a huge collection of specimens (37 crates), despite losing most of his possessions in the Brazilian Civil War, and deposited 12,293 birds and around 24,000 insects with Naturhistorisches Museum Wien, where they can still be seen. He died of a lung ailment. He published only a few short accounts, and unfortunately his notebooks and diary were destroyed by fire (1848). He never received the credit he was due in Austria but was held in the highest esteem abroad. Among the many taxa named after him are 2 mammals, 11 birds, and a piranha.

Neang Thy

Gecko sp. *Cnemaspis neangthyi* Grismer, **Grismer,** and Chav, 2010

Dr. Neang Thy (b. 1970) is a Cambodian herpetologist

and conservationist who works at the Ministry of Environment, where he is head of the Botanical Garden Office and of Fauna and Flora International's Cardamom Mountains Research Group. He co-wrote "A New Species of *Chiromantis* Peters 1854 (Anura: Rhacophoridae) from Phnom Samkos in the Northwestern Cardamom Mountains, Cambodia" (2007).

Nečas

Necas' Chameleon *Chamaeleo necasi* Ullenbruch, Krause, and **Böhme,** 2007

Petr Nečas (b. 1969)—not the Czech politician of the same name—is a herpetologist particularly interested in chameleons. He has written over 100 articles and published (in German) *Chameleons: Nature's Hidden Jewels* (1995). A chameleon parasite is named after him.

Neill

Florida Crowned Snake *Tantilla relicta neilli* Telford, 1966

Neill's Snail Sucker *Sibon sanniolus neilli* **Henderson,** Hoevers, and Wilson, 1977

Wilfred Trammell Neill Jr. (1922–2001) was a herpetologist, linguist, archeologist, and author who described some new reptile species. During WW2 he served in the U.S. Army Air Force in the South Pacific and New Guinea. He was Research Director, Florida's Ross Allen Reptile Institute (1949–1962), and an Associate Curator, University of Florida (1964). He died of pulmonary pneumonia, his health never having recovered from a near-fatal snake bite 23 years earlier. The holotype of the snail sucker was discovered preserved in a jar of alcohol in a chemist's shop. Neill's best-known book is *The Last of the Ruling Reptiles: Alligators, Crocodiles, and Their Kin* (1971).

Neiva

Bahia Snail-eater *Dipsas neivai* **Amaral,** 1926

Dr. Arthur Neiva (1880–1943) was an epidemiologist and biologist. He qualified in Rio de Janeiro in medicine (1903) and did entomological research at the Institute of Manghuinos. He organized the Medical Section of Zoology and Parasitology, Institute Bacteriológico, Buenos Aires, for the government of Argentina (1915). He returned to Brazil (1916), becoming Director of Public Health for São Paulo State, a member of the staff at Instituto Butantan, São Paulo, and then Director of Museu Nacional, Rio de Janeiro (1923). He was first Director, Institute Bacteriológico (1928–1932). After the 1930 revolution he held a number of appointments, including being Director-General of Research, Ministry of Agriculture. He entered politics (1933–1937) and then gave it up to resume his original research at the Institute of Manghuinos.

Nelly Carrillo

Sun Tegu sp. *Euspondylus nellycarillae* Köhler and Lehr, 2004

Nelly Carrillo de Espinoza (b. 1932) is a zoologist and herpetologist at Universidad Nacional Mayor de San Marcos, Peru. She wrote, with Icochea, "Lista taxonomica preliminar de los reptiles vivientes del Perú" (1995). The original etymology says the lizard is named "in recognition of her contributions to the knowledge of Peruvian herpetology."

Nelson, E. W.

Nelson's Anole *Anolis nelsoni* **Barbour,** 1914

Nelson's Milk Snake *Lampropeltis triangulum nelsoni* **Blanchard,** 1920

Nelson's Tree Lizard *Urosaurus bicarinatus nelsoni* **Schmidt,** 1921

Nelson's Lizard *Sceloporus nelsoni* **Cochran,** 1923

Nelson's Spotted Box Turtle *Terrapene nelsoni nelsoni* **Stejneger,** 1925

Dr. Edward William Nelson (1855–1934) was founding President of the American Ornithologists' Union. His family became homeless after the Chicago Fire (1871). He traveled to the Rockies (1872) and also collected in Alaska and Mexico. While taking part in the search for the missing Arctic exploration vessel *Jeanette,* his expedition became the first to reach and explore Wrangell Island (1881). Then he joined the Biological Survey of the Department of Agriculture and explored in Arizona and the southwestern USA (1882). He was on the Death Valley expedition (1890–1891) and collected in Mexico (1892–1902). He was Chief Field Naturalist (1907–1912) in charge of biological investigations (1912–1913), Assistant Chief (1914–1915), then Chief of the U.S. Biological Survey (1916–1927), finally becoming Principal Biologist (1927–1929). His greatest contribution was the establishment of the Migratory Bird Treaty, which is still in force today. Six birds and 15 mammals are named after him.

Nelson, G.

Florida Redbelly Turtle *Pseudemys nelsoni* **Carr,** 1938

George Nelson (b. 1873) was a botanist, lecturer, zoologist, and photographer who became Chief Taxidermist at the Museum of Comparative Zoology, Harvard (1901). He specialized in the fauna of Florida, where he spent his winters studying Brown Pelicans. He acquired land and built a house there (1910). He often collected turtles and moccasin snakes in the marshes. He wrote a study of Pelican Island (1911) and the changes to its ecology after the 1910 hurricane.

Nelson Jorge

Burrowing Snake sp. *Apostolepis nelsonjorgei* **De Lema** and Renner, 2004

Dr. Nelson Jorge da Silva Jr. of Universidade Católica de Goiás, Goiânia, Brazil, is a herpetologist who researches snake venoms. He began a degree in history at Universidade Católica de Goiás (1983) before switching to biology at the same university, where his bachelor's (1986) and master's degrees (1987) were awarded. His doctorate (1995) was awarded by Brigham Young University. He wrote "Novas ocorrências de *Micrurus brasiliensis* Roze, 1967 (Serpentes: Elapidae) em áreas de tensão ambiental no centro-oeste Brasileiro" (2007).

Nepthys

Eungella Leaf-tailed Gecko *Phyllurus nepthys* **Couper, Covacevich,** and **Moritz,** 1993

Nepthys was a goddess of ancient Egypt and the sister of Isis. She was known as the "Useful Goddess" or the "Excellent Goddess" because she represented divine assistance on a number of levels. This gecko was named at the same time as *Phyllurus isis* (see **Isis**), and the two names seem to have been applied fancifully.

Neuhauss

Forest Skink sp. *Sphenomorphus neuhaussi* **Vogt,** 1911

Professor Richard Neuhauss (1855–1915) was a leading German expert on the island of New Guinea. He published *Unsere Kolonie Deutsch-Neu-Guinea* (1910).

Neumann

Neumann's Orangetail Lizard *Philochortus neumanni* **Matschie,** 1893

Neumann's Agama *Agama neumanni* **Tornier,** 1905 [Junior syn. of *Pseudotrapelus sinaitus* Heyden, 1827]

Neumann's Sand Lizard *Heliobolus neumanni* Tornier, 1905

Professor Oskar Rudolph Neumann (1867–1946) was a German ornithologist who collected in East Africa (1892–1894). He was in Somaliland and Ethiopia with Carlo von Erlanger (1899–1901). In the early 1900s he studied the birds and mammals of Rothschild's collection. Later in his life he moved to Chicago to escape Nazi persecution and worked at the Field Museum.

Neuwied

Neuwied's Lancehead *Bothrops neuwiedi* **Wagler,** 1824

Neuwied's False Boa *Pseudoboa neuwiedii* **Duméril, Bibron,** and **Duméril,** 1854

Neuwied's Polemon *Polemon neuwiedi* **Jan,** 1858

Neuwied's False Fer-de-Lance *Xenodon neuwiedii* **Günther,** 1863

Neuwied's Tree Snake *Sibynomorphus neuwiedi* **Ihering,** 1930

See **Maximilian.**

Nevermann

Dunn's Road Guarder *Crisantophis nevermanni* **Dunn,** 1937

The original description has no etymology, but we believe the reptile is named after Wilhelm Heinrich Ferdinand Nevermann (1881–1938), a friend of Dunn's. He was a German coleopterist who owned Hamburg Farm, Costa Rica, from at least 1922. He was collecting at night and was accidentally killed by a hunter who mistook him for a large animal.

Nevin

Skink sp. *Lerista nevinae* **L. A. Smith** and Adams, 2007

Anne F. Nevin was Secretary to the Director, Natural Sciences Department, Western Australian Museum (1982–2006), where Smith and Adams worked at the same time.

Newman, R. J. and M.

Newman's Knob-scaled Lizard *Xenosaurus newmanorum* **Taylor,** 1949

Newman's Earth Snake *Adelphicos quadrivirgatus newmanorum* Taylor, 1950

Dr. Robert J. Newman and his wife, Marcella, are honored in the scientific names of these reptiles. He was a zoologist and ornithologist at the Zoological Museum, Louisiana State University, which also awarded his doctorate. See also **Marcella.**

Newton

Newton's Beaked Snake *Rhinotyphlops newtonii* **Bocage,** 1890

Newton's Leaf-toed Gecko *Hemidactylus newtoni* **Ferreira,** 1897

Colonel Francesco Newton (1864–1909) was a Portuguese botanist who collected in São Tomé (1888) and Timor (1896). His record keeping and accounts of his findings were exceptionally meticulous for his time. Two birds are named after him.

Ngo Van Tri

Ngo Van Tri's Lady Butterfly Lizard *Leiolepis ngovantrii* **Grismer** and Grismer, 2010

Ngo Van Tri (b. 1969) is a scientist at the Institute of Tropical Biology, Ho Chi Minh City. He graduated in biology (1994) at Hue General University and worked for Fauna and Flora International (1994–2000). He has carried out extensive fieldwork in southern Vietnam resulting in the discovery of this and many other new species of lizards.

Nguyen Van Sang

Colubrid snake sp. *Colubroelaps nguyenvansangi* Orlov et al., 2009

See **Sang.**

Niceforo

Loveridge's Large-scaled Lizard *Ptychoglossus nicefori* Loveridge, 1929

Northern Ground Snake *Atractus nicefori* **Amaral,** 1930

Niceforo's Lizard *Pantodactylus nicefori* **Burt** and Burt, 1931 [Said to be a syn. of *Psammodromus algirus*]

Amazon Burrowing Snake *Apostolepis niceforoi* Amaral, 1935

Dunn's Ameiva *Ameiva niceforoi* **Dunn,** 1943

Niceforo's Andes Anole *Anolis nicefori* Dunn, 1944

Santander Blind Snake *Leptotyphlops nicefori* Dunn, 1946

Brother Niceforo Maria (1888–1980), né Antoine Rouhaire, became a missionary in Colombia under his monastic name. He went from France to Medellin (1908) and was given the task of forming a natural history museum (1913). Primarily a herpetologist, he was an excellent taxidermist. Many taxa are named after him, including amphibians, birds, and mammals.

Nicholls

Nicholls' Lerista *Lerista nichollsi* **Loveridge,** 1933 [Alt. Inland Broad-blazed Slider]

Professor Gilbert Ernest Nicholls (b. 1893) of the Western Australia Museum, Perth, collected the holotype of this skink. His specialty was crustaceans. He wrote the report on Crustacea Amphipoda from the Australian Antarctic expeditions (1911–1914).

Nichols

Nichols' Least Gecko *Sphaerodactylus nicholsi* **Grant,** 1931 [Alt. Nichols' Dwarf Sphaero]

Snail-eating Snake sp. *Dipsas nicholsi* **Dunn,** 1933

John Treadwell Nichols (1883–1958) was an ichthyologist and ornithologist who co-described the rediscovered Bermuda Petrel (1916), believed extinct since the 1620s. He founded *Copeia* (1913), which became the official journal of the American Society of Ichthyologists and Herpetologists (1923). He was Associate Curator of the Ichthyology Department, American Museum of Natural History (1920–1942). He published over 1,000 articles and books.

Nicosia

Chameleon sp. *Furcifer nicosiai* Jesu, Mattioli, and Schimmenti, 1999

Guido Nicosia is an Italian novelist and diplomat. He entered the Italian diplomatic service (1961) and was Italian Ambassador to Madagascar and Mauritius (1996–1999). His novels are mostly thrillers with an ambassadorial figure as the hero. He was very helpful to Jesu and Schimmenti during their visits to Madagascar (1995–1999).

Nieden

Nieden's Dwarf Skink *Panaspis megalurus* Nieden, 1913

Fritz Nieden (1883–1942) was a zoologist at Museum für Naturkunde Berlin. He concentrated on African herpetofauna. He wrote *Die Reptilien (außer den Schlangen) und Amphibien Kamerum* (1910).

Nieuwenhuis

Nieuwenhuis' Skink *Lamprolepis nieuwenhuisii* Lidth de Jeude, 1905

Anton Willem Nieuwenhuis (1864–1953) was a Dutch explorer and ethnologist who traversed central Borneo with Büttikofer (q.v.) (1893–1894). He wrote *In Central Borneo* (1900).

Nikhil

Nikhil's Kukri Snake *Oligodon nikhili* **Whitaker** and Dattatri, 1982

Nikhil Whitaker (b. 1979) is Curator of the Madras Crocodile Bank Trust, where he works with his father, Romulus. See **Whitaker, R.**

Nikolsky

Nikolsky's Rock Agama *Laudakia erythrogastra* Nikolsky, 1896 [Alt. Redbelly Rock Agama; Syn. *Stellio erythrogaster*]

Nikolsky's Bow-fingered Gecko *Cyrtopodion kirmanensis* Nikolsky, 1900

Nikolsky's Iranian Gecko *Cyrtopodion agamuroides* Nikolsky, 1900

Nikolsky's Middle-toed Gecko *Cyrtopodion sagittifer* Nikolsky, 1900

Racerunner (lacertid lizard) sp. *Eremias nikolskii* Nikolsky, 1905

Nikolsky's Blind Snake *Leptotyphlops hamulirostris* Nikolsky, 1907

Nikolsky's Adder *Vipera nikolskii* Vedmederja, Grubant, and Rudajewa, 1986

Nikolsky's Tortoise *Testudo graeca nikolskii* Ckhikvadze and Tunijev, 1986 [Alt. Nikolsky's Spur-thighed Tortoise; Syn. *T. ibera nikolskii*]

Dr. Alexander Mikhailovich Nikolsky (1858–1942) was a herpetologist and zoologist. He studied at the university in St. Petersburg (1877–1881), taking his doctorate in 1887, in which year he became Assistant Professor at the university and a Curator of the zoological collection. He became Director, Department of Herpetology, Natural History Museum, Russian Academy of Sciences (1895). He resigned to become Professor at Kharkov University, Ukraine (1903). He made a number of expeditions to such destinations as the Caucasus Mountains, Iran, Siberia, and Japan (1881–1891). Today in Russia the A. M. Nikolsky Herpetological Society commemorates him.

Nilson

Nilson's Snake Skink *Ophiomorus nuchalis* Nilson and Andrean, 1978

Fringe-fingered Lizard sp. *Acanthodactylus nilsoni* Rastegar-Pouyani, 1998

Dr. Göran Nilson (b. 1948) is a herpetologist, an Associate Professor at the Department of Zoology, Göteborgs Universitet. He co-wrote "A New Subspecies of the Subalpine Meadow Viper, *Vipera ursinii* (Bonaparte) (Reptilia, Viperidae), from Greece" (1988).

Nilsson

Softshell Turtle genus *Nilssonia* **Gray,** 1872

Dr. Sven Nilsson (1787–1883) was a naturalist, zoologist, and archeologist. He began studying for the priesthood (1806) at Lunds Universitet, but was persuaded by Anders Jahan Retzius, Professor of Zoology, to switch to natural history. He was Director, Naturhistoriska Riksmuseet, Stockholm, and endeavored to assemble a complete collection of the vertebrates of Sweden (1822–1831). He was Professor of Zoology and Curator of the Museum, Lunds Universitet (1831–1856), where he had taken his doctorate. He wrote *Illuminerade figurer till Skandinaviens fauna* (1832–1840). A mammal is named after him.

Nitsche

Nitsche's Bush Viper *Atheris nitschei* **Tornier,** 1902

Heinrich Nitsche (1845–1902) was a German zoologist and entomologist who specialized in forest habitats.

Noble

Noble's Bachia *Bachia intermedia* Noble, 1921

Noble's Anole *Anolis altavelensis* Noble and **Hassler,** 1933 [Alt. Hassler's Anole]

Noble's Anole *Anolis noblei* **Barbour** and **Shreve,** 1935

Dr. Gladwyn Kingsley Noble (1894–1940) was a biologist and herpetologist. His father was a co-founder of the publishers Barnes and Noble. He made trips to Peru, Newfoundland, and Guadeloupe (1914–1916). His bachelor's (1917) and master's degrees (1919) were awarded by Harvard. Colombia University, New York, awarded his doctorate (1922). He served in the U.S. Navy (1918–1919). He joined the American Museum of Natural History, New York (1919), as an Assistant, becoming Assistant Curator (1922), and was Curator of Herpetology (1923–1940). He was also Curator of Experimental Biology (1928) and Visiting Professor, University of Chicago (1931). He led expeditions to Santo Domingo (1922) and Cuba (1937). Among his publications is *The Experimental Animal from the Naturalist's Point of View* (1939). He died from a streptococcal infection. Two different lizards have been given the same vernacular name of Noble's Anole, so note the different scientific names.

Nogge

Nogge's Water Skink *Tropidophorus noggei* Ziegler, **Thanh,** and Thanh, 2005

Dr. Gunther Nogge (b. 1942) was the Director of the Cologne Zoological Garden (1981–2007). He studied biology at Friedrich-Wilhelms-Universität Bonn and was in Afghanistan, where he lectured at the university and worked at the Kabul Zoo (1969–1973). He returned to Bonn for further study and became Director of the Cologne Zoo (1981), where he developed extensive and successful captive breeding programs. In 1985 he was attacked by two escaped chimpanzees and severely injured, needing a seven-hour operation to save his life.

Norris

Norris' Dragon Lizard *Amphibolurus norrisi* **Witten** and **Coventry,** 1984 [Alt. Mallee Heath Lashtail]

Kenneth Charles Norris was a zoologist who worked for the Victorian Fisheries and Wildlife Department Survey Team. He co-authored *Sites of Zoological Significance in Central Gippsland* (1982).

Norvill

Alcock's Flying Dragon *Draco norvillii* **Alcock,** 1859

F. H. Norvill of Upper Assam (now Arunachal Pradesh) collected the holotype.

Nuaulu

Bent-toed Gecko sp. *Cyrtodactylus nuaulu* **Oliver** et al., 2009

Named after the Nuaulu people of south Seram (Indonesia).

Núñez

Núñez' Tree Iguana *Liolaemus curis* Núñez and Labra, 1985

Herman Núñez Cepeda is a herpetologist at Museo Nacional de Historia Natural, Santiago de Chile. See also **Herman Núñez.**

Nurse

Nurse's Blind Snake *Leptotyphlops nursii* **Anderson,** 1896

Lieutenant Colonel C. G. Nurse of the Indian Service Corps was an entomologist who divided his time between northwest India and Aden (which garrison was part of the Western Army Corps of India). He was in Ferozepore, India (1902), and in Baluchistan (1906). He sold his private collection of Lepidoptera to John James Joicey (1919).

Nutaphand

Nutaphand's Narrowhead Softshell *Chitra chitra* Nutaphand 1986

Gecko sp. *Gekko nutaphandi* **Bauer, Sumontha,** and **Pauwels,** 2008

Wing Commander Ajarn Wirot Nutaphand (1932–2005) was a herpetologist and an officer in the Royal Thai Airforce. He took an arts degree but did postgraduate study of anatomy and histology. He wrote *The Turtles of Thailand* (1979). See also **Wirot.**

O

Oates, E. W.

Oates' Worm Snake *Typhlops oatesii* **Boulenger,** 1890 [Alt. Andaman Island Worm Snake]

Eugene William Oates (1845–1911) was a civil servant in British Colonial India and Burma (now Myanmar) and an amateur naturalist. When he returned to England he was Secretary of the British Ornithologists' Union (1898–1901). He wrote *The Fauna of British India* (1889). A bird is named after him.

Oates, F.

Oates' Savanna Vine Snake *Thelotornis capensis oatesii* **Günther,** 1881

Francis "Frank" Oates (1840–1875) was a naturalist who was in Matabeleland (1874) and became one of the first Europeans, after Livingstone, to sight the Victoria Falls. He traveled in Central and North America (1871–1872), and he and his brother William Edward Oates, a renowned traveler and hunter, left England for Africa (1873) to trek to the Zambesi. Francis died of malaria, and another brother, Charles George Oates, collected his diaries and letters, edited them, and published them as *Matabele Land and the Victoria Falls: A Naturalist's Wanderings in the Interior of South Africa* (1881). His nephew was Captain Lawrence Oates, who died on Scott's last expedition to the Antarctic.

Oberon

Royal Lesser Spiny Lizard *Sceloporus minor oberon* **H. M. Smith** and **Brown,** 1941 [Syn. *Sceloporus ornatus oberon*]

Oberon, King of the Fairies and Shadows, is best known from Shakespeare's *A Midsummer Night's Dream*.

Obst

Obst's Pond Turtle *Emys orbicularis fritzjuergenobsti* Fritz, 1993

Keeled Box Turtle ssp. *Cuora mouhotii obsti* Fritz, **Andreas,** and Lehr, 1998

Obst's Rock Gecko *Pristurus obsti* Rösler and Wranik, 1999

Professor Dr. Fritz Jürgen Obst (b. 1939) is a German herpetologist. He studied psychology and biology at Universität Heidelberg and Universität Hohenheim, Stuttgart, which awarded his doctorate (1996). He was Curator of Lower Vertebrates and Insects at the Zoological Gardens, Stuttgart (1990–1996), and he became herpetologist at the Staatliche Museum für Tierkunde, Dresden, becoming Deputy Director (1997), then Director (2001). Since 1997 he has also taught at Universität Leipzig. He wrote *Turtles, Tortoises, and Terrapins* (1988) and has co-authored books with Fritz and Andreas.

Ochoterena

Guerreran Skink *Eumeces ochoterenae* **Taylor,** 1933

Ochoterena's Lizard *Sceloporus ochoterenae* **H. M. Smith,** 1934

Northern Chiapas Arboreal Alligator Lizard *Abronia ochoterenai* **Martín del Campo,** 1939

Dr. Isaac Ochoterena Mendieta (1885–1950) was Professor of Histology and Embryology at, and Director of, Instituto de Biologia, Universidad Nacional de Mexico, Mexico City. He was a Lieutenant Colonel in the Mexican army, as he had been Professor of Histology at the Mexican Army Medical School.

Ocoa

Peravia Least Gecko *Sphaerodactylus ocoae* **Schwartz** and **Thomas,** 1977

Named after Sierra de Ocoa, a mountainous area in the Dominican Republic.

Oelofsen

Oelofsen's Girdled Lizard *Cordylus oelofseni* Mouton and Van Wyk, 1990

Dr. Burger W. Oelofsen is Director of Resource Management, Ministry of Fisheries and Marine Resources, Walvis Bay, Namibia. He has worked with Mouton, with whom he co-wrote "A Model Explaining Patterns of Geographic Character Variation in *Cordylus cordylus* (Reptilia: Cordylidae) in the South-western Cape, South Africa" (1988).

Oenpelli

Oenpelli's Rock Python *Morelia oenpelliensis* **Gow,** 1977

Oenpelli is an Aboriginal community in the Northern Territory of Australia.

Oertzen

Oertzen's Lizard *Lacerta oertzeni* **Werner,** 1904 [Syn. *Anatololacerta oertzeni*]

E. von Oertzen was an entomologist who visited Crete, Greece, and Asia Minor (Turkey) (1884–1887), during which travels he collected reptiles including the holotype of this lizard. He wrote *Verzeichnis der Coleopteren Griechenlands und Cretas, nebst einigen Bemerkungen uber ihre geographische Verbreitung und die Zeit des Vorkommens einiger Arten betreffenden Sammelberischten* (1886).

Ogilby

Ogilby's Knobtail Gecko *Nephrurus sphyrurus* Ogilby, 1892

James Douglas Ogilby (1853–1925) was an Irish ichthyologist and taxonomist who migrated to Australia (1884) after having worked at the British Museum and in the USA. He was appointed to the Australian Museum, Sydney (1885), but was sacked (1890) for being drunk on the job. The contemporary report criticized his "extreme and undiscriminating affinity for alcohol." Though sacked as a permanent employee, he went on working on a contract basis. He worked for the Queensland Museum (1901–1904 and 1913–1920).

Okada

Okada's Skink *Plestiodon okadae* Stejneger, 1907 [Alt. Okada's Five-lined Skink; Syn. *Eumeces okadae, Eumeces latiscutatus okadae*]

Shigefumi Okada was the author of *Catalogue of Vertebrated Animals of Japan* (1891). Stejneger does no more than attribute the species to "S. Okada" in his etymology. He is not to be confused with his famous namesake Professor Dr. Yaichiro Okada (1892–1976), a herpetologist and ichthyologist who wrote more than 400 books and papers dealing with nearly all branches of zoology, including *Ecology and Evolution of Reptiles* (1932).

Oldham

Oldham's Leaf Turtle *Cyclemys oldhami* **Gray,** 1863

Oldham's Bow-fingered Gecko *Cyrtodactylus oldhami* **Theobald,** 1876

Richard Oldham (1837–1864) was a gardener at Kew. He was sent to the Far East as a collector (1861). He was in Japan, the Bonin Islands, and Formosa (Taiwan) (1862–1864). He died in China.

Olfers

Lichtenstein's Green Racer *Philodryas olfersii* **Lichtenstein,** 1823

Ignaz Franz Werner Maria von Olfers (1793–1871) was a naturalist, historian, and diplomat. He was posted to Brazil (1816). He became Director of Museum für Naturkunde Berlin (1838).

Olive

Banded Gecko *Gymnodactylus olivii* **Garman,** 1901 [Alt. Ring-tailed Gecko]

Edmund Abraham Cumberbatch Olive (1844–1921) emigrated from England to Australia. He lived in Cooktown, Queensland (1875–1921), at the peak of the Palmer gold rush. He became an auctioneer and commission agent (1875) and developed his interest in natural history from his home, Mount Olive, outside Cooktown, with the help of an Aboriginal man known as Billy Olive. Mount Olive was a good starting point for their excursions to nearby areas of dense, undisturbed vegetation, and Olive accumulated impressive collections of native fauna at his home. He also sent many specimens to Australian, European, and American collectors and museums. A bird is named after him.

Oliver, J. A.

Oliver's Parrot Snake *Leptophis nebulosus* Oliver, 1942

Juventud Least Gecko *Sphaerodactylus oliveri* **Grant,** 1944

Oliver's Bronzeback *Dendrelaphis oliveri* **Taylor,** 1950

Oliver's Coral Snake *Micrurus distans oliveri* **Roze,** 1967

Dr. James "Jim" Arthur Oliver (1914–1981) was a zoologist and herpetologist. All three of his degrees, from bachelor's to doctorate, were awarded by the University of Michigan. He became Assistant Curator, Herpetology Department, American Museum of Natural History (1942). During WW2 he served in the U.S. Navy, then returned to the museum, later resigning (1948) to become Assistant Professor of Biology, University of Florida (1948–1950). He returned to New York as Curator of Reptiles and Director of the Bronx Zoo (1951). He was Director of the American Museum of Natural History (1959–1969) and, at one time, Director of the New York Aquarium. He wrote *The Natural History of North American Amphibians and Reptiles* (1955). The etymology for the coral snake states that it was "named for Dr. James A. Oliver, the distinguished herpetologist, who collected two of the paratypes." Taylor's bronzeback etymology says that Oliver "has contributed to the stabilization of the generic name of this form."

Oliver, W. R. B.

Oliver's Skink *Cyclodina oliveri* **McCann,** 1955 [Alt. Marbled Skink]

Dr. Walter Reginald Brook Oliver (1883–1957) was a New Zealand ornithologist and paleontologist. He wrote *New Zealand Birds* (1930). All his notes and papers are at Te Papa Tongarewa, the Museum of New Zealand, Wellington, where he was Director (1928–1947). An extinct bird is named after him.

Olivier

Olivier's Sand Lizard *Mesalina olivieri* Audouin, 1829

Guillaume-Antoine Olivier (1756–1814) was a botanist, entomologist, and malacologist and one of the great French naturalists. He qualified as a physician, practicing in his hometown, but was bored and badly paid. A rich patron paid him to collect in Europe, particularly Great Britain and the Netherlands, and he was employed as a naturalist on a major expedition (1792–1798) to Persia (now Iran). He returned home with a significant natural history collection from Turkey, Asia Minor, Persia, Egypt, and some Mediterranean islands, now in Muséum National d'Histoire Naturelle, Paris. He wrote *Voyage dans l'Empire Othoman, l'Egypte et la Perse* (1807). His grandson was Ernest Olivier, a French herpetologist who presented his grandfather's collection of insects to the French Academy. A mammal is named after him.

Olson

Olson's Cleft Lizard *Sceloporus mucronatus olsoni* **Webb, Lemos-Espinal,** and **H. M. Smith,** 2002

Dr. Rupert Earl Olson (b. 1934) is a herpetologist who specializes in Mexican herpetofauna. He wrote "A New Subspecies of *Sceloporus torquatus* from the Sierra Madre Oriental, Mexico" (1986).

Olsson

Desert Grass Anole *Anolis olssoni* **Schmidt,** 1919

Dr. Axel Adolph Olsson (1889–1977) was a paleontologist and malacologist who supplied fossil crustacean collections to learned institutions including the Smithsonian. He was a consultant to the Creole Petroleum Corporation's activities in Colombia and Venezuela. He wrote "The Miocene of Northern Costa Rica with Notes on Its General Stratigraphic Relations" (1922).

Omman

Omman's Lizard *Lacerta cyanura* **Arnold,** 1972 [Alt. Blue-tailed Oman Lizard; Syn. *Omanosaura cyanura*]

Named after the (wrongly spelled) Oman Mountains, Arabian Peninsula.

O'Neill

O'Neill's Tree Snake *Sibynomorphus oneilli* **Rossman** and **Thomas,** 1979

John P. O'Neill (b. 1942) is an American field ornithologist and artist. The etymology reads "During the summer of 1975, John P. O'Neill made a small collection of herpetological specimens. As luck would have it, the one snake he collected represents a previously undescribed species." A bird is named after him.

Oort

Indonesian Scaly-toed Gecko *Lepidodactylus oortii* **Kopstein,** 1926

Professor Eduard Daniel Van Oort (1876–1933) was a Dutch zoologist who collected in the East Indies and was Director of the bird collections at Nationaal Natuurhistorisch Museum, Leiden. He wrote *Ornithologia Neerlandica. De Vogels van Nederland* (1918). Four birds are named after him.

Orbigny

Orbigny's Slider *Trachemys dorbigni* **Duméril** and **Bibron,** 1835

See **D'Orbigny.**

Orcés

Peters' Ameiva *Ameiva orcesi* **Peters,** 1964

Orcés' Coral Snake *Micrurus steindachneri orcesi* Roze, 1967

Orcés' Andes Anole *Phenacosaurus orcesi* **Lazell,** 1969

Gymnophthalmid lizard sp. *Riama orcesi* Kizirian, 1995

Rough Teiid sp. *Echinosaura orcesi* **Fritts,** Almendariz, and Samec, 2002

Professor Gustavo Orcés (1902–1999) was a zoologist and herpetologist who worked at the Polytechnic, Quito. Part of his collection is housed at Fundación Herpetológica Gustavo Orces at Museo Ecuatoriano de Ciencias Naturales, Quito, founded in 1989. Roze, who described the coral snake, wrote that Orces "made available to me his large coral snake collection." Two birds and a mammal are named after him.

Orcutt

Granite Spiny Lizard *Sceloporus orcutti* **Stejneger,** 1893

Charles Russell Orcutt (1864–1929) was primarily a botanist and malacologist. He combined collecting with publishing scientific journals. He had no formal schooling, being taught on the farm by his parents. The family moved from Vermont to San Diego (1879). In 1884 he began publishing *The West American Scientist* to get his own work and notes before the public. It continued to appear, sporadically, until 1919. Orcutt accumulated a large, if eclectic, collection, which eventually finished up with the San Diego Society of Natural History. He collected for the Smithsonian (1927–1929) in Baja California, Mexico, Central America, and the Caribbean, including Haiti, where he died.

Orestes

Sadlier's Caledonian Skink *Caledoniscincus orestes* **Sadlier,** 1986

Pampas Snake sp. *Tomodon orestes* **Harvey** and Muñoz, 2004

Orestes was the son of King Agamemnon and his wife, Clytemnestra. Clytemnestra murdered Agamemnon, and then Orestes and his sister Electra murdered Clytemnestra to avenge their father. At least seven plays by three of the great Athenian classical dramatists have survived and tell this story. However, "Orestes" means "he who stands/dwells on the mountain," and Sadlier perhaps had the meaning in mind, as the holotype was taken in a mountainous region; Harvey and Muñoz certainly did, as they say so in their description.

Orlov

Orlov's Viper *Vipera orlovi* Tuniyev and Ostrovskikh, 2001

Orlov's Forest Lizard *Bronchocela orlovi* Hallermann, 2004

Dr. Nikolai Lusteranovich Orlov (b. 1952) is a Russian zoologist, herpetologist, and Senior Research Scientist, Herpetology Department, Zoological Institute, Russian Academy of Sciences, St. Petersburg. He has made over 60 field trips, mainly to countries of the former USSR, but also to the Indian subcontinent and Southeast Asia. His publications include, as co-author, "A New Species of Mountain Stream Snake, Genus *Opistotropis* Günther, 1872 (Serpentes: Colubridae: Natricinae), from the Tropical Rain Forests of Southern Vietnam" (1998).

Ornelas

Cerra Baul Alligator Lizard *Abronia ornelasi* **Campbell,** 1984

Julio Ornelas Martinez first met Campbell when he helped the latter extract his vehicle from the bottom of a muddy Mexican ravine. Over many years they became friends and often collected together.

Orsini

Orsini's Viper *Vipera ursinii* Bonaparte, 1835 [Alt. Ursini's Viper, Meadow Viper]

Antonio Orsini (1788–1870) was an Italian pharmacist and naturalist who collected the viper holotype (1833). He made large collections of artifacts, minerals, fossils, shells, and plants, all of which are housed at the Natural History Museum, Ascoli Piceno (where he lived). A number of plants are named after him.

Ortiz

Ortiz's Tree Iguana *Liolaemus ortizi* **Laurent,** 1982

Tree Iguana sp. *Liolaemus juanortizi* Young-Downey and Moreno, 1992

Professor Dr. Juan Carlos Ortiz Zapata (b. 1945) is a zoologist and herpetologist at Universidad de Concepcion, Chile. His initial degree in biology (1972) and his doctorate (1981) were both from the Sorbonne.

Orton

Orton's Anole *Anolis ortonii* **Cope,** 1868 [Alt. Amazon Bark Anole, Ortoni's Anole (in error)]

Orton's Boa *Boa constrictor ortonii* Cope, 1877

Professor James Orton (1830–1877) was a zoologist who collected in Latin America (1860s). He taught at Vassar College, New York (1866–1877). Cope makes clear in the title of his article describing the Anole—"An Examination of the Reptilia and Batrachia Obtained by the Orton Expedition to Equador and the Upper Amazon, with Notes on Other Species"—that Professor Orton is the man intended to be honored. Orton collected the holotype of the boa. Among his publications is *The Andes and the Amazon* (1876). Two birds are named after him.

Osborn

Osborn's Dwarf Crocodile *Osteolaemus osborni* **Schmidt,** 1919 [Alt. Congo Dwarf Crocodile; Syn. *O. tetraspis osborni*]

Dr. Henry Fairfield Osborn (1857–1935) was a zoologist, paleontologist, humanist, and evolutionist. He was Professor of Natural Sciences, Princeton (1881–1891), where he had graduated (1880). He was Professor of Biology and Zoology, Columbia University, New York (1891–1907), and worked at the American Museum of Natural History (1908–1933), where he became a Trustee. He was a leading proponent of the theory of evolution and was called as an expert witness at the famous "Monkey Trial." However, this clearly led him down some dark avenues because of ignorant assumptions, as he was a confirmed racist, once saying, "The Negroid stock is even more ancient than the Caucasian and Mongolians as may be proved by an examination not only of the brain, of the hair, of the bodily characteristics, but of the instincts, the intelligence. The standard intelligence of the average adult Negro is similar to that of the eleven-year-old youth of the species Homo sapiens." Three mammals are named after him.

Osborne

Osborne's Lancehead *Bothrops osbornei* Freire-Lascano, 1991

Probably named after Steven T. Osborne, an American who has bred snakes in captivity in Escondido in California for over 20 years. He published "Life History Note—*Rhinocheilus lecontei antonii* (Mexican Long-nosed Snake)—Behavior" (1984).

Osella

Osella's Skink *Leptoseps osellai* **Böhme,** 1981

Dr. Giuseppe Osella is an entomologist with a particular interest in Coleoptera who has worked at Museo Regionale di Scienze Naturali di Torino. He was, before the 2009 earthquake, Professor of Zoology and Entomology, Università degli Studi dell'Aquila. He was Curator of Natural History, Verna Museum (1976). He has collected beetles all over the world and described over 400 species, and is now concentrating on weevils. He was collecting in Thailand when he found the holotype of the skink. Among his publications is "Taxonomy, Ecology and Distribution of Curculionoidea (Coleoptera: Polyphaga)" (1998).

O'Shaughnessy

O'Shaughnessy's Galliwasp *Diploglossus bilobatus* O'Shaughnessy, 1874
O'Shaughnessy's Anole *Anolis gemmosus* O'Shaughnessy, 1875
O'Shaughnessy's Keeled Iguana *Ophryessoides aculeatus* O'Shaughnessy, 1879
O'Shaughnessy's Lightbulb Lizard *Riama simoterus* O'Shaughnessy, 1879
O'Shaughnessy's Skink *Amphiglossus gastrostictus* O'Shaughnessy, 1879
O'Shaughnessy's Chameleon *Calumma oshaughnessyi* **Günther,** 1881
O'Shaughnessy's Dwarf Iguana *Enyalioides oshaughnessyi* **Boulenger,** 1881
O'Shaughnessy's Gecko *Gonatodes concinnatus* O'Shaughnessy, 1881
O'Shaughnessy's Banded Gecko *Pachydactylus oshaughnessyi* Boulenger, 1885
White-striped Eyed Lizard *Cercosaura oshaughnessyi* Boulenger, 1885

Arthur William Edgar O'Shaughnessy (1844–1881) is probably best remembered today as a minor poet. He began work (1861) as a transcriber in the Library of the British Museum, transferring (1863) to the Zoology Department, where he became an expert in herpetology. He was not thought to have any literary talents and must have surprised all his friends and relations when he published the first of his four anthologies, *An Epic of Women* (1870). He was an associate of the Pre-Raphaelite Brotherhood.

Osman Hill

Colombo Wolf Snake *Lycodon osmanhilli* **Taylor,** 1950

Professor William Charles Osman Hill (1901–1975) was a physician, an anthropologist, and a primatologist at London University. His collection of skeletons and tissue is held by the Royal College of Surgeons, London. The Primate Society of Great Britain awards an Osman-Hill Medal named in his honor. He published *The Primates—Comparative Anatomy and Taxonomy* (1949). A monkey is named after him.

Osten-Sacken

Osten-Sacken's Ribbon Snake *Thamnophis sauritus sackenii* **Kennicott,** 1859 [Alt. Peninsula Ribbon Snake]

Baron Carl Robert Romanovich von der Osten-Sacken (1828–1906) was a Russian aristocrat and entomologist of German ancestry. He first became interested in entomology at the age of 11 while on a visit to Baden-Baden, Germany. He entered the Russian Imperial Diplomatic service in 1849 and seven years later was appointed secretary to the Russian legation in Washington, DC. He became Consul-General of Russia in New York City (1862). He resigned in 1871 and for the next two years journeyed back and forth between Europe and America. He was in the USA during 1873–1875, collaborating on a study of the Diptera of the Americas north of Panama. He donated his collection of holotypes to the Museum of Comparative Zoology, Harvard. He left the Americas and lived in Heidelberg, Germany (1877–1906). He published *Catalog of the Described Diptera of N. America* (1858).

Osvaldo

Gymnophthalmid lizard sp. *Leposoma osvaldoi* Avila-Pires, 1995

Osvaldo Rodrigues da Cunha. See **Cunha.**

Ota

Ota's Japalure *Japalura makii* **Ota,** 1989
Ota's Wolf Snake *Lycodon bibonius* Ota and **Ross,** 1994
Ota's Rock Gecko *Cnemaspis otai* **Das** and **Bauer,** 1998
Japalure sp. *Japalura otai* Mahony, 2009

Professor Hidetoshi Ota (b. 1959) is a Japanese herpetologist formerly of the Tropical Biosphere Research Center, University of the Ryukyus, Okinawa.

Oudemans

Oudemans' Dtella *Gehyra interstitialis* Oudemans, 1894
Oudemans' Four-fingered Skink *Lygisaurus laevis* Oudemans, 1894

Dr. Johannes Theodorus Oudemans (1862–1934) was a Dutch entomologist. He wrote "Étude sur la position de repos chez les lépidoptères" (1903).

Oudri

Oudri's Fan-footed Gecko *Ptyodactylus oudrii* **Lataste,** 1880 [Alt. Algerian Fan-fingered Gecko]

General Émile Oudri (1843–1919) was a soldier who spent his entire career in the French Colonies, commanding troops of the French Foreign Legion. He was a Colonel (1895) in the Second Foreign Regiment, which was sent to Madagascar to help put down the revolt (1895–1896). He was promoted to Brigadier General (1896) and to full General (1900). At the end of his career, in which he had fought in 36 campaigns, he was in command of the Fourth Army Corps. He was a member of the French Zoological Society (1879).

Ours

Malagasy Night Snake sp. *Ithycyphus oursi* **Domergue,** 1986

Dr. Jacques de Saint-Ours (1924–1968) was a French geologist and, like C. A. Domergue, who described the snake, an enthusiastic speleologist. He undertook a geological survey of Madagascar (1951). In his description of the snake Domergue wrote that Saint-Ours was killed in a plane crash in the course of his work: "Jacques de Saint-Ours, qui fût directeur du Service d'Hydrogéologie de Madagascar, mort en service commandé, dans un accident d'avion, à Nouakchott, en 1968." He wrote, with Pavlovsky, *Étude géologique de l'archipelago des Comores* (1953).

Oustalet

Oustalet's Chameleon *Furcifer oustaleti* **Mocquard,** 1894 [Alt. Giant Madagascar Chameleon]

Dr. Jean-Frédéric Émile Oustalet (1844–1905) was a zoologist. He succeeded Jules Verreaux as Assistant Naturalist, Muséum National d'Histoire Naturelle, Paris (1873), and succeeded Alphonse Milne-Edwards as Professor of Mammalogy (1900). He wrote *Les oiseaux de la Chine* (1877), with Père Armand David as co-author. Six birds and a mammal are named after him.

Owen, R.

Owen's Galliwasp *Diploglossus oweni* **Duméril** and **Bibron,** 1839

Professor Sir Richard Owen (1804–1892) was a British anatomist and paleontologist. He was a midshipman in the Royal Navy and later became a surgeon, having studied at Edinburgh. His fame as a scientist led to his appointment to teach Natural History to Queen Victoria's children. He was largely responsible for the creation of the Natural History Museum, London, having separated that department from the British Museum. In 1863 he reported on an unusual fossil from Germany: it was the now-famous primitive bird *Archaeopteryx lithographica*. He was the first person to give fossil reptiles, found in southern England, the name Dinosauria, "terrible lizards," so creating the present-day dinosaur industry. He was originally friendly with Darwin but violently disagreed with Darwin's ideas on evolution, and they became lifelong enemies. A species of kiwi is named after him.

Owen, W. F. W.

Owen's Three-horned Chameleon *Chamaeleo oweni* **Gray,** 1831

Vice Admiral William Fitz-William Owen (1774–1857) was born and died in Canada. As a child he was taken to Wales and brought up there, and at age 10 he became a midshipman in the Royal Navy, in which he served for 43 years. He served with all the great naval officers of his day, and among his friends was Nelson. He commissioned HMS *Leven* (1821) and equipped her for a voyage he thought would last about four years. Another brig, HMS *Barracouta*, was also commissioned and placed under his command. They set off for a voyage to southern Africa (1822), and in 1823 Owen famously reported a sighting of the mysterious *Flying Dutchman*. He published *Narrative of Voyages to Explore the Shores of Africa, Arabia, and Madagascar; Performed in HM Ships Leven and Barracouta under the Direction of Captain W F W Owen RN* (1833). He was the sole owner of the island of Campobello in New Brunswick, to which he retired (1835) and ruled as a benevolent despot. Owen collected the holotype of this chameleon.

Owen Stanley

Owen Stanley Forest Snake *Toxicocalamus stanleyanus* Boulenger, 1903

See Stanley.

Ozorio

Skink sp. *Mabuya ozorii* **Bocage,** 1893

Dr. Balthazar Osorio (sometimes Ozorio) (1855–1926) was a Portuguese ichthyologist and naturalist. He was the third Director, Zoological Section, Museu Bocage, Lisbon. He wrote *Memorias do Museu Bocage* (1909).

P

Pacheco-Gil

Zapotitlan Coral Snake *Micrurus pachecogili* **Campbell,** 2000

E. Pacheco-Gil collected the holotype. Oddly, the original text says that the snake is being named after his children, though the scientific name gives no indication of that.

Pagaburo

Tree Iguana sp. *Liolaemus pagaburoi* **Lobo,** 1999

Omar Pagaburo is described in the text as "having collected extensively throughout Argentina for over 20 years."

Pagenstecher

Southern Grass Tussock Skink *Pseudemoia pagenstecheri* Lindholm, 1901

Dr. Heinrich Alexander Pagenstecher (1825–1889) was a physician, zoologist, and comparative anatomist. He practiced as a physician before becoming a teacher at Universität Heidelberg (1856), then Professor of Zoology there (1865). In retirement he wrote the four-volume *Allgemeine Zoologie* (1875–1881). He was persuaded to come out of retirement and became Director, Zoologischen Museum und Institut, Hamburg (1882–1889).

Pails

False King Brown Snake *Pseudechis pailsei* Hoser, 1998 [Syn. *Pailsus pailsei*]

Roy Pails (b. 1956) is a reptile breeder from Ballarat, Victoria, Australia. He collected the holotype (1984) and kept it captive until it died (1990). He froze the corpse and eventually sent it to the Victoria Museum (1998).

Palacios

Palacios' Bunchgrass Lizard *Sceloporus palaciosi* Lara-Góngora, 1983

Dr. Prococo Palacios works for Comunidad Rancho el Capulin, Mexico.

Palfreyman

Palfreyman's Window-eyed Skink *Niveoscincus palfreymani* **Rawlinson,** 1974 [Alt. Pedra Branca Cool-skink]

A. E. Palfreyman and M. Forster made the first recorded landing (1947) on Pedra Branca, off southern Tasmania. Palfreyman returned (1956) and collected specimens of this skink. He was a keen sailor; a racing boat (Dragon class) was built for him (1955), and he was the founding Vice President, Tasmanian International Dragon Association (1958).

Pallas

Pallas' Glass Lizard *Ophisaurus apodus* Pallas, 1775 [Alt. Scheltopusik; Syn. *Pseudopus apodus*]

Pallas' Viper *Gloydius halys* Pallas, 1776

Dagestani Tortoise *Testudo graeca pallasi* Chkikvadze and Bakradze, 2002

Peter Simon Pallas (1741–1811) was a German zoologist, geographer, and traveler, and one of the greatest 18th-century naturalists. He was born in Berlin and arrived in Russia in 1767, but on retiring returned to Berlin. His doctorate (1760) was from Universiteit Leiden. He went to London (1761) to study the English hospital system and was enchanted by the countryside. Empress Catherine II summoned him to Russia (1767) to become Professor of Natural History, St. Petersburg Academy of Sciences, and to investigate Russia's natural environment and lesser known areas. He led an expedition that studied many regions of Russia (1768–1774). Among his publications is *A Journey through Various Provinces of the Russian State* (1771). In 1772 he found a mass of iron weighing 700 kilograms (1,500 pounds) that proved to be a new kind of meteorite; it was named pallasite after him. A volcano on the Kurile Islands is also named after him, as are 14 birds and 7 mammals.

Palmer, M. G.

Palmer's Anole *Anolis palmeri* **Boulenger,** 1908

Mervyn George Palmer (1882–1954) was an English naturalist and traveler who collected the holotype of this lizard. After graduating he became an analytical chemist, then decided upon a career as a freelance collector and naturalist. He collected for the Natural History Museum, London (1904–1910), in Colombia, Ecuador, and Nicaragua. During this time he discovered over 60 species new to science, learned to speak Spanish and two South American Indian languages, married a South American woman, undertook archeological digs, and explored and mapped the Río Segovia between Nicaragua and Honduras. He worked for commercial concerns in Ecuador (1910–1918) before moving to London with the same company. Having suffered from malaria and yellow fever, he was declared unfit for overseas army service during WW1. He was in Venezuela from 1919 to 1921, later being based in London but making frequent visits to South America. He lived in Ilfracombe, Devon, England (1932–1954), where he founded and was curator of a museum and ran a library and field club, as he "wanted something to do." He was also at one time the editor of the *Natural Science Gazette*. He wrote *Through Unknown Nicaragua—The Adventures of a Naturalist on a Wild-Goose Chase* (1945). Two amphibians and a bird are named after him.

Palmer, T. S.

San Pedro Martir Side-blotched Lizard *Uta palmeri* **Stejneger,** 1890

Sierra Alligator Lizard *Elgaria coerulea palmeri* Stejneger, 1893

Dr. Theodore Sherman Palmer (1868–1958) was a botanist. He worked for the U.S. Biological Survey as Assistant Chief for 15 years (1889–1933) and was also the Law Enforcement Officer of the U.S. Fish and Wildlife Service (1900–1916). He led an expedition (1891) to study the flora and fauna of Death Valley. Following a broken hip he was confined to his house for his last 21 years. He wrote *Index generum mammalium* (1904). A mammal is named after him.

Pamela

Seychelles Skink genus *Pamelaescincus* **Greer,** 1970

Greer's etymology informs us, "The genus is named after Pamela, the older of my two sisters."

Pamela K.

Arnhemland Watercourse Dtella *Gehyra pamela* **King,** 1982

Pamela King is the author's wife, who has played a key role in the collection of specimens.

Pan

Pan's Box Turtle *Cuora pani* Song, 1984

Unfortunately Song does not give an etymology in his description, so we cannot be sure who he had in mind. It is probably Pan Lei, a fellow Chinese herpetologist. He co-wrote "Studies on Genus *Cuora* of Testudoformes" (1988).

Paniai

Coastal Emo Skink *Emoia paniai* Brown, 1991

Named after the Lake Paniai area in New Guinea.

Papenfuss

Papenfuss' Rock Agama *Laudakia papenfussi* **Zhao,** 1998

Leaf-toed Gecko sp. *Phyllodactylus papenfussi* **Murphy, Blair,** and de la Cruz, 2009

Dr. Theodore Johnstone Papenfuss (b. 1941) is a Research Specialist in herpetology, University of California, Berkeley. He and Zhao jointly presented a paper at the First Asian Herpetological Meeting (1993). He co-wrote "Karyotypes of Chinese Species of the Genus *Teratoscincus* (Gekkonidae)" (1998).

Parish

Parish's Fanged Snake *Ialtris parishi* **Cochran,** 1932

Lee H. Parish (d. 1931) was on the 1930 Parish-Smithsonian Expedition to Haiti, led by Alexander Wetmore and financed by Semmes W. Parish. A "Captain Semmes Parish" commanded a company of infantry during the Spanish-American War (1898), and we think this is the same man. Lee may have been his son. The snake holotype was collected during that expedition.

Parker, F.

Parker's Snake-necked Turtle *Chelodina parkeri* Rhodin and **Mittermeier,** 1976

Fred Parker (b. 1941) of Kirwan, Queensland, is an Australian naturalist and explorer. In the 1970s he worked for the Wildlife Section, District Administration, Konebobu, Papua New Guinea. He co-wrote "A New Species of *Cyrtodactylus* (Gekkonidae) from New Guinea with a Key to Species from the Island" (1973). See also **Fred Parker.**

Parker, H. W.

Parker's Blind Snake *Typhlops ocularis* Parker, 1927

Parker's Ground Snake *Atractus carrioni* Parker, 1930

Parker's Pholidobolus *Pholidobolus annectens* Parker, 1930

Parker's Leaf-toed Gecko *Hemidactylus megalops* Parker, 1932

Parker's Banded Snake *Chamaelycus parkeri* **Angel,** 1934

Parker's Whorl-tailed Iguana *Stenocercus carrioni* Parker, 1934

Parker's Zonure Lizard *Zonurus parkeri* Cott, 1934

Parker's Dwarf Gecko *Lygodactylus keniensis* Parker, 1936

Parker's Green Tree Skink *Prasinohaema parkeri* **M. A. Smith,** 1937

Parker's Least Gecko *Sphaerodactylus parkeri* **Grant,** 1939

Parker's Dwarf Boa *Tropidophis parkeri* Grant, 1940

Parker's Day Gecko *Phelsuma parkeri* **Loveridge,** 1941

Parker's Many-fingered Teiid *Cercosaura parkeri* **Ruibal,** 1952

Parker's Keelback *Tropidonophis parkeri* Malnate and **Underwood,** 1988

Parker's Emo Skink *Emoia parkeri* Brown, Pernetta, and Watling, 1980

Parker's Worm Snake *Leptotyphlops parkeri* **Broadley,** 1999

Dr. Hampton Wildman Parker (1897–1968) was an English zoologist. He took both his bachelor's (1923) and master's degrees (1935) at Cambridge. He joined the Natural History Museum, London (1923), interrupted by war service in the Admiralty, returning (1945) to become Keeper of Zoology (1947–1957). He wrote *Snakes of the World—Their Ways and Means of Living* (1963).

Parona

Peracca's Iguana *Leiosaurus paronae* **Peracca,** 1897

Dr. Corrado Parona (1842–1922) graduated as a physician in Pavia, Italy, before moving to Genoa and becoming Professor of Zoology, Università degli Studi di Genova (1883). His brother, Carlo, was a Professor, Università degli Studi di Torino.

Parrhasius

Fire-tailed Rainbow-skink *Carlia parrhasius* **Couper, Covacevich,** and Lethbridge, 1994

Parrhasius of Ephesus was an ancient Greek artist who was a master of artistic deception. There is a tale recorded of him describing his contest with Zeuxis. Zeuxis painted some grapes so perfectly that birds came to peck at them. He then called on Parrhasius to draw aside the curtain and show his picture. However, his rival's picture *was* the curtain. Zeuxis acknowledged himself to be surpassed, for he had deceived birds but Parrhasius had deceived Zeuxis.

Parson/Parsons

Parson's Giant Chameleon *Calumma parsonii* **Cuvier,** 1824

James Parsons (1705–1770) was a physician, author, and antiquary who was brought up in Ireland, studied medicine in Paris, received his degree from Université de Reims Champagne-Ardenne, and moved to England (1836). Here he was elected to the Royal Society and became Assistant Foreign Corresponding Secretary (1741). He described this chameleon (1768), which really ought to be called Parsons' (not Parson's) Chameleon. It was only later fully identified and named.

Partello

Partello's Waterside Skink *Tropidophorus partelloi* **Stejneger,** 1910

Colonel Joseph McDowell Trimble Partello (1851–1934) was an officer in the U.S. Army. He served under General Custer and survived Custer's disastrous campaign against the Sioux. He knew the Sioux Chief Sitting Bull. He was also an amazing shot with a rifle, setting a world record (1878) of 224 out of 225 shots at ranges of up to 900 meters (1,000 yards). He was commanding officer of Fort Bliss, El Paso, Texas (1906). He was stationed in Seattle, Washington (1910), and was a leading light of the Seattle Philatelic Club. He served in the Philippines, and his wife sent woven basketry from there to the Burke Museum, Washington State.

Pastaza

Shreve's Keelback *Helicops pastazae* **Shreve,** 1934

Pastaza is a tributary of the Amazon in Ecuador.

Pasteur

Pasteur's Cylindrical Skink *Chalcides atlantis* Pasteur, 1962 [Junior syn. of *C. mionecton* **Boettger,** 1873]

Pasteur's Dwarf Gecko *Lygodactylus arnoulti* Pasteur, 1964

Pasteur's Lizard *Mesalina pasteuri* **Bons,** 1960

Pasteur's Day Gecko *Phelsuma (v-nigra) pasteuri* Meier, 1984

Georges Pasteur (b. 1930) is a French biologist who is Honorary Director, École Pratique des Hautes Études and Centre de Récherches sur l'Evolution et ses Mecanismes and of Muséum National d'Histoire Naturelle, Paris. He worked at Institut Chérifien, Rabat, Morocco (1960).

Pathfinder

Pathfinder Short-legged Skink *Brachymeles pathfinderi* **Taylor,** 1925

This skink was named after the steamship *Pathfinder.*

Patrick Couper

Patrick Couper's Python *Broghammerus reticulatus patrickcouperi* Hoser, 2004

See **Couper, P.**

Patton

Colubrid snake sp. *Liophidium pattoni* Vieites et al., 2010

Dr. James "Jim" L. Patton (b. 1941) is a mammalogist who works at the Museum of Vertebrate Zoology, University of California, Berkeley, USA. The University of Arizona awarded his doctorate (1969). He was appointed as an Assistant Curator at the Museum of Vertebrate Zoology (1969), becoming Full Professor and Curator, Zoology Department (1979), from where he retired (2001) as Emeritus Professor of Integrative Biology. He has published over 160 scientific papers or contributions to larger works, including descriptions of a number of new species. Three mammal species and a mammal genus are named after him.

Paulian

Paulian's Five-toed Skink *Leptosiaphos pauliani* **Angel,** 1940

Ornate Dwarf Gecko *Lygodactylus pauliani* **Pasteur** and **Blanc,** 1991

Dr. Renaud Paulian (1913–2003) was the leading European expert on scarab beetles. He became Assistant, Laboratory of Entomology, Muséum National d'Histoire Naturelle, Paris (1937). He was at Institut de Recherche Scientifique de Madagascar, Tananarive (1947–1966), eventually becoming Director, and founded the publication *Faune de Madagascar* (1956). He was Director, Institut Scientifique de Congo-Brazzaville (1961–1966), and headed the local university, moving to become head of Université d'Abidjan, Ivory Coast (1966–1969). He returned to France (1969), becoming, successively, Rector of Académie d'Amiens and Académie de Bordeaux. Among the taxa named after him are a bird and an amphibian.

Paulina

Paulina's Tree Iguana *Liolaemus paulinae* **Donoso-Barros,** 1961

Paulina is one of the describer's daughters.

Paulson

Paulson's Bevel-nosed Boa *Candoia paulsoni* Stull, 1956 [Syn. *C. carinata paulsoni*]

John Paulson was a Swedish herpetologist from Gothenburg.

Pauwels

Blind Snake sp. *Letheobia pauwelsi* **Wallach,** 2005

Olivier Sylvain Gérard Pauwels (b. 1971) is a herpetologist. He took his master's degree and doctorate in zoology at Université Libre de Bruxelles. He is a Research Associate, Koninklijk Belgisch Instituut voor Natuurwetenschappen, Brussels. Wallach's description states that Pauwels collected the snake holotype in Gabon. He co-wrote "*A New Leptodactylodon Species from Gabon (Amphibia: Anura: Astylosternidae)*" (2003).

Peal

Peal's Keelback *Amphiesma pealii* Sclater, 1891

Samuel E. Peal (d. 1897) was an ethnographer and a tea planter in Assam, where he lived for many years. He wrote "Note on the Origin and Orthography of River Names in Further India" (1889).

Peers

Peers' Girdled Lizard *Cordylus peersi* **Hewitt,** 1932 [Alt. Hewitt's Spiny-tailed Lizard]

Victor Peers and his son, Bertram (Bertie), discovered Peers Cave at Fish Hoek near False Bay, South Africa (1926), and started excavating it (1927). They found nine burials (one became known as "Fish Hoek Man"), dating from 12,000 years ago, and many stone tools. We do not know which Peers is honored but think Bertram is more probable, as he was described as a fine amateur scientist and a dedicated naturalist whose enthusiasm proved fatal; he was bitten by a puff adder.

Peltier

Chameleon sp. *Calumma peltierorum* **Raxworthy** and Nussbaum, 2006

The Peltier family, including Valerie and Jeffrey, have a family trust that has given considerable support to the herpetological and conservation work carried out at the American Museum of Natural History.

Pemba

Pemba Worm Snake *Leptotyphlops pembae* **Loveridge,** 1941

Named after Pemba Island, Tanzania.

Peña, L. E.

Coquimbo Marked Gecko *Homonota penai* **Donoso-Barros,** 1966

Dr. Luis E. Peña (d. 1995) was an entomologist and a lepidopterist who collected zoological specimens (1950s–1960s) in Chile for the American Museum of Natural History. He co-wrote *Las mariposas de Chile* (1996).

Peña, Z. U.

Peña's Knob-scaled Lizard *Xenosaurus penai* Pérez Ramos, de la Riva, and **Campbell,** 2000 [Alt. Pico de Aguila Knob-scaled Lizard]

Zeferino Uribe Peña is a Mexican herpetologist. He was President, Sociedad Herpetológica Mexicana (1988–1990).

Peracca

Peracca's Large-scaled Lizard *Ptychoglossus festae* Peracca, 1896

Peracca's Iguana *Leiosaurus paronae* Peracca, 1897

Peracca's Whorl-tailed Iguana *Stenocercus festae* Peracca, 1897

Peracca's Anole *Anolis peraccae* **Boulenger,** 1898

Peracca's Teiid *Alopoglossus festae* Peracca, 1904

Peracca's Scaly-eyed Gecko *Lepidoblepharis peraccae* Boulenger, 1908

Peracca's Ground Snake *Liophis pseudocobella* Peracca, 1914

Count Dr. Mario Giancinto Peracca (1861–1923) was a herpetologist who kept a collection of Galapagos Tortoises and iguanas in his greenhouse in Turin. He abandoned medical studies in favor of zoology. He took his doctorate (1886) and joined the Institute of Zoology, Università degli Studi di Torino, as an Assistant, serving there until his retirement (1920). He visited the Cape Verde Islands (1891), bringing back to Italy 40 living specimens of the now extinct Cape Verde Giant Skink.

Percival

Percival's Lance Skink *Acontias percivali* **Loveridge,** 1935 [Alt. Percival's Legless Skink]

Arthur Blayney Percival (1874–1940) was a British game warden in East Africa (1901–1928). He went to Arabia as part of a Royal Society expedition (1899), then was appointed Assistant Collector (a public administrative job) in Kenya (1900). He was made Ranger (1900) for Game Preservation in Kenya and served in that post until his retirement (1923). He was instrumental in the establishment of two big game reserves and was largely the author of the codified game laws, East African Game Ordinance (1906). He was a founding member, East Africa and Uganda Natural History Society (1909). He wrote *A Game Ranger's Notebook* (1924). Three mammals and two birds are named after him.

Père David

Père David's Rat Snake *Elaphe davidi* **Sauvage,** 1884

Jean Pierre Armand David (1826–1900) was a French Lazarist priest and zoologist. He taught science at Savona College, Italy, before going as a missionary to China (1862). He became the first Westerner to observe such species as the Giant Panda. He co-authored *Les oiseaux de Chine* (1877). He collected thousands of specimens, and many taxa are named after him, including 5 mammals, 12 birds, and a giant salamander that he also discovered.

Perinet

Perinet Chameleon *Calumma gastrotaenia* **Boulenger,** 1888

Fandrefiala *Ithycyphus perineti* **Domergue,** 1986 [Alt. Perinet Night Snake]

Perinet Leaf Chameleon *Brookesia therezieni* Brygoo and Domergue, 1970

Named after Périnet, a forest reserve in eastern Madagascar.

Péringuey

Péringuey's Adder *Bitis peringueyi* **Boulenger,** 1888 [Alt. Dwarf Puff Adder]

Peringuey's Leaf-toed Gecko *Cryptactites peringueyi* Boulenger, 1910

Dr. Louis Albert Péringuey (1855–1924) was a French entomologist and naturalist. He left France (1879) for South Africa, where he became a Scientific Assistant, South African Museum (1884), was in charge of the Invertebrates Collection (1885), and became the museum's Director (1906–1924). He dropped dead as he was walking home from the museum.

Perkins, C. B.

Spotted Leaf-nosed Snake *Phyllorhynchus decurtatus perkinsi* **Klauber,** 1935

Clarence Basil Perkins (1888–1955) was better known as C.B. or "Si." He graduated from Princeton University (1912). He was a collector for the San Diego Zoo (1926–1931) and thereafter the zoo's Curator of Reptiles.

Perkins, G. A.

Perkins' Short-headed Snake *Oligodon perkinsi* **Taylor,** 1925

Dr. Granville A. Perkins collected the holotype and presented it to Taylor.

Peron

Lowlands Earless Skink *Hemiergis peronii* **Gray,** 1831

Skink sp. *Carlia peronii* **Duméril** and **Bibron,** 1839

Peron's Sea Snake *Acalyptophis peronii* Duméril, 1853

François Péron (1775–1810) was a French voyager and naturalist. He was a member of Baudin's scientific expedition with the ships *Geographe* and *Naturaliste* (1800–1804), which visited New Holland (Australia), Van Diemen's Land (now Tasmania), and Timor, Indonesia. He died of tuberculosis. Two mammals, a bird, and two amphibians are named after him, as is the Peron Peninsula, Western Australia.

Perret

Perret's Mole Viper *Atractaspis coalescens* Perret, 1960

Blind Snake sp.. *Leptotyphlops perreti* **Roux-Estéve,** 1979

Perret's Nigerian Gecko *Cnemaspis gigas* Perret, 1986

Perret's Chameleon *Chamaeleo wiedersheimi perreti* Klaver and **Böhme,** 1992

Jean-Luc Perret (b. 1925) is a herpetologist who concentrates on African amphibians. He is Honorary Curator, Department of Herpetology and Ichthyology, Museum d'Histoire Naturelle, Geneva, where he was previously a Research Assistant (1970–1987).

Perrotet

Teita Mabuya *Trachylepis perrotetii* **Duméril** and **Bibron,** 1839 [Alt. Red-sided Skink; Syn. *Mabuya perrotetii*]

Bronze-headed Vine Snake *Ahaetulla perroteti* Duméril and Bibron, 1854

Perrotet's Mountain Snake *Xylophis perroteti* Duméril and Bibron, 1854

Perrotet's Shieldtail *Plectrurus perroteti* Duméril and Bibron, 1854

Gustave Samuel Perrotet (1793–1867) was an explorer and collector. He was the naturalist (1819) on board the *Rhône.* The expedition was sent to Cayenne (French Guiana) to introduce plants that they thought would be useful. While there Perrotet made large mineralogical and botanical collections, returning to France in 1821. He made a number of voyages to Africa and South America (1822–1832), including one circumnavigation, and wrote *Souvenirs d'un voyage autour du monde* (1831).

Peter

Gymnophthalmid lizard sp. *Riama petrorum* Kizirian, 1996

The binomial honors three people called Peter or Peters (see **Peters, J. A.**, and **Peters, W. K. H.**, for two of them). The third is Peter D. Spoecker, who collected the holotype. He studied for a doctorate at the Museum of Vertebrate Zoology, University of California, Berkeley (1964–1966), but completed it at the University of Kansas (1994). He wrote "Movements and Seasonal Activity Cycles of the Lizard *Uta stansburiana* Stejneger" (1967).

Peters, G.

Peters' Rock Agama *Acanthocercus trachypleurus* Peters, 1982
Peters' Butterfly Lizard *Leiolepis guentherpetersi* **Darevsky** and Kupriyanova, 1993
Agama sp. *Acanthocercus guentherpetersi* **Largen** and Spawls, 2006

Dr. Günther Peters (b. 1932) was a herpetologist and onetime Director, Institute for Systematic Zoology, Museum für Naturkunde, Humboldt-Universität, Berlin. He studied for his doctorate in St. Petersburg (then called Leningrad) (1957). He wrote "Eine neue Wirtelschwanzagame aus Ostafrika (Agamidae: Agama)" (1987).

Peters, J. A.

Peters' Black-striped Snake *Coniophanes piceivittis frangivirgatus* Peters, 1950
Peters' Ameiva *Ameiva orcesi* Peters, 1964
Peters' Coral Snake *Micrurus petersi* **Roze,** 1967
Peters' Gecko *Gonatodes petersi* **Donoso-Barros,** 1967
Peters' Anadia *Anadia petersi* Oftedal, 1974
Spiral Keelback *Helicops petersi* **Rossman,** 1976
Peters' Black-headed Snake *Tantilla petersi* Wilson, 1979
Neotropical Tree Snake sp. *Sibynomorphus petersi* **Orces** and Almendariz, 1989
Gymnophthalmid lizard sp. *Riama petrorum* Kizirian, 1996

Dr. James Arthur Peters (1922–1972) was a zoologist who specialized in Ecuadorean herpetofauna. He attended the University of Michigan and was awarded a bachelor's degree (1948), a master's (1950), and a doctorate (1952). He taught at Brown University as an Associate Professor (1952–1958), leaving to become a Fulbright Lecturer at Universidad Centrale de Ecuador (1958–1959). He was a Professor at San Fernando Valley State College (1959–1964) and was at the Smithsonian as Assistant Curator, Reptiles and Amphibians (1965–1972), and as Curator for the last few years of his life. He co-wrote *Catalogue of the Neotropical* Squamata (1970). (Regarding *Riama petrorum*, see also **Peter.**)

Peters, W. K. H.

Peters' Lava Lizard *Tropidurus hispidus* **Spix,** 1825
Peters' Philippine Earth Snake *Rhinophis philippinus* **Cuvier,** 1829
Peters' Eyelid Skink *Lygosoma afrum* Peters, 1854 [Sometimes given as *L. afer*]
Peters' Writhing Skink *Lygosoma ater* Peters, 1854
Peters' Thread Snake *Leptotyphlops scutifrons* Peters, 1854
Peters' Tree Skink *Lankascincus fallax* Peters, 1860
Peters' Burrowing Skink *Scelotes caffer* Peters, 1861
Peters' Cobra *Naja samarensis* Peters, 1861 [Alt. South-eastern Philippine Cobra]
Peters' Leaf-toed Gecko *Phyllodactylus reissii* Peters, 1862
Peters' Sand Lizard *Pseuderemias striata* Peters, 1862
Peters' Blind Snake *Ramphotyphlops bituberculatus* Peters, 1863 [Syn. *Austrotyphlops bituberculatus*]
Peters' Forest Racer *Dendrophidion nuchale* Peters, 1863
Peters' Pholidobolus *Pholidobolus affinis* Peters, 1863
Peters' Running Snake *Coniophanes dromiciformis* Peters, 1863
Peters' Banded Skink *Scincopus fasciatus* Peters, 1864
Peters' Swamp Skink *Egernia luctuosa* Peters, 1866
Peters' Dasia *Dasia semicincta* Peters, 1867
Peters' Keeled Cordylid *Tracheloptychus petersi* **Grandidier,** 1869 [Alt. Malagasy Keeled Plated Lizard]
Peters' Odd-scaled Snake *Achalinus spinalis* Peters, 1869
Peters' Brazilian Lizard *Placosoma glabellum* Peters, 1870
Peters' Side-necked Turtle *Phrynops tuberosus* Peters, 1870
Peters' Bow-fingered Gecko *Cyrtodactylus consobrinus* Peters, 1871
Peters' Forest Dragon *Gonocephalus doriae* Peters, 1871
Peters' Smooth Snake *Gongylosoma longicauda* Peters, 1871
Peters' Sea Snake *Hydrophis bituberculatus* Peters, 1872
Peters' Anole *Anolis petersii* **Bocourt,** 1873
Peters' Giant Blind Snake *Rhinotyphlops schlegelii petersii* **Bocage,** 1873
Peters' Lidless Skink *Panaspis breviceps* Peters, 1873
Peters' Bright Snake *Liophidium mayottensis* Peters, 1874
Peters' Longtail Lizard *Latastia carinata* Peters, 1874
Peters' Lateral Fold Lizard *Coloptychon rhombifer* Peters, 1877
Indian Eyed Turtle *Morenia petersi* **Anderson,** 1879
Peters' Burrowing Snake *Apostolepis erythronota* Peters, 1880
Peters' Spotted Gecko *Geckolepis maculata* Peters, 1880

Peters' Worm Lizard *Leposternon petersi* **Strauch,** 1881
Peters' Snake Eater *Polemon notatus* Peters, 1882
Peters' Keelback *Amphiesma petersi* **Boulenger,** 1893
Peters' Earth Snake *Geophis petersii* Boulenger, 1894
Gymnophthalmid lizard sp. *Riama petrorum* Kizirian, 1996

Wilhelm Karl Hartwig Peters (1815–1883) was a zoologist and traveler who made important collections in Mozambique. For many years he headed Museum für Naturkunde Berlin. He was elected as a corresponding member of the Russian Academy of Sciences (1876). At first sight two of these reptiles appear to have been described much too early to be associated with Peters, but in both cases his name has stuck to the vernacular. The lava lizard, described by Spix (1825), was the subject of a redescription by Peters (1877), and the Philippine earth snake was the core of his 1861 thesis *De serpentum familia Uropeltaceorum*. (Regarding *Riama petrorum*, see also **Peter.**)

Peterson

Pale Broad-blazed Slider *Lerista petersoni* **Storr,** 1976

Magnus Peterson is an Australian herpetologist. He co-wrote "Overwintering in a Southwestern Long-necked Tortoise *Chelodina oblonga*" (2004).

Petit

Eyeless Skink sp. *Voeltzkowia petiti* **Angel,** 1924
Angel's Petite Gecko *Paragehyra petiti* Angel, 1929
Angel's Gecko *Geckolepis petiti* Angel, 1942

Louis Petit (1856–1943) was in the Congo and Angola (1876–1884). He was primarily an ornithologist and should not be confused with his father, also an ornithologist and with the same name. Two birds are named after him.

Petronella

Kukri Snake sp. *Oligodon petronellae* **Roux,** 1914

See **De Rooij.**

Petter, F.

Petter's Short Skink *Pygomeles petteri* **Pasteur** and **Paulian,** 1962

Professor Dr. Francis Petter works at Laboratoire de Zoologie, Mammifères et Oiseaux, Muséum National d'Histoire Naturelle, Paris. He was elected an honorary member of the American Society of Mammalogists (1987). He has been on collecting trips to several African countries including Ivory Coast, where he collected a number of specimens of *Pachybolus laminaria* (a millipede); microbiologists found parasites on them that appear to be the source of Ebola fever. He co-wrote with his wife, Dominique, a microbiologist, *Les félins* 1993. Three mammals are named after him.

Petter, J.-J.

Petter's Chameleon *Furcifer petteri* **Brygoo** and **Domergue,** 1966

Professor Jean-Jacques Petter (1927–2002) was a French zoologist, primatologist, and population biologist who first qualified as a physician. He was Professor of Ecology and Ethnology at Muséum National d'Histoire Naturelle, Paris, and Director of the Vincennes Zoo. He wrote extensively on Madagascan mammals in general and lemurs in particular, and was awarded the World Wildlife Fund Gold Medal (1980) for his conservation work there.

Peyrieras

Peyrieras' Chameleon *Calumma peyrierasi* **Brygoo, Blanc,** and **Domergue,** 1974 [Alt. Brygoo's Chameleon]
Peyrieras' Dwarf Chameleon *Brookesia peyrierasi* Brygoo and Domergue, 1974

Dr. André Peyrieras is a French biologist who became one of Madagascar's most respected naturalists. He has discovered over 3,000 insects new to science. He runs Mandraka Nature Farm, which contains his private collection and where a wide variety of Madagascar's rare reptiles, frogs, and insects are bred in captivity. A mammal is named after him.

Pfeffer

Pfeffer's Chameleon *Chamaeleo pfefferi* **Tornier,** 1900
Pfeffer's Reed Snake *Calamaria pfefferi* **Stejneger,** 1901
Pfeffer's Worm Lizard *Leposternon pfefferi* **Werner,** 1910

Dr. Georg Johann Pfeffer (1854–1931) was a zoologist who became Curator, Naturhistorisches Museum zu Hamburg.

Phayre

Phayre's Tortoise *Manouria emys phayrei* **Blyth,** 1854 [Alt. Burmese Mountain Tortoise, Black Mountain Tortoise]

Lieutenant General Sir Arthur Purves Phayre (1812–1885) was Commissioner in Burma (now Myanmar) (1862–1867) and Governor of Mauritius (1871–1878). He wrote *History of Burmah* (1883). Three birds and three mammals are named after him.

Phelps

Phelps' Sun Tegu *Cercosaura phelpsorum* **Lancini,** 1968 [Syn. *Euspondylus phelpsorum*]

The Phelps family had their own specially equipped yacht, *Ornis*, in which they made 49 trips to the Caribbean islands in addition to many trips to the hinterland of Venezuela. William H. Phelps (1875–1965) was an American-born Venezuelan ornithologist. He first visited Venezuela as a Harvard student (1896). The Phelps

Ornithological Collection, Caracas, was built up by his son, William "Billy" H. Phelps Jr. (1902–1988), and Billy's wife, Kathleen. A bird is named after William Sr.

Phelsum

Day Gecko genus *Phelsuma* **Gray,** 1825

Dr. Murk van Phelsum (1730–1779) was a physician in general practice in Holland and was also a helminthologist (studying parasitic worms and their effects on their hosts).

Philby

Philby's Spiny-tailed Lizard *Uromastyx ornata philbyi* **Parker,** 1938

Philby's Arabian Skink *Scincus philbyi* Schmidt, 1941 [Junior syn. of *S. mitranus* Anderson, 1871]

Harry St. John Bridger Philby (1885–1960) was a noted Arabist and an explorer who was among the first Europeans to travel in the southern Arabian provinces. He collected the holotype of the skink. He resigned from the Foreign Service in 1925, then was an adviser for 30 years to King ibn Saud of Saudi Arabia. Philby became a Muslim, renaming himself Hajj Abdullah. His son was Kim Philby, who infamously spied for the Soviet Union in Britain. He wrote *Heart of Arabia* (1923). A bird is named after him.

Philippen

Philippen's Stripeneck Turtle *Ocadia philippeni* **McCord** and **Iverson,** 1992

Dr. Hans-Dieter Philippen (b. 1957) is a German herpetologist who specializes in turtles.

Philippi

Philippi's Tree Iguana *Liolaemus bisignatus* Philippi, 1860

Philippi's Steppe Iguana *Urostrophus torquatus* Philippi, 1861 [Syn. *Pristidactylus torquatus*]

Philippi's Snail Sucker *Tropidodipsas philippii* **Jan,** 1863

Rodolfo Amando Philippi (1808–1904) was a German-Chilean zoologist and paleontologist. He left Germany as a young man, ostensibly because he thought he was gravely ill and wanted to end his days in a mild Mediterranean climate. However, the decision to leave may have been influenced by his political activities on the losing side of the movement to unify Germany. He recovered his health and was invited to go to Chile by his brother, who worked for the Chilean government. Rodolfo became Director, Museo Nacional de Historia Natural, Santiago de Chile (1853–1883), and is noted for having described thousands of plant species. A species of fur seal is named after him.

Phillips, B.

Burrowing Snake sp. *Apostolepis phillipsae* **Harvey,** 1999

Barbara Phillips collected the holotype (1998).

Phillips, E. L.

Phillips' Agama *Acanthocercus phillipsii* **Boulenger,** 1895

Phillips' Rock Gecko *Pristurus phillipsii* Boulenger, 1895

Phillips' Orangetail Lizard *Philochortus phillipsii* Boulenger, 1898

Ethelbert Lort Phillips (1857–ca. 1926) was a British traveler, hunter, and collector who shot big game in many parts of the world. He traveled in East Africa (1884–1895) and explored parts of Somaliland (now Somalia). He is remembered in Norway as the man who developed an estate called Vangshaugen, Lake Storvatnet, where in the late 19th century he planted a garden of rhododendrons—then virtually unknown in Norway. He became Vice President, Zoological Society, London. Two mammals and a bird are named after him.

Phillips, J. C.

Phillips' Mole Viper *Atractaspis phillipsi* **Barbour,** 1913

Phillips' Blind Snake *Leptotyphlops phillipsi* Barbour, 1914

Dr. John Charles Phillips (1876–1938) was an American physician and traveler. He commanded a field hospital during WW1 before and after which he traveled widely making zoological collections for Harvard's Museum of Comparative Zoology. He co-authored *Extinct and Vanishing Mammals of the Old and New World* (in sections 1942–1945). An antelope is named after him.

Phillips, J. S.

Phillips' Sand Snake *Psammophis phillipsi* **Hallowell,** 1844 [Alt. West African Olive Sand Racer]

John S. Phillips. All we know about him is Hallowell's statement that the snake is "named after my friend John S. Phillips, Esq."

Phillips, W. W. A.

Phillips' Earth Snake *Uropeltis phillipsi* **Nicholls,** 1929

Major William Watt Addison Phillips (1892–1981) was a tea and rubber planter and naturalist in Ceylon (now Sri Lanka). While a prisoner-of-war in Turkey during WW1 he developed an interest in zoology. He was Secretary and subsequently Chairman of the Ceylon Bird Club. He also collected herpetofauna. He wrote *Check List of the Birds of Ceylon* (1952). He returned to England in 1956.

Phipson

Phipson's Shieldtail *Uropeltis phipsonii* Mason, 1888

Herbert Musgrave Phipson (1849–1936) was one of the founders of the Bombay Natural History Society (1883) and Honorary Secretary (1886–1906). He lent space in his wine shop, Phipson and Co., Forbes Street, as office accommodation for the society. He married Mary Edith Pechey (1889), a pioneering Victorian physician who was Senior Officer, Cama Hospital for Women and Children, Bombay.

Pianka

Pianka's Ctenotus *Ctenotus piankai* **Storr,** 1969 [Alt. Course Sand Ctenotus]

Professor Dr. Eric R. Pianka (b. 1939) is a herpetologist who spent seven years in Australia, most of it in the desert, where he studied the *Ctenotus* genus in great detail. He was awarded his bachelor's degree by Carleton College, Minnesota (1960), and his doctorate by the University of Washington, Seattle (1965). He works at the University of Texas, being Assistant Professor (1968). He lost part of his left leg and part of a finger as a result of a bazooka blast in his garden (1952). He is proud of the fact that a tapeworm parasite has been named after him. He co-edited *Varanoid Lizards of the World* (2004).

Picado

Picado's Pit-viper *Atropoides picadoi* **Dunn,** 1939

Dr. Clodomiro Picado Twight (1887–1944) was a leading Costa Rican botanist, zoologist, and toxicologist. The Sorbonne awarded his doctorate (1913). He was internationally recognized for his studies of venomous snakes and his development of antivenins. The scientific research institute Instituto Clodomiro Picado, Costa Rica, is named after him.

Pickering

Pickering's Garter Snake *Thamnophis sirtalis pickeringii* **Baird** and **Girard,** 1853 [Alt. Puget Sound Garter Snake]

Dr. Charles F. Pickering (1805–1878) was an ornithologist and ethnologist who was a friend of Audubon. He explored the White Mountains, New Hampshire (1825). He qualified as a physician at Harvard (1826) and set up practice in Philadelphia (1827). He was Librarian of the Philadelphia Academy of Natural Sciences (1828–1833) and, thereafter, Curator. He was on the United States Exploring Expedition (1838–1842) as naturalist. He helped Holbrook with the publication of *North American Herpetology* (1842). He went abroad to study ethnology (1844), traveling from Egypt to Zanzibar and then on to India. He wrote *The Races of Man and Their Geographical Distribution* (1847). A bird is named after him.

Pictet

Jan's Snake *Elapotinus picteti* **Jan,** 1862

Professor François Jules Pictet de la Rive (1809–1872) was a Swiss zoologist and paleontologist. He studied at Muséum National d'Histoire Naturelle, Paris, under Cuvier (1829–1830) before becoming Professor of Zoology, Université de Genève (1835). He retired from teaching (1859) to devote his time to Museum d'Histoire Naturelle, Geneva. Jan, in his description of the snake, wrote that the holotype, collected in 1840, came from Museum d'Histoire Naturelle, Geneva. We surmise that Pictet may have lent Jan the specimen, though Jan gives no explicit etymology.

Pietschmann

Cork Bark Leaftail Gecko *Uroplatus pietschmanni* Böhle and Schönecker, 2004 [Alt. Spiny Leaftail Gecko]

Jürgen Pietschmann (1949–2005) was a German gecko breeder.

Pilona

Stuart's Graceful Brown Snake *Rhadinaea pilonaorum* **Stuart,** 1954

Antonio and Marta Piloña lived on, and Antonio was administrator of, Finca La Gloria, a coffee plantation in Guatemala where Stuart had been a guest. Antonio obtained the holotype.

Pilsbry

Spiny Lizard sp. *Sceloporus pilsbryi* **Dunn,** 1936 [Junior syn. of *S. grammicus disparilis* **Stejneger,** 1916]

Pilsbry's Dwarf Boa *Tropidophis pilsbryi* **Bailey,** 1937

Dr. Henry Augustus Pilsbry (1862–1957) was a conchologist and malacologist, an expert on barnacles, and a leading light in the Philadelphia Academy of Sciences. He was awarded his bachelor's degree (1882) and an honorary doctorate (1899) at the University of Iowa. He worked as a newspaper reporter in Iowa (1883–1887). He was a conservator of conchology at Philadelphia (1888–1895). He edited *Manual of Conchology* (1889–1932), and he founded *Nautilus* (1889), a journal he edited until his death. He collected in much of the Americas and in Australia, Japan, and the Pacific islands.

Pinchot

Crab Cay Anole *Anolis pinchoti* **Cochran,** 1931

Gifford Pinchot (1865–1946) went to Yale (1885) but found no suitable course there, and after taking an arts degree, he went to Nancy, France, to study forestry. After returning to the USA he worked as resident forester for the Vanderbilt family. He was in government service (1898–1910) as Chief of the Division of Forestry, Department of the Interior, transferring as first head of a new Forestry Service (1905). During his administration the number of national forests in the USA grew enormously. President Taft sacked him (1910) in a controversy over coal claims in Alaska. He entered politics and was Governor of Pennsylvania (1923–1927 and 1931–1935). He was in the Caribbean (1929) on an epic voyage he made with his son, on board the *Mary Pinchot*, from New York to the Society Islands via Key West, Colombia, the Galapagos, and the Marquesas.

Piraja

Piraja's Lancehead *Bothrops pirajai* **Amaral,** 1923

Dr. Manuel Augusto Piraja da Silva (1873–1961) was a Brazilian physician and medical researcher. He qualified as a physician at the College of Medicine, Bahia (1896), becoming an Assistant Professor there (1902). His main early work was in parasitology. He was in Europe for a time (1909–1912), studying at tropical medicine institutes in Hamburg and Paris, then became Professor of Natural History and Parasitology at the Gymnasium, Bahia (1914–1935).

Pitman

Pitman's Shovelsnout Snake *Prosymna pitmani* **Battersby,** 1951

Lieutenant Colonel Charles Robert Senhouse Pitman (1890–1975) was an ornithologist and a herpetologist. He was an officer in the Indian army (1909–1921), serving with distinction in WW1 in France, Egypt, Mesopotamia (now Iraq), and Palestine. He settled in Kenya and farmed until becoming a game warden in Uganda (1925). He was in Northern Rhodesia (now Zambia) (1931–1932) and was an intelligence officer (1941–1946). He retired to England (1951). He wrote *A Guide to the Snakes of Uganda* (1938). A mammal is named after him.

Plate

Braided Tree Iguana *Liolaemus platei* **Werner,** 1898

Dr. Ludwig Hermann Plate (1862–1937) was a German zoologist and geneticist. He studied under Haeckel at Friedrich-Schiller-Universität Jena, which awarded his doctorate (1886). He qualified in zoology at Philipps-Universität Marburg (1888), then taught at the Veterinary High School in Berlin (1898–1905) and at the Agricultural College (1905–1908) as Professor of Zoology. He was Haeckel's successor as Professor at Jena (1909–1935). He became embroiled in an unpleasant case when he accused Haeckel and his circle of slandering him. He was a convinced Darwinist. He was also a virulent anti-Semite.

Plee

Anguilla Bank Ameiva *Ameiva pleei* **Duméril** and **Bibron,** 1839

Puerto Rican Galliwasp *Diploglossus pleii* Duméril and Bibron, 1839

Plee's Tropical Racer *Mastigodryas pleei* Duméril, Bibron, and **Duméril,** 1854 [Alt. Plee's Forest Racer; Syn. *Dryadophis pleei*]

Martinique Spectacled Tegu *Gymnophthalamus pleei* **Bocourt,** 1881

August Plée (1787–1825) was a collector for Muséum National d'Histoire Naturelle, Paris, in Colombia and the Antilles. Primarily a botanist, he went to Canada and the USA (1819) and from there to Puerto Rico (1822). He traveled to the Virgin Islands and Leeward Islands (1823), finally arriving in Martinique where "he contracted some tropical malady and died."

Pleske

Pleske's Racerunner *Eremias pleskei* **Nikolsky,** 1905 [Syn. *Rhabderemias pleskei*]

Fedor Dimitrievich Pleske (1858–1932) was a zoologist, geographer, and ethnographer. From childhood he collected birds and insects in European Russia. He graduated from St. Petersburg University (1882). He became a Fellow of the Russian Imperial Academy of Science, St. Petersburg, when appointed (1886) as Scientific Keeper, Department of Ornithology, Zoological Museum, Imperial Academy of Sciences, and later became Director (1892–1896). From 1918 he was an active member of the USSR Zoological Museum. He wrote *Ornithological Fauna of Imperial Russia* (1891). Two birds are named after him.

Plowes

Plowes' Legless Skink *Typhlosaurus plowesi* **Fitzsimons,** 1943

Darrel Charles Herbert Plowes (b. 1925) is a citizen of Zimbabwe. He has over 50 years of field experience in southern Africa. He was an agricultural director and is now retired, but he remains active in photography of the natural world and is still collecting plants.

Plummer

Dwarf Gecko sp. *Sphaerodactylus plummeri* **Thomas and Hedges,** 1992

Dr. Michael V. Plummer (b. 1945) is a biologist and zoologist at Harding University, Arkansas, where he gained his bachelor's degree (1967). He earned his master's at Utah State University (1969) and his doctorate at the University of Kansas (1976). He was a Psychiatric Technician in the U.S. Army (1969–1971), avoiding, as he put it, the "herps of Vietnam" in favor of those of California and Texas. He returned to Harding University in 1971 as an Instructor in Biology, becoming Professor of Biology in 1985.

Poeppig

Basin Ground Snake *Atractus poeppigi* **Jan,** 1862

Professor Eduard Friedrich Poeppig (1798–1868) was a German naturalist and collector. He studied medicine and natural science at Universität Leipzig, leaving to undertake an expedition to Cuba and the USA. He was on an expedition to Brazil and Peru (1829–1832). When he returned to Germany he became Professor of Zoology, Universität Leipzig. He wrote *Reise nach Chili, Peru, und auf dem Amazonen-Flusse* (1835). A mammal and an amphibian are named after him.

Poilane

Skink sp. *Leptoseps poilani* **Bourret,** 1937

Laotian False Bloodsucker *Pseudocalotes poilani* Bourret, 1939

Eugène Poilane (1887–1964) was a French botanist who was in Cochin-China (now a part of Vietnam) and neighboring areas of French Indochina from 1909 until he was assassinated by Viet Cong troops. He started a coffee plantation (1918). He was notable for having fathered 10 children, 5 of them after the age of 60.

Poinsett

Crevice Spiny Lizard *Sceloporus poinsettii* **Baird** and **Girard,** 1852

Dr. Joel Robert Poinsett (1779–1851) was an American physician and politician but is best remembered as the botanist after whom the popular plant *Poinsettia* is named. He traveled in Europe, Russia, and Asia in the early 19th century but was back in the USA by 1809. He was a "special agent" in South America (1810–1814), investigating the prospects of colonial revolutions against Spanish rule, so he sounds like a pre-CIA spy, working under diplomatic cover. He was a member of the House of Representatives (1821–1826) and the first U.S. Minister to Mexico (1822–1830). He was Secretary of War (1837–1841) during Martin van Buren's presidency. He was one of the founders (1840) of the National Institute for the Promotion of Science and the Useful Arts, which evolved into the Smithsonian.

Pollen

Madagascar Coastal Skink *Amphiglossus polleni* **Grandidier,** 1869

Mayotte Chameleon *Furcifer polleni* **Peters,** 1874

François P. L. Pollen (1842–1886) was a Dutch naturalist who collected in Madagascar (1863–1866). From his first name, the French colony where he collected, and the language in which he published, one might think he was French, but he wasn't. He wrote *Récherches sur la faune de Madagascar et de ses dépendances—d'après les déscouvertes de F. P. L. Pollen et D. C. van Dam* (1868). A bird is named after him.

Polyphemus

Gopher Tortoise *Gopherus polyphemus* **Daudin,** 1802

Polyphemus was a mythical cave-dwelling giant, one of the Cyclopes.

Poncelet

Poncelet's Helmet Skink *Tribolonotus ponceleti* **Kinghorn,** 1937

False Poncelet's Helmet Skink *Tribolonotus pseudoponceleti* **Greer** and **Parker,** 1968

Rev. Jean-Baptiste Poncelet (1884–1958) headed the Catholic Mission at Buin, Bougainville, Solomon Islands, in the 1930s. He also collected natural history specimens for V. Danis, who worked at Muséum National d'Histoire Naturelle, Paris. A mammal is named after him.

Poole

Shark Bay Sea-snake *Aipysurus pooleorum* **L. A. Smith,** 1974

Named after W. and W. Poole, Fremantle fishermen who collected many sea-snake specimens, including the holotype of this species.

Pootipong

Pootipong's Ground Skink *Scincella pootipongi* **Taylor,** 1962 [Perhaps synonymous with *S. reevesii* **Gray,** 1838]

Professor M. R. Pootipongse Nupartpat Varavudhi (d. 2005) was a zoologist and biologist with the Department of Biology, Chulalongkorn University, Thailand, who wrote on medical zoological matters. He was of great help to Taylor, who said that Pootipongse "accompanied me on numerous journeys into the hinterlands of Thailand."

Pope

Pope's Writhing Skink *Lygosoma popae* **Shreve,** 1940

The common name is a misnomer, as the skink is named after Mount Popa in Burma (now Myanmar).

Pope, C. H.

Pope's Keelback *Amphiesma popei* **Schmidt,** 1925

Pope's Bamboo Viper *Trimeresurus popeiorum* **M. A. Smith,** 1937 [Alt. Pope's Pit-viper; Syn. *Popeia popeiorum*]

Pope's Emo Skink *Emoia popei* **Brown,** 1953

Pope's Skink *Eumeces popei* **Hikida,** 1989

Clifford Hillhouse Pope (1899–1974) took a bachelor's degree at the University of Virginia (1921). He was in

China (1921–1926) as an employee of the American Museum of Natural History, New York, becoming a fluent speaker of Chinese after having collected in Fujian Province (1925–1926). He was Assistant Curator of Herpetology, American Museum (1928–1935), and was sacked after a disagreement with the Director, Gladwyn Kingsley Noble (q.v.).Unemployed during the Great Depression, he wrote a number of popular books on herpetology to support himself. He was Curator of Reptiles, Field Museum (1941–1954). He took early retirement and went to live in California, where he continued to write. *Eumeces popei* was eventually described and named over 60 years after he collected the holotype. *Trimeresurus popeiorum* is named after Clifford and his wife, Sarah H. Pope (q.v.), who was also a herpetologist. They worked together and co-wrote "A Study of the Green Pit-vipers of Southeastern Asia and Malaysia, Commonly Identified as *Trimeresurus gramineus*" (1933). The viper is sometimes placed in the genus *Popeia*, also named after Pope.

Pope, S. H.

Pope's Bamboo Viper *Trimeresurus popeiorum* **M. A. Smith,** 1937 [Alt. Pope's Pit-viper; Syn. *Popeia popeiorum*]

Sarah Haydoc Pope, née Davis, was a herpetologist. See **Pope, C. H.**

Popov

Saudi Rock Gecko *Pristurus popovi* **Arnold,** 1982

Dr. George B. Popov (1922–1998) was an entomologist who was born of Russian parents in Iran. He was recruited to the Middle East Anti-Locust Unit (1943). He worked for the rest of his life in research on locusts, visiting nearly every desert in the world where locusts breed. He visited Yemen (1948) and worked for the World Food Programme.

Porras

White-tailed Hognose Pit-viper *Porthidium porrasi* **Lamar,** 2003

Louis William Porras (b. 1948) is a Costa Rican herpetologist, now based in Florida. He went into publishing (1995) with *Fauna Magazine* and is President of Eagle Mountain Publishing. He wrote (and published) "Island Boa Constrictors *(Boa constrictor)*" (1999).

Portschinski

Portschinski's Rock Lizard *Darevskia portschinskii* Kessler, 1878

Professor Josef Aloizievitsch Portschinski (1848–1916) was a Russian entomologist. He was associated with the entomological service of the Ministry of Agriculture, St. Petersburg. He collected the holotype near Tiflis, Georgia, while on an entomological collecting expedition there. He wrote "Diptera Europae et Asiatica nova aut minus cognita (cum notis biologicis)" (1884).

Posadas

Posadas' Graceful Brown Snake *Rhadinaea posadasi* **Slevin,** 1936

Juan Zenon Posadas had a coffee plantation on the slopes of Volcan Zunil (Guatemala). Slevin was his guest there and collected reptiles and amphibians in the area.

Potanin

Skink sp. *Scincella potanini* **Günther,** 1896

Shansi Toadhead Agama *Phrynocephalus potanini* Bedriaga, 1909

Grigory Nikolayaevich Potanin (1835–1920) was a Russian explorer of Inner Asia. He served in a Cossack regiment in Siberia (1850s) and was a supporter of Siberian separatism. He studied Physics in St. Petersburg (1858–1861). He was arrested and imprisoned (1861) for participating in a student demonstration. Expelled from the university, on release he returned to Siberia to work as a publisher. Arrested again (1867) for political activities, he was sentenced to three years in prison followed by five years of hard labor. He led expeditions to Mongolia (1876–1877) and Northern China (1884–1886). He was a founder of Tomsk State University (1889). He supported the 1905 Revolution and was arrested. He was chairman of Siberia's short-lived Provisional Council (1917–1918). He wrote *The Tangut-Tibetan Borderlands of China and Central Mongolia* (1893). An asteroid is named after him.

Potsch

Ground Snake sp. *Atractus potschi* Fernandes, 1995

Dr. Sergio Potsch de Carvalho e Silva is a Brazilian herpetologist in the Department of Vertebrates, Museu Nacional, Rio de Janeiro.

Prado

Prado's Ground Snake *Atractus manizalesensis* Prado, 1940

Prado's Coastal House Snake *Thamnodynastes rutilus* Prado, 1942

Prado's Lancehead *Bothrops pradoi* **Hoge,** 1948 [Junior syn. of *Bothrops leucurus* Wagler, 1824, according to some]

Professor Dr. Alcides Prado was a Brazilian parasitologist, entomologist, and herpetologist who worked at Instituto Butantan, São Paulo (1930–1949). He may have worked in Mexico in the early 1940s, as he began publishing in Spanish in Mexican scientific journals at that time.

Prashad

Bombay Leaf-toed Gecko *Hemidactylus prashadi* **M. A. Smith,** 1935

Dr. Baini Prashad (1894–1969) was an Indian zoologist and malacologist who was Director of the Zoological Survey of India, Indian Museum, Calcutta. He was the senior collector of the type series of this gecko.

Pratt

Pratt's Snail-eater *Dipsas pratti* **Boulenger,** 1897

Forest Skink sp. *Sphenomorphus pratti* Boulenger, 1903

Antwerp Edgar Pratt (ca. 1850–ca. 1920) was an explorer and naturalist, as were his sons, Felix and Charles. Antwerp and Felix collected in the Colombian Andes in the 1890s and later were in New Guinea. Antwerp wrote *To the Snows of Tibet through China* (1892).

Preston

Preston's Torquate Lizard *Sceloporus torquatus mikeprestoni* **H. M. Smith** and **Alvarez,** 1976

Dr. Michael J. Preston is Professor of English at the University of Colorado, where he teaches Middle English and early Renaissance literature. He took his bachelor's degree at Gonzaga University (1965), and his master's (1972) and doctorate (1975) at the University of Colorado. He gave the describers of this lizard vital aid in processing and accessing data on Mexican herpetological literature.

Pretre

Rio Grande Worm Lizard *Amphisbaena pretrei* **Duméril** and **Bibron,** 1839

Jean Gabriel Prêtre (1800–1840) was a highly regarded French artist who was employed by Muséum National d'Histoire Naturelle, Paris. He illustrated a number of classic natural history publications. Four birds are named after him.

Preuss

Preuss' Forest Snake *Toxicocalamus preussi* **Sternfeld,** 1913

Professor Paul Preuss (1861–1926), although born in Poland, was a German naturalist, botanist, and horticulturist. He collected in West Africa (1886–1898 and 1910) and in New Guinea (1903). He was a member of Zintgraff's military expedition to explore the hinterland of Cameroon, then a German colony (1888–1891). While storming a native village the troop commander was killed and the second-in-command severely wounded. Preuss took charge and led the remaining troops back to the coast. He constructed the botanical gardens of Victoria (now Limbe), Cameroon (1901). Three birds and a monkey are named after him.

Prevost

Gerard's Water Snake *Gerarda prevostiana* **Eydoux** and **Gervais,** 1822 [Alt. Cat-eyed Water Snake]

Florent Prévost (or Prevot) (d. 1870) was a French artist and writer who illustrated works by Temminck, Bonaparte, and Buffon. He was Assistant Naturalist at Muséum National d'Histoire Naturelle, Paris. He wrote *Histoire naturelle des oiseaux d'Europe* (1864). A squirrel and a bird are named after him.

Price

Price's Rattlesnake *Crotalus pricei* **Van Denburgh,** 1895 [Alt. Twin-spotted Rattlesnake]

William Wightman Price (1871–1922) appears to have been a wild child, as he ran away from home at age 8 and for a few days lived with a band of Native Americans in Wisconsin. The result was that his father moved him to California (1880). When his father died (1885), William departed for Arizona and spent 18 months exploring and living rough. He returned to California (1887) and entered Oakland High School, paying his fees by selling bird and mammal skins. He collected in California, Nevada, and Arizona for Stanford (1892–1895). He founded a camp for boys (1897) near Lake Tahoe and named it after Agassiz. He took a bachelor's degree in economics at Stanford (1898) and a master's (1899). He worked for the Red Cross as Assistant Field Director in charge of the Palo Alto Base Hospital (1917–1919).

Prince Ruspoli

Prince Ruspoli's Gecko *Hemidactylus ruspolii* **Boulenger,** 1896 [Alt. Farm Leaf-toed Gecko]

Prince Ruspoli's Shovel-snout *Prosymna ruspolii* Boulenger, 1896

Prince Eugenio Ruspoli (1866–1893) was an Italian explorer. His family were eminent Roman aristocrats in the 19th century who intermarried with the Bonaparte dynasty. He traveled in Ethiopia (1891–1893), where he was killed in "an encounter with an elephant" that he had wounded. A bird is named after him. See also **Ruspoli.**

Pritchard

Pritchard's Snake-necked Turtle *Chelodina pritchardi* Rhodin, 1994

Pritchard's Pond Turtle *Mauremys pritchardi* **McCord,** 1997

Dr. Peter Charles Howard Pritchard (b. 1943) is a British herpetologist and conservationist who founded, and is head of, the privately funded Chelonian Research Institute. He read chemistry at Oxford and then switched to zoology, taking his doctorate at the University of Florida (1969). He lives and works in Florida, where he has a private collection of chelonians. He co-wrote *The Turtles of Venezuela* (1984).

Procter

Loveridge's Forest Snake *Geodipsas procterae* **Loveridge,** 1922

Tortoise sp. *Testudo procterae* Loveridge, 1923 [Junior syn. of *Kinixys spekii* **Gray,** 1863]

Joan Beauchamp Procter (1897–1931) was a zoologist and herpetologist at the Natural History Museum, London, which she joined in 1917 as an assistant to Boulenger (q.v.) whom she succeeded in 1920. She became Curator of Reptiles, London Zoo (1923), where she had a "pet" Komodo Dragon called Sumbawa that used to accompany her on her strolls around the zoo. She wrote "On a Living Komodo Dragon, *Varanus komodensis* Ouwens, Exhibited at the Scientific Meeting, October 23rd, 1928."

Pronk

Pronk's Day Gecko *Phelsuma pronki* **Seipp,** 1994

Olaf Pronk is a Dutch photographer, naturalist, and animal dealer who is now a resident of Madagascar, where he breeds reptiles for sale. He also has a nursery in Antananarivo where he grows Malagasy orchids. He co-wrote and also illustrated "The Ghost Geckos of Madagascar: A Further Revision of the Malagasy Leaf-toed Geckos (Reptilia, Squamata, Gekkonidae)" (1998).

Pruthi

Skink sp. *Lygosoma pruthi* **Sharma,** 1977

Dr. Hem Singh Pruthi (1897–1969) was an Indian entomologist who was one of the founders of the Division of Entomology of the Imperial Indian Agricultural Research Institute established in 1905. He became Imperial Entomologist to the Viceroy (1935–1947). The skink holotype was collected (1929) by Pruthi but named long afterward.

Pryer

Pryer's Keelback *Amphiesma pryeri* **Boulenger,** 1887

Henry James Stovin Pryer (1850–1888) was a lepidopterist who collected specimens in Japan in the late 19th century, including the holotype of this snake. His collection was later donated to the British Museum. He wrote *Rhopalocera niponica* (1886). A bird is named after him.

Przewalski

Gobi Racerunner *Eremias przewalskii* **Strauch,** 1876

Przewalski's Toad-headed Agama *Phrynocephalus przewalskii* Strauch, 1876

Przewalski's Wonder Gecko *Teratoscincus przewalskii* Strauch, 1887

Xinjiang Even-fingered Gecko *Alsophylax przewalskii* Strauch, 1887

Przewalski's Dwarf Skink *Scincella przewalskii* **Bedriaga,** 1912

General Nikolai Mikhailovitch Prjevalsky (1839–1888) was a Russian Cossack naturalist who explored Central Asia. Undoubtedly one of the greatest explorers the world has ever seen, he made five major expeditions: one to the Russian Far East and the others to Mongolia and northern China. He is best known for having discovered the Mongolian wild horse, which was named after him. There are a least half a dozen different spellings of his name, but he signed himself as Prjevalsky (pronounced "she-val-ski"). He died of typhus while preparing for a sixth expedition. Tsar Alexander II decreed that the town where he died, Karakol, should immediately have its name changed to Prjevalsk. Five mammals, including Przewalski's Horse, and seven birds are named after him. He wrote *Mongolia, and the Tangut Country* (1875).

Puccioni

Zanzibar Leaf-toed Gecko *Hemidactylus puccionii* Calabresi, 1927

Professor Dr. Nello Puccioni (1881–1937) was a physician, paleontologist, and naturalist in Florence who collected mainly ichthyological specimens (1924).

Puelche

Tree Iguana sp. *Liolaemus puelche* Avila, Morando, Fulvio Perez, and Sites, 2007

Named after the Puelche, a group of aboriginal people living in the mountainous regions of Mendoza Province, Chile.

Puiseux

Israeli Fan-fingered Gecko *Ptyodactylus puiseuxi* Boutan, 1893

Victor Alexandre Puiseux (1820–1883) was an astronomer. On graduating (1837) he entered École Normale Supérieure, where he was awarded his doctorate (1841). He was Professor of Mathematics, Collège Royal de Rennes (1841–1844), and held the same position at Université de Franche-Comté, Besançon (1844–1849). He was at Bureau de Longitudes (1868–1872). He was also a keen mountaineer and was the first to scale the Alpine peak that is now named after him.

Purcell

Thick-toed Gecko sp. *Pachydactylus purcelli* **Boulenger,** 1910

Dr. William Frederick Purcell (1866–1919) was a South African entomologist and naturalist particularly interested in spiders and scorpions. He was born in England but was educated in part in Germany. The University of Berlin awarded his doctorate (1895). He worked at the South African Museum (1896–1905) and collected the gecko holotype.

Putjata

Toadhead Agama sp. *Phrynocephalus putjatae* **Bedriaga,** 1909

Dmitrji Wassiljewitsch Putjata was a Russian botanist who led an expedition to Chingan (1891). He was a member of the staff at the Imperial Institute of Sciences, St. Petersburg, and was one of those who had written up the results of Prjevalsky's expeditions, during one of which the agama holotype was collected. Putjata, like many 19th-century explorers, was probably a soldier, as after his expedition he wrote a paper on the Chinese army, a translation of which was published in Vienna (1895). A number of plants are named after him.

Putnam

Ridgehead Snake *Manolepis putnami* **Jan,** 1863

Frederic Ward Putnam (1839–1915) was an anthropologist and naturalist. He went to Harvard (1856) and studied under Louis Agassiz. He worked as Agassiz's assistant (1857–1864) and became immersed in ichthyology and herpetology. He worked in a variety of museums and institutions (1864–1875). He was Curator, Peabody Museum of American Archaeology and Ethnology, Harvard (1875–1909). He was made Peabody Professor (1888), then Emeritus Curator and Emeritus Professor (1909–1915).

Pyburn

Pyburn's Earth Snake *Geophis pyburni* **Campbell** and **Murphy,** 1977

Pyburn's Tropical Forest Snake *Umbrivaga pyburni* Markezich and **Dixon,** 1979

Pyburn's Bachia *Bachia pyburni* **Kizirian** and Mc-Diarmid, 1998

Emeritus Professor Dr. William Frank "Billy" Pyburn (1927–2007) founded in 1956 the specimen collection at the University of Texas, Arlington, that became the Amphibian and Reptile Diversity Research Center. His trips to the jungles of Colombia during the 1970s were not for the fainthearted, as he would climb into his motor vehicle in Arlington and drive all the way to Colombia. He was primarily a specialist in the biology of amphibians, and at least one is named after him.

Pylzow

Pylzow's Toadhead Agama *Phrynocephalus pylzowi* **Bedriaga,** 1909

Lieutenant Mikhail A. Pylzow (b. 1850) was a Russian who was one of Przewalski's traveling companions in the exploration of Central Asia in the early 1870s. A bird is named after him.

Pym

Pym's Burrowing Snake *Apostolepis pymi* **Boulenger,** 1903

J. Pym owned the single male specimen upon which Boulenger based his description. Pym's collection was acquired by the Natural History Museum, London, after his death.

Pyrrhus

Desert Death Adder *Acanthophis pyrrhus* **Boulenger,** 1898

Pyrrhus (318–272 B.C.) was a King in ancient Greece, but the snake is so named for the literal meaning of the word, which is "flamelike."

Q

Quadras

Quadras' Flying Lizard *Draco quadrasi* **Boettger,** 1893
José Florencio Quadras was a Spanish forester who made a collection of Philippine molluscs that is now in the Field Museum. He worked closely with Möllendorff (q.v.) when he visited the Philippines. A number of plants are named after him.

Quedenfeldt

Gecko genus *Quedenfeldtia* **Boettger,** 1883
Max Quedenfeldt (1851–1891) was a German naturalist and entomologist who mainly worked on beetles.

Quilmes

Tree Iguana sp. *Liolaemus quilmes* **Etheridge** 1993
Named after the Quilmes, an indigenous tribe of northwestern Argentina

Quim

Quim's Mussurana *Clelia quimi* Franco, Marques, and Puorto, 1997
Joaquim "Quim" Cavalheiro is a Laboratory Assistant, Laboratório de Herpetologia do Instituto Butantan. The original description says that "his knowledge, gathered diligently over the years, has greatly contributed to the studies of snakes at this Institute."

Quiroga

Quiroga's Burrowing Snake *Apostolepis quirogai* Giraudo and **Scrocchi,** 1998
Horacio Quiroga (1878–1937) was, according to the etymology, "a famous Uruguayan writer" who also collected snakes in the early 1900s. After a sojourn in Paris, he returned to teach in Argentine schools and tour the wilds of Argentina as a photographer. He settled (1904) in Chaco Province, tried and failed to grow cotton, so returned to teaching in Buenos Aires. He was registrar in the San Ignacio district of Misiones (1909–1915). He returned to Buenos Aires and worked (1919–1925) in the Uruguayan consulate, then went again to San Ignacio, where he became Honorary Uruguayan Consul (1935). Now regarded as one of the greatest Uruguayan writers, he wrote works dealing with anthropomorphic, intelligent animals, a jungle that seems to be alive, fate, and bizarre coincidences set against a backdrop of despair—which is perhaps understandable, given his famously unhappy life. His father was killed in an accidental shooting when Quiroga was young. His stepfather committed suicide (1900). His first wife poisoned herself (1915), and his second marriage failed. He found he had cancer and swallowed cyanide. Later both his children also committed suicide.

Quoy

Quoy's Australian Skink *Eulamprus quoyii* **Duméril** and **Bibron,** 1839 [Alt. Eastern Water Skink; Syn. *Sphenomorphus quoyii*]
Jean René Constant Quoy (1790–1869) was a French zoologist and naval surgeon. He went on a number of voyages of discovery, including two circumnavigations of the globe (1817–1820 and 1826–1829). Many taxa, including a bird, are named after him.

R

Rabino

Rabino's Tree Iguana *Liolaemus rabinoi* **Cei,** 1974

M. Rabino collected the holotype (1972). He may have been a member of a field trip led by Cei, but no further information is given in the original text.

Rabor

Rabor's Lipinia *Lipinia rabori* **Brown** and **Alcala,** 1956

Rabor's Short-legged Skink *Brachymeles cebuensis* Brown and Rabor, 1967

Dr. Dioscoro S. "Joe" Rabor (1911–1996) was the preeminent Philippine ornithologist, mammalogist, herpetologist, ichthyologist, and conservationist of the 20th century. He graduated from the University of the Philippines (1934) and studied for a Ph.D. at Yale (1958). Sillman University, where he held a number of teaching posts and professorships (1947–1967), awarded him an honorary doctorate (1974). He led over 50 expeditions (1935–1977), normally accompanied by his wife, two sons, and four daughters; all his children became physicians. He produced the most thorough documentation of the birds and mammals of the Philippines and wrote *Philippine Birds and Mammals: A Project of the U. P. Science Education Center* (1977) and *Philippine Reptiles and Amphibians* (1981). Four mammals and a bird are named after him.

Racenis

Roze's Neusticurus *Neusticurus racenisi* **Roze,** 1958

Dr. Janis Racenis (1915–1980) was a Latvian entomologist, geographer, ornithologist, and ecologist who emigrated from the Ukraine to Venezuela and worked at the Institute of Tropical Zoology, Faculty of Sciences, Universidad Central de Venezuela. The holotype was collected during a Phelps-Tate expedition for the American Museum of Natural History, about 20 years before it was described.

Rackham

Rackham's Knob-scaled Lizard *Xenosaurus grandis rackhami* **Stuart,** 1941

Horace H. Rackham (1858–1933) was a Detroit lawyer who invested $5,000 in Henry Ford's new venture to build "horseless carriages." Ford bought out minority shareholders (1919), and Rackham's $5,000 turned into $12.5 million on top of the $2 million in dividends he had already been paid. He spent the rest of his life giving his money away. He sponsored and paid for expeditions, he was President of the Detroit Zoological Commission (1924–1928), and his will provided for a foundation in his and his wife's names to ensure graduate education and student loan facilities. Stuart's studies of Guatemalan herpetofauna were financed by this foundation.

Radde

Toadhead Agama sp. *Phrynocephalus raddei* **Boettger,** 1888

Radde's Viper *Vipera raddei* Boettger, 1890

Azerbaijan Lizard *Darevskia raddei* Boettger, 1892

Gustav Ferdinand Richard Radde (1831–1903) was an apothecary. Born in Danzig, Prussia (Gdansk, Poland), he moved to Russia (1852). He was on numerous expeditions through Siberia, the Crimea, the Caucasus, Trans-Caucasus, and other regions of Russia, and also Iran and Turkey. He settled in Georgia (1863) and founded the Caucasian Museum, Tbilisi (1867). Toward the end of the 19th century he made two further journeys, accompanying Russian Imperial family members to India and Japan (1895) and to North Africa (1897). He wrote *Reisen im Süden von Ost-Sibirien in den Jahren 1855–1859*. Two birds and two mammals are named after him.

Raddi

Raddi's Lizard-eating Snake *Elapomorphus quinquelineatus* Raddi, 1820

Giuseppe Raddi (1770–1829) was an Italian botanist who worked in various gardens in Florence (1785–1795). He was Curator, Museo di Storia Naturale, Florence (1795–1817), except for a period (1807–1814) during which the post was abolished and he worked wherever he could. He was one of the party accompanying Leopoldina, Archduchess of Austria, on her journey to marry Dom Pedro, and he traveled in Brazil (1817–1818). He died after returning from a visit to Egypt.

Rafael

Rafael's Alligator Lizard *Mesaspis moreleti rafaeli* **Hartweg** and Tihen, 1946

See **Del Campo** and **Martin del Campo.**

Raffone

Raffone's Wall Lizard *Podarcis raffoneae* **Mertens,** 1952 [Alt. Aeolian Wall Lizard]

Dr. Antonia Trischitta, née Raffone. Dr. Antonino Trischitta collected the holotype of the lizard (1951), and at his request it was named after his wife.

Raffrey

Raffrey's Hook-nosed Snake *Scaphiophis raffreyi* **Bocourt,** 1875 [Alt. Ethiopian Hook-nosed Snake]

Marie Jacques Achille Raffray (b. 1844)—Bocourt misspelled his name—was a French traveler, civil servant, and collector. The French Ministry of Public Instruction charged him with making scientific collections during the voyage of the French navy ship *Correze*. He collected in Zanzibar and Abyssinia (Ethiopia) (1874–1875) and later in New Guinea. In the 1890s he was in the Philippines. He wrote *Tour du monde* (1879). Two mammals are named after him.

Ragazzi

Ragazzi's Cylindrical Skink *Chalcides ragazzii* **Boulenger,** 1890

Sahelian Fan-toed Gecko *Ptyodactylus ragazzii* **Anderson,** 1898

Dr. Vincenzo Ragazzi (1856–1929) of the Modena Natural History Society explored and collected in Ethiopia. He was a physician and was posted to the Italian research station, Let Marefia, Ethiopia (1884), later serving as its Director for several years. He was on good terms with the Emperor Menilek and accompanied his military expeditions to Harrar (1886–1887) to make geographical surveys. Menilek selected Ragazzi as his emissary to go to Italy (1887).

Rahm

Rahm's Sun Tegu *Euspondylus rahmi* de Grijs, 1936

Peter Rahm collected the holotype.

Raithma

Skink sp. *Ophiomorus raithmai* **Anderson** and **Leviton,** 1966

The name is formed from the words *raith mai*, meaning "sand fish," the Sindhi name for the skink.

Rajery

Blind Snake sp. *Typhlops rajeryi* Renoult and Raselimanana, 2009

Emile Rajeriarison is a Malagasy experimental naturalist and nature guide working at Ranomafana National Park. An amphibian is named after him.

Rakiura

Stewart's Sticky-toed Gecko *Hoplodactylus rakiurae* **Thomas,** 1981

Named after Rakiura National Park.

Ramanantsoa

Ramanantsoa's Leaf Chameleon *Brookesia ramanantsoai* **Brygoo** and **Domergue,** 1975

Professor Guy A. Ramanantsoa is a Malagasy zoologist and herpetologist at the University of Madagascar. He was Chief Engineer (1970) dealing with water and forestry resources in the National Parks, and he was one of the originators of the Beza Mahafaly Project to create a nature reserve for use in training in agronomy and zoology (1975).

Ramsay

Ramsay's Copperhead *Austrelaps ramsayi* **Krefft,** 1864

Ramsay's Python *Aspidites ramsayi* **Macleay,** 1882 [Alt. Woma]

Edward Pearson Ramsay (1842–1916) was an Australian naturalist, oologist, and ornithologist who was Curator of the Australian Museum (1874–1894) and remained the museum's consulting ornithologist (1895–1916). He also worked on marine zoology, mammalogy, botany, and herpetology. He corresponded at length with Gould on oology. A bird is named after him.

Ramirez, A.

Ramirez's Alligator Lizard *Abronia ramirezi* **Campbell,** 1994

Antonio Ramirez Velazquez is a herpetologist and Curator of Reptiles and Amphibians, Instituto de Historia Natural, Chiapas, Mexico. He collected the holotype (1990).

Ramirez, J.

Ramirez's Hook-nosed Snake *Ficimia ramirezi* **H. M. Smith** and Langebartel, 1949

Juan Ramirez collected the holotype (1949).

Ramirez, M.

Tree Iguana sp. *Liolaemus ramirezae* **Lobo** and **Espinoza,** 1999

Professor Dr. Martha Patricia Ramirez-Pinilla is a biologist at Universidad Nacional de Colombia, Bogotá. Universidad Nacional de Tucumán, Argentina, awarded her doctorate (1992). She co-wrote "Annual Reproductive Activity of *Mabuya mabouya* Squamata, Scincidae" (2002).

Ramsden

Ramsden's Least Gecko *Sphaerodactylus ramsdeni* **Ruibal,** 1959

Dr. Charles Theodore Ramsden (1876–1951) was an entomologist, herpetologist, and naturalist who received his doctorate from Universidad de La Habana (1917). He collected mainly in eastern Cuba. He co-wrote *The Herpetology of Cuba* (1919). The museum at Universidad de Oriente, Santiago de Cuba, is named after him.

Ranawana

Ranawana's Golden Cat Snake *Boiga ranawanei* Samarawickrama et al., 2006

Dr. Kithsiri Bandara Ranawana is a Sri Lankan zoologist and ecologist, Department of Zoology, Faculty of Science, University of Peradeniya, Sri Lanka. He also acts as a consulting ecologist for commercial companies. He co-wrote "Species Composition, Status, and Feeding Ecology of Avifauna in High Altitude Forests of Sri Lanka" (1998).

Rand

Pedernales Least Gecko *Sphaerodactylus randi* **Shreve,** 1968

Dr. Austin Stanley Rand (1932–2005) was a herpetologist at the Smithsonian. His father was the famous Canadian ornithologist Austin Loomer Rand (1905–1982). His mother was a herpetologist, as was his wife, Patricia Grubbs Rand, whom he met when they catalogued salamanders together at the Smithsonian as a student summer job. Harvard awarded his doctorate (1961). He was an Assistant, Division of Amphibians and Reptiles, Field Museum (1950). He served in the U.S. armed forces in Germany (1955–1957). He worked at the Smithsonian's Tropical Research Institute, Panama (1964–1997), and after retiring continued to visit Panama annually.

Raney

Gymnophthalmid lizard sp. *Riama raneyi* **Kizirian,** 1996

Richard H. Raney graduated from the University of Kansas (1949) and was Mayor of Lawrence, Kansas (1967–1968). Kizirian's etymology says that the name honors "Richard H. Raney of Lawrence, Kansas, in recognition of his generous support of the Panorama Society, Museum of Natural History, The University of Kansas."

Range

Namib Sand Gecko *Palmatogecko rangei* **Andersson,** 1908

Dr. Paul Range (1879–1952) was born in Berlin. He worked as a government geologist in German South-West Africa (Namibia) (1906–1912), where he also collected plants. He later taught at the University of Berlin. He wrote *Beiträge und Ergänzungen zur Landeskunde des deutschen Namalandes* (1914).

Rankin

Rankin's Spiny-tailed Gecko *Strophurus rankini* **Storr,** 1979 [Alt. Exmouth Spiny-tailed Gecko]

Rankin's Elf Skink *Nannoscincus rankini* **Sadlier,** 1987

Rankin's Dragon *Pogona rankini* **Manthey** and Schuster, 1999

Peter Rankin (1955–1979) was a herpetologist who died when he fell from a tree while collecting lizards at night in New Caledonia. He had just completed his bachelor's degree in science at Macquarie University, Sydney. His visit to New Caledonia was partly holiday and partly preparation for his future work and studies.

Ranwella

Ranwella's Day Gecko *Cnemaspis ranwellai* Mendis Wickramasinghe, 2006

Dr. Sanjeewa Ranwella (1965–2003) was a Sri Lankan physician, zoologist, and herpetologist. He, his wife, and their young son were all drowned in a boating accident on Lake Bolgoda. His brother, Dimuth Ranwella, is a Sri Lankan wildlife photographer who has provided photographs of geckos to illustrate articles written by Mendis Wickramasinghe, who described this species and collected the holotype in the year that the Ranwella family died.

Rapp

Himalayan Stripe-necked Snake *Liopeltis rappi* **Günther,** 1860

Professor Dr. Wilhelm Ludwig von Rapp (1794–1868) was a German physician, naturalist, and ichthyologist. He practiced medicine in Stuttgart (1818–1819) and taught anatomy at Eberhard Karls Universität Tübingen (1819–1858). He wrote *Neue Batrachier* (1842).

Rasmussen

Rasmussen's Gecko *Urocotyledon rasmusseni* **Bauer** and Menegon, 2006

Sulu Water Monitor *Varanus rasmusseni* Koch, Gaulke, and **Böhme,** 2010

Jens Bødtker Rasmussen (1947–2005) was Curator of Herpetology, Zoologisk Museum, Copenhagen (1977–2005). His research focused on phylogeny and biogeography of African snakes.

Raun

Raun's Spotted Whiptail *Aspidoscelis gularis rauni* **Walker,** 1967

Dr. Gerald George Raun (b. 1932) graduated from Texas Tech University (1954) with a bachelor's degree in geology. He served in the U.S. Army (1954–1956). His master's degree (1958) and doctorate (1961), both in zoology, are from the University of Texas. He was employed in the university's Texas Memorial Museum as Curator of Vertebrates (1960–1966). He was Assistant Professor of Biology, North Texas State University (1967–1970), and then Professor of Biology, Department of Biology, Angelo State University, San Angelo (1970–1978). He became the publisher of *Devil's River News* in Sonora (1979–1989) and of *Alpine Avalanche* (1989–1993). After 1993 he wanted a change and so began conducting research into cacti and the Mexican Revolution (1910–1920). He wrote *Snakes of Texas* (1972).

Ravergier

Ravergier's Whip Snake *Coluber ravergieri* **Ménétries,** 1832 [Alt. Spotted Whip Snake; Syn. *Hemorrhois ravergieri*]

Ravergier was an attaché at the French embassy in St. Petersburg. He reported he had seen this snake in Georgia.

Rawlinson

Rawlinson's Ctenotus *Ctenotus rawlinsoni* **Ingram,** 1979

Rawlinson's Window-eyed Skink *Pseudemoia rawlinsoni* **Hutchinson** and Donnellan, 1988 [Alt. Glossy Grass Skink]

Dr. Peter Alan Rawlinson (1942–1991) was a biologist, a conservationist, and a senior lecturer in zoology at LaTrobe University, Melbourne. He was Treasurer and Vice President of the Australian Conservation Foundation; after his death, during a field trip to Indonesia, they set up an annual award in his honor. He wrote "Biogeography and Ecology of the Reptiles of Tasmania and the Bass Strait Area" (1974).

Raxworthy

Raxworthy's Leaf Chameleon *Brookesia valerieae* Raxworthy, 1991 [Alt. Valerie's Leaf Chameleon]

Dr. Christopher John Raxworthy (b. 1964) is an English herpetologist who is Curator-in-Charge and Associate Curator, Department of Herpetology, Division of Vertebrate Zoology, American Museum of Natural History, New York, which he joined in 2000. The University of London awarded his bachelor's degree (1985) and the Open University, Milton Keynes, England, his doctorate in biology (1989). He is Assistant Professor, Center for Environmental Research and Conservation, Columbia University, and an external lecturer, University of Antananarivo, Madagascar, where he has carried out fieldwork.

Rebentisch

Dwarf/Reed Snake sp. *Calamaria rebentischii* **Bleeker,** 1860

J. H. A. B. Sonnemann Rebentisch collected specimens of fish for Bleeker (1858–1859). He wrote *Reptiliën van Borneo* (1859). A fish is named after him.

Rebouch

Rebouch's Mabuya *Mabuya ficta* Rebouças-Spieker, 1981

Regina Rebouças-Spieker is a herpetologist, Museum of Zoology, Universidade de São Paulo.

Rechinger

Dwarf Racer sp. *Eirenis rechingeri* **Eiselt,** 1971

Dr. Karl Heinz Rechinger (1906–1998) was an Austrian botanist. He worked in Naturhistorisches Museum Wien (1929–1971) in the Botanical Department and became the museum's Director (1963). He was a Visiting Professor in Baghdad (1956–1967). Documenting the wild plants of Iran was his life's work.

Reeves

Reeves' Terrapin *Chinemys reevesii* **Gray,** 1831

Reeves' Butterfly Lizard *Leiolepis reevesii* Gray, 1831

Reeves' Smooth Skink *Scincella (Leiolopisma) reevesi* Gray, 1838

John Reeves (1774–1856) was an English amateur naturalist and collector who served in China, chiefly Macao, for the East India Company as an "Inspector of Tea" (1812–1831). He sent specimens of Chinese fauna back to England, and various other taxa are named after him.

Regina

Plain-backed Two-line Dragon *Diporiphora reginae* **Glauert,** 1959

Western Soil-crevice Skink *Proablepharus reginae* Glauert, 1960

Named after a location in Australia, Queen Victoria Spring, rather than directly after the British monarch.

Reich

Reich's Tree Iguana *Phrynosaura reichei* **Werner,** 1907

Dr. Karl Friedrich Reiche (1860–1929) was a botanist. He took his doctorate at Universität Leipzig. He was a Professor, Universität Dresden (1886–1889). He worked at Museo Nacional de Historia Natural, Santiago de Chile, being Director, Botany Department (1896–1911). He held a professorship in Mexico (1912–1924), then lived in Monaco (1924–1926), went back briefly to Mexico, and returned to Germany (1928). An amphibian is named after him.

Reichenow

Reichenow's Lidless Skink *Lacertaspis reichenowi* **Peters,** 1874

Reichenow's Spiny-tail Lizard *Cordylus vittifer* Reichenow, 1887

Dr. Anton Reichenow (1847–1941) dominated German ornithology for many years. He worked at the Zoological Museum, Humboldt-Universität, Berlin (1874–1921), starting as an Assistant, then becoming Curator of Birds (1888) and Vice Director (1906), succeeding Cabanis (his father-in-law). He was regarded as the leading expert of his time on African birds. He traveled to Africa only once on a collecting expedition, visiting Gold Coast (now Ghana), Gabon, and Cameroon (1869). After retiring he moved to Hamburg and was active in the local natural history museum. He wrote *Die Vogel Africas* (3 vols., 1900–1905). Nineteen birds are named after him.

Reid

Reid's Arboreal Alligator Lizard *Abronia reidi* Werler and **Shannon,** 1961

Jack Robert Reid (b. 1933) of San Antonio, Texas, was a member of the group that collected the holotype (1953).

Reimann

Reimann's Snakeneck Turtle *Chelodina reimanni* **Philippen** and **Grossman,** 1990

Dr. Michael J. Reimann is a herpetologist. He read biology and zoology at Johannes Gutenberg-Universität Mainz. He went to Turkey (1987) to study turtles and in 2003 established a farm at Foca that exports turtles to Europe; he also has a turtle-breeding farm in Germany. In 2007 he announced that he intended to expand production and also breed snakes and iguanas at Foca.

Reimschiissel/Reimschisel

Reimschisel's Emo Skink *Emoia reimschiisseli* **Tanner,** 1950

Ernest F. Reimschiissel (b. 1917) was a graduate of Brigham Young University who became Assistant Professor of Horticulture there (1942). After joining the army (1943) he was sent to New Caledonia, where as Private First Class he met Dr. Elden Beck (then a First Lieutenant), who interested him in collecting reptiles and insects. He continued to make collections at various Pacific locations until his return to the USA (1945).

Reinhardt

Reinhardt's Lined Snake *Cyclocorus lineatus* Reinhardt, 1843

Reinhardt's Shovelsnout Snake *Prosymna meleagris* Reinhardt, 1843

African Burrowing Python *Calabaria reinhardtii* **Schlegel,** 1848

Reinhardt's Snake-eater *Polemon acanthias* Reinhardt, 1860

Reinhardt's Lava Lizard *Tropidurus hygomi* Reinhardt and Lutken, 1861

Professor Dr. Johannes Theodor Reinhardt (1816–1882) was a Danish zoologist who was Curator of Terrestrial Vertebrates at Det Kongelige Naturhistoriske Museum, Copenhagen. He was an ardent supporter of Darwin's theories. He co-authored *Bidrag til vestindiske Öriges og navnligen til de danskvestindiske Öers Herpetologie* (1862). A bird and an amphibian are named after him.

Reisinger

Reisinger's Tree Monitor *Varanus reisingeri* Eidenmüller and Wicker, 2005

Manfred Reisinger is a German naturalist and reptile breeder.

Reiss

Peters' Leaf-toed Gecko *Phyllodactylus reissii* **Peters,** 1862

Carl Reiss was a German living in Ecuador. He regularly sent specimens, including this gecko, to Peters, who mentions him as the source of material from Guayaquil in the early 1860s.

Remacle

Witte's Worm Lizard *Monopeltis remaclei* **Witte,** 1933

David L. Remacle collected fish (1932) from River Lukulu, the site where the lizard's holotype was collected.

Renard

Renard's Viper *Vipera renardi* Christoph, 1861

Dr. Charles Renard was a Russian naturalist and a Councillor of State. He was elected a member of the Imperial Society of Naturalists, Moscow (1840), becoming Secretary (before 1854) and President (1884). The American Philosophical Society elected him a member (1854). Christoph met with Renard (1860), who, as editor of the Moscow society's *Bulletin*, published Christoph's paper.

Rendahl

Rendahl's Agama *Salea brachydactyla* Rendahl, 1937

Rendahl's Wolf Snake *Lycodon paucifasciatus* Rendahl and **M. A. Smith,** 1943

Dr. Carl Hialmar Rendahl (1891–1969) was a zoologist, ichthyologist, and artist at Naturhistoriska Riksmuseet, Stockholm (1912–1958), where he also was Professor of Natural History (1933–1958), retiring as Professor Emeritus. Stockholms Universitet awarded his bachelor's degree (1918) and his doctorate in zoology (1924).

Renevier

Renevier's New Caledonian Skink *Caledoniscincus renevieri* **Sadlier, Bauer,** and Colgan, 1999

Alain Renevier is a French resident of Noumea, New Caledonia, where he is involved in atoll research. He and his family were generous with hospitality and support for the describers.

Rengger

Striped Kentropyx *Kentropyx renggerii* Peters, 1869 [Junior syn. of *K. striata* Daudin, 1802]

Dr. Johann Rudolph Rengger (1795–1832) was a

German-born Swiss physician and naturalist who studied at Université de Lausanne and Eberhard Karls Universität Tübingen. He traveled extensively and lived in Paraguay (1819–1825). He wrote *Natural History of the Mammals of Paraguay* (1830).

Renjifo

Renjifo's Coral Snake *Leptomicrurus renjifoi* **Lamar,** 2003 [Alt. Ringed Slender Coral Snake]

Dr. Juan Manuel Renjifo (b. 1948) is a biologist, herpetologist, and wildlife photographer at Instituto Nacional de Salud, Bogotá, Colombia, and is also an Associate Professor of Biology, Universidad del Magdalena.

Rensch

Skink sp. *Cryptoblepharus renschi* **Mertens,** 1928

Professor Bernhard Rensch (1900–1990) was a biologist, philosopher, and artist. He was at Museum für Naturkunde Berlin (1925–1936) and was Director, Zoological Institute, Westfälische Wilhelms-Universität Münster (1947–1968). He was in the Lesser Sunda Islands (Indonesia) with his wife, the botanist Ilse Rensch-Maier (1927).

Reuss

Reuss' Water Snake *Enhydris alternans* Reuss, 1834

Dr. Adolph Reuss (1804–1878) was a physician, zoologist, arachnologist, and herpetologist at Naturmuseum Senckenberg, Frankfurt. His brother-in-law, who was involved in the attempted 1848 revolution, was imprisoned in the state prison fortress of Mayence. Reuss apparently became a political refugee and emigrated. He and his family settled in Shiloh, Illinois, in the early 1850s.

Reuter

Reuter's Worm Snake *Typhlops reuteri* **Boettger,** 1881

C. Reuter collected the holotype on Nossi-Bé, Madagascar, and gave it to Dr. H. Lenz, who forwarded it Boettger. A person called Reuter is reported as late as 1887 as being on Nossi-Bé collecting arachnids, and is presumably the same person. Boettger chose to write the description and etymology in abbreviated Latin: "Hab. in insula Nossi-Bé, spec. unicum ab ill. C. Reuter collectum et ab ill. Dr. H. Lenz mihi communicatum."

Revoil

Revoil's Short Snake *Brachyophis revoili* **Mocquard,** 1888

Georges Emmanuel Joseph Révoil (1852–1894) was a French naturalist who made several expeditions to Somaliland (1877–1883) and wrote *La vallée du Darro: Voyage aux pays Somalis* (1882). A bird is named after him.

Reyes

Reyes' Caribbean Gecko *Aristelliger reyesi* Diaz and Hedges, 2009

Ernesto Reyes works for Estacion de Investigaciones Integrales de la Montana, Cuba. He photographed this newly discovered Cuban species (2007).

Reynolds

Florida Sand Skink *Neoseps reynoldsi* **Stejneger,** 1910 [Syn. *Plestiodon reynoldsi*]

A. G. Reynolds of Gulfport, Florida, dealt in insects, fish, reptiles, amphibians, shells, and invertebrates. He collected the holotype (1910).

Rhode

Rhode's Snouted Snake *Simophis rohdei* **Boettger,** 1885

The common name reflects a misspelling. See **Rohde, Ricardo.**

Ricardini

São Paulo Sharp Snake *Uromacerina ricardinii* **Peracca,** 1897

Dr. Ricardini presented the holotype to Museo Civico di Storia Naturale, Turin. Unfortunately that is all Peracca wrote, and we have been unable to trace him.

Richard, A. and L. C. M.

Richard's Anole *Anolis richardi* **Duméril** and **Bibron,** 1837

Richard's Worm Snake *Typhlops richardii* Duméril and Bibron, 1844

Louis Claude Marie Richard (1754–1821) was a botanist who was in the West Indies and Central America (1780–1784). He became Professor of Botany, School of Medicine, Paris (1790). He was a specialist in orchids and invented the special terminology used when describing them. His son, Achilles Richard (1794–1852), was also a noted botanist. The snake may be named after either or both of them.

Richard S.

Richard's Skink *Egernia richardi* **Peters,** 1869 [Alt. Bright Crevice-Skink]

See **Schomburgk.**

Richardson

Broad-banded Sand-swimmer *Eremiascincus richardsonii* **Gray,** 1845

Richardson's Leaf-toed Gecko *Hemidactylus richardsonii* Gray, 1845

Richardson's Least Gecko *Sphaerodactylus richardsonii* Gray, 1845

Richardson's Mangrove Snake *Myron richardsonii* Gray, 1849 [Alt. Gray's Mangrove Snake]

Dr. Sir John Richardson (1787–1865) was a naval surgeon and Arctic explorer. He was a friend and relation by marriage of Sir John Franklin, and took part in Franklin's expeditions (1819–1822 and 1825–1827). He participated from 1847 in the vain search for Franklin and his colleagues; their fate was not discovered until Rae's expedition (1853–1854). He co-edited *The Zoology of the Voyage of H.M.S. Erebus and Terror, under the Command of Capt. J. C. Ross during 1839–1845* (1845). Five mammals and seven birds are named after him, as are Canada's Richardson Mountains.

Richard T.

Richard's Banded Sphaero *Sphaerodactylus richardi* Hedges and **Garrido,** 1993

Professor Dr. John Paul Richard Thomas (b. 1938) works at the Department of Biology, Universidad de Puerto Rico. His doctorate was awarded at Louisiana State University (1976). He co-wrote "A New Amphisbaenian from Cuba" (1998).

Richmond

Richmond's Worm Snake *Typhlops capitulatus* Richmond, 1964

Neil Dwight Richmond (1912–1992) was a herpetologist and zoologist who was an Instructor in Zoology, Syracuse University (1935–1938), Marshall College (1938–1939), and Shackelford Farms (1939–1940). He worked for the U.S. Fish and Wildlife Service, Pennsylvania (1947–1948). He became Curator, Amphibians and Reptiles, Carnegie Museum (1951). An amphibian is named after him.

Richter

Richter's Andes Anole *Phenacosaurus richteri* **Dunn,** 1944

Dr. Leopold Richter was a friend of Dunn's and provided him with photographs to illustrate an article on this genus.

Ricord

Haitian Green Anole *Anolis ricordi* **Duméril** and **Bibron,** 1837

Ricord's Ground Iguana *Cyclura ricordi* Duméril and Bibron, 1837

Alexandre Ricord (1798–1876) was a French naval surgeon who qualified as a physician in Paris (1824), and became a Corresponding Member, Académie Nationale de Médecine, Paris (1838). He collected in Latin America (1826–1834). A bird is named after him.

Ridgeway

Derafshi Snake *Lytorhynchus ridgewayi* **Boulenger,** 1887

Colonel Sir Joseph West Ridgeway (1844–1930) was the senior British officer on the Russian-Afghan Boundary Commission (1885–1899), when it seemed probable that Russia would attack British India via Afghanistan. Afghanistan's northwestern frontier with Russia, delineated by the commission, was called the Ridgeway Line. He was Governor of Ceylon (Sri Lanka) (1895–1903). The snake holotype was collected by Dr. Aitchison, another member of the commission.

Ridley

Olive Ridley Turtle *Lepidochelys olivacea* **Eschscholtz,** 1829

Pernambuco Teiid *Stenolepis ridleyi* **Boulenger,** 1887

Ridley's Worm Lizard *Amphisbaena ridleyi* Boulenger, 1890

Henry Nicholas Ridley (1855–1956) was a British botanist and collector who was working on the island of Fernando de Noronha (1887), when he first reported sightings of Olive Ridley Turtles in Brazil. However, it seems unlikely that the "Ridley" in the turtle's name refers to him. There are several theories, including one that it was a "riddle" as to where the turtles came from, and "riddle" came to be pronounced "riddlie" and so "ridley." Ridley was known as "Mad Ridley" or "Rubber Ridley," because he was keen to get the rubber tree transplanted to British territory. He was Superintendent of the Tropical Gardens, Singapore (1888–1912), where early experiments in growing the tree outside Brazil took place. He wrote "The Habits of Malay Reptiles" (1889). Two mammals and a bird are named after him.

Riebeck

Riebeck's Leaf-toed Gecko *Haemodracon riebeckii* **Peters,** 1882

Dr. Emil Riebeck (1853–1885) was a German ethnologist, mineralogist, and explorer. He traveled in Arabia (1880) and Socotra (1881). He visited (1862) the hill country near Chittagong in East Bengal (Bangladesh) and wrote *Chittagong Hill Tribes* (1825). He discovered the mineral named after him, riebeckite.

Rieppel

Chameleon genus *Rieppeleon* Matthee, **Tilbury,** and **Townsend,** 2004

Dr. Oliver Cedric Rieppel (b. 1951) is a Swiss-born American scientist at the Field Museum, where he has been Curator of Geology, with special responsibility for fossil amphibians and reptiles, since 1990. His early education in zoology resulted in a diploma from Universität Basel (1974), followed by a master's degree in vertebrate paleontology from University College, London (1975), and a doctorate in zoology, again from Universität Basel (1978).

Riggenbach

Riggenbach's File Snake *Mehelya riggenbachi* **Sternfeld,** 1910

F. W. Riggenbach (1864–1944) was a zoologist and collector employed on German Central African expeditions led by Adolf Friedrich, Duke of Mecklenburg, in the first two decades of the 20th century. A mammal is named after him.

Riise

Ameiva sp. *Ameiva riisei* **Reinhardt** and Lütken, 1862

Albert Heinrich Riise (1810–1882) was a Danish pharmacist, botanist, and collector who supplied much material to Zoologisk Museum, Copenhagen. He was an apprentice pharmacist (1824–1830), then moved to Copenhagen where he graduated (1832) and worked until 1838. The Danish King appointed him to be a pharmacist in the Danish West Indies (now U.S. Virgin Islands) with an exclusive licence to open a retail shop, A. H. Riise in St. Thomas, which is still in business. The year 1868 saw epidemics of cholera, smallpox, and yellow fever in St. Thomas, so he returned to Denmark with his family and never returned to St. Thomas.

Rijgersma

Anguilla Racer *Alsophis rijgersmai* **Cope,** 1869 [Alt. Leeward Islands Racer; scientific name often given as *A. rijersmai*]

Dr. Henrik Elingsz Van Rijgersma (1835–1877) qualified as a physician (1858). He was the government physician on the island St. Martins, Netherlands Antilles (1863–1877). He was an enthusiastic amateur naturalist and collector, especially of molluscs. He sent specimens of this snake to the Philadelphia Academy of Natural Sciences.

Riley

San Salvador Iguana *Cyclura rileyi* **Stejneger,** 1903

Joseph Harvey Riley (1873–1941) was a biologist and ornithologist at the Smithsonian (1896–1941) who became Associate Curator of Birds (1932). He traveled in Cuba (1900) and the Bahamas (1905). He collected the holotype in 1900.

Riva

Ethiopian Girdled Lizard *Cordylus rivae* **Boulenger,** 1896

Dr. Domenico Riva was traveling with Prince Eugenio Ruspoli (q.v.) in Somaliland and Gallaland (1893) when the holotype of this lizard was collected. He had previously traveled in Eritrea (1889–1891).

Rivero, J. A.

Rivero's Ground Snake *Atractus riveroi* **Roze,** 1961

Dr. Juan Arturo Rivero-Quintero (b. 1923) gained his bachelor's degree (1945) from Universidad de Puerto Rico, and both his master's (1951) and doctorate (1953) from Harvard. He is the founder of the Dr. Juan A. Rivero Zoo, Universidad de Puerto Rico, which he joined in 1945, becoming Professor of Biology in 1958. He was a Research Associate at Harvard and Visiting Scientist, Instituto Venezolano de Investigaciones Científicas (1966–1968).

Rivero, R.

Neotropical House Snake sp. *Thamnodynastes ramonriveroi* Manzanilla and **Sánchez,** 2005

Ramón A. Rivero is Curator and Collections Manager, Estación Biológica Rancho Grande, Maracay, Venezuela.

Rivers

Island Night Lizard *Klauberina riversiana* **Cope,** 1883 [Syn. *Xantusia riversiana*]

Dr. James John Rivers (1824–1913), who qualified as a physician in London, had a taste for entomology. He knew such eminent persons as Thomas Huxley, Charles Darwin, and Alfred Wallace. He visited the USA (1867), settling there (1870). He traveled (1875–1880), becoming associated with the California Academy of Science, and until 1895 was Curator of Natural History, University of California. There is mystery about the origin of *riversiana*; although Cope described this Night Lizard in 1883, an anonymous article appeared (1879) in *American Naturalist* with the comment that the zoologist/describer wanted "to commemorate Mr Rivers of the University of California."

Rivet

Despax's Parrot Snake *Leptophis riveti* **Despax,** 1910

Professor Dr. Paul Rivet (1876–1958) originally trained as a physician. He was on the second French Geodesic Mission to Ecuador (1901), staying until 1906. After returning to Paris he worked at Muséum National d'Histoire Naturelle, Paris, and was one of the founders of Institut d'Ethnologie (1926). He became Director, Muséum National d'Histoire Naturelle, Paris (1928), and founded Musée de l'Homme (1937). He went to Colombia, where he founded Instituto Colombiano de Antropología e Historia (1942–1945). Despax published an article (1911) on the collection of reptiles and amphibians that Rivet had made in Ecuador.

Robecchi

Chameleon sp. *Rhampholeon robecchii* **Boulenger,** 1892

Robecchi's Agama *Agama robecchii* Boulenger, 1892

Stumptail Chameleon sp. *Rieppeleon kerstenii robecchii* Boulenger, 1892

Luigi Robecchi Brichetti (1855–1926) was an Italian geologist and explorer who visited Africa often (1895–1905). He campaigned strongly against the slavery that still existed in Somalia in his day, and his efforts are commemorated in several Italian cities having streets named after him.

Robert

Robert's Worm Lizard *Cercolophia roberti* **Gans,** 1964

See **Mertens.**

Robert Mertens

Robert's Tree Iguana *Liolaemus robertmertensi* **Hellmich,** 1964

Robert Mertens' Day Gecko *Phelsuma robertmertensi* **Meier,** 1980

See **Mertens.**

Roberto

Roberto's Lizard *Liolaemus robertoi* Pincheira-Donoso and **Núñez,** 2004

See **Donoso-Barros.**

Roberts, G.

Grey-bellied Sunskink *Lampropholis robertsi* **Ingram** 1991

Blind Snake sp. *Ramphotyphlops robertsi* **Couper, Covacevich,** and **Wilson,** 1998 [Syn. *Austrotyphlops robertsi*]

Gregory Roberts is a Queensland naturalist.

Roberts, J. A.

Roberts' Girdled Lizard *Pseudocordylus robertsi* **Van Dam,** 1921

Large-scaled Gecko *Pachydactylus robertsi* **Fitzsimons,** 1938

Dr. J. Austin Roberts (1883–1948) was a South African zoologist who in his day was the most prominent ornithologist in southern Africa. He had little formal zoological training but was awarded an honorary doctorate by the University of Pretoria (1935). He worked at the Transvaal Museum (1910–1948). He died in a car accident in the Transkei region. He wrote *Birds of South Africa* (1940). In his etymology Van Dam refers only to "A. Roberts," but there is no more likely candidate. One bird and two mammals are named after him.

Robertson

Robertson Dwarf Chameleon *Bradypodion gutturale* **Andrew Smith,** 1849 [Alt. Robertson's Dwarf Chameleon]

Named after the Robertson region of Cape Province, South Africa.

Robinson, D.

Robinson's Spiny-tailed Gecko *Strophurus robinsoni* **L. A. Smith,** 1995

David Robinson is an amateur naturalist who is a Technical Officer, Western Australian Department of Agriculture. He has helped the Department of Herpetology, Western Australian Museum, both in the field and in the laboratory, for over 25 years, taking part in many reptile surveys. He co-wrote *A Guide to Reptiles and Frogs of the Perth Region* (1995).

Robinson, H. C.

Robinson's Forest Dragon *Gonocephalus robinsonii* **Boulenger,** 1908

Robinson's Keeled Skink *Tropidophorus robinsoni* **M. A. Smith,** 1919

Herbert Christopher Robinson (1874–1929) was a British zoologist and ornithologist. He went to Switzerland for his health (1894) and visited Queensland (1896). He was assistant to Dr. H. O. Forbes (q.v.) at the Liverpool Museum (1897–1900), and in later years he sent many specimens to that museum. He spent the rest of his life in the tropics, in the Malay Peninsula (1900–1902), and as Curator of the Federated Malay States Museum, Selangor (1903–1926). Cecil Boden Kloss (q.v.) joined him in exploring the Indo-Malay region (1908). He wrote *The Birds of the Malay Peninsula* (1927). Four mammals and two birds, among other taxa, are named after him.

Robison

Robison's Crevice Spiny Lizard *Sceloporus poinsettii robisoni* **Tanner,** 1987

Dr. Wilbur Gerald Robison Jr. (1933–2004) was awarded his doctorate by the University of California, Berkeley. The holotype was collected by Robison and Tanner (1958). He co-wrote "New and Unusual Serpents from Chihuahua, Mexico" (1960).

Roborovski/Roborowski

Roborovski's Toad-headed Agama *Phrynocephalus roborowskii* **Bedriaga,** 1906

Roborowski's Wonder Gecko *Teratoscincus roborowskii* Bedriaga, 1906

Captain Vladimir Ivanovich Roborovski (1856–1910) was an explorer of parts of China and Tibet. He accompanied Przewalski on his third and fourth expeditions

(1879–1880 and 1883–1885). He led his own expedition for the Imperial Russian Geographical Society to Eastern Tien-Shan, Nanshan, and Northern Tibet (1893–1896), and during it was hit by paralysis, despite which he continued the handling of field material. He wrote *Ekspeditisii v storonu ot pugey Tibetskoy ekspeditsii* (1896). One bird and one mammal are named after him.

Rodeck

Rodeck's Whiptail *Aspidoscelis rodecki* **McCoy** and **Maslin,** 1962

Dr. Hugo George Rodeck (1903–2005) took his bachelor's degree in chemistry and biology (1928) and his master's (1929) at the University of Colorado, where he worked at the university's museum (1922–1971), becoming Curator (1933) and then Director (1939). He was also Professor of Natural History, University of Colorado, from 1934. His doctorate in entomology was from the University of Minnesota (1944). He served in the U.S. Army Air Corps (1942–1945), returning to his old job after WW2. For April Fool's Day 2005, at the age of 101, he arranged for his definitive collection of lavatory papers of the world to arrive in a shoebox as a legacy to the Whole World Toilet Paper Museum. He had traveled extensively from the 1960s onward, and wherever he went he collected a sample and stored it in an envelope with a note of where it had been collected. He lived to see his 102nd birthday, and we are sure he had a laugh.

Rodenburg

Rodenburg's Mabuya *Mabuya rodenburgi* **Hoogmoed,** 1974

Willem F. Rodenburg, who collected the holotype, worked at the Center for Environmental Studies, Leiden (1987), and previously for the United Nations Food and Agriculture Organisation. He wrote *The Trade in Wild Animal Furs in Afghanistan* (1977).

Rodhain

Rodhain's Purple-glossed Snake *Amblyodipsas rodhaini* **Witte,** 1930

Dr. Jérome Alphonse Hubert Rodhain (1876–1956) was a physician, zoologist, and parasitologist at the School of Tropical Medicine, Université Libre de Bruxelles. He was Honorary Professor of Biological and Medical Sciences, Université Catholique de Louvain, and Honorary Director, Institut de Médecine Tropicale, Antwerp (1948–1949). He was in the Katanga region of the Belgian Congo (Zaire) (1910–1911). He wrote "Les petits crocodiles du District des Bangala" (1926).

Rodrigues

Rodrigues' Day Gecko *Phelsuma edwardnewtoni* **Vinson** and Vinson, 1969 EXTINCT

Named after the island of Rodrigues.

Rodrigues, M. T. U.

Rodrigues' Lava Lizard *Tropidurus nanuzae* Rodrigues, 1981

Rodrigues' Four-fingered Teiid *Procellosaurinus tetradactylus* Rodrigues, 1991

Rodrigues' Microteiid *Calyptommatus nicterus* Rodrigues, 1991

Rodrigues' Red Teiid *Procellosaurinus erythrocercus* Rodrigues, 1991

Gymnophthalmid lizard sp. *Pantepuisaurus rodriguesi* Kok, 2009

Dr. Miguel Trefaut Urbano Rodrigues (b. 1953) is a zoologist and herpetologist. He took a degree in quantitative biology, Université Paris VII (1978), and obtained his doctorate from Universidad de São Paulo (1984). Since 1996 he has been Professor of Biological Sciences, Universidad de São Paulo, and was Director of the university's Zoological Museum (1997–2001). A fish is named after him. See also **Trefaut.**

Rodriguez

Rodriguez's Anole *Anolis rodriguezi* **Bocourt,** 1873

Dr. Juan J. Rodriguez Luna (1840–1916) was Director, Museo Nacional de Historia Natural, Guatemala (1896–1916). See also **Luna.**

Roedinger

Roedinger's Lancehead *Bothrops roedingeri* **Mertens,** 1942

Hermann Rödinger (1879–1957) was a German businessman and naturalist. His father apprenticed him to a trading house (1895). After working for a time at his father's tobacco factory, he worked for a trading house in Guatemala (1900–1903), then went to Peru, where he was first employed as a clerk (1903–1911) and from 1911 owned his own business. His property was confiscated, and he was forced to live outside of Peru (1942–1947). On his return he was unable to recover his assets. He gave all his collections to museums in Hamburg. He collected the holotype during the Hamburg South Peru expedition (1936).

Roesler

Day gecko sp. *Phelsuma roesleri* Glaw et al., 2010

Roesler's Bent-toed Gecko *Cyrtodactylus roesleri* Ziegler et al., 2010

Herbert Rösler (b. 1952) is a German herpetologist in

Dresden who has studied and bred geckoes over a long period. He wrote *Geckoes* (1991).

Roger Roman

Graceful Brown Snake sp. *Rhadinaea rogerromani* Köhler and McCranie, 1999

Roger Roman is in charge of Nicaragua's Reserva de la Biosfera Bosawas, Managua.

Rogers

Rogers' Racer *Coluber rogersi* **Anderson,** 1893 [Syn. *Platyceps rogersi*]

Sir John Rogers was a physician who was Director-General, Sanitary Department, Egypt. He took swift and firm measures when there was an outbreak of the plague (1896), thus limiting the duration to six months and the death toll to 45 persons. He is probably the same person as the Sir John Rogers Pasha who was a Director of the Cairo Electric Railways and Heliopolis Oasis Society that was established (1906) with its headquarters in Shepheard's Hotel. Anderson wrote, "I have much pleasure in connecting Dr. Rogers's name with this new form."

Rohan

Rohan's Blind Dart Skink *Typhlacontias rohani* **Angel,** 1923

Comte Jacques de Rohan-Chabot (1889–1958) was a French explorer who led the Rohan-Chabot expedition to Angola and Rhodesia (Zimbabwe/Zambia) (1912–1914).

Rohde, Reinhold

Gaboon Lidless Skink *Lacertaspis rohdei* **Müller,** 1910

Reinhold Theodor Rohde, an Australian Lutheran missionary, arrived in Cameroon (then a German colony) for the Swiss Basel Mission prior to 1904. He collected fish and ants for Naturmuseum Senckenberg, Frankfurt (1909), and sent anthropological material to Goethe-Universität Frankfurt am Main. Despite being a British subject, he was ejected from his farm and mission by British troops at the outbreak of WW1. He was forced to live with fellow missionaries elsewhere in Cameroon (1915–1917), then he was arrested and taken to England with his wife and three children. The charges were that his passport was out of date, and it was suspected that he was not a British citizen but something more sinister. He was prevented from going to the mission headquarters in Basel or returning to Australia but was given shelter by Quakers. The family (five children by this time) returned to Australia only in 1921. Rhode was Pastor, Hatton Vale Lutheran Church, Queensland (1924–1948).

Rohde, Ricardo

Rhode's Snouted Snake *Simophis rohdei* **Boettger,** 1885

Ricardo Rohde was a collector in Paraguay (1885–1886). He joined the German New Guinea Company (1889). The difference between the spellings of "Rhode" and "Rohde" in the vernacular and the scientific names is typical of the ability of us Anglophones to get it wrong. A bird and an amphibian are named after him.

Roig

Mountain Lizard sp. *Phymaturus roigorum* **Lobo** and Abdala 2007

This unusual dedication honors a number of Brazilian naturalists—Fidel Roig, Virgilio Roig, Sergio Roig, Fidel Roig Juñent, and Arturo Roig Alsina, all of whom made inestimable contributions to the knowledge of Brazilian biodiversity and were great role models.

Rolfe

Skink sp. *Lerista rolfei* **L. A. Smith** and Adams, 2007

James "Jim" K. Rolfe is a herpetologist and Senior Technical Officer, Woodvale Wildlife Research Centre, Western Australian Department of Conservation and Land Management, where he has worked since 1984. He co-wrote "The Reptiles and Amphibians of Kimberley Rainforests" (1991).

Roman

Roman's Saw-scaled Viper *Echis leucogaster* Roman, 1972

Roman's Tropical Snake *Crotaphopeltis acarina* Roman, 1974

Brother Dr. Benigno Roman Gonzalez was a Venezuelan herpetologist, ichthyologist, and Jesuit monk whose doctorate was awarded by Universidad de Barcelona (1969). He wrote "Deux sous-espèces de la vipère *Echis carinatus* (Schneider) dans les territoires de Haute-Volta et du Niger" (1972). The Oceanologic Museum, Fundación La Salle, Venezuela, is named after him.

Ronaldo

Colubrid snake sp. *Calamodontophis ronaldoi* Franco, De Carvalho Cintra, and **De Lema,** 2006

Dr. Ronaldo Fernandes (b. 1966) is a herpetologist, comparative biologist, and Associate Professor at Museu Nacional, Rio de Janeiro. He graduated from Universidade Federal do Rio Grande do Sul with a master's degree (1993), having previously been awarded his bachelor's by Universidade Federal do Rio de Janeiro (1991). The University of Texas awarded his doctorate in quantitative biology (1995).

Roosevelt

Roosevelt's Giant Anole *Anolis roosevelti* **Grant,** 1931 [Alt. Culebra Island Giant Anole]

Roosevelt's Least Gecko *Sphaerodactyous roosevelti* Grant, 1931

Major General Theodore Roosevelt Jr. (1887–1944) fought in both world wars and was awarded America's highest decoration, the Congressional Medal of Honor. He was the eldest son of President Theodore Roosevelt. He was Governor, Puerto Rico (1929–1932), and Governor-General, the Philippines (1932–1933), after which he became a businessman, being Chairman of American Express. He and his brother, Kermit, have a mammal named after them.

Roper

Roper's Banded Snake *Simoselaps roperi* **Kinghorn,** 1931 [Alt. Northern Shovel-nosed Snake]

Named after the Roper River in Northern Territory, Australia.

Roper, G.

Chameleon sp. *Chamaeleo roperi* **Boulenger,** 1890

G. D. Trevor-Roper collected the holotype and presented it to the Natural History Museum, London.

Rosado

Rosado's Bevel-nosed Boa *Candoia paulsoni rosadoi* **H. M. Smith** et al., 2001

José P. O. Rosado joined the Harvard Museum of Comparative Zoology (1977) and is now Collection Manager, Herpetology Department.

Rosaura

Bay Island Least Gecko *Sphaerodactylus rosaurae* **Parker,** 1940

This gecko is named after Lord Moyne's yacht *Rosaura*, which was cruising near Honduras in 1937. Lord Moyne (1880–1944) and his chauffeur were murdered by members of the extremist Zionist Stern Gang when he was Resident British Minister in Cairo.

Rosén

Rosén's Desert Skink *Egernia inornata* Rosén, 1905

Rosén's Snail-eater *Dipsas infrenalis* Rosén, 1905

Rosén's Snake *Suta fasciata* Rosén, 1905

Nils Rosén (1882–1970) was a zoologist, ichthyologist, and herpetologist. He worked at the Biology Museum, Lunds Universitet (1910–1914). He edited *Svenskt Fiskelexicon* (1955).

Rosenberg

Rosenberg's Heath Monitor *Varanus rosenbergi* **Mertens,** 1957

Baron Carl (originally Karl) Benjamin Hermann von Rosenberg (1817–1888) was a German naturalist, ornithologist, and geographer who collected in the East Indies and mapped outlying districts of the archipelago. He enlisted as a common soldier in the Dutch colonial army in the Malay Archipelago and served for 30 years, the first 16 as a topographic draftsman on Sumatra, then as a civil servant in the Moluccas and around New Guinea. He traveled there in a Dutch warship, the *Etna*, and met Alfred Wallace. He wrote *Reistochten naar de Geelvinkbaai op Nieuw-Guinea in de jaren 1869 en 1870*. One mammal and five birds are named after him.

Rosenmann

Tree Iguana sp. *Liolaemus rosenmanni* **Núñez** and **Navarro,** 1992

Dr. Mario Rosenmann Abramovich (1933–2004) was a zoologist who, after early education in Chile, went to the University of Alaska, which awarded his doctorate. He returned to Chile (1974) and became Professor, Department of Ecological Sciences, Science Faculty, Universidad de Chile, Santiago de Chile. He co-wrote "Comparative Diet Activity of Pristidactylus Lizards from Forest and Scrubland Habitats" (1992).

Ross

Ross' Wolf Snake *Lycodon chrysoprateros* **Ota** and Ross, 1993

Ross' Calayan Gecko *Gekko rossi* Brown et al., 2010

Charles A. "Andy" Ross is a herpetologist and paleontologist who is retired from the Department of Vertebrate Zoology, the Smithsonian, where he was a specialist on crocodilians. He co-wrote "Four New Species of *Lycodon* (Serpentes: Colubridae) from the Northern Philippines" (1994).

Rossignon

Yucatan Snapping Turtle *Chelydra rossignoni* **Bocourt,** 1868 [Alt. Central American / Mexican Snapping Turtle]

Jules (or Julio) Rossignon (d. 1883) was a Frenchman who arrived in Guatemala in 1843 as Director of Scientific Affairs, Belgian Colonization Company. He started coffee cultivation at Finca "Las Victorias." He was internationally influential as a planter and grower of coffee, exhibiting at the Exhibition in Paris (1867). Before leaving Europe he had been Professor, Natural Sciences, Université de Paris. He wrote "Manual of Coffee, Cocoa, Vanilla, and Snuff in Spanish America and All Their Applications" (1859).

Rossikow

Uzbekistan Toadhead Agama *Phrynocephalus rossikowi* **Nikolsky,** 1898

Konstantin Nikolaevich Rossikow (often Rossikov) (b. 1854) was a zoologist, ichthyologist, arachnologist, and "zoographer." He graduated from the University of St. Petersburg and traveled in the Northern Caucasus region (1890), where he mentioned seeing "multi-colored" vipers. He wrote *a Trip to Zaakdan and the Bolshaya Laba River Head for the Purpose of Zoogeographic Research* (1890) and collected the agama holotype.

Rossman

Rossman's Garter Snake *Thamnophis rossmani* **Conant,** 2000

Professor Dr. Douglas "Dag" Athon Rossman (b. 1936) is a herpetologist specializing in garter snakes. He was awarded his doctorate by the University of Florida (1961). He is retired from his posts as Professor of Zoology, Louisiana State University, Baton Rouge, and Director, Museum of Natural Sciences, Louisiana State University. He co-wrote *Amphibians and Reptiles of Louisiana* (1996).

Rostombekov

Rostombekov's Lizard *Darevskia rostombekovi* **Darevsky,** 1957 [Syn. *Lacerta rostombekowi*]

Dr. V. N. Rostombekov was a Georgian biologist. He wrote "On the Herpetofauna of Abkhasia" (1939).

Roth, J.

Roth's Dwarf Racer *Eirenis rothii* **Jan,** 1863

Johannes Rudolph Roth (1814–1858) was a naturalist, botanist, entomologist, and malacologist, and a member of Akademie der Bildenden Künste München. He was on the 1840 Major Harris expedition to the Schoa area of Abyssinia (Ethiopia). He worked at Zoologische Staatssammlung München (1848). He visited Lebanon and Palestine and collected the holotype of this snake. He wrote "List of Mammalia, Observed and Partly Collected, in Abyssinia" (1843). Other taxa are named after him, including a species of garlic.

Roth, S.

Roth's Tree Iguana *Liolaemus rothi* **Koslowsky,** 1898

Dr. Santiago Roth (1850–1924) was a Swiss-born Argentine paleontologist, geologist, and naturalist whose baptismal names were Kaspar Jacob. His family moved to a Swiss community in Buenos Aires Province (1866). By 1870 he had collected sufficient fossils and plants to sell to finance his continued explorations. He surveyed the provinces of Entre Ríos and Corrientes (1890–1892), making a vertebrate collection from the "Conglomerado Osífero" of the Ituzaingó Formation, part of which was deposited in Museo de La Plata, where he was appointed Curator of Paleontology (1895). He later became Director, Bureau of Geology and Mines, for the province of Buenos Aires.

Rothschild

Rothschild's Skink *Paracontias rothschildi* **Mocquard,** 1905

Baron Maurice de Rothschild (1881–1957) was a member of the French branch of the famous banking family. In his youth he was a well-known playboy and quarreled with his relations over an investment they regarded as risky but that turned out to be highly profitable, as did many of his subsequent ventures. He became a politician and was one of the few senators of France to vote against giving Marshall Pétain full powers (1940). He was instrumental in helping De Gaulle become the leader of the Free French in exile in England during WW2, but he later upset De Gaulle and was virtually banished from France to the Bahamas. He traveled in East Africa (1904–1905).

Roule

Roule's Ground Snake *Atractus roulei* **Despax,** 1910
Chonburi Snake Skink *Isopachys roulei* **Angel,** 1920

Dr. Louis Roule (1861–1942) was a zoologist, ichthyologist, and herpetologist. He was at the Marine Biology Station, Faculty of Sciences, Université de Toulouse (1885–1910), becoming a Professor and gaining a doctorate of medicine at Paris (1902). He became Professor of Zoology, Muséum National d'Histoire Naturelle, Paris (1910).

Roux, J.

Roux's Blind Dart Skink *Typhlacontias gracilis* Roux, 1907
Roux's Sea Snake *Parahydrophis mertoni* Roux, 1910 [Alt. Merton's Sea Snake]
Roux's Emo Skink *Emoia loyaltiensis* Roux, 1913
Roux's Gecko *Bavayia crassicollis* Roux, 1913
Roux's Giant Gecko *Rhacodactylus sarasinorum* Roux, 1913
Roux's Lipinia *Lipinia rouxi* **Hediger,** 1934

Dr. Jean Roux (1876–1939) was a zoologist who gained his doctorate at Université de Genève (1899) and studied protozoa in Berlin until 1902. He was Curator, Naturhistorisches Museum Basel (1902–1930). He traveled in New Guinea and Australia (1907–1908) and in New Caledonia and the Loyalty Islands (1911–1912) with Fritz Sarasin. He wrote "Les reptiles de la Nouvelle-Calédonie et des Îles Loyalty" (1913).

Roux, P.

Roux's Forest Lizard *Calotes rouxii* **Duméril** and **Bibron,** 1837 [Alt. Bombay Agama]

(Jean Louis Florent) Polydore Roux (1792–1833) was a naturalist and painter. He studied under Cuvier at Muséum National d'Histoire Naturelle, Paris. He became Curator, Muséum d'Histoire Naturelle de Marseille (1819). Roux accompanied Carl Alexander Anselm Freiherr von Hügel (1796–1870) for the first part of his expedition, meeting him in Egypt and traveling on to India, but died in Bombay after they had argued and parted company. His death is a mystery; some sources say it was from plague. A mammal is named after him.

Roux-Estève

Amphisbaena sp. *Cynisca rouxae* Hahn, 1979

Roux-Estève's Worm Snake *Typhlops rouxestevae* **Trape** and Mane, 2004

Mme. Rolande Roux-Estève is a French herpetologist and ichthyologist who was Curator of Reptiles, Muséum National d'Histoire Naturelle, Paris. In the early 1950s she described fishes that Jaques Cousteau had collected in the Red Sea and off Aldabra during his expeditions with the *Calypso*. She wrote "Révision systématique des typhlopidae d'Afrique reptilia-serpentes" (1974).

Rowley

Rowley's Palm Pit-viper *Bothriechis rowleyi* **Bogert,** 1968

John Stuart Rowley (1907–1968) was an ornithologist who was a Research Associate, Department of Ornithology and Mammalogy, California Academy of Sciences. He took his degree at Berkeley and went to work (1933) in a weighing machine company, which he sold (1957) to devote the rest of his life to birds. Bogert and Rowley collected the pit-viper together (1967). He was killed in a fall from a cliff.

Roxane

Roxane's Blind Snake *Typhlops roxaneae* **Wallach,** 2001

Roxane Coombs is an artist and a librarian who retired from Harvard in 2003. Wallach waxed lyrical in his etymology: "Roxane Coombs, a special friend, talented artist, and excellent librarian . . . proofread my manuscripts, and supported my research in numerous ways."

Roze

Roze's Green Racer *Philodryas carbonelli* Roze, 1957

Roze's Ground Snake *Liophis breviceps canaima* Roze, 1957

Roze's Coastal House Snake *Thamnodynastes chimanta* Roze, 1958

Roze's Neusticurus *Neusticurus racenisi* Roze, 1958

Roze's Worm Lizard *Amphisbaena rozei* **Lancini** 1963

Roze's Hog-nosed Pit-viper *Porthidium lansbergii rozei* **Peters,** 1968

Dr. Janis Arnolds Roze (b. 1926) was born in Latvia. He is Emeritus Professor of Biology, City College and Graduate School, City University of New York, and was associated with the American Museum of Natural History and the United Nations. He was a founder of the International Center for Integrative Studies. He wrote "On the Synonymy and Holotypes of the Coral Snakes" (1966).

Rozella

Rozella's Lesser Galliwasp *Diploglossus rozellae* **H. M. Smith,** 1942

Rozella's Dwarf Short-tailed Snake *Tantillita lintoni rozellae* Perez-Higareda, 1985

Mrs. Rozella Blood Smith was the wife of Hobart Smith (q.v.), who named the lizard after her. She worked at the University of Kansas (1937). They married (1940) and spent a romantic honeymoon hunting snakes in Mexico. She has worked closely with him in research, and they co-wrote *The Synopsis of the Herpetofauna of Mexico* (in sections, 1968–1993).

Ruben

Ruben's Sand Lizard *Pedioplanis rubens* **Mertens,** 1955

Not an eponym. Someone has taken the binomial, which is Latin for "tinged with red," and assumed it to be a person's name; hence the common name, complete with spurious apostrophe.

Rudebeck

Skink sp. *Typhlacontias rudebecki* **Haacke,** 1997

Gustaf Rudebeck (1913–2005) was a zoologist, ornithologist, and entomologist at Lunds Universitet, where he became Professor Emeritus in Ecology. He was attached to the Transvaal Museum, South Africa (1950–1951 and 1954–1956). He was on the Swedish Lund University expedition to South Africa (1956).

Rueda

Rueda's Anole *Anolis megalopithecus* Rueda-Almonacid, 1989

José Vicente Rueda-Almonacid is a Colombian herpetologist. He co-wrote *Libro rojo de los anfibios de Colombia* (2004). An amphibian is named after him.

Ruhana

Deraniyagala's Earth Snake *Uropeltis ruhunae* **Deraniyagala,** 1954

Thorntail Snake sp. *Platyplectrurus madurensis ruhanae* Deraniyagala, 1954

Ruhana (or Ruhuna) was an ancient kingdom in Sri Lanka in the early centuries of the first millennium.

Ruhstrat

Formosa Wolf Snake *Lycodon ruhstrati* **Fischer,** 1886

Ernst Konrad A. Ruhstrat (d. 1913) was a German who joined the Imperial Chinese Customs Service (1881). He was stationed at Takow, Formosa (now Taiwan), from 1884. He wrote "Geschichtliche Notiz über die Insel Formosa" (1886) and collected the snake holotype. He died while on home leave.

Ruibal

Ruibal's Brown Tegu *Leposoma annectans* Ruibal, 1952

Ruibal's Least Gecko *Sphaerodactylus ruibali* **Grant,** 1959

Ruibal's Tree Iguana *Liolaemus ruibali* **Donoso-Barros,** 1961

Cabo Cruz Pallid Anole *Anolis ruibali* **Navarro** and **Garrido,** 2004

Dr. Rodolfo Ruibal (b. 1931) is a Cuban-born American biologist and herpetologist whose family emigrated to the USA when his father, a sugar trader, moved the family to Wall Street (1938). He worked at the University of California Riverside (1954–1995), retiring as Emeritus Professor. His bachelor's degree is from Harvard and his doctorate from Columbia University, New York. He was in the Medical Corps of the U.S. Army (1947–1948). He visited Cuba regularly until Castro took power. He collected in Argentina (1960–1961).

Ruiz

Anole sp. *Anolis ruizi* **Rueda** and **Williams,** 1986

Professor Pedro Miguel Ruiz-Carranza (1932–1998) was a herpetologist and Curator of Herpetology, Instituto de Ciencias Naturales, Museo de Historia Natural, Bogotá, Colombia. He co-wrote "Two New Species of *Hyla* from the Andes of Central Colombia and Their Relationships to Other Small Andean *Hyla*" (1957). Six amphibians are named after him.

Ruiz Leal

Ruizleal's Tree Iguana *Liolaemus ruizleali* **Donoso-Barros** and **Cei,** 1971

Dr. Adrián Ruiz Leal (1898–1980) was an Argentine botanist. A street in Mendoza is named after him.

Rüppell/Rueppell

Rüppell's Desert Chameleon *Chamaeleo affinis* Rüppell, 1845

Arboreal Agama *Agama rueppelli* **Vaillant,** 1882

Wilhelm Peter Eduard Simon Rüppell (1794–1884) was a collector in the broadest sense, of coins and manuscripts as well as natural history specimens. He went to Egypt and ascended the Nile to Aswan (1817) and later made two extended expeditions to eastern Africa, Sudan (1821–1827), and Ethiopia (1830–1834), bringing back large zoological and ethnographical collections. He wrote *Reise in Abyssynien* (1838–1840). Nine birds and five mammals are named after him.

Rurk

Rurk's Ristella *Ristella rurkii* **Gray,** 1839

Dr. Rurk sent the holotype of this Indian skink to Gray. That is all Gray says about Rurk, and we can find nothing to add.

Ruschenberger

Central American Tree Boa *Corallus ruschenbergerii* **Cope,** 1876

Commodore Dr. William Samuel Waithman Ruschenberger (1807–1895) was a physician in both the U.S. Amy and Navy and was fleet surgeon on several cruises. He was a member of the Philadelphia Academy of Natural Sciences. At the outbreak of the American Civil War he became Chief Surgeon at the Boston Navy Yard. Cope might have regretted the eponym, as later he had disagreements with Ruschenberger, who reported "an unpleasant run-in" with Cope in the foyer of Philadelphia's St. George Hotel (1883). He went on to say, "My conviction is that Mr Cope is ambitious, selfish, unscrupulous and wholly unreliable and unfaithful." He wrote *Three Years in the Pacific* (1834).

Ruspoli

Ruspoli's Chameleon *Chamaeleo ruspolii* **Boettger,** 1893

See **Prince Ruspoli.**

Russel

Russel's Sea Snake *Hydrophis obscurus* **Daudin,** 1803

See **Russell, P.**

Russell, A.

Russell's Gecko *Cyrtodactylus russelli* **Bauer,** 2003

Professor Dr. Anthony Patrick Russell (b. 1947) is a zoologist who was awarded his bachelor's degree by the University of Exeter (1969) and his doctorate in zoology by the University of London (1972). He teaches biological sciences at the University of Calgary, Canada. He wrote *The Amphibians and Reptiles of Alberta* (1993).

Russell, P.

Russell's Viper *Daboia russelii* **Shaw,** 1797

Russell's Sand Boa *Eryx conicus* **Schneider,** 1801 [Alt. Rough-scaled Sand Boa; Syn. *Gongylophis conicus*]

Russell's Kukri Snake *Oligodon taeniolatus* **Jerdon,** 1853 [Alt. Streaked Kukri Snake]

Dr. Patrick Russell (1726–1805) was a British surgeon and naturalist. He first went to India in 1781 to look after his brother, who was employed by the Honourable East India Company in Vizagapatnam. He became fascinated by the plants in the region and was appointed to be the company's Botanist and Naturalist, Madras Presidency (1785). He spent six years in Madras (Chennai) and sent a large collection of snakes to the British Museum (1791). One of his major concerns was snakebite, and he tried to find a way for people to identify poisonous snakes without first getting bitten and seeing what happened. The sand boa has his name attached to it because it appears to mimic Russell's Viper, something he commented on in his *A Continuation of an Account of Indian Serpents* (1801).

Russell Train

Russell Train's Marble Gecko *Gekko russelltraini* Ngo, **Bauer**, Wood, and **Grismer**, 2009

Dr. Russell Errol Train (b. 1920) is a lawyer and public servant who earned a bachelor's degree from Princeton (1941), after which he served in the U.S. armed forces (1941–1946). He was a judge for the U.S. Tax Court (1957–1965); President of the Conservation Foundation (1965–1969); Under Secretary of the Department of the Interior (1969–1970); Chairman of the Council on Environmental Quality (1970–1973); Administrator of the Environmental Protection Agency (1973–1977); and President (1978–1985), Chairman (1985–1994), and Emeritus Chairman of the World Wildlife Fund in the USA.

Russow

Russow's Bent-toed Gecko *Cyrtopodion russowii* **Strauch,** 1887

Valerian von Russow (1842–1879) was an Estonian naturalist. He became Curator of the Natural History Museum of the University of Tartu. He gave an overview of the birds of Matsalu Bay (1870), which was later designated a nature reserve. He was Head of the Ornithological Department of the Imperial Academy Science in St. Petersburg (where Strauch became Director of the Zoological Museum [1879]). He went on an expedition to Turkmenistan (1878–1879). He died of smallpox.

Ruthven

Ruthven's Kingsnake *Lampropeltis ruthveni* **Blanchard,** 1921

Ruthven's Macropholidus *Macropholidus ruthveni* **Noble,** 1921

Ruthven's Anole *Anolis bonairensis* Ruthven, 1923

Ruthven's Whipsnake *Masticophis schotti ruthveni* Ortenburger, 1923

Ruthven's Bachia *Bachia talpa* Ruthven, 1925

Ruthven's Earth Snake *Geophis ruthveni* **Werner,** 1925

Ruthven's Anadia *Anadia pulchella* Ruthven, 1926

Ruthven's Scaly-eyed Gecko *Lepidoblepharis ruthveni* **Parker,** 1926

Ruthven's Burrowing Snake *Apostolepis tenuis* Ruthven, 1927

Louisiana Pine Snake *Pituophis ruthveni* Stull, 1929 [Syn. *Pituophis melanoleucus ruthveni*]

Bleached Earless Lizard *Holbrookia maculata ruthveni* **H. M. Smith,** 1943

Dr. Alexander Grant Ruthven (1882–1971) was a herpetologist who was awarded his doctorate in zoology by the University of Michigan (1906) and worked at that university's Museum of Zoology (1906–1929), first as Curator and then as Director (1913) and Professor of Zoology (1915). He was President of the University of Michigan (1921–1951) and Emeritus Professor thereafter. He was President of the American Society of Ichthyologists and Herpetologists (1962). He co-wrote *Herpetology of Michigan* (1912).

Rutten

Venezuela Leaf-toed Gecko *Phyllodactylus rutteni* **Hummelinck,** 1940

Louis Martin Robert Rutten (1884–1946) was a geologist and paleontologist. He mapped large parts of the Dutch East Indies (Indonesia), Cuba, and the Netherlands Antilles. He graduated from Universiteit Utrecht (1909), joined a company later known as Royal Dutch Shell, and was sent oil prospecting in Borneo. He traveled on business to Argentina, Cuba, Mexico, and Peru (1919–1921). He became Professor of Crystallography, Geology, and Paleontology, Universiteit Utrecht (1921). He led expeditions by his students to the Netherlands Antilles (1930) and to Cuba (1933 and 1938).

S

Sack

Sack's Spotted Whiptail *Aspidoscelis sackii* **Wiegmann,** 1834

Baron Sebastian Albert von Sack was a traveler and chamberlain to the King of Prussia. He wrote *A Narrative of a Voyage to Surinam. Of a Residence There during 1805, 1806 and 1807 and of the Author's Return to Europe by the Way of North America* (1801). He traveled in Cyprus and Egypt (1817–1821), during which he made a small collection of birds. He decided to travel to Mexico and recruited Ferdinand Deppe (q.v.) to accompany him. The expedition (1824–1829) was marred by the death of the Baron's footman from yellow fever during the voyage from Jamaica to Veracruz.

Sadlier

Sadlier's Caledonian Skink *Caledoniscincus orestes* Sadlier, 1986

Sadlier's Skink *Graciliscincus shonae* Sadlier, 1987

Sadlier's New Caledonian Gecko *Bavayia septuiclavis* Sadlier, 1988

Sadlier's Menetia *Menetia sadlieri* **Greer,** 1991

Sadlier's Bevel-nosed Boa *Candoia paulsoni sadlieri* **H. M. Smith** and **Chiszar,** 2001

Northern Rough-scaled Snake *Tropidechis sadlieri* Hoser, 2003

Ross Allen Sadlier (b. 1955) has been the Collection Manager of the Herpetological Section of the Australian Museum, Sydney, since 1996. He came to the museum as a volunteer in 1978 and joined the staff as a Technical Officer in 1980. His bachelor's degree (1986) is from Macquarie University. His main interest is the rainforest fauna of eastern Australia and New Caledonia. He co-wrote *The Herpetofauna of New Caledonia* (2000).

Saengsom

Banded Cat Snake *Boiga saengsomi* **Nutaphand,** 1985

Buntot Saengmahasom is an animal collector. He sent a consignment of snakes from Krabi Province to Bangkok. It included a specimen of this snake, which spent the rest of its life in the Pata Zoo. Why Nutaphand contracted Saengmahasom's full name for the binomial is unknown.

Sage

Tree Iguana sp. *Liolaemus sagei* **Etheridge** and Christie, 2003

Dr. Richard David Sage is a zoologist at the Museum of Vertebrate Zoology, University of California, Berkeley. He co-wrote "Taxonomy of the House Mouse" (1981). The etymology notes "his important contribution to Argentine herpetology." A mammal is named after him.

Sagre

Sagré's Anole *Anolis sagrei* **Duméril,** 1837

See **De la Sagra.**

Saint Girons

New Caledonian Sea Krait *Laticauda saintgironsi* **Cogger** and **Heatwole,** 2005

Kukri Snake sp. *Oligodon saintgironsi* David, **Vogel,** and **Pauwels,** 2008

Dr. Hubert Saint Girons (1926–2000) was a distinguished French herpetologist. The vipers he saw on the grounds of his chateau fascinated him. Université de Paris rewarded his studies of them with a doctorate (1951). He then worked at Centre National de la Recherche Scientifique and in 1962 became France's youngest Directeur de Recherche. He kept snakes in large outdoor enclosures, allowing him to study them in near-natural environments. He worked for many years at Centre d'Ecologie, Brunoy, then in Laboratoire d'Evolution des Etres Organisés, Université Pierre et Marie Curie, Paris. He published over 200 scientific papers, including the co-written "Reproductive Cycle of the Male Tuatara, *Sphenodon punctatus*, on Stephens Island, New Zealand" (1987).

Sajdak

Great Bird Island Racer *Alsophis sajdaki* **Henderson,** 1990

Richard A. Sajdak (b. 1945) was Curator of Reptiles at Milwaukee County Zoo, Milwaukee, Wisconsin. He and Henderson co-wrote "Status of West Indian Racers in the Lesser Antilles" (1991).

Salensky

Salensky's Toadhead Agama *Phrynocephalus salenskyi* **Bedriaga,** 1907

Dr. Vladimir Vladimirovich Zalenski (1847–1918) was an embryologist, anatomist, and zoologist who was Director of the Zoological Institute and Museum of the Imperial Academy of Sciences, St. Petersburg (1897–1906). He studied the embryology of neural systems in invertebrates and fishes. He wrote *Equus Przewalski* (1902). A mammal is named after him.

Salgueiro

Espírito Santo Blind Snake *Leptotyphlops salgueiroi* **Amaral,** 1955

W. S. Salgueiro collected the holotype (1934). Amaral wrote the description when it was rediscovered, naming it after the earlier collector.

Sallé

Sallé's Anole *Anolis sericeus sallaei* **Günther,** 1859

Sallé's Earth Snake *Geophis sallaei* **Boulenger,** 1894

Auguste Sallé (1820–1896) was a French taxonomist and entomologist who collected in tropical parts of the Americas (1846–1856). Many taxa are named after him.

Salvadori

Crocodile Monitor *Varanus salvadorii* **Peters** and **Doria,** 1878

Conte Adelardo Tommaso Paleotti Salvadori (1835–1923) was an eminent physician, author, educator, and ornithologist. He was Vice Director of the Museum of Zoology at Università degli Studi di Torino (1879–1923). He was Medical Officer in Garibaldi's battalion during his second "expedition" in Sicily. His collection is at the Natural Science Museum of Villa Vitali. He wrote *Catalogo sistimatico degli uccelli di Borneo* (1874). Eighteen birds are named after him.

Salvin

Giant Musk Turtle *Staurotypus salvinii* **Gray,** 1864

Salvin's Anole *Anolis salvini* **Boulenger,** 1885

Salvin's Spiny Lizard *Sceloporus salvini* **Günther,** 1890

Huamantlan Rattlesnake *Crotalus scutulatus salvini* Günther, 1895

Osbert Salvin (1835–1898) was a naturalist and ornithologist. He was the first European to record an observation of a Resplendent Quetzal. He pronounced it "unequalled for splendour among the birds of the New World"—and promptly shot it. Salvin contributed to and co-edited the 40-volume *Biologia Centrali Americana* (1879), a near-complete catalogue of Central American species. Two mammals and about 20 birds are named after him.

Sanchez

Sanchez's Night Lizard *Xantusia sanchezi* **Bezy** and Flores, 1999 [Alt. Zacatecas Night Lizard]

Dr. Oscar Sanchez Herrera (b. 1954) is a herpetologist who was at Instituto de Biologia, Universidad Nacional Autónoma de México (1999). Presently he is Director for Flora and Woodland Fauna Sub-Secretariat of Forest and Woodland Fauna, Secretariat for Agriculture and Water Resources. He co-wrote "Another Suggested Case of Ophidian Deceptive Mimicry" (1981).

Sandford

Sandfords' Ethiopian Mountain Snake *Pseudoboodon sandfordorum* Spawls, 2004

Named after a whole family. Spawls wrote, "Three generations of Sandfords have served Ethiopia, their adopted country, in the military field, in education and development."

Sanford

Sanford's Emo Skink *Emoia sanfordi* **Schmidt** and **Burt,** 1930

Dr. Leonard Cutler Sanford (1868–1950) was a zoologist and Trustee of the American Museum of Natural History. He co-wrote *The Waterfowl Family* (1924). Five birds, a mammal, and other taxa are named after him.

Sang

Sang's Reed Snake *Calamaria sangi* Truong, Koch, and Ziegler, 2009

Dr. Nguyen Van Sang of the Institute of Ecology and Biological Resources, Hanoi, is a zoologist and expert on the squamate reptiles of Vietnam. He collected the holotype of the colubrid snake. He co-wrote *Herpetofauna of Vietnam* (2009). See also **Nguyen Van Sang.**

Sanoja

Sanoja's Canyon Lizard *Sceloporus merriami sanojae* **H. M. Smith, Lemos-Espinal,** and **Chiszar,** 2003

Susy Sanoja Sarabia is a herpetologist and is also Mrs. Lemos-Espinal. She has co-written a number of papers, such as "Ecological Observations of the Lizard *Xenosaurus grandis* in Cuautlapan, Veracruz" (1995).

Sapper

Sapper's Rustyhead Snake *Amastridium veliferum sapperi* **Werner,** 1903

Sapper's Variable Coral Snake *Micrurus diastema sapperi* Werner, 1903

Karl Theodor Sapper (1886–1945) was a traveler, antiquarian, vulcanologist, linguist, and explorer who spent more than a dozen years exploring southern Mexico and Central America at the end of the 19th century. Despite suffering from malaria, he recorded Mayan languages; over 100 years later, his notes and records are all we know of several Mayan dialects. After returning to Germany he taught at Université de Strasbourg (then in Germany), then at Universität Würzburg, and established the Institute for American Studies. His publications include *The Volcanoes of Central America* (1925).

Sara

Sierra de Coalcomán Striped Snake *Coniophanes sarae* Ponce-Campos and **H. M. Smith,** 2001

Sara M. Huerta-Ortega is a biologist who works for the organization "Tropical Forest" in Jalisco, where she is head of Ecology and Conservation Biology. She co-wrote "Range Extensions and Variational Notes on Some Amphibians and Reptiles of Jalisco and Michoacán, México" (2003). The original description says that she is honored "for her contributions to the knowledge of the herpetology and crocodile conservation in Jalisco."

Sarasin

Müller's Nessia *Nessia sarasinorum* **Müller,** 1889
Sarasins' Reed Snake *Pseudorabdion sarasinorum* Müller, 1895
Keelback Snake sp. *Amphiesma sarasinorum* **Boulenger,** 1896
Forest Skink sp. *Sphenomorphus sarasinorum* Boulenger, 1897
Roux's Giant Gecko *Rhacodactylus sarasinorum* **Roux,** 1913

Paul Benedict Sarasin (1856–1929) and Fritz Carl Frederich Sarasin (1859–1942) were cousins. They were Swiss zoologists, explorers, and collectors who wrote *Reisen in Celebes* (1905). Two birds and a mammal are named after them.

Sarg

Sarg's Earth Snake *Adelphicos sargii* **Fischer,** 1885
[Syn. *A. quadrivirgatus sargii*]

Franz Sarg (1840–1920) was a businessman who had coffee and sugar plantations in the Alta Verapaz, Guatemala, where he was the German Consul and from whence he sent specimens of this snake to Staatlisches Museum für Naturkunde Stuttgart.

Sargent

Sargent's Graceful Brown Snake *Rhadinaea sargenti* **Dunn** and **Bailey,** 1939

W. M. Sargent was a surveyor employed by the Panama Canal Company. He was head of a surveying party that collected 78 snake specimens, including the type of this one (1937).

Sarmiento

Tree Iguana sp. *Liolaemus sarmientoi* **Donoso-Barros,** 1973

Pedro Sarmiento de Gamboa (1532–1592) was a Spanish explorer, navigator, scientist, and author. He was in Mexico (1555–1557) and then lived in Peru for over 20 years. He took part in an expedition that discovered the Solomon Islands (1568). As the first person to pay serious scientific attention to the area round the Strait of Magellan, he was commemorated in the name of this lizard.

Sartorius

Sartorius' Snail-sucker *Sibon sartorii* **Cope,** 1863
[Syn. *Tropidodipsas sartorii*]

Dr. Christian Carl Wilhem Sartorius (1796–1872) was a naturalist who lived and collected in Mexico (1826–1872). His hacienda, El Mirador, was a magnet for German scientists, especially botanists. Sartorius collected everything and is mentioned in connection with, inter alia, botany, herpetology, and ornithology.

Sauter

Kosempo Keelback *Amphiesma sauteri* **Boulenger,** 1909
Koshun Grass Lizard *Takydromus sauteri* **Van Denburgh,** 1909
Taiwan Coral Snake *Sinomicrurus sauteri* **Steindachner,** 1913

Dr. Hans Sauter (1871–1943) was an entomologist who developed an interest in herpetology. He studied biology at Ludwig-Maximilians-Universität München and Eberhard Karls Universität Tübingen. He was in Formosa (Taiwan, then Japanese owned), collecting insects (1902–1904). He was in Tokyo (1905) and then returned to Taiwan for the rest of his life. He worked for a British trading company but spent as much time as he could on entomology. Though Japan and Germany were enemies during WW1, he kept his job and continued collecting, though he was kept under observation. He was the first person to offer private piano lessons in Taiwan and gave German and English lessons.

Sauvage

Sauvage's New Caledonian Gecko *Bavayia sauvagii* **Boulenger,** 1883
Sauvage's Snail-eater *Dipsas albifrons* Sauvage, 1884

Dr. Henri-Èmile Sauvage (1844–1917) was a French paleontologist, herpetologist, and ichthyologist. He wrote "Note sur les geckotiens de la Nouvelle-Calédonie" (1878).

Savage

Savage's Sand Snake *Chilomeniscus savagei* Cliff, 1954
Yellow-spotted Pilbara Gecko *Diplodactylus savagei* **Kluge,** 1963
Racerunner (lizard) sp. *Pseuderemias savagei* **Gans, Laurent,** and Pandit, 1965
Savage's Least Gecko *Sphaerodactylus savagei* **Shreve,** 1968
Savage's Earth Snake *Geophis downsi* Savage, 1981

Dr. Jay Mathers Savage (b. 1928), a herpetologist, is Emeritus Professor of Biology at the University of Miami and Adjunct Professor at San Diego State University. He took all his degrees at Stanford (1950,1954, and 1955). He was a member of the faculty at the University of Southern California before taking up his post in Florida. He has published nearly 200 papers and 3 books including *The Amphibians and Reptiles of Costa Rica: A Herpetofauna between Two Continents between Two Seas* (2002).

Savigny

Savigny's Fringe-fingered Lizard *Acanthodactylus savignyi* Audouin, 1809
Savigny's Agama *Trapelus savignii* **Duméril** and **Bibron,** 1837

Marie Jules Cesar Lelorgne de Savigny (1778–1851) was a

zoologist and artist. He studied medicine, but influenced by Étienne Geoffroy Saint-Hilaire, turned to zoology. He was in Egypt during Napoleon's occupation (1798–1800) and undertook several expeditions, including one to Lake Manzala, where he studied the birds, assisting Saint-Hilaire, with whom he eventually quarreled so badly that Saint-Hilaire prevented him becoming Professor at Muséum National d'Histoire Naturelle, Paris. He wrote *Description d'Egypte; ou Recueil des observations et des recherches qui ont été faites en Egypte pendant l'expédition de l'armée française* (1798–1801).

Sawin

Sawin's Tropical Night Lizard *Lepidophyma sawini* **H. M. Smith,** 1973

Professor H. Lewis Sawin was an expert on the fictional works of George Meredith and a computer expert in the English Department, University of Colorado. He started a program of using computers to assist in integrating conflicting bibliographies (1964). His techniques were adopted and adapted by herpetologists. He appears as co-author of a number of books and papers, such as "A Summary of Snake Classification (Reptilia, Serpentes)" (1977).

Say

Say's Pine Snake *Pituophis melanoleucus sayi* **Schlegel,** 1837 [Alt. Bull Snake; Syn. *P. catenifer sayi*]

Thomas Say (1787–1843) was a self-taught naturalist, zoologist, and entomologist who described over 1,000 new species of beetles and over 400 other new insects. He originally trained as a pharmacist and was a charter member and first Secretary of the Philadelphia Academy of Natural Sciences (1812). He was Professor of Natural History at the University of Pennsylvania (1822–1828). He was chief zoologist with Major Long's expeditions, and explored the Rocky Mountains with him (1819–1820) and the sources of the Missouri River (1823). He was a Quaker and lived at the utopian village of New Harmony (1825–1843) in the Indiana wilderness, where he was the Superintendent of Literature, Science, and Education. He wrote *American Entomology* (1824).

Schach

Schach's Ground Snake *Atractus schach* **Boie,** 1827

We think this is not named after a person and that a wrong assumption was made when coining the vernacular name. Schach means "chess" in German, so could this refer to the appearance of the snake, which when coiled could look like the checkered board used in chess?

Schadenberg

Southern Burrowing Skink *Brachymeles schadenbergi* **Fischer,** 1885

Alexander Schadenberg (1851–1896) was a German chemist who went to the Philippines to join a wholesale pharmaceutical company. With Otto Koch he went on a scientific expedition to Mindanao (1881), staying for six months, and climbed the volcano Mount Apo (1882). Their collection of botanical and zoological specimens included thousands of butterflies. A mammal is named after him.

Schaefer, C.

Schaefer's Spinejaw Snake *Xenophidion schaeferi* **Günther** and **Manthey,** 1995

Christian Schäfer, an amateur herpetologist from Berlin, collected the holotype of this snake.

Schaefer, H.

Cameroon Worm Lizard *Cynisca schaeferi* **Sternfeld,** 1912

Brothers Hans Schäfer (b. 1884) and Dr. Fritz Schäfer (d. 1911) both collected natural history specimens in Africa. In early 1900 one at least was in Namibia, and later that year one or the other was in Cameroon. Their botanical specimens from both areas are in the collection of Universität Berlin. One of them collected the holotype of this lizard in Cameroon, and as Hans was definitely located there, we believe it is named after him.

Schatti

Schatti's Racer *Coluber messanai* **Schätti** and **Lanza,** 1989

Beat Schätti was Curator of the Department of Herpetology and Ichthyology at Museum d'Histoire Naturelle, Geneva. While he was Curator, he and his girlfriend were arrested (1991) after a complaint that he had beaten up a man with whom he was in dispute. He later complained that he was beaten by the police and that he had been unlawfully imprisoned. His case was taken up and published by Amnesty International. He co-wrote "The Herpetofauna of Southern Yemen and the Sokotra Archipelago" (1999).

Scheben

Skink sp. *Scelotes schebeni* **Sternfeld,** 1910

Leonard Scheben lived in Windhoek, Namibia, and collected natural history specimens including reptiles, amphibians, and ants during at least the first three decades of the 20th century.

Scheffler

Scheffler's Dwarf Gecko *Lygodactylus scheffleri* **Sternfeld,** 1912

Georg Scheffler was a German collector working in East Africa at the end of the 19th century and the beginning of the 20th century. It is known that he was still collecting in Tanganyika (Tanzania) in 1911. A number of African shrubs and trees are named after him.

Scherz

Gecko sp. *Pachydactylus scherzi* **Mertens,** 1954

Dr. Ernst Rudolph Scherz (1906–1981) was a chemical engineer who was educated in Berlin. He hated Nazism and in 1933 emigrated to South-West Africa (Namibia) and was Managing Director of the Karakul Breeders Association (1946–1963), karakul being a breed of sheep. He was interested in rock art and accompanied Abbé Breuil on expeditions to Brandberg (1947–1950) and produced unrivaled documentation of Namibian rock art. He and his wife, a well-known photographer, returned to Germany in 1980.

Schevill

Black-soil Rises Ctenotus *Ctenotus schevilli* **Loveridge,** 1933

William Edward Schevill (1906–1994) was a cetologist and biologist who was the Librarian of the Museum of Comparative Zoology and Assistant Curator of Invertebrate Paleontology at Harvard (1938). He was at the Woods Hole Oceanographic Institution, Massachusetts (1950s), when he and his wife, the zoologist Barbara Lawrence, made the first hydrophone recordings of cetaceans in the wild. He collected the holotype of this skink (1932).

Schiede

Schiede's Anole *Anolis schiedei* **Wiegmann,** 1834

Dr. Christian Julius Wilhelm Schiede (1798–1836) was a German physician, naturalist, and botanist who accompanied Ferdinand Deppe (q.v.) when he was in Mexico (1828–1836). He practiced medicine in Mexico after Deppe left.

Schinz

Schinz's Beaked Blind Snake *Rhinotyphlops schinzi* **Boettger,** 1887

Dr. Heinrich Rudolph Schinz (1777–1861) was an ornithologist, herpetologist, and physician who gave up medicine to teach natural history at the Medical Institute and Universität Zürich. He published *Die Vögel des Schweiz* with Meisner (1815).

Schlegel

Schlegel's Japanese Gecko *Gekko japonicus* **Schlegel,** 1836

Schlegel's Golden Snake *Cercophis auratus* Schlegel, 1837

Schlegel's Thread Coral Snake *Leptomicrurus collaris* Schlegel, 1837

False Gharial *Tomistoma schlegelii* **Müller,** 1838

Schlegel's Blind Snake *Leptotyphlops nigricans* Schlegel, 1839

Eyelash Viper *Bothriechis schlegelii* **Berthold,** 1846

Schlegel's Beaked Blind Snake *Rhinotyphlops schlegelii* Bianconi, 1847

Pink-headed Reed Snake *Calamaria schlegeli* **Duméril, Bibron,** and **Duméril** 1854

Schlegel's Adder *Aspidomorphus schlegelii* **Günther,** 1872

Forest Skink sp. *Sphenomorphus schlegeli* **Dunn,** 1927

Professor Hermann Schlegel (1804–1884) was a German zoologist, the first person to use trinomials to describe separate races of animals. In his youth C. L. Brehm, the father of Alfred Brehm, tutored him. He assisted at Universität Wien (1824–1825), was recruited by Temminck as a preparator at Nationaal Natuurhistorisch Museum, Leiden (1825), and succeeded Temminck as Director there (1858). Although primarily an ornithologist, he wrote extensively on herpetology. He co-wrote *Fauna Japonica* (1845–1850). Ten birds, two mammals, and many other taxa are named after him.

Schleich

Agamid lizard sp. *Sitana schleichi* Anders and **Kästle,** 2002

Dr. Hans Hermann Schleich (b. 1952) is a herpetologist. He studied zoology, paleontology, and geology at Ludwig-Maximilians-Universität München, where he is now a Professor. He is active in the society Amphibian and Reptile Conservation of Nepal and co-wrote *Amphibians and Reptiles of Nepal* (2002).

Schlueter

Schlueter's Snake-eyed Lizard *Ophisops elegans schlueteri* **Boettger,** 1880

Schlueter's Sipo *Chironius schlueteri* **Werner,** 1899

Wilhelm Schlüter (1828–1919) founded the eponymous firm of natural history dealers in Halle (1853). His father, Friedrich, was an entomologist and malacologist. His brother, Julius, emigrated to Brazil and was one of a network of collectors, including the Geisler brothers in Australia and New Guinea, who kept the company supplied with specimens. The company was preeminent in the trade and until 1914 dealt with all the major museums and private collectors. It is likely that Werner and Boettger needed to keep in with the firm of Schlüter. Werner's description specifically mentions the company as having supplied the specimens. The company was later run by Wilhelm's sons, both of whom had studied natural sciences: Wilhelm "Willy" Schlüter Jr. (1886–1938) and Curt (1881–1944). After WW2 Schlüter relocated from the Russian Zone (later East Germany) to West Germany.

Schmeltz

Robust Rainbow-skink *Carlia schmeltzii* **Peters,** 1867

Dr. Johannes Dietrich Eduard Schmeltz (1839–1909) had no formal training but became one of the leading experts on Pacific Islands ethnography. He was the Director of the Godeffroy Museum, Hamburg (1861–1883), which specialized in the natural history and ethnography of the South Seas. The Godeffroy Shipping and Trading Company, which was eventually bankrupted, owned it, and the collections were all sold off (1886). Schmeltz was at Rijksmuseum voor Volkenkunde, Leiden (1883–1909), first as Curator and then as Director (1897).

Schmidt, A.

Schmidt's Mastigure *Uromastyx alfredschmidti* Wilms and **Böhme,** 2001 [Alt. Ebony Mastigure]

Alfred A. Schmidt was a herpetologist from Frankfurt. He wrote *The Concept of Nature in the Theory of Karl Marx* (1971). See also **Alfred Schmidt.**

Schmidt, K. P.

Schmidt's Orange-throated Whiptail *Aspidoscelis hyperythrus schmidti* **Van Denburgh** and **Slevin,** 1921

Schmidt's Ground Skink *Scincella schmidti* **Barbour,** 1927

Spiny Lizard sp. *Sceloporus schmidti* Jones, 1927 [Junior syn. of *S. smaragdinus* Bocourt, 1873]

Schmidt's Helmet Skink *Tribolonotus schmidti* **Burt,** 1930

Schmidt's Bold-eyed Tree Snake *Thrasops schmidti* **Loveridge,** 1936

Faded Black-striped Snake *Coniophanes schmidti* **Bailey,** 1937

Schmidt's Anole *Anolis schmidti* **H. M. Smith,** 1939

Lined Tree Lizard *Urosaurus ornatus schmidti* **Mittleman,** 1940

Schmidt's Green Racer *Philodryas tachymenoides* Schmidt and **Walker,** 1943

Schmidt's Monitor *Varanus karlschmidti* **Mertens,** 1951

Schmidt's Emo Skink *Emoia schmidti* **Brown,** 1954

Schmidt's Reed Snake *Calamaria schmidti* **Marx** and **Inger,** 1955

Beaked Blind Snake sp. *Rhinotyphlops schmidti* **Laurent,** 1956

Schmidt's Fringe-fingered Lizard *Acanthodactylus schmidti* **Haas,** 1957

Schmidt's Tree Iguana *Liolaemus schmidti* Marx, 1960

Schmidt's Worm Lizard *Amphisbaena schmidti* **Gans,** 1964

Karl Patterson Schmidt (1890–1957) was a herpetologist. He graduated from Cornell (1916) and worked as a Scientific Assistant in herpetology at the American Museum of Natural History until 1922. He was Assistant Curator of the newly founded Department of Amphibians and Reptiles, Field Museum (1922–1940), then Curator of Zoology (1941–1955), becoming Emeritus Curator in the latter year. He undertook many expeditions, to destinations including Santo Domingo (1916), Puerto Rico (1919), Central America (1923), Brazil (1926), and Guatemala (1933). He edited the journal *Copeia* (1937–1949). He was an avid collector of herpetological specimens and wrote books on them. He co-wrote *Field Book of Snakes of the US and Canada* (1941). He was bitten by a boomslang. Believing that the juvenile snake could not inject a fatal dose of venom, he went home to his wife and received no medical treatment. He kept a careful note of the development of the symptoms he experienced until he died. See also **Karl Schmidt.**

Schmidt, P. M. P. F.

Schmidt's Sea Snake *Praescutata viperina* Schmidt, 1852 [Alt. Viperine Sea Snake; Syn. *Thalassophis viperina*]

Philipp Moses Paul Frederich Schmidt (1800–1869/1873) was a physician in Hamburg who wrote *Beiträge zur ferneren Kenntniss der Meerschlangen* (1852).

Schmidt, P. Y.

Schmidt's Whip Snake *Dolichophis schmidti* **Nikolsky,** 1909 [Syn. *Coluber schmidti, Hierophis schmidti*]

Professor Petr Yulevich Schmidt (1872–1949) was a specialist in ichthyology at the Zoological Museum and Institute, St. Petersburg. After Nikolsky described a new species of fish (1889), Schmidt gave its status wider credence in *Fishes of Eastern Seas of the Russian Empire* (1904).

Schmidtler

Zagros Whip Snake *Coluber schmidtleri* **Schätti** and McCarthy, 2001 [Syn. *Platyceps najadum schmidtleri*]

Josef Friedrich Schmidtler (b. 1942) is a German herpetologist. He co-wrote "A New Dwarf-snake (Eirenis) from Lake Van in Eastern Turkey" (1990). He often worked with his father, Josef Johann Schmidtler (1910–1983).

Schmitz

Schmitz' Agama *Trapelus schmitzi* Wagner and **Böhme,** 2006

Dr. Andreas Schmitz is a research officer at the Department of Herpetology and Ichthyology, Muséum d'Histoire Naturelle, Geneva. The holotype was collected in 1954 but waited over 50 years to be described. Schmitz co-wrote "A New Polytypic Species of the Genus *Uromastyx* Merrem 1820 (Reptilia: Squamata: Agamidae: Leiolepidinae) from Southwestern Arabia" (2007). In the etymology the authors state that Schmitz "is a good friend and colleague."

Schmutz

Schmutz's Worm Snake *Typhlops schmutzi* **Auffenberg,** 1980

Father Erwin Schmutz is a missionary, ornithologist, and herpetologist. He trained as a pharmacist and, having studied theology (1956–1962), went to Flores, Indonesia (1963). He co-wrote "Living Space of *Varanus (Odatria) t. timorensis* (Gray, 1931) (Sauria: Varanidae" (1986) and has written on the birds of Indonesia.

Schneider, J. G. T.

Schneider's Bevel-nosed Boa *Candoia carinata carinata* Schneider, 1801

Schneider's Dwarf Caiman *Paleosuchus trigonatus* Schneider, 1801 [Alt. Smooth-fronted Caiman]

Schneider's Earth Snake *Rhinophis oxyrhynchus* Schneider, 1801

Schneider's Skink *Eumeces schneideri* **Daudin,** 1802

Pond Turtle sp. *Emys schneideri* Schweigger, 1812 [Junior syn. of *Terrapene carolina* Linnaeus, 1758]

Python sp. *Python schneideri* **Merrem,** 1820 [Junior syn. of *Python (Broghammerus) reticulatus*]

Johann Gottlob Theaenus Schneider (1750–1822) was a scholar in the days when scholars were expected to be polymaths and scholarship covered everything from the natural sciences to dead languages. He wrote a great deal, and his most important work is considered to be *Kritisches griechisch-deutsches Handwörterbuch* (1797–1798).

Schneider, O.

Schneider's Adder *Bitis schneideri* **Boettger,** 1886 [Alt. Namaqua Dwarf Adder]

Dr. Oskar Schneider (1841–1903) of Dresden, where he was a Professor, was a conchologist and one of Boettger's friends. He wrote a study of the use of shells as money, published posthumously (1905).

Schneider, W.

Rock Gecko sp. *Pristurus schneideri* Rösler, Köhler, and **Böhme,** 2008

Dr. Wolfgang Schneider is a zoologist and herpetologist at Hessisches Landesmuseum, Darmstadt. He collected the type series of this gecko and has made many contributions to the study of Middle Eastern zoology.

Schoede

Vogt's Forest Dragon *Hypsilurus schoedei* **Vogt,** 1932

Hermann Schoede collected reptiles in Japan (1913) and in the Solomon Islands and New Guinea. He was a wealthy man who leased a schooner and explored the islands of German New Guinea (1909–1910).

Schomburgk

Schomburgk's Ctenotus *Ctenotus schomburgkii* **Peters,** 1863

Moritz Richard Schomburgk (1811–1891) was working as a gardener at the Palace of Sans Souci (1840) when he had the opportunity to accompany his more famous brother, Sir Robert Schomburgk (1804–1865), on an expedition to British Guiana (Guyana) and Venezuala (1841–1844). Encouraged by Alexander von Humboldt, a family friend, he wrote *Reisen in Britisch-Guiana in den Jahren 1840–1844*. He decided to emigrate (1848) to escape the political turmoil in Europe and, with another brother, Otto, arrived in Adelaide in 1849. He was the second Director, Adelaide Botanical Gardens (1866–1891). He died from a heart attack. See also **Richard S.**

Schott

Schott's Whipsnake *Masticophis schotti* **Baird** and **Girard,** 1853 [Syn. *M. taeniatus schotti*]

Schott's Tree Lizard *Urosaurus ornatus schotti* Baird, 1859

Arthur Carl Victor Schott (1814–1875) was born in Stuttgart, Germany, where he was apprenticed at the Royal Gardens. He studied at the Institute of Agriculture, Hohenheim. He spent 10 years in Hungary, managing a mining property and studying geology, botany, and zoology. He traveled in Europe and the Near East, then went to the USA (1850) where the Corps of Topographical Engineers in Washington employed him. He was a member of the U.S.-Mexican border survey (1853–1855) and collected animals, fossils, and minerals in the Rio Grande valley. He was naturalist and geologist on Michler's survey of the Isthmus of Darien (1857) and surveyed in the Yucatan Peninsula (1864–1866).

Schouteden

Middle Congo Worm Lizard *Monopeltis schoutedeni* **Witte,** 1933

Schouteden's Sun Snake *Helophis schoutedeni* Witte, 1942

Schouteden's Chameleon *Chamaeleo schoutedeni* **Laurent,** 1952

Henri Eugene Alphonse Hubert Schouteden (1881–1972) was a zoologist who undertook many expeditions to the Congo. He published on both ornithology and entomology, for example, *De Vogels van Belgisch-Congo en van Ruanda-Urundi* (1948).

Schrader

Hooded Scalyfoot *Pygopus schraderi* **Boulenger,** 1913

Dr. Carl Wilhelm Otto Schrader (1852–1930) was a German explorer, naturalist, herpetologist, and astronomer. He was at the O-Gyalla Observatory in Hungary (1876) and in Hamburg (1878). He went on expeditions to South Georgia and New Guinea (1880s). He became Imperial Inspector for nautical examinations (1889).

Schreiber

Schreiber's Green Lizard *Lacerta schreiberi* **Bedriaga,** 1878

Schreiber's Fringe-fingered Lizard *Acanthodactylus schreiberi* **Boulenger,** 1918

Dr. Egid Schreiber (1836–1913) was an Austrian zoologist. He wrote *Herpetologia Europaea* (1875).

Schreibers

Schreibers' Many-fingered Teiid *Cercosaura schreibersii* **Weigmann,** 1834 [Syn. *Pantodactylus schreibersii*]

Schreibers' Curly-tailed Lizard *Leiocephalus schreibersii* **Gravenhorst,** 1837 [Alt. Red-sided Curly-tailed Lizard]

Dr. Carl Franz Anton Ritter von Schreibers (1775–1852) was a zoologist who collected in Brazil (1817). He qualified as a physician and studied botany and mineralogy. He became Director of the Viennese Natural History Collections (1806) and worked for decades to overhaul them, including documenting the expeditions of Natterer. His main interest was meteorites. Most of his papers literally went up in smoke, when the Imperial Army bombarded the revolutionaries (1848). He retired in 1851, a broken man.

Schrenck

Schrenck's Rat Snake *Elaphe schrenckii* **Strauch,** 1873 [Alt. Amur Rat Snake]

Leopold Ivanovich von Schrenck (sometimes transcribed as Schenk or Shrenk) (1826–1894) was a Russo-German zoologist, geographer, and ethnographer who was Director (1879) of the Imperial Academy of Sciences, St. Petersburg. He explored the Amur River and Sakhalin Island (1854–1856), the results of which he published in *Reisen und Forschungen im Amur-Lande in den Jahren 1854–1856* (4 vols., 1860–1900). He coined the term "Paleoasiatic nations." He believed that the mammoths he found preserved in the permafrost must have died recently, and that they were subterranean animals that ate earth. Among other taxa, two birds are named after him.

Schroeder

Schroeder's Tree Iguana *Liolaemus schroederi* **Müller** and **Hellmich,** 1938

William C. Schroeder (1894–1977) was an oceanographer and ichthyologist. He joined the Woods Hole Oceanographic Institution (1932), initially as a business manager in connection with a ship they had acquired, the *Atlantis*. He was also an Associate Curator of Ichthyology at the Harvard Museum of Comparative Zoology (1937) and went with the *Atlantis* on a collecting trip to waters off Central and South America. Müller says in his etymology, "Wir benennen diese Eidechse nach Herrn. W. Schröder, dem wir eine große Bereicherung unserer südamerikanischen Sammlungen verdanken."

Schubart

Sao Paulo Blind Snake *Liotyphlops schubarti* **Vanzolini,** 1948

Dr. Otto Schubart was a German biologist who emigrated to Brazil (1934). He was regarded as the grand old man of Brazilian diplopodology. He worked at Museu Paulista, São Paulo, where a street is named after him.

Schubert

Schubert's Least Gecko *Sphaerodactylus schuberti* **Thomas** and Hedges, 1998

Andreas Schubert is a German zoologist who lives in the Dominican Republic and works for the National Park Service. The etymology says the gecko is named "in honor of his efforts towards the conservation of biodiversity in the Dominican Republic."

Schubotz

Schubotz's Chameleon *Chamaeleo schubotzi* **Sternfeld,** 1912 [Alt. Mt Kenya Stripe-sided Chameleon]

Dr. Johann G. Hermann Schubotz (1881–1955) was a zoologist. He was in East Africa (1907–1908) with the Duke of Mecklenburg's expedition, and he acted as overall editor of *Ergebnisse der deutschen Zentral-Afrika-Expedition 1907/08 unter Fuehrung Ad. Friedrichs, Herzog zu Mecklenburg*, in which Sternfeld, who described this chameleon, wrote the section on reptiles. Schubotz traveled from Ubangi to the Nile (1910–1911). He became Professor at Naturhistorisches Museum zu Hamburg (1914).

Schultze

Schultze's Pit-viper *Trimeresurus schultzei* **Griffin,** 1909 [Syn. *Parias schultzei*]

Forest Skink sp. *Sphenomorphus schultzei* **Vogt,** 1911

Northern New Guinea Snapping Turtle *Elseya schultzei* Vogt, 1911

Schultze's Blunt-headed Tree Snake *Boiga schultzei* **Taylor,** 1923

Dr. Leonhard Schultze-Jena (1872–1955) was a geographer, zoologist, botanist, philologist, and ethnographer. He was Professor of Geography at Philipps-Universität

Marburg (1913–1937). He led a number of expeditions that collected in New Guinea early in the 20th century. He traveled in Mexico and Guatemala studying Mayan and Aztec culture and languages, and published translations from and dictionaries of these languages (1930s).

Schultze-Westrum

Forest Dragon sp. *Hypsilurus schultzewestrumi* Urban, 1999

Dr. Thomas Schultze-Westrum is a German zoologist, ecologist, and filmmaker based in Greece. He has produced articles and films on Greek populations of the endangered Monk Seal and has written about reefs off West Papua.

Schulz

Schulz's Mountain Reed Snake *Macrocalamus schulzi* **Vogel** and David, 1999

Klaus-Dieter Schulz is a herpetologist. With co-authors he wrote *Amphibians and Reptiles of Peninsular Malaysia and Thailand: An Illustrated Checklist* (1999). According to Vogel, this snake was named after Schulz "for his major contribution to the knowledge of the snake fauna of Southeast Asia, especially of the genus *Elaphe*."

Schunke

Schunke's Snail-eater *Dipsas schunkii* **Boulenger,** 1908

Carlos Schunke (d. 1923) and his brother, José, were professional collectors. The holotype of this snake was part of a collection made by Carlos at Chanchamayo, Peru, where the brothers also collected mammals and botanical specimens for the Field Museum.

Schwartz

Cayman Brac Trope *Tropidophis (caymanensis) schwartzi* **Thomas,** 1963

Schwartz's Island Racer *Arrhyton landoi* **Schwartz,** 1965

Schwartz's Anole *Anolis schwartzi* **Lazell,** 1972 [Syn. *A. wattsi schwartzi*]

Acklins Island Boa *Epicrates chrysogaster schwartzi* Buden, 1975

Schwartz's Worm Snake *Typhlops schwartzi* Thomas, 1989

Schwartz's Dwarf Gecko *Sphaerodactylus schwartzi* Thomas, Hedges, and **Garrido,** 1992

Schwartz's Wall Gecko *Tarentola albertschwartzi* Sprackland and Swinney, 1998

Albert Schwartz (1923–1992) was a biologist and entomologist. He was Professor Emeritus of Biology, Miami-Dade Community College, and was associated with the Florida Museum of Natural History, the National Museum of Natural History, and Museo Nacional de Historia Natural, Santo Domingo. He was a specialist in West Indies fauna, writing extensively on amphibians, reptiles, and Lepidoptera of the region. He co-wrote *Amphibians and Reptiles of the West Indies: Descriptions, Distributions, and Natural History* (1991).

Schwartze

Schwartze's Skink *Eumeces schwartzei* **Fischer,** 1884 [Syn. *Mesoscincus schwartzei*]

Dr. E. W. E. Schwartze was Chairman of the Board of Directors of the company that managed the affairs of the zoo in Hamburg.

Schweizer

Schweizer's Viper *Macrovipera schweizeri* **Werner,** 1935 [Alt. Milos Viper]

Hans Schweizer (1891–1975) was a herpetologist who wrote "Über Vipera lebetina lebetina und Natrix natrix" (1932).

Sclater

Colombian Longtail Snake *Enuliophis sclateri* **Boulenger,** 1894

Dr. Philip Lutley Sclater (1829–1913) was a graduate of Oxford and practiced law for many years. He was the founding editor (1858–1865 and 1877–1903) of *The Ibis*, the journal of the British Ornithologists' Union, and Secretary of the Zoological Society of London (1860–1903). Sclater's study of bird distribution resulted in the classification of the biogeographical regions of the world into six major categories. He later adapted his scheme for mammals, and it is still the basis for work in biogeography. He wrote *Exotic Ornithology* (1866). Six mammals and five birds are named after him.

Scortecci

Scortecci's Agama *Agama cornii* Scortecci, 1928

Scortecci's Blind Snake *Leptotyphlops braccianii* Scortecci, 1929 [Alt. Bracciani's Worm Snake]

Scortecci's Whole-toed Gecko *Holodactylus cornii* Scortecci, 1930

Scortecci's Orange-tailed Lizard *Philochortus zolii* Scortecci, 1934

Scortecci's Mole Viper *Atractaspis scorteccii* **Parker,** 1949

Scortecci's Mastigure *Uromastyx princeps scorteccii* **Cherchi,** 1954

Scortecci's Dwarf Gecko *Lygodactylus scorteccii* **Pasteur,** 1959

Scortecci's Racer *Coluber scortecci* **Lanza,** 1963

Scortecci's Sand Gecko *Tropiocolotes scortecci* Cherchi and Spano, 1963

Scortecci's Diadem Snake *Spalerosophis josephscorteccii* Lanza, 1964
Scortecci's Blind Snake *Letheobia scorteccii* **Gans** and **Laurent,** 1965 [Syn. *Rhinotyphlops scorteccii*]
Scortecci's Wall Lizard *Latastia doriai scorteccii* Arillo, Balletto, and Spano, 1967

Professor Dr. Giuseppe Scortecci (1898–1973) was an Italian zoologist and herpetologist. After taking his doctorate at Università degli Studi di Firenze (1921), he joined the staff of its Institute of Comparative Anatomy. He became Professor of Zoology, Università degli Studi di Genova (1942). Before WW2 he explored the Sahara, Italian Somaliland (Somalia), and Ethiopia. He produced around 50 publications on herpetology, particularly that of desert regions. An amphibian is named after him.

Scrocchi

Scrocchi's Ground Snake *Atractus canedii* Scrocchi and **Cei,** 1991
Tree Iguana sp. *Liolaemus scrocchii* Abdala and **Lobo,** 2008

Dr. Gustavo José Scrocchi Manfrini is an Argentine herpetologist who works at Fundación Miguel Lillo and is Secretary of the Argentine Herpetological Association. His doctorate (1956) is in biology and zoology. He has written widely on South American herpetology and co-wrote "A New Species of the Genus *Atractus* from Northwestern Argentina (Serpentes, Colubridae)" (1991).

Seba

African Rock Python *Python sebae* **Gmelin,** 1788
Redback Coffee Snake *Ninia sebae* **Duméril, Bibron,** and **Duméril,** 1854
Seba's False Coral Snake *Oxyrhopus petola sebae* Duméril, Bibron, and Duméril, 1854

Albertus Seba (1665–1736) was an extremely wealthy collector and apothecary who formed the Seba Museum, Amsterdam, regarded as the richest museum of his time. He sold a huge collection to the Russian Tsar, Peter the Great (1717), then started collecting again. Linnaeus, who visited him (1735), found that Seba's 1734 book on animals was extremely useful to him in the development of his ideas. Many of Seba's animals became holotypes for Linnaeus' descriptions.

Seiglie

Estados Sucre Gecko *Gonatodes seigliei* **Donoso-Barros,** 1966

Dr. George Alfredo Seiglie (1926–1988) was Cuban who went to live in Venezuela (1960) rather than remain under Castro's regime. He was an engineer, geologist, and micropaleontologist who worked in the petroleum industry. He became an academic, being an Associate Professor at Instituto Oceanografico, Universidad de Oriente, Venezuela. He moved to Universidad de Puerto Rico, Mayaguez (1966). He was enticed back to the petroleum industry (1978) and worked in Houston until his death from cancer. He published on paleontology. His last co-written article, "Parkiella n. gen., a Late Cretaceous Turrilinacid Foraminifer from Gabon and Cameroon, West Africa" (1993), appeared posthumously.

Seipp

Seipp's Day Gecko *Phelsuma seippi* **Meier,** 1987 [Alt. Olafian Day Gecko]

Robert Seipp is a herpetologist from Frankfurt, and a gecko specialist. He wrote "Rhacodactylus—Biologie, Haltung und Zucht; mit einem Anhang der Geckoarten Neukaledoniens" (2000).

Semon

Semon's Green Tree Skink *Prasinohaema semoni* **Oudemans,** 1894

Dr. Richard Wolfgang Semon (1859–1918) was a German embryologist, evolutionary biologist, and physiologist. He was particularly interested in memory and whether it could be hereditary. He took doctorates at Friedrich-Schiller-Universität Jena in zoology (1883) and medicine (1896). He was an Associate Professor at Friedrich-Schiller-Universität Jena (1892–1897) and led an expedition to Australia (1892–1893) to investigate monotreme reproduction (science had been shaken by the revelation that some mammals were egg-layers). His party discovered 207 new species and 24 new genera. He used native Queenslanders as trappers and collectors to help him study the platypus and the Australian lungfish. He was forced to resign (1897) because he had an affair with the Professor of Pathology's wife, whom he married after moving to Munich, where he worked as a private scholar. He committed suicide, depressed by the death of his wife and by Germany's role and defeat in WW1.

Semper

Mindoro Forest Dragon *Gonocephalus semperi* **Peters,** 1867
Semper's Lipinia *Lipinia semperi* Peters, 1867
Luzon Sea Snake *Hydrophis semperi* **Garman,** 1881 [Alt. Garman's Sea Snake, Lake Taal Snake]

Dr. Carl Gottfried Semper (1832–1893) was an anatomist, ethnologist, and zoologist. Universität Würzburg awarded his doctorate in zoology (1856). He visited the Philippines and China (1858–1865) and published *Travels in the Philippine Archipelago* (1886). He lectured (1877) at the Lowell Technological Institute (now part of the University of Massachusetts), and some of his 12 lectures were published as *Animal Life as Affected by the Natural Conditions of Existence*. A bird is named after him.

Seoane

Seoane's Viper *Vipera seoanei* **Lataste,** 1879 [Alt. Portuguese Viper, Iberian Viper]

Dr. Victor Lopez Seoane y Pardo-Montenegro (1832–1900) was a naturalist. His early education was at Santiago de Compostela, after which he studied medicine in Madrid, becoming Professor of Botany there. He returned to Galicia and became Professor of Physics, Chemistry, and Natural History, Universidade da Coruña (1869), later eschewing medicine for the law. His publications include *Review of Natural History of Galicia* (1870).

Serrano

Anole sp. *Anolis serranoi* Kohler, 1999

Dr. Francisco Serrano is a biologist and conservationist in El Salvador. As Director of the National Parks Authority, he has been working to recreate areas with a pre-Columbian environment and ecology.

Serventy

Plains Ctenotus *Ctenotus serventyi* **Storr,** 1975 [Alt. Northwestern Sandy-loam Ctenotus]

Dr. Dominic Louis Serventy (1904–1988) was an outstanding ornithologist, interested in all aspects of ornithology from biogeography and speciation to breeding seasons, and had a long-term influence on conservation and government policies. After a bachelor's degree from the University of Western Australia, he took a doctorate at Cambridge (1933). He was Assistant Lecturer in Zoology, University of Western Australia (1934–1937). He worked for the Fisheries Division of the Commonwealth Scientific and Industrial Research Organisation (1937–1951) and the Wildlife Survey Section in Perth (1951). He was a major contributor to scientific journals and *Western Australian Naturalist* (1947–1980), was President, Royal Australasian Ornithologists' Union (1947–1949). He helped his brother Vincent, a well-known naturalist, and sister Lucy to revive the Western Australian Naturalists' Club after WW2. He co-wrote *Birds of Western Australia*.

Setaro

Setaro's Dwarf Chameleon *Bradypodion setaroi* **Raw,** 1976

Gordon Setaro is described by Raw as "an enthusiastic field collector of southern African herpetofauna." They collected the holotype (1974).

Seydel

Skink sp. *Afroablepharus seydeli* **Witte,** 1933

Charles Henri Victor Seydel (1873–1960) was a lepidopterist and botanical collector for Jardin Botanique National de Belgique, notably in the Belgian Congo (1924–1938). He was the only member of the Lepidopterist Society living (1951) in Elisabethville, Congo. He advertised that he had specimens for sale, including the words, "Prices are low."

Shannon

Arizona Brush Lizard *Urosaurus graciosus shannoni* **Lowe,** 1955

Shannons' Spiny Lizard *Sceloporus shannonorum* Langebartel, 1959 [Syn. *S. heterolepis shannonorum*]

Dr. Frederick Albert Shannon (1921–1964) was a physician and zoologist. He collected herpetological specimens in Mexico for the Field Museum (1940–1941). His bachelor's degree in zoology (1943) and his medical degree (1947) were both from the University of Illinois. He returned to the university (1948) for postgraduate work in herpetology before moving to Arizona (1949). He was a Lieutenant in the U.S. Army on active service in Korea (1951–1953) but managed to collect 600 herpetological specimens there. He died from the bite of a Mojave rattlesnake, despite being treated in several hospitals. He wrote extensively on herpetology, particularly on venomology (venomous snake bites). The spiny lizard is named after him and his wife; they financed the collecting trip on which the holotype was acquired.

Sharma

Sharma's Mabuya *Mabuya nagarjuni* Sharma, 1969 [Alt. Sharma's Skink; Syn. *Eutropis nagarjuni*]

Sharma's Racer *Coluber bholanathi* Sharma, 1976

Dr. Ramesh Chandra Sharma is a herpetologist. He wrote *Handbook—Indian Snakes* (2003).

Shaw, C. R.

Potosi Centipede Snake *Tantilla shawi* **Taylor,** 1949

Dr. Charles R. Shaw is a zoologist and ornithologist. He took part in several collecting expeditions to Mexico (1940s and 1950s), on one of which he collected the snake holotype (1947) while with a field party that included Robert and Marcella Newman (q.v.). He worked for the Louisiana Wildlife and Fisheries Commission to try to restore the Brown Pelican as a breeding bird along that coast (1967). He co-wrote articles such as "Genetic Variation, Selection, and Speciation in *Thomomys talpoides* Pocket Gophers."

Shaw, G. K.

Shaw's Dark Ground Snake *Liophis melanotus* Shaw, 1802

Shaw's Sea Snake *Lapemis curtus* Shaw, 1802

Dr. George Kearsley Shaw (1751–1813) was a physician,

botanist, and zoologist. He lectured on botany at Oxford (1786–1791), then was Assistant Keeper (1791–1807) and Keeper (1807–1813) at the Natural History Section, British Museum. He was a co-founder of the Linnean Society (1788). His works include *General Zoology* (1800–1812).

Shea

Kimberley Rough Knob-tail *Nephrurus sheai* **Couper** and Gregson, 1994

Dr. Glenn Michael Shea (b. 1961) is a veterinary surgeon and herpetologist and a Research Associate at the Australian Museum, Sydney. He qualified at Sydney University (1983), after which he taught anatomy while completing his doctoral thesis (1992) on Blue-tongued Lizards. He is now a Senior Lecturer in veterinary science at Sydney University. He served on the New South Wales Non-Indigenous Animals Advisory Committee (2003–2009). He co-wrote *A Field Guide to Reptiles of New South Wales* (2004).

Sheba

Guadalcanal Scaly-toed Gecko *Lepidodactylus shebae* **Brown** and **Tanner,** 1949

It is frustrating that the etymology gives no explanation of the binomial. The holotype was collected (1944) by John Chattin and given to D. Elden Beck, who later deposited the specimen in the herpetological collections at Brigham Young University. Brown died in 2002 and Tanner in 1989, so there is no one to clear up the mystery.

Shelford

Forest Skink sp. *Sphenomorphus shelfordi* **Boulenger,** 1900

Shelford's Skink *Lamprolepis vyneri* Shelford, 1905

Robert Walter Campbell Shelford (1872–1912) was a Singapore-born English-educated naturalist and Curator of the Sarawak Museum (1897–1905). He became an Assistant Curator at the Oxford University Museum. His special interest was entomology, particularly insect mimicry. A number of plants are named after him.

Sheplan

Cabral Anole *Anolis sheplani* **Schwartz,** 1974

Bruce R. Sheplan is a herpetologist who worked with Schwartz. They co-wrote "Hispaniolan Boas of the Genus *Epicrates* (Serpentes, Boidae) and Their Antillean Relationships" (1974).

Sherbrooke

Sherbrooke's Night Lizard *Xantusia sherbrookei* **Bezy,** Bezy, and Bolles, 2008

Dr. Wade Cutting Sherbrooke (b. 1941) is Director Emeritus, Southwestern Research Station, American Museum of Natural History, Arizona, where he moved to study lizards. The University of Arizona awarded his doctorate (1988). He was Visiting Professor at the University of Normal Schools, Tingo Maria, Peru (1966–1968). His publications include *Introduction to Horned Lizards of North America* (2003).

Sherman

Sherman's Blind Snake *Typhlops madgemintonae shermani* **Khan,** 1999

See **Minton, S. A.**

Shibata

Gecko sp. *Gekko shibatai* Toda, Sengoku, **Hikida,** and **Ota,** 2008

Yasuhiko Shibata is a Japanese herpetologist who was on the staff of the Osaka Museum of Natural History. He co-wrote "Two Snake Records for the Koshiki Islands, Kyushu, Japan (Reptilia: Serpentes)" (1989).

Shin

Lizard family Shinisauridae [A single species: *Shinisaurus crocodilurus* **Ahl,** 1930]

Shu-szi Sin led an expedition to southern China. His name was constantly mispronounced as "Shin" by the other expedition participants, who were all German.

Shine

Shine's Australian Whipsnake *Demansia shinei* **Shea,** 2007

Dr. Richard "Rick" Shine (b. 1950) is a herpetologist. The Australian National University in Canberra awarded his bachelor's degree in zoology (1971). He did a doctorate at the University of New England, New South Wales (1976), and to obtain a D.Sc. he presented a thesis to the University of Sydney (1988). He was a Postdoctoral Fellow at the University of Utah (1975–1978) and at the University of Sydney (1978–1980), becoming a Lecturer in Biology at the University of Sydney (1980), where he is still, having risen to the position of University Chair, specially created for excellence in research. He became a Research Associate at the Australian Museum, Sydney, and the Carnegie Museum of Natural History, Pittsburgh (1988). He was President of the Australian Society of Herpetologists (1983–1985).

Shona

Sadlier's Skink *Graciliscincus (Graciliscincus) shonae* **Sadlier** 1987

Shona Sadlier, née von Sturmer, who we presume is the describer's wife.

Shreve

Shreve's Marbled Tree Snake *Dipsadoboa shrevei* **Loveridge,** 1932
Shreve's Keelback *Helicops pastazae* Shreve, 1934
Shreve's Lightbulb Lizard *Riama shrevei* **Parker,** 1935
Shreve's Anole *Anolis shrevei* **Cochran,** 1939
Shreve's Least Gecko *Sphaerodactylus shrevei* **Lazell,** 1961

Benjamin Shreve (1908–1985) was a volunteer herpetologist at the Harvard Museum of Comparative Zoology (being from a wealthy family of jewelers). He co-wrote articles such as "Concerning Some Bahamian Reptiles" (1935).

Shropshire

Shropshire's Puffing Snake *Pseustes shropshirei* **Barbour** and **Amaral,** 1924

James B. Shropshire was Sanitary Inspector in the U.S. Army in the Panama Canal Zone.

Siebenrock

Siebenrock's Snake-necked Turtle *Chelodina siebenrocki* **Werner,** 1901
Siebenrock's Longtail Lizard *Latastia siebenrocki* **Tornier,** 1905
Siebenrock's Caspian Turtle *Mauremys caspica siebenrocki* Wischuf and Fritz, 1997

Dr. Friedrich Siebenrock (1853–1925) was a naturalist whose major interest was chelonians. He attended Leopold-Franzens-Universität Innsbruck and Universität Wien. He was a friend of Steindachner (q.v.) and succeeded him as Curator of Herpetology and Ichthyology at Naturhistorisches Museum Wien (1919–1920). He retired and lived in considerable poverty thereafter.

Siebold, K. T. E.

Siebold's Earth Snake *Geophis sieboldi* **Jan,** 1862

Dr. Karl Theodor Ernst von Siebold (1804–1885) studied at Universität Berlin and Georg-August-Universität Göttingen, becoming a physician who practiced in East Prussia (1831–1834) at Heilsberg (Lidzbark Warmiński, Poland) and Königsberg (Kaliningrad, Russia). He was headmaster in Danzig (Gdansk, Poland) (1834–1840) and subsequently Professor of Zoology, Comparative Anatomy, and Veterinary Science, Friedrich-Alexander-Universität Erlangen (1840–1845); Zoology and Physiology, Albert-Ludwigs-Universität Freiburg (1845–1850); Physiology, Universität Breslau (Wroclaw, Poland) (1850–1853); and Zoology and Comparative Anatomy, Maximilians-Universität, Munich (1853). He co-wrote *Lehrbuch der Vergleichenden Anatomie* (1845–1848). He was a cousin of P. F. B. Siebold (q.v.).

Siebold, P. F. B.

Siebold's Smooth Water Snake *Enhydris sieboldii* **Schlegel,** 1837
Siebold's Keelback *Amphiesma sieboldii* **Günther,** 1860

Dr. Philipp Franz Balthazar von Siebold (1796–1866) was a German physician, biologist. and botanist. He was Medical Officer to the Dutch East Indian Army in Batavia and at the Dutch Trading Post, Dejima Island, Nagasaki, Japan. He taught Western medicine and treated Japanese patients, accepting ethnographic and art objects as payment. Using local Japanese agents he collected in the interior (1823–1829). With the connivance of the Imperial librarian and astronomer, he copied a map of the northern regions of Japan, so upsetting the government that all his known Japanese contacts were imprisoned, his house was searched, and many possessions were confiscated. He packed all of his manuscripts, maps, and books in a large lead-lined chest, which was then hidden. Banished from Japan (1829) and forced to leave behind his young Japanese mistress and a two-year-old daughter (shades of *Madame Butterfly*), he returned to Holland, prepared his Japanese materials for publication, and was appointed by the King to advise on Japanese affairs (1831). The Japanese ban was eventually lifted (1859), and he was chief negotiator for all European nations who were trying to establish trade links with Japan (1861). His mission was a failure, and he was pensioned off (1863). The Bavarian government bought his ethnographical, botanical, and zoological collections at Nationaal Natuurhistorisch Museum, Leiden. He wrote extensively about Japan, in works including *Fauna Japonica—Aves* (1844). In Nagasaki the Siebold Memorial Museum was founded to honor his contributions to the modernization of Japan. Two birds are named after him.

Sievers

Three-horn-scaled Pit-viper *Triceratolepidophis sieversorum* Ziegler et al., 2000

Moritz and Julian Sievers are brothers. The peculiar etymology states that the name is "in recognition of the efforts of their father, Dr. J.-H. Sievers, in financially supporting zoological research and nature conservation."

Sikora

Southern Flat-tail Gecko *Uroplatus sikorae* **Boettger,** 1913

Franz Sikora (1863–1902) was an Austrian explorer and collector who was based in Réunion and collected in Madagascar for seven years (1890s). He discovered fossil remains of giant lemurs and early human settlers at Andrahomana Cave, Madagascar (1899).

Silva

Tree Iguana sp. *Liolaemus silvai* **Ortiz,** 1989

Professor Francisco Silva G. is a zoologist at Universidad de Concepción, Chile.

Silvana

Donoso's Forest Iguana *Liolaemus silvanae* **Donoso-Barros** and **Cei,** 1971 [Syn. *Vilcunia silvanae*]

Silvana Cei is the widow of Dr. José Miguel Cei, who was one of the describers.

Silvestri

Silvestri's Worm Lizard *Amphisbaena silvestrii* **Boulenger,** 1902

Professor Dr. Filippo Silvestri (1873–1949) was a zoologist and entomologist. He attended Università degli Studi di Roma "La Sapienza" (1892), later moving to Università degli Studi di Palermo, where he graduated (1896). He worked at the Institute of Comparative Anatomy, Rome, until 1902, then went to Laboratorio di Zoologia Generale e Agraria della R. Scuola Superiore d'Agricoltura, Portico, becoming Director (1904–1949). He visited South America around 1900.

Silvia

Coral Snake sp. *Micrurus silviae* Di Bernardo, Borges-Martins, and Da Silva, 2007

Silvia Di Bernardo (1966–2002) was a Brazilian biologist and herpetologist and the senior author's wife.

Simmons

Simmons' Anole *Anolis simmonsi* Holman, 1964

Dr. Robert Stanley Simmons (1924–1985) was a herpetologist working with the U.S. Fish and Wildlife Service, Baltimore, Maryland. He kept a large herpetological collection, particularly rattlesnakes. He co-wrote articles such as "A Checklist of the Rattlesnakes of South America" (1972).

Simon

Simon's Beaked Snake *Rhinotyphlops simonii* **Boettger,** 1879

Simon's Desert Racer *Mesalina simoni* Boettger, 1881

Chameleon sp. *Chamaeleo simoni* Boettger, 1885 [Junior syn. of *C. gracilis* Hallowell, 1844]

Hans Simon (d. 1898) was a collector for Naturmuseum Senckenberg, Frankfurt. He traveled in Haifa and Jerusalem (1879), Morocco (1880), and West Africa (1884).

Simonetta

Coastal Rock Gecko *Pristurus simonettai* **Lanza** and Sassi, 1968

Simonetta's Writhing Skink *Lygosoma simonettai* Lanza, 1979

Dr. Alberto Mario Simonetta (b. 1930) traveled with Lanza on many expeditions to Somalia (1959–1970). He graduated in both medicine and natural sciences. He worked at the Department of Animal Biology and Genetics, Università degli Studi di Firenze, and retired (1997) from the Chair of Zoology. His publications include "An Outline of the Status of the Somali Fauna and Its Conservation and Management Problem" (1983).

Simons

Simons' Whorltail Iguana *Stenocercus simonsii* **Boulenger,** 1899

Simons' Green Racer *Philodryas simonsi* Boulenger, 1900

Simons' Sun Tegu *Euspondylus simonsii* Boulenger, 1901

Simons' Tree Iguana *Liolaemus simonsi* Boulenger, 1902 [Syn. *L. ornatus*]

Perry O. Simons (1869–1901) was an American who collected in the Neotropics. He collected herpetology specimens in Peru (1899–1900) and birds in Bolivia (1901), and was murdered by his guide while crossing the Andes. Two birds and a mammal are named after him. F. A. Simons was collecting in Colombia a decade earlier, and we are curious to know if the two were related.

Simony

Simony's Lizard *Lacerta simonyi* **Steindachner,** 1889

East Canary Islands Skink *Chalcides simonyi* Steindachner, 1891

Skink sp. *Hakaria simonyi* Steindachner, 1899

Dr. Oskar Simony (1852–1915) was a mathematician, physicist, and naturalist. He was Professor of Mathematics, Physics, and Mechanics at an academy in Vienna. He went to the Canary Islands (1888) and made a botanical collection on Socotra Island (Gulf of Aden) (1899).

Sison

Skink sp. *Parvoscincus sisoni* Ferner, **Brown,** and **Greer** 1997

Rogelio V. Sison is a herpetologist who was a taxidermist at the Philippine National Museum and now teaches taxidermy. He co-wrote "The Amphibians and Reptiles of Panay Island, Philippines" (2001).

Sita

Sita's Lizard *Sitana ponticeriana* **Cuvier,** 1829

In Hindu mythology, Sita is the wife of Rama, the seventh avatar of Vishnu. She is considered to be an example of womanly and wifely virtue.

Sjöstedt

Sjöstedt's Five-toed Skink *Lacertaspis gemmiventris* Sjöstedt, 1897

Bror Yngve Sjöstedt (1866–1948) was an entomologist and ornithologist. He was in Cameroon collecting for the Uppsala Universitet Zoological Department (1890–1891) and for Naturhistoriska Riksmuseet, Stockholm, which he joined (1897) and traveled for (1898), visiting U.S. and Canadian entomological stations in order to study their methods. He went on the Swedish zoological expedition to Mount Kilimanjaro (1905–1906). His publications include *Zur ornithologie Kameruns*, *Wissenschaftliche ergenbisse der schwedischen expedition nach dem Kilimanjaro.*

Skilton

Skilton's Skink *Eumeces skiltonianus* **Baird** and **Girard,** 1852 [Alt. Western Skink]

Dr. Avery Judd Skilton (1802–1858) was a physician with an interest in natural history, particularly geology, mineralogy, botany, and conchology. He practiced medicine in Connecticut (1828–1858). He was Curator of the Troy Lyceum of Natural History, New York, and regularly sent specimens to Baird at the Smithsonian.

Skoog

Desert Plated Lizard *Gerrhosaurus skoogi* **Andersson,** 1916

Hilmer Nils Erik Skoog (1870–1927) started as an errand-boy at Göteborgs Naturhistoriska Museum, Sweden (1884). He advanced through the ranks, becoming Assistant Curator (1900) and Curator (1904–1927). He was on Nordenskiöld's expedition to Greenland to collect for the museum (1909) and collected in South Africa (1911–1913).

Slater, K.

Slater's Egernia *Egernia slateri* **Storr,** 1968 [Alt. Centralian Floodplains Desert Skink]

Kenneth R. Slater (1923–1999) was a zoologist and herpetologist. He worked in New Guinea and collected (1950) and described (1956) the holotype of the Papuan Taipan. He then worked for the Animal Industry Branch, Northern Territory Administration, Alice Springs (1960s). The skink was named after Slater, "who was first to demonstrate its morphological and ecological distinctiveness. Mr Slater kindly donated the holotype to the Western Australian Museum." He wrote *A Guide to the Snakes of Papua* (1956).

Slater, T.

Slater's Worm Lizard *Amphisbaena slateri* **Boulenger,** 1907

Thomas Slater collected a single specimen of this amphisbaenid, presenting it to the British Museum. Boulenger said no more, and we cannot find out more about him.

Slavens

Slavens' Centipede Snake *Tantilla slavensi* Pérez-Higareda, **H. M. Smith,** and R. B. Smith, 1985

Frank Leo Slavens (b. 1947) of the Washington Fish and Wildlife Service and Reptile Curator of the Woodland Park Zoo, Seattle, is an expert on the captive breeding of reptiles and amphibians. The original etymology says the snake is "named in honor of Frank Slavens . . . recognizing his unique contributions to captive breeding." He and his wife, Kate, are involved in the Western Pond Turtle recovery program in Washington State and Oregon. They publish (annually) *Reptiles and Amphibians in Captivity: Breeding—Longevity*

Slevin

Banded Galapagos Snake *Antillophis slevini* **Van Denburgh,** 1912

Slevin's Chuckwalla *Sauromalus slevini* Van Denburgh, 1922 [Alt. Monserrate Island Chuckwalla]

Slevin's Lizard Eater *Dryadophis melanolomus slevini* **Stuart,** 1933 [Syn. *Mastigodryas melanolomus slevini*]

Slevin's Worm Lizard *Amphisbaena slevini* **Schmidt,** 1936

Slevin's Bunch Grass Lizard *Sceloporus slevini* **H. M. Smith,** 1937

Slevin's Tropical Ground Snake *Trimetopon slevini* **Dunn,** 1940

Slevin's Dwarf Skink *Nannoscincus slevini* **Loveridge,** 1941

Baja California Night Snake *Eridiphas slevini* **Tanner,** 1943

Slevin's Banded Gecko *Coleonyx variegatus slevini* **Klauber,** 1945

San Esteban Island Whipsnake *Masticophis slevini* **Lowe** and **Norris,** 1955

Slevin's Short-fingered Gecko *Stenodactylus slevini* **Haas,** 1957

Slevin's Skink *Emoia slevini* Brown and Falanruw, 1972

Dr. Joseph Richard Slevin (1881–1957) came from a family tradition of naturalists: his father was an ornithologist and a member of the California Academy of Sciences. Joseph was employed by the California Academy of Sciences (1904) and went to the Galapagos (1905–1906). He missed the San Francisco earthquake (1906), which destroyed much of the academy's museum—Van Denburgh, then Curator of Herpetology, managed to save some specimens and books from the fire—but established the foremost collection of Galapagos flora and fauna. Apart from WW1 service as a submarine officer, he worked with Van Denburgh until succeeding him as Curator (1928–1957). He wrote, among other works, *The Amphibians of Western North America* (1928).

Sloan

Sloan's Leaf-toed Gecko *Phyllodactylus xanti sloani* Bostic, 1971

Allan John Sloan took his master's degrees at San Diego State College (1967) while working as a herpetologist at the San Diego Natural History Museum, where he had been an assistant since 1961. He co-wrote "Biogeography and Distribution of the Reptiles and Amphibians on Islands in the Gulf of California, Mexico" (1966).

Sloane

Sloane's Skink *Mabuya sloanii* **Daudin,** 1803

Sir Hans Sloane (1660–1753) was a physician and avid collector. His collections, with the contents of George II's royal library, became the basis of today's British Museum, British Library, and Natural History Museum. He graduated in medicine at the University of Orange (1683). He traveled to Jamaica (1687–1688), where he made a natural history collection and invented the practice of drinking chocolate by mixing it with milk instead of water. His investments including buying Chelsea, London, where Sloane Square is named after him. He wrote *Natural History of Jamaica* (1707). Daudin reworked many of his herpetological specimens and admired his Latin descriptions and illustrations.

Slowinski

Slowinski's Corn Snake *Pantherophis slowinskii* Burbrink, 2002 [Syn. *Elaphe slowinskii*]

Slowinski's Gecko *Cyrtodactylus slowinskii* **Bauer,** 2002

Red River Krait *Bungarus slowinskii* Kuch et al., 2005

Dr. Joseph Bruno Slowinski (1962–2001) was a herpetologist who reportedly caught his first snakes and frogs at the age of four. His bachelor's degree was from the University of Kansas (1984) and his doctorate from the University of Miami (1991). He worked at universities and museums in Louisiana (1992–1997). He became Associate Curator of Herpetology at the California Academy of Sciences. During 11 trips to Myanmar he discovered 18 new species of reptiles and amphibians. Though allergic to antivenin, he survived eight of nine venomous snake bites, the first being from a rattlesnake in Nebraska in 1977. He was also bitten by two copperheads and a Monocled Cobra, and a spitting cobra once envenomed his eyes, but local villagers successfully advised rinsing with water and tamarind leaf juice. The ninth, and fatal, bite was from a krait. He died despite the heroic efforts of his team, who gave mouth-to-mouth respiration for 26 hours.

Smallwood

Smallwood's Anole *Anolis smallwoodi* **Schwartz,** 1964

James D. Smallwood collected the holotype of the anole (1960) when assisting Schwartz, who acknowledged him in his book on the mammals of the West Indies.

Smith, Alexander

Smith's Ground Skink *Oligosoma smithi* **Gray,** 1845 [Syn. *Leiolopisma smithi*]

Lieutenant Alexander Smith was a British naval officer who was Gray's nephew. He discovered this skink and presented the holotype. It was probably collected in New Zealand during Ross' expedition with HMS *Erebus* and HMS *Terror* to the southern ocean and Antarctic (1839–1843).

Smith, Andrew

Smith's Dwarf Chameleon *Bradypodion taeniabronchum* Andrew Smith, 1831

Smith's Sand Lizard *Meroles ctenodactylus* Andrew Smith, 1838

Smith's Green-eyed Gecko *Gekko smithii* **Gray,** 1842

Smith's Burrowing Skink *Scelotes inornatus* Andrew Smith, 1845

Smith's Worm Snake *Typhlops verticalis* Andrew Smith, 1846

Smith's Thick-toed Gecko *Pachydactylus formosus* Andrew Smith, 1849

Brown Roofed Turtle *Kachuga smithii* **Gray,** 1863 [Syn. *Batagur smithii*]

Smith's Tropical Night Lizard *Lepidophyma smithii* **Bocourt,** 1876

Smith's Red-sided Skink *Mabuya homalocephala smithii* **FitzSimons,** 1943

Dr. Sir Andrew Smith (1797–1872) was a Scotsman who joined the Army Medical Service (1819) after graduating from Edinburgh University. He was a zoologist and herpetologist and famous for his scrupulous accuracy. He was in Cape Colony, South Africa (1820–1837), and was the first Superintendent of the South African Museum of Natural History, Cape Town (1825). He visited Namaqualand to discover more about its inhabitants (1828), publishing a paper on the history and lives of "Bushmen" (1831). He led the first scientific expedition into the South African interior (1834–1836). He returned to Britain, becoming Principal Medical Officer at Fort Pitt, Chatham (1841), and later Director General of the Army Medical Services (1853), a post that included organizing medical services in the Crimean War (an enquiry cleared him of charges of inefficiency and incompetence instigated by the *Times* newspaper), before ill health brought about his retirement (1858). He wrote the five-volume *Illustrations of the Zoology of South Africa* (28 parts, 1838–1850). An amphibian, four mammals, and four birds are named after him.

Smith, Arthur

Smith's Leaf-toed Gecko *Hemidactylus smithi* **Boulenger,** 1895
Smith's Racer *Coluber smithi* Boulenger, 1895
Smith's Racerunner *Pseuderemias smithii* Boulenger, 1895

Dr. Arthur Donaldson-Smith (1864–1939) was an American physician, traveler, naturalist, and big game hunter who spent much time in East Africa. He visited Lake Rudolph (Lake Turkana) (1895 and 1899). He was in Ethiopia (1896–1897) and may have witnessed the Ethiopian victory over the Italians at the Battle of Adwa. He published *Through Unknown African Countries* (1897). A mammal and three birds are named after him.

Smith, C.

Smith's Water Snake *Grayia smythii* **Leach** 1818

Professor Dr. Christen Smith (1785–1816) was a botanist, geologist, and physician, qualifying in medicine at the University of Copenhagen. He became Professor of Botany and Land Economy at the newly founded University of Christiania (Oslo) (1814). Sometimes stated to have been Danish, he was Norwegian; during his lifetime Norway was Danish-owned. He visited England (1814) and met Sir Joseph Banks. He studied all aspects of natural history and science in the Canary Islands (1815), in which year Banks persuaded him to be the botanist and geologist on Captain James Tuckey's expedition to the Cape Verde Islands and the Congo River, where he died. Why Leach chose the spelling *smythii* for the binomial is a mystery. A bird is named after him.

Smith, H. M.

Smith's Black-headed Snake *Tantilla hobartsmithi* **Taylor,** 1936
Smith's Rose-bellied Lizard *Sceloporus smithi* **Hartweg** and **Oliver,** 1937
Smith's Blue Spiny Lizard *Sceloporus serrifer plioporus* H. M. Smith, 1939
Smith's Two-spotted Snake *Coniophanes bipunctatus biseriatus* H. M. Smith, 1940
Smith's Yellowbelly Snake *Coniophanes fissidens dispersus* H. M. Smith, 1941
Smith's Garter Snake *Thamnophis vicinus* H. M. Smith, 1942 [Syn. *T. cyrtopsis collaris*]
Smith's Short-horned Lizard *Phrynosoma douglasii brachycercum* H. M. Smith, 1942
Smith's Bunchgrass Lizard *Sceloporus scalaris unicanthalis* H. M. Smith and Taylor, 1950
Smith's Earth Snake *Uropeltis smithi* **Gans,** 1966
Smith's Milk Snake *Lampropeltis triangulum smithi* **Williams,** 1978
Smith's Arboreal Alligator Lizard *Abronia smithi* **Campbell** and **Frost,** 1993

Dr. Hobart Muir Smith, né Frederick William Stouffer (b. 1912), was adopted (1916) and his names were changed. His bachelor's degree (1932), master's (1933), and doctorate (1936) were awarded by the University of Kansas. He worked for various institutions, including the Field Museum and the Smithsonian (1937–1941), and undertook a number of field trips to Mexico, collecting over 20,000 specimens. He was Professor of Zoology, University of Rochester, New York (1941–1945), then returned to Kansas (1946). He became Associate Professor, Wildlife Management, Texas A&M University (1946). He was Professor of Zoology, University of Illinois (1947–1968). After supposedly retiring, he became Professor of Biology, Colorado University, Boulder, then Chairman, Department of Ecology and Evolutionary Biology (1972). He retired again (1983) as Emeritus Professor. At age 95 he was still writing and researching. He has become the most published herpetologist ever. His first major publication was *Handbook of Lizards: Lizards of the US and Canada* (1945). See also **Hobart.**

Smith, L. A.

Smith's Legless Lizard *Aprasia smithi* **Storr,** 1970

Lawrence Alec Smith (b. 1944) was Storr's assistant. He was a co-author of several books, including the series *Lizards of Western Australia*.

Smith, M. A.

Smith's Water Snake *Enhydris smithi* **Boulenger,** 1914
Smith's Mountain Keelback *Opisthotropis spenceri* M. A. Smith, 1918 [Alt. Spencer's Stream Snake]
Smith's Agama *Oriocalotes paulus* M. A. Smith, 1935
Smith's Japalure *Japalura kaulbacki* M. A. Smith, 1937
Smith's Bent-toed Gecko *Cyrtodactylus malcolmsmithi* Constable, 1949 [Alt. Malcolm's Bow-fingered Gecko]
Smith's Blind Skink *Dibamus smithi* **Greer,** 1985
Bearded Snake sp. *Fimbrios smithi* Ziegler et al., 2008

Dr. Malcolm Arthur Smith (1875–1958) was a herpetologist. He practiced medicine in Bangkok (1902–1924), including five years as Court Physician. He visited French Indochina (1918). He was founding President of the British Herpetological Society (1949–1954). He published *A Physician at the Court of Siam* (1947). See also **Malcolm.**

Sneidern

Saphenophis Snake *Saphenophis sneiderni* **Myers,** 1973

Kjell von Sneidern (1910–2000) was a Swedish naturalist and taxidermist who worked as a collector in Colombia for the Philadelphia Academy of Natural Sciences. In 1946 he became Deputy Director of the Natural History Museum at Universidad del Cauca. He settled in South America. His son, Erik, runs shooting lodges in Colombia and Paraguay.

Snell

Pilbara Bandy-bandy *Vermicella snelli* **Storr,** 1968

Charles Snell donated the holotype of this snake, and many others, to the Western Australian Museum.

Snethlage

Ground Snake sp. *Atractus snethlageae* **Da Cunha** and Do Nascimento, 1983

Gymnophthalmid lizard sp. *Leposoma snethlageae* Avila-Pires, 1995

Dr. Maria Elizabeth Emilia Snethlage (1868–1929) was a German ornithologist and former assistant in zoology at Museum für Naturkunde Berlin. She collected in the Amazon Basin (1905–1929), having been recommended by Reichenow to Museu Paraense Emilio Goeldi, Belém. She succeeded Goeldi and was head of the museum's Zoological Section (1914–1929) except for a suspension (1917–1918) when Brazil entered WW1 against Germany. She was the first woman scientist to direct a Brazilian museum and to work in Amazonia. She wrote *Catalogo das Aves Amazonicas* (1914) and various works on local languages.

Sochurek

Sochurek's Saw-scaled Viper *Echis sochureki* **Stemmler,** 1969 [Alt. Stemmler's Saw-scaled Viper]

Erich Sochurek (1923–1987) was a herpetologist and animal dealer. His collection of photographic slides is at Naturhistorisches Museum Wien. He worked with Stemmler, and together they published *Die Sandrasselotter von Kenya: Echis carinatus leakeyi* (1969). Among his unusual pets were wolverines.

Soini

Anole sp. *Anolis soinii* Poe and Yañez-Miranda, 2008

Pekka Soini (1941–2004) was a Finnish naturalist and herpetologist at Iquitos (Peru). His doctorate was honorary, awarded by Universidad Nacional de la Amazonia Peruana. He co-wrote *The Reptiles of the Upper Amazon Basin, Iquitos Region* (1977). He died from lung cancer.

Sokolov

Sokolov's Glass Lizard *Ophisaurus sokolovi* **Darevsky** and Sang, 1983

Vladimir Evgenevich Sokolov (1928–1998) was a mammalogist and member of the Russian Academy of Sciences. He co-wrote *Guide to the Mammals of Mongolia* (1980). Two mammals are named after him.

Sommer

Terrenueve Least Gecko *Sphaerodactylus sommeri* **Graham,** 1981

William W. Sommer was an entomologist at the Allyn Museum of Entomology, Sarasota, Florida (now part of the University of Florida). He and Graham collected together in Haiti (1978). He published "A New Species of Atlantea (Nymphalidae) from Hispaniola, West Indies" (1980).

Somsak

Somsak's Blind Lizard *Dibamus somsaki* Honda et al., 1997

Dr. Somsak Panha is a zoologist and Associate Professor, Biology Department, Chulalonkorn University, Thailand. His bachelor's degree was awarded by Srinakarinwirote University (1977), his master's by Chulalonkorn (1981), and his doctorate by the University of Kyoto (1988). He is particularly interested in malacology and mollusc taxonomy and phylogeny. The etymology honors "his great contribution to the progress in zoological researches in Thailand." He has written over 70 scientific papers and articles.

Sons

Sons' Lizard *Liolaemus filiorum* Ramirez Leyton and Pincheira-Donoso, 2005

Filiorum means "of the sons." We don't know if Pincheira-Donoso has any sons, but Leyton has two, and perhaps they are among those after whom this lizard is named. Unfortunately the description gives no etymology.

Sophia

Negros Forest Dragon *Gonocephalus sophiae* **Gray,** 1845

Unfortunately Gray does not identify Sophia. One hundred sixty-five years later, all we can do is speculate that she was one of his relatives.

South

Southern Leposoma *Leposoma southi* **Ruthven** and **Gaige,** 1924 [Alt. Northern Spectacled Lizard]

Dr. John Glover South (1873–1940) was a physician and diplomat. He qualified as a physician at the University of Louisville (1897) and practiced medicine in Kentucky until he became a diplomat (1921) and was the U.S. Minister to the Republic of Panama (1921–1930). He mediated between the Panamanian government and the Kuna Indians of the San Blas archipelago (off Panama's Caribbean coast), who were in armed rebellion against the government (1925). He was U.S. Ambassador to Portugal (1930–1933). The name "Southern" Leposoma apparently arises from a misunderstanding of the binomial *southi*.

Sowerby

Skink sp. *Lygosaurus sowerbyi* **Stejneger,** 1924

Arthur de Carle Sowerby (1885–1954) was a zoologist, naturalist, explorer, and artist who was born in China, where his father was a Baptist missionary. He went to Bristol University but stayed only a short time before returning to China and beginning to collect specimens for the Natural History Museum in Tai-yuan Fu. He collected mammals (1907) for the British Natural History Museum during the expedition to the Ordos Desert in Mongolia and (1908) was part of the Clark expedition to Shansi and Kansu provinces; jointly with Robert Sterling Clark, he wrote *Through Shên Kan, the Account of the Clark Expedition in North China 1908–09* (1912). Clark was a very wealthy man, and he financed a number of collecting trips for Sowerby. There was a revolution in China (1911), and Sowerby led an expedition to evacuate foreign missionaries from Shensi and Sianfu provinces. During WW1 he was a technical officer in the Chinese Labour Corps and saw service in France. After the war Sowerby settled in Shanghai and established *The China Journal of Science and Arts*, which he edited until the Japanese occupied Shanghai during WW2. The Japanese army in Shanghai interned him for the duration, but despite that he appears to have been able to go on writing and publishing, as evidenced by "Birds Recorded from or Known to Inhabit the Shanghai Area" (1943). He emigrated to the USA (1949) and lived the rest of his life in Washington DC, spending his time in genealogical research that resulted in a family history, *The Sowerby Saga*. Two birds and a reptile are named after him.

Spalding

Spalding's Ctenotus *Ctenotus spaldingi* **Macleay,** 1877

Edward Spalding (1836–1900) was an entomologist, taxidermist, and collector of aboriginal artifacts, many of which are now in the Macleay Museum Ethnographic Collection. Macleay employed him as a collector (1870s) and took him on the *Chevert* expedition (1875) to New Guinea. He was a taxidermist at the Queensland Museum (1880–1894).

Spannring

Spannring's Gecko *Matoatoa spannringi* Nussbaum, **Raxworthy,** and **Pronk,** 1998

Jürgen Spannring is a botanist and an entomological trader. He lives in Diego, Madagascar, and collected the gecko holotype.

Spear

Spear's Prairie Lizard *Sceloporus undulatus speari* **H. M. Smith, Chiszar, Lemos-Espinal,** and **Bell,** 1995

Dr. Norman E. Spear (b. 1937) became Professor of Psychology, State University of New York, Binghamton (1974). His doctorate was awarded by Northwestern University (1963). He is co-Director of the Center for Development of Psychobiology and was Chiszar's adviser. He has studied animal behavior, and the original description recognizes him for "fostering interest in developmental psychobiology that persists to the present in our collaborative work with zoo professionals aimed at studying the ontogeny of behavior in captive-reared amphibians and reptiles."

Spegazzini

Spegazzini's Diadem Snake *Elapomorphus spegazzinii* **Boulenger,** 1913

Carlos Luigi Spegazzini (1858–1926) was an Italian-born Argentinean mycologist and naturalist. He was trained in oenology but his main interest was fungi. He traveled to Brazil (1879), swiftly moving from there to Argentina to escape a yellow fever epidemic. He went on the Italo-Argentine expedition to Patagonia and Tierra del Fuego (1881), but they were shipwrecked, and Spegazzini had to swim for it, bearing all his notes on his shoulder to keep them dry. He took up permanent residence in Argentina (1884). He was Professor at Universidad de la Plata and Universidad de Buenos Aires, Curator of the National Department of Agriculture Herbarium, first head of the Herbarium of Museo de La Plata, and founder of the Institute of Mycology La Plata. A mammal is named after him.

Speke

Speke's Hinge-back Tortoise *Kinixys spekii* **Gray,** 1863

Speke's Sand Lizard *Heliobolus spekii* **Günther,** 1872

Captain John Hanning Speke (1827–1864) was an explorer. He was the first European to see Lake Victoria (Lake Nyanza), proving it to be the source of the Nile. Speke joined Burton's expedition to discover the Nile's source because he wanted the chance to hunt big game. By the time he parted from Burton, who went on to Lake Tanganyika, Speke too had caught the source-location obsession. His own shotgun killed him when he stumbled over a stile while out shooting in England; some believe he committed suicide. Two mammals and a bird are named after him.

Spencer, F. D.

Spencer's Stream Snake *Opisthotropis spenceri* **M. A. Smith,** 1918 [Alt. Smith's Mountain Keelback]

F. D. Spencer collected the holotype. Unfortunately Smith

gives no personal details about Spencer. He may be the same F. D. Spencer who was a prominent Freemason in Bombay (1908), where he was Master of his Lodge.

Spencer, W. B.

Spencer's Window-eyed Skink *Pseudemoia spenceri* Lucas and Frost, 1894

Spencer's Monitor *Varanus spenceri* Lucas and Frost, 1903

Sir Walter Baldwin Spencer (1860–1929) was an explorer, biologist, anthropologist, zoologist, and patron of the arts. He went to Australia to become the first Professor of Biology, University of Melbourne (1887–1919), and Director of the Natural History Museum, Melbourne. He became Honorary Director of the Natural Museum of Victoria (1899). He was the photographer and zoologist on the Horn expedition, the first major scientific expedition to the center of Australia (1904). He resigned his position at the museum (1928) to return to England, but died at Navarin Island, Tierra del Fuego, while on the journey home. He mainly wrote on ethnology, publishing such works as *The Northern Tribes of Central Australia* (1904).

Spengler

Black-breasted Leaf Turtle *Geoemyda spengleri* **Gmelin,** 1789

Lorentz Spengler (1720–1807) was a zoologist employed by Det Kongelige Danske Kunstkammer, Copenhagen, as Assistant to the Keeper (1765). He became Keeper (1777), being then "Master Turner and Conchologist" to the King. He wrote a series of scientific papers on bivalves and other molluscs. In the 1780s his son Johan Conrad (1767–1839) became his assistant, taking over as Keeper (1807).

Spirrelli

Spirrelli's Worm Lizard *Amphisbaena spurrelli* **Boulenger** 1915

"Spirrelli" is a transcription error. See **Spurrell.**

Spix

Amazonian Coral Snake *Micrurus spixii* **Wagler,** 1824

Spix's Kentropyx *Kentropyx calcarata* Spix, 1825

Spix's Whiptail Lizard *Cnemidophorus ocellifer* Spix, 1825

Spix's Sideneck Turtle *Acanthochelys spixii* **Duméril** and **Bibron,** 1835

Spix's Sipo *Chironius spixii* **Hallowell,** 1845

Dr. Johann Baptist Ritter Von Spix (1781–1826) was a naturalist working in Brazil (1817–1820). He gained his doctorate (1800) at the age of 19. He studied theology for three years at Universität Würzburg, then medicine and natural sciences, qualifying as a physician (1806). The King of Bavaria awarded him a scholarship to study zoology at Muséum National d'Histoire Naturelle, Paris (1808), then *the* center for the natural sciences, with renowned scientists such as Cuvier, Lamarck, and Étienne Geoffroy Saint-Hilaire at the height of their reputations. The King appointed him Assistant to the Bavarian Royal Academy of Sciences with special responsibility for the natural history exhibits (1810). A group of academicians was invited (1816) to travel to Brazil, and King Maximilian agreed to two members of the Bayerische Akademie der Wissenschaften going. Spix was in South America (1817–1820), returning with specimens of 85 mammals, 350 birds, 130 amphibians, 116 fish, 2,700 insects, and 6,500 plants, plus 57 species of living animals; this formed the basis for Zoologische Staatssammlung München. The King awarded him a knighthood and a pension for life. He catalogued and published his findings despite extremely poor health contracted in Brazil. The expedition report was published in three volumes (1823–1831).

Spooner

Spooner's Mud Turtle *Kinosternon flavescens spooneri* P. W. Smith, 1951 [Alt. Yellow Mud Turtle]

Dr. Charles Stockman Spooner (b. 1885) was an entomologist. He was awarded his bachelor's degree by Cornell (1905), returning for graduate study (1910), but was awarded a doctorate (by the University of Illinois, 1936) only when he had achieved professorial status. Early in his career he worked as a field entomologist for the U.S. government and for various state boards. He taught in an Illinois state teachers college, becoming Professor of Zoology and departmental head (1920–1948).

Spurrell

Colombian Coral Snake *Micrurus spurrelli* **Boulenger,** 1914

Mud Turtle sp. *Kinosternon spurrelli* Boulenger, 1913

Spurrell's Worm Lizard *Amphisbaena spurrelli* Boulenger, 1915

Professor Dr. Herbert George Flaxman Spurrell (1877–1918) was a British physician and zoologist who collected in the Gold Coast (Ghana) and Colombia. He wrote *Modern Man and His Forerunners, a Short Study of the Human Species Living and Extinct* (1917). He served in the Royal Army Medical Corps during WW1 as a Captain. He died in Egypt of pneumonia. Three mammals and two amphibians are named after him.

Stadelman

Stadelman's Worm Snake *Typhlops stadelmani* **Schmidt,** 1936

Stadelman's Graceful Brown Snake *Rhadinaea stadelmani* **Stuart** and **Bailey,** 1941

Raymond Edward Stadelman (1907–1991) was Curator of a serpentarium at Tela, Honduras (1930s). He may also have been an agronomist, since he was probably the same Stadelman who wrote a report entitled *Maize Cultivation in Northwestern Guatemala* (1940).

Standing

Standing's Day Gecko *Phelsuma standingi* **Methuen** and **Hewitt,** 1913

Dr. Herbert F Standing was a medical missionary in Madagascar and also Headmaster of the Boys' High School of the Friends' Foreign Mission Association, Antananarivo. He was interested in paleontology and described a subfossil lemur (1905). He wrote *The Children of Madagascar* (1897).

Stanger

Skink sp. *Mabuya stangeri* **Gray,** 1845 [Syn. *Chioninia stangeri*]

Dr. William Stanger (1811–1854) was a physician, geologist, and explorer. He took part in the Niger expedition (1841), which used three ships to sail up the Niger River. He wrote the expedition's geological report. He suffered from fever (presumably malaria) intermittently after returning to England. He was the first Surveyor-General of Natal (1845–1854). He was instrumental in establishing the Durban Botanical Gardens (1848). A Durban writer remarked, "Stanger who had come to Natal fresh from exploring the Niger River and was well pickled with tropical diseases died in Durban." He was buried in England (1857).

Stanjorger

Stanjorger's Mabuya *Mabuya stanjorgeri* **Gray,** 1845

This is almost certainly a misspelling based on a corruption of *Mabuya stangeri*. See **Stanger.**

Stanley

Stanley's Lobulia *Papuascincus stanleyanus* **Boulenger,** 1897 [Syn. *Lobulia stanleyana*]

Owen Stanley Forest Snake *Toxicocalamus stanleyanus* Boulenger, 1903

Named after the Owen Stanley Range, Papua New Guinea.

Stanley, A.

Stanley's Slug Snake *Pareas stanleyi* **Boulenger,** 1914

Dr. Arthur Stanley resided in China until at least 1919, when he reported on the very high death rate in Zhejiang Province as a result of the influenza pandemic. He was Curator of the Shanghai Museum (1914).

Stansbury

Stansbury's Swift Lizard *Uta stansburiana* **Baird** and **Girard,** 1852 [Alt. Northern Side-blotched Lizard]

Major Howard Stansbury (1806–1863) was a civil engineer. He was in charge (1828) of a survey to link lakes Erie and Michigan by canal to the Wabash River. He surveyed the Illinois and Kaskaskia rivers (1837) and a route for a railway from Milwaukee to Dubuque. He joined the U.S. Army (1838) as a Lieutenant in the Corps of Topographical Engineers and was promoted to Captain (1840). He was in charge (1847) of building an iron lighthouse in Florida. He commanded the Great Salt Lake expedition (1849–1851); his report on it, *An Expedition to the Valley of the Great Salt Lake of Utah*, was greatly admired. He was promoted to Major (1861). He died in Wisconsin as a result of "disease contracted in the Rocky Mountains."

St. Croix

Saint Croix Racer *Alsophis sanctaecrucis* **Cope,** 1863 EXTINCT

St. Croix Ground Lizard *Ameiva polops* **Cope,** 1863 [Alt. St. Croix Ameiva]

Saint Croix's Sphaero *Sphaerodactylus beattyi* **Grant,** 1937 [Alt. Beatty's Least Gecko]

Named after St. Croix, U.S. Virgin Islands.

Steele-Scott

Northern Hooded Scalyfoot *Pygopus steelescotti* James, Donnellan, and **Hutchinson,** 2001

Dr. Colin Steele-Scott was a keen supporter of the South Australian Museum. He took part in a number of expeditions east of Lake Eyre.

Steere

Steere's Sphenomorphus *Sphenomorphus steerei* **Stejneger,** 1908

Dr. Joseph Beal Steere (1842–1940) was a zoologist and botanist whose bachelor's degree in natural history (1868) and bachelor of law (1870) were both from the University of Michigan. He made a round-the-world trip (1870–1875) during which he went up the Amazon as far as he could by boat, crossed the Andes to Peru, and sailed for China. He was thrice in Formosa (Taiwan) (1871–1874). He visited China, the Moluccas, and the Philippines and collected there (1874–1875 and 1887–1888). His collection from his first round-the-world trip was huge. According the *Michigan County Histories*, it included 3,000 birds, 100,000 seashells, 12,000 insects, 300 fishes, 200 reptiles, and 1,000 corals, not to mention Chinese bronzes and the like. He joined the faculty of the University of Michigan (1876) and became Full Professor (1879–1893). He ceased active involvement with the university (1893) and took up farming and private research, but undertook a final expedition to the Amazon (1901) for the Smithsonian. *A List of the Birds and Mammals Collected by the Steere Expedition to the Philippines* was published in 1890. Five birds and two mammals are named after him.

Stehlin

Gran Canaria Giant Lizard *Gallotia stehlini* **Schenkel,** 1901

Hans Georg Stehlin (1870–1941) was a paleontologist based at Naturhistorisches Museum Basel. He coined the expression *Grande Coupure* (Great Break) to describe the sudden extinctions of fauna at the end of the Eocene period. He collected the lizard holotype (1895). He wrote "Remarques sur les faunules de mammifères des couches eocenes et oligocenes du Bassin de Paris" (1910).

Steinbach

Colubrid snake sp. *Liophis steinbachi* **Boulenger,** 1905

Dr. Jose Steinbach (1856–1929) was a botanical and zoological collector in Argentina and Bolivia for the Field Museum. Many of the plants he collected are in Instituto de Botánica Darwinion, San Isidro, Argentina. A bird is named after him.

Steindachner

Steindachner's Emo Skink *Emoia adspersa* Steindachner, 1870

Steindachner's Ground Skink *Lioscincus steindachneri* **Bocage,** 1873

Steindachner's Worm Lizard *Cercolophia steindachneri* **Strauch,** 1881

Steindachner's Gecko *Diplodactylus steindachneri* **Boulenger,** 1885 [Alt. Box-patterned Gecko; Syn. *Lucasium steindachneri*]

Steindachner's Coral Snake *Micrurus steindachneri* **Werner,** 1901

Florida Mud Turtle *Kinosternon subrubrum steindachneri* **Siebenrock,** 1906

Steindachner's Soft-shelled Turtle *Palea steindachneri* Siebenrock, 1906

Steindachner's Toadhead Agama *Phrynocephalus steindachneri* **Bedriaga,** 1907

Striped Galapagos Snake *Antillophis steindachneri* **Van Denburgh,** 1912

Steindachner's Turtle *Chelodina steindachneri* Siebenrock, 1914 [Alt. Dinner Plate Turtle]

Franz Steindachner (1834–1919) was a zoologist, herpetologist, and ichthyologist. He originally planned to become a lawyer but became interested in fossil fishes. He worked at Naturhistorisches Museum Wien (1860–1919), becoming a Curator (1861), head of the Zoology Department (1874), and Director of the museum (1898). Unlike many curators he traveled actively and collected in the Americas, including the Galapagos, and the Middle East. He wrote the herpetological sections of the report of the results of the circumnavigation by the Austrian frigate *Novara*. A bird and seven amphibians are named after him.

Steiner

Lacertid lizard sp. *Darevskia steineri* **Eiselt,** 1995

Hans M. Steiner is an Austrian herpetologist who collected the lizard holotype (1968). He and Eiselt published on the Persian Brook Salamander (1970).

Steinhaus

Steinhaus' Worm Snake *Typhlops steinhausi* **Werner,** 1909

Dr. Carl Otto Steinhaus (1870–1919), an expert on marine worms, was an assistant at Naturhistorisches Museum zu Hamburg (1887), where the holotype resided when Werner described it and where Steinhaus was fully employed from 1900. His doctorate was awarded by Kiel University (1896). Werner and Steinhaus both contributed (1915) to a publication on the Geographical Society's expedition to German East Africa (1911–1912). He died of a nervous disease brought on by injuries sustained while serving in the German army during WW1.

Stejneger

Stejneger's Spiny Lizard *Sceloporus clarkii boulengeri* Stejneger, 1893

Coastal Whiptail *Aspidoscelis tigris stejnegeri* **Van Denburgh,** 1894

Stejneger's Bamboo Snake *Pseudoxenodon stejnegeri* **Barbour,** 1908

Stejneger's Snail Sucker *Sibon longifrenis* Stejneger, 1909

Stejneger's Grass Lizard *Takydromus stejnegeri* Van Denburgh, 1912

Long-tailed Rattlesnake *Crotalus stejnegeri* **Dunn,** 1919

Desert Side-blotched Lizard *Uta stansburiana stejnegeri* **Schmidt,** 1921

Stejneger's Worm Lizard *Amphisbaena stejnegeri* **Ruthven,** 1922

Stejneger's Rock Racer *Platyplacopus intermedius* Stejneger, 1924

Stejneger's Bamboo Viper *Trimeresurus stejnegeri* Schmidt, 1925

Central Antillean Slider *Trachemys stejnegeri* Schmidt, 1928

Stejneger's Beaked snake *Rhinotyphlops stejnegeri* **Loveridge,** 1931

Stejneger's Least Gecko *Sphaerodactylus cinereus stejnegeri* **Cochran,** 1931
Gobi Pit-viper *Gloydius intermedius stejnegeri* **Rendahl,** 1933
Stejneger's Blackcollar Spiny Lizard *Sceloporus stejnegeri* **H. M. Smith,** 1942
Stejneger's Leaf-toed Gecko *Hemidactylus stejnegeri* **Ota** and **Hikida,** 1989

Dr. Leonhard Hess Stejneger (1851–1943) was a zoologist and herpetologist who grew up in Bergen, Norway. He studied philosophy and law at Universitetet i Christiana (now Universitetet i Oslo). He went on an expedition (1881) to the USA; to Bering Island, Kamchatka; and to Commander Island in the North Pacific. He joined the Smithsonian as Assistant Curator of Birds. He became Curator of Reptiles (1889) and was Head Curator of Biology until 1911. He was laden with honors, among them being made Permanent Commander of the International Zoological Congress. His extensive publications include *Herpetology of Japan and Adjacent Territories* (1907). Two mammals and a bird are named after him.

Stemmler

Stemmler's Saw-scaled Viper *Echis sochureki* Stemmler, 1969 [Alt. Sochurek's Saw-scaled Viper]

Othmar Stemmler (b. 1934) was a herpetologist who worked at Naturhistorisches Museum Basel. He worked closely with the Austrian herpetologist Erich Sochurek, from whose name the binomial is derived. His publications are mainly concerned with the herpetology of Switzerland and surrounding countries. He co-wrote "Nos Reptiles" (1992).

Stephen

Stephen's Sticky-toed Gecko *Hoplodactylus stephensi* **Robb,** 1980 [Alt. Stephens Island Gecko]

Named after Stephens Island, New Zealand.

Stephens, F.

Panamint Rattlesnake *Crotalus stephensi* **Klauber,** 1930 [Syn. *Crotalus mitchellii stephensi*]

Frank Stephens (1849–1937) was an ornithologist and mammalogist. He was Curator Emeritus of the San Diego Society of Natural History and a member of the Death Valley Expedition (1891). His wife, Kate (who lived to be over 100), was a conchologist, and because Stephens was hard of hearing and had a reputation as a careless driver, she insisted on traveling with him. They were both members of the Alexander expedition to southeastern Alaska (1907), and they accompanied Joseph Grinnell on the Colorado River (1910). Stephens gave the San Diego Society of Natural History 2,000 bird and mammal specimens (1910). He was knocked down by a tram and died 10 days later. Two mammals are named after him.

Stephens, W. J.

Stephens' Banded Snake *Hoplocephalus stephensii* **Krefft,** 1869

William John Stephens (1829–1890) was a teacher and scholar who, after taking a B.A. (1852) and M.A. (1855) at Oxford, was appointed to be the first Headmaster of Sydney Grammar School. He quarreled with the governors, resigned, and opened his own private school at Darlinghurst (1867). This school (Eaglesfield, as of 1879) was an immediate success. He became the first Professor of Natural History, University of Sydney (1882), despite having qualifications only in mathematics and classics. He was involved in the planning and development of the Macleay Museum. He was a founding member of a number of eminent bodies, such as the Entomological Society of New South Wales (1862), the Linnean Society of New South Wales (1874), and the Zoological Society of New South Wales (1879). He died of nephritis.

Sternfeld

Sternfeld's Sand Lizard *Pedioplanis breviceps* Sternfeld, 1911
Sternfeld's Beaked Snake *Rhinotyphlops graueri* Sternfeld, 1912 [Alt. Grauer's Blind Snake]
Sternfeld's Gecko *Cnemaspis quattuorseriata* Sternfeld, 1912 [Alt. Nocturnal Forest Gecko]
Chameleon sp. *Chamaeleo sternfeldi* **Rand,** 1963

Dr. Richard Sternfeld (1884–1943) was a herpetologist and zoologist. He was in charge of the Herpetological Section of Naturmuseum Senckenberg, Frankfurt, but was dismissed (1920). He wrote *Die Reptilien und Amphibien Mitteleuropas* (1912). He was murdered during the Holocaust.

Steudner

Algerian Sand Gecko *Tropiocolotes steudneri* **Peters,** 1869

Dr. Hermann Steudner (1832–1863) was a physician and explorer. He studied botany and mineralogy at Universität Berlin. He joined Heuglin's expedition (1861–1862) to the Nile, traveling from the Red Sea to Lake Tana and back to Khartoum. In 1863 Steudner and Heuglin joined the Dutch adventuress Alexina Tinné (1839–1869) in her exploration of the White Nile, but Steudner fell ill and died in Sudan.

Stevenson

Stevenson's Dwarf Gecko *Lygodactylus stevensoni* **Hewitt,** 1926

Major James Stevenson-Hamilton (1867–1957) is

considered to be the father of South Africa's Kruger National Park. The army posted him to be warden of the Sabie Nature Reserve (1902). In the early days there were many problems, including renegade Boers who had not accepted that the war was over, and fevers that killed European horses. He returned to the army for WW1 and afterward was employed by the Sudan Civil Service and was influential in drafting Sudan's game protection laws (1921). He returned to the Nature Reserve (late 1920s) and retired (1946), having expanded it into today's Kruger National Park.

Stewart

Stewart's Sticky-toed Gecko *Hoplodactylus rakiurae* **Thomas,** 1981

Named after Stewart Island. In the binomial, *rakiurae* refers to Rakiura National Park.

Stewart, T.

Panamanian Coral Snake *Micrurus stewarti* **Barbour** and **Amaral,** 1928

Captain Thomas H. Stewart Jr. was part of the U.S. Army Medical Corps. He made a collection of reptiles while stationed in Panama.

Steyer

Steyer's Anadia *Anadia steyeri* **Nieden,** 1914

Dr. Steyer was connected with Naturhistorisches Museum, Lübeck. The etymology gives no substantial information.

Steyer, L.

Gymnophthalmid lizard sp. *Cercosaura steyeri* Tedesco, 1998

Dr. Ligia Steyer Krause is a Brazilian herpetologist. Her bachelor's degree in biological sciences (1971), her master's (1976), and her doctorate in geosciences (1983) were all awarded by Universidade Federal do Rio Grande do Sul, where she is now a Professor in the Faculty of Biological Sciences and Zoology. She wrote "Notes on Biological Aspects and Reproductive Behavior of *Dromicus poecilogyrus* in Captivity (Serpentes, Colubridae)" (1988).

Steyermark

Steyermark's Ground Snake *Atractus steyermarki* **Roze,** 1958

Dr. Julian Alfred Steyermark (1909–1988) was a botanist, explorer, taxonomist, and plant collector who was Curator of the Missouri Botanical Gardens. His speciality was the flora of Missouri, Guatemala, and Venezuela. His bachelor's degree (1929), his master's (1930), and his doctorate (1933) were all awarded by Washington University, St. Louis. During his life he collected more than 132,000 plants in 26 countries, thus earning himself an entry in the *Guinness Book of Records*. This ground snake was first collected during a Field Museum expedition (1958).

Steyn

San Steyn's Gecko *Pachydactylus sansteynae* **Steyn** and **Mitchell,** 1967

San Steyn is the senior author's wife. The description recognizes her voluntary contribution to curatorial duties at the South African Museum.

Stickel

Forest Skink sp. *Sphenomorphus stickeli* **Loveridge,** 1948

William Henson Stickel (1912–1996) was a wildlife research scientist with the U.S. Fish and Wildlife Service. He collected in New Guinea, having served with the U.S. Army there and in the Philippines during WW2. He wrote "The Snakes of the Genus *Sonora* in the United States and Lower California" (1938). His wife, Dr. Lucille Farrier Stickel (1915–2007), was also a herpetologist and the first woman to be Director of the U.S. Geological Survey Patuxent Wildlife Research Center.

Stimson

Stimson's Worm Lizard *Anops bilabialatus* **Stimson,** 1972

Stimson's Python *Antaresia stimsoni* **L. A. Smith,** 1985

Parrot Snake sp. *Leptophis stimsoni* Harding, 1995

Andrew Francis Stimson (b. 1940) was a herpetologist at the Natural History Museum, London. He wrote "A New Species of *Anops* from Mato Grosso, Brazil (Reptilia: Amphisbaenia)" (1972).

St. Johann / St. John

St. Johann's Tree Snake *Lycodryas sanctijohannis* **Günther,** 1879

Comoro Ground Gecko *Paroedura sanctijohannis* Günther, 1879

Named after Johanna (Anjouan) Island, Comoro Islands.

St. John

Tropical Snail-eater *Dipsas sanctijoannis* **Boulenger,** 1911

Named after the San Juan River, Colombia.

St. John, O. B. C.

St. John's Keelback Water Snake *Xenochrophis sanctijohannis* **Boulenger,** 1890

Colonel Sir Oliver Beauchamp Coventry St. John (1837–1891) joined the Bengal Army of the Honourable East India Company (1856). He was on a special mission (1860) to Persia (Iran) to try to improve the speed of communication between London and India. He was Chief Commissioner in Baluchistan (1877–1887) and Chief Commissioner of Mysore (1889–1891). He returned to Baluchistan but died of pneumonia after only a couple of months. See also **Johan.**

St. Marta

Gecko sp. *Lepidoblepharis sanctaemartae* **Ruthven** 1916
St. Marta's Ground Snake *Atractus sanctaemartae* **Dunn,** 1946
Santa Marta Anole *Anolis santamartae* **Williams,** 1982

Named after the Sierra Nevada de Santa Marta, Colombia.

Stoddart

Horned Agama *Ceratophora stoddarti* **Gray,** 1834

Lieutenant Colonel Charles Stoddart (1806–1842) was a diplomat and a spy for the British Empire. He was at the siege of Herat (1838) (then in Afghanistan, now in Iran) and later that year was sent to Bokhara to try to form an alliance with Nasrullah Khan, Emir of Bokhara. Instead he was imprisoned (1839) and, with his companion and would-be rescuer, Captain Arthur Conolly of the Honourable East India Company's Bengal Army, was beheaded after having been kept in a pit of vipers for months (1842).

Stokes

Stokes' Egernia *Egernia stokesii* **Gray,** 1845 [Alt. Gidgee Skink]
Stokes' Sea Snake *Astrotia stokesii* Gray, 1846

Admiral John Lort Stokes (1811–1885) joined the navy at the age of 12 (normal for midshipmen in those days). He was on HMS *Beagle*, commanded by Captain Philip Parker King (q.v.), engaged in surveying South American waters. He later commanded HMS *Acheron*, surveyed in New Zealand, and spent much of his time exploring in Australasia. He wrote *Discoveries in Australia* (1846). It was said of him that "in the course of his voyages around both the Australian mainland and Tasmania, [he] had taken the trouble to get to know some Aboriginal people personally, and the fact that he had once been speared did not affect his opinion that actions of the colonists, some of which he had himself witnessed, were atrocious in the extreme." A bird is named after him.

Stoliczka

Frontier Bow-fingered Gecko *Cyrtodactylus stoliczkai* **Steindachner,** 1867 [Syn. *Altigekko stoliczkai*, *Cyrtopodion stoliczkai*]
Stoliczka's Gecko *Cnemaspis affinis* Stoliczka, 1870 [Alt. Penang Rock Gecko]
Stoliczka's Pit-viper *Ovophis convictus* Stoliczka, 1870 [Syn. *Ovophis monticola convictus*]
Colubrid snake genus *Stoliczkia* **Jerdon,** 1870
Stoliczka's Xenodermid *Stoliczkia khasiensis* Jerdon 1870 [Alt. Khasi Red Snake]
Stoliczka's Rock Agama *Laudakia agrorensis* Stoliczka, 1872
Mongolia Rock Agama *Laudakia stoliczkana* **Blanford,** 1875
Stoliczka's Stripe-necked Snake *Liopeltis stoliczkae* Sclater, 1891
Stoliczka's Tawny Cat Snake *Boiga ochracea stoliczkae* **Wall,** 1909

Dr. Ferdinand Stoliczka (1838–1874) was a paleontologist and zoologist born in Moravia (Czech Republic). He was educated at Prague and Universität Wien, where he obtained his doctorate. He was an Assistant Superintendent of the Geological Survey of India and took part in the Second Yarkand Mission (1873–1874) but died of spinal meningitis while "returning loaded with the spoils and notes of nearly a year's research in one of the least-known parts of Central Asia." He wrote many papers on Indian zoology (1863–1872). Among other taxa named after him are three birds and three mammals.

Stolzmann

Stolzmann's Tree Iguana *Liolaemus stolzmanni* **Steindachner,** 1891 [Alt. Stolzmann's Pacific Iguana; Syn. *Microlophus stolzmanni*, *Phrynosaura stolzmanni*]

Jean Stanislas Stolzmann (or Szrolcman) (1854–1928) was a Polish zoologist who went to Peru (1871). He worked as a collector there (1875–1883). Oldfield Thomas described him as "one of the best known and most successful of Peruvian collectors." Among his publications is *On the Ornithological Researches of M. Jean Kalinowski in Central Peru* (1896). Five birds and a mammal are named after him.

Storer

Storer's Snake *Storeria occipitomaculata* **Storer,** 1839
Brown Snake genus *Storeria* **Baird** and **Girard,** 1853

Dr. David Humphreys Storer (1804–1891) qualified in obstetrics at Harvard Medical School (1825) and founded the Tremont Street Medical School (1837). He was a physician at the Massachusetts General Hospital (1849–1858) and Professor (later Dean), Harvard Medical School (1854–1868). He was President of the American Medical Society (1866). The Massachusetts legislature wanted a new look at the state's natural resources, and Storer was put in charge of the Department of Zoology and Herpetology. He also collected and described molluscs. He wrote *Ichthyology and Herpetology of Massachusetts* (1839).

Storey

Isle of Pines Sphaero *Sphaerodactylus storeyae* **Grant,** 1944

Margaret Hamilton Storey (1900–1960) was an ichthyologist and herpetologist who researched and collected in Cuba. She worked in the Natural History Museum, Stanford, California.

Storm

Sulawesi Wolf Snake *Lycodon stormi* **Boettger,** 1892

Captain Hugo Storm was a seaman from Lübeck who was captain of the *Lübeck* (1887–mid-1890s). He collected in Singapore, Borneo, and other parts of Indonesia where his vessel traded. He supplied zoological specimens from Asia to the museum in Lübeck and continued to do so after he left the sea and emigrated to the USA. A bird is named after him.

Storr

Storr's Knob-tailed Gecko *Nephrurus vertebralis* Storr, 1963
Storr's Monitor *Varanus storri* **Mertens,** 1966 [Alt. Dwarf Monitor]
Storr's Two-pored Dragon *Diporiphora lalliae* Storr, 1974
Storr's Scrub Skink *Cyclodomorphus maxima* Storr, 1976 [Alt. Giant Slender Bluetongue]
Storr's Ctenotus *Ctenotus storri* **Rankin,** 1978
Storr's Whipsnake *Demansia rufescens* Storr, 1978
Storr's Morethia *Morethia storri* **Greer,** 1980
Storr's Lerista *Lerista storri* Greer, MacDonald and Lawrie, 1983
Storr's Cross-banded Snake *Suta ordensis* Storr, 1984 [Alt. Ord Curl Snake]
Storr's Carlia *Carlia storri* **Ingram** and **Covacevich,** 1989
Storr's Prickly Gecko *Heteronotia planiceps* Storr, 1989

Dr. Glen Milton Storr (1921–1990) was a biologist, ornithologist, and herpetologist. He started training as a surveyor, but WW2 interrupted his studies. He served in the Australian infantry in Queensland and New Guinea (1942–1945). Afterward he qualified as a surveyor (1947) and worked in the South Australia Lands Department (1946–1952). He became interested in natural history and entered the University of Western Australia (1953), earning a bachelor's degree (1957) and a doctorate (1960). He joined the Western Australian Museum (1962), becoming Curator of Ornithology and Herpetology (1965).

Strahm

Strahm's Anole *Anolis strahmi* **Schwartz,** 1979

Michael H. Strahm is a herpetologist. He was originally Schwartz's pupil and became his friend. They co-wrote "Osteoderms in the Anguid Subfamily Diploglossinae and Their Taxonomic Importance" (1977).

Strauch

Strauch's Racerunner *Eremias strauchi* Kessler, 1878
Turkish Worm Lizard *Blanus strauchi* **Bedriaga,** 1884
Strauch's Worm Lizard *Leposternon strauchi* **Boettger,** 1885
Strauch's Ctenotus *Ctenotus strauchii* **Boulenger,** 1887
Strauch's Even-fingered Gecko *Alsophylax loricatus* Strauch, 1887
Strauch's Toad Agama *Phrynocephalus strauchii* **Nikolsky,** 1899
Strauch's Pit-viper *Gloydius strauchi* Bedriaga, 1912

Professor Dr. Alexander Alexandrovich Strauch (1832–1893) was a Russian-German naturalist. He qualified as a physician in Estonia (1859), but his doctorate was in zoology. He was in Algeria (1859–1860). He became Director of the Zoological Museum, St. Petersburg (1879), and Permanent Secretary of the Library of the Academy of Science (1890). He wrote *Essai d'une exploration de l'Algérie* (1862).

Strecker

Strecker's Snake *Ficimia streckeri* **Taylor,** 1931 [Alt. Hook-nosed Snake]
Western Pygmy Rattlesnake *Sistrurus miliarius streckeri* **Gloyd,** 1935

John Kern Strecker (1875–1933) was a naturalist and herpetologist. He was also interested in conchology and folklore, being President of the American Folklore Society. He is regarded as the father of Texan herpetology, starting field surveys there (1895). He was Head Librarian, Baylor University, Waco, Texas (1919–1933), and Curator of the university's museum (1903), which after his death was renamed the Strecker Museum (1940). The John K. Strecker Herpetological Society was formed in 1964 but lasted just two years. He wrote *Reptiles and Amphibians of Texas* (1915). A mammal is named after him.

Street

Street's Snake Skink *Ophiomorus streeti* **Anderson** and **Leviton,** 1966

William S. Street (1904–2000), an executive of Marshall Field Co., Chicago, and his wife (Janice Kergan Street) were big game hunters and benefactors of the Field Museum. They sponsored and led at least five expeditions to collect mammals for the museum. They visited Australia once, and Iran and Afghanistan twice each. They collected large mammals themselves, employing graduate students to gather small mammals. They would collect amphibians and reptiles too, but only casually, and often they acquired them from villagers who brought them specimens. They wrote, with Richard Sawyer, *Iranian Adventure—The First Street Expedition* (1986). A mole is named after them.

Stresemann

Bent-toed Gecko sp. *Cyrtodactylus stresemanni* Rösler and **Glaw,** 2008

Erwin Friedrich Theodor Stresemann (1889–1972) was a German ornithologist and collector in the Far East. He was President of the German Ornithological Society and Chairman of the Standing Committee on Ornithological Nomenclature, International Ornithological Congress (1954), and Curator of Birds, Museum für Naturkunde Berlin. He wrote *Aves* (1927). He collected the holotype of the gecko (1910), which took 98 years to get described. Eleven birds are named after him.

St. Rita

Burrowing Snake sp. *Apostolepis sanctaeritae* **Werner,** 1924

Worm Lizard sp. *Amphisbaena sanctaeritae* **Vanzolini,** 1994

Named after a place in Brazil.

St. Thomas

St. Thomas Beaked Snake *Rhinotyphlops feae* **Boulenger,** 1906

Named after the island of São Tomé (St. Thomas) in the Gulf of Guinea.

Stuart

Point Stuart Ctenotus *Ctenotus stuarti* **Horner,** 1995

The original etymology refers to both Point Stuart, Northern Territory, Australia, and also to the explorer John McDouall Stuart (1815–1866). Stuart led the second expedition to cross Australia south to north, and the route is today called Stuart Highway. He was trained as a civil engineer and arrived in Adelaide from Scotland (1839). He worked as a surveyor for the government (1839–1842) and freelance thereafter. He was on Charles Sturt's 1844 expedition to the interior, during which he nearly died of scurvy. He led six expeditions (1858–1962), his last being from Adelaide to the coast near Darwin. Alice Springs was originally named Stuart after him.

Stuart, L. C.

Stuart's Forest Skink *Sphenomorphus incertus* Stuart, 1940

Stuart's Brown Forest Skink *Sphenomorphus cherriei stuarti* **H. M. Smith,** 1941

Stuart's Burrowing Snake *Adelphicos veraepacis* Stuart, 1941

Stuart's Anole *Anolis cobanensis* Stuart, 1942

Stuart's Lizard Eater *Dryadophis melanolomus stuarti* H. M. Smith, 1943

Stuart's Black-nosed Lizard *Sceloporus melanorhinus stuarti* H. M. Smith, 1948

Stuart's Graceful Brown Snake *Rhadinaea pilonaorum* Stuart, 1954

Stuart's Coral Snake *Micrurus stuarti* **Roze,** 1967

Dr. Laurence Cooper Stuart (1907–1983) of the University of Michigan's Museum of Zoology was an expert on Guatemalan herpetofauna. He collected (1938) in the Yucatan Peninsula. He co-wrote "A New *Hyla* from Guatemala" (1934).

Stuart Bigmore

Stuart Bigmore's Python *Broghammerus reticulatus stuartbigmorei* Hoser, 2003

See **Bigmore.**

Stuhlmann

East African Shovel-snout *Prosymna stuhlmannii* **Pfeffer,** 1893 [Syn. *P. ambigua stuhlmanni*]

Professor Dr. Franz Stuhlmann (1863–1928) was a zoologist, anthropologist, and collector in German East Africa (now Rwanda, Burundi, and Tanzania) (1888–1900). He traveled with Emin Pasha, after whose murder he and others, survivors of a smallpox outbreak, returned from Lake Albert with a large collection and cartographic material from which the first comprehensive map of German East Africa was made. About 1,800 years ago Ptolemy's map showed the Mountains of the Moon (Ruwenzori Mountains), snow-covered peaks that fed the Nile, and Stuhlmann was the first European (1891) to confirm their existence. He wrote *Dr. Franz Stuhlmann: Mit Emin Pasha ins Herz von Africa* (1894). Three birds and two mammals are named after him.

Stumpff

Stumpff's Ground Gecko *Paroedura stumpffi* **Boettger,** 1879

Yellow-striped Water Snake *Thamnosophis stumpffi* Boettger, 1881 [Syn. *Bibilava stumpffi*]

Stumpff's Skink *Amphiglossus stumpffi* Boettger, 1882

Plated Leaf Chameleon *Brookesia stumpffi* Boettger, 1894

Anton Stumpff was a traveler and collector who accompanied Ebenau (q.v.) on his visit to Nossi-Bé Island, Madagascar. Boettger split the specimens gathered between Ebenau and Stumpff and wrote *Diagnoses Reptilium et Batrachiorum novorum ab ill. Antonio Stumpff in insula Nossi Bé Madagascariensi lectorum* (1881).

Styan

Chinese Mountain Snake *Plagiopholis styani* **Boulenger,** 1899

Frederick William Styan (1838–1934) was a tea trader and collector in Kiukiang, China for 27 years. He was a Fellow of the Zoological Society and a Member of the British Ornithologists' Union. Three mammals and two birds are named after him.

Sumichrast

Sumichrast's Garter Snake *Thamnophis sumichrasti* **Cope,** 1866
Sumichrast's Skink *Eumeces sumichrasti* Cope, 1867
Sumichrast's Longtail Snake *Enulius flavitorques sumichrasti* **Bocourt,** 1883

Adrien Jean Louis François de Sumichrast (1828–1882) was a naturalist who traveled with Saussure in the West Indies, the USA, and Mexico (1854–1856). Saussure took their considerable collection back to Geneva (1856) while Sumichrast stayed on in Mexico until his death. He used various parts of his name, depending upon with whom he was dealing. To the Smithsonian, which employed him for an expedition in Mexico, he was Professor François Sumichrast, a Frenchman. Four birds and three mammals are named after him.

Sumontha

Gecko sp. *Cyrtodactylus sumonthai* **Bauer, Pauwels,** and **Chanhome,** 2002

Montri Sumontha is a biologist at the Ranong Marine Fisheries Station, Thailand, and collected the type series of this gecko.

Sundberg

Seychelles Giant Day Gecko *Phelsuma sundbergi* **Rendahl,** 1939

Henrik Sundberg was a Swedish trader and amateur ichthyologist who was in India and the Indian Ocean area prior to WW2.

Sundevall

Sundevall's Garter Snake *Elapsoidea sundevalli* **Andrew Smith,** 1848
Sundevall's Shovel-snout Snake *Prosymna sundevalli* Andrew Smith, 1849
Sundevall's Writhing Skink *Mochlus sundevalli* Andrew Smith, 1849
Sundevall's Blind Snake *Leptotyphlops sundewalli* **Jan,** 1861

Dr. Carl Jakob Sundevall (sometimes Sundewall) (1801–1875) was a zoologist, ornithologist, and arachnologist. His doctorate in zoology was from Lunds Universitet (1823). He traveled to East Asia before returning to Lund and qualifying as a physician (1830). He was employed by Naturhistoriska Riksmuseet, Stockholm (1833–1871), being Professor and Keeper of the Vertebrates from 1839. He wrote *Svenska fåglarna* (1856). Two mammals and a bird are named after him.

Supriatna

Flying Dragon sp. *Draco supriatnai* McGuire et al., 2007

Dr. Jatna Supriatna is a zoologist, herpetologist, and primatologist who is Director of Conservation International, Indonesia. He made an important study and investigation into the herpetology of the Togian Islands.

Suter

Suter's Ground Skink *Oligosoma suteri* **Boulenger** 1906

Hans Heinrich "Henry" Suter (1841–1918) was a Swiss analytical chemist, zoologist, paleontologist, and malacologist whose collection is in the Museum Te Papa Tongarewa, Wellington, New Zealand. He joined the family firm of silk manufacturers, which failed (ca. 1885). Wanting a new start, he emigrated to New Zealand with his wife and seven living children (three others died in childhood). He was unable to clear the land he had taken up, so became (1881) Assistant Manager at the Hermitage Hotel, Mount Cook. He took New Zealand citizenship (1890) and published his first paper, describing new species of snails. He lived in Auckland (1900–1910) and in Christchurch (1911–1918), earning an uncertain living by identifying and arranging the mollusc collections of the major museums. He wrote *Manual of the New Zealand Mollusca* (1913).

Swanson

Swanson's Burrowing Skink *Anomalopus swansoni* **Greer** and **Cogger,** 1985

Stephen Swanson is a herpetologist who wrote *Lizards of Australia* (1987).

Swinhoe

Peking Gecko *Gekko swinhonis* **Günther,** 1864
Swinhoe's Lizard *Japalura swinhonis* Günther, 1864 [Alt. Swinhoe's Japalure]
Swinhoe's Grass Snake *Rhabdophis swinhonis* Günther, 1868 [Alt. Taiwan Keelback]
Swinhoe's Soft-shelled Turtle *Rafetus swinhoei* **Gray,** 1873 [Alt. Shanghai Soft-shell Turtle]

Robert Swinhoe (1836–1877) was born in Calcutta, India, but educated at London University (1852). He joined the China Consular Corps (1854). As a diplomat in China he had great opportunities for natural history research; he explored a vast area never previously open to any collector. He discovered new species at the rate of about one per month for 19 years. Although the majority of his discoveries were birds, he is also associated with many other Chinese taxa. He first returned to London in 1862, taking part of his huge collection of specimens to meetings of the zoological societies in England, France, and Holland. He was surprised at having to allow someone else to name the 200-plus new bird species he had discovered. He related, "I have been blamed by some naturalists for allowing Mr. Gould to reap the fruits of my labours, in having the privilege of describing most of my novelties. I must briefly state, in explanation, that I returned to England elated with the fine new species I had discovered, and was particularly anxious that they should comprise one entire part of Mr. Gould's fine work on the Birds of Asia, still in progress. On an interview with Mr. Gould, I found that the only way to achieve this was to consent to his describing the entire series to be figured, as he would include none in the part but novelties, which he should himself name and describe. I somewhat reluctantly complied; but as he has done me the honour to name the most important species after me, I suppose I have no right to complain." Four mammals and 15 birds are named after him.

Switak

Switak's Banded Gecko *Coleonyx switaki* **Murphy,** 1974 [Alt. Switak's Barefoot Gecko]

Karl-Heinz Switak (b. 1938) is a German herpetologist, author, and wildlife photographer. He was the herpetologist at the Steinhart Aquarium, San Francisco (1963–1974). He became Curator of Reptiles at the Transvaal Snake Park, South Africa (1987). He wrote *Adventures in Green Python Country* (2006).

Sworder

Johore Bow-fingered Gecko *Cyrtodactylus sworderi* **M. A. Smith,** 1925 [Alt. Kota-Tinggi Forest Gecko]

Hope Sworder sent a collection of reptiles to Smith, who wrote, "I am indebted to Mr. Hope Sworder for the pleasure of examining a small collection of reptiles and amphibians collected by him in Johore."

Swynnerton

Swynnerton's Worm Lizard *Chirindia swynnertoni* **Boulenger,** 1907

Charles Francis Massy Swynnerton (1877–1938) was principally an entomologist. He was born in India and worked in Africa, becoming the first game warden in Tanganyika (Tanzania) (1919–1929). He was head of tsetse research in East Africa (1929–1938). Among his publications is "On the Birds of Gazaland, Southern Rhodesia" (1907). He was killed in an air crash.

T

Taczanowski

Taczanowski's Dwarf Boa *Tropidophis taczanowskyi* **Steindachner,** 1880

Dr. Wladyslaw Taczanowski (1819–1890), a zoologist and ornithologist, was Curator, Zoological Cabinet, Royal University of Warsaw (later Branicki Museum), which he transformed into a scientific center. He collected in North Africa (1866–1867) and with Kalinowski in South America (1884). He wrote *Ornithologie du Peru* (3 vols., 1884–1886). Among the taxa named after him are eight birds and two mammals.

Tamessar

Ground Snake sp. *Atractus tamessari* Kok, 2006

Michael Tamessar (b. 1937) is a retired Senior Scientific Officer, Department of Biology, University of Guyana, and an expert on biodiversity in Guyanese forests. He was on the 1973 expedition to Mount Roraima and made the first successful ascent of it from the north.

Tana

Loveridge's Writhing Skink *Lygosoma tanae* **Loveridge,** 1935

Tana Worm Snake *Leptotyphlops tanae* **Broadley** and **Wallach,** 2007

Named after the Tana River in Kenya.

Tancredi

Boulenger's Dwarf Skink *Afroablepharus tancredi* **Boulenger,** 1909

Captain Alfonso Mario Tancredi (d. 1942) was an Italian explorer in East Africa (1899–1908). He led an expedition to Lake Tsana, Ethiopia, during which the type of this skink was collected (1908). He wrote *La missione della Società Geografica Italiana in Etiopia settentrionale* (1908).

Tania

Estado Aragua Gecko *Gonatodes taniae* **Roze,** 1963

Tania Cobo was a biology student when she collected the gecko holotype and paratype.

Tanner, C.

Tanner's Brown Snake *Pseudonaja tanneri* Worrell, 1961

Forest Skink sp. *Sphenomorphus tanneri* **Greer** and **Parker,** 1967

Tanner's Four-fingered Skink *Lygisaurus tanneri* **Ingram** and **Covacevich,** 1988

Charles Tanner (1911–1996) was an Australian herpetologist who milked the most poisonous snakes of their venom for the production of antivenin. He survived many bites, including one from a taipan; the antivenin worked, and he said it was "like mother's milk." He died of old age. He published occasionally with Covacevich, who wrote his obituary.

Tanner, W. W.

Tanner's Spiny Lizard *Sceloporus tanneri* **H. M. Smith** and Larsen, 1975

Tanner's Blind Snake *Leptotyphlops dulcis supraorbicularis* Tanner, 1985

Dr. Wilmer Webster Tanner (b. 1909) is an American zoologist who was influenced by his herpetologist elder brother, Dr. Vasco Myron Tanner, from whom he eventually took over at Brigham Young University, Utah, where he started the University Natural History Museum (1978). He served two years as a Mormon missionary in the Netherlands (1929–1931). He attended Brigham Young University, taking a bachelor's degree (1936) and a master's in zoology (1937). After WW2 service, he worked as assistant to E. H. Taylor at the University of Kansas (1946), which awarded his doctorate (1948). He edited *Herpetologica* for 18 years and co-wrote *Snakes of Utah* (1995).

Tanzer

Tanzer's Night Snake *Hypsiglena tanzeri* **Dixon** and Lieb, 1972

Ernest Claude Tanzer (d. 1971) was a Texas-based herpetologist. He was President of the Texas Herpetological Society (1965). Color polymorphism was his particular study; he published a number of articles on the subject in relation to reptile species, including showing that *Lampropeltis alterna* and *L. blairi* were two distinct morphs of the Grey-banded Kingsnake. He co-wrote "New Locality Records for Amphibians and Reptiles in Texas" (1966).

Taphorn

Taphorn's Ground Snake *Atractus taphorni* Schargel and García-Pérez, 2002

Dr. Donald Charles Taphorn Baechle (b. 1951) is an American zoologist and ichthyologist who lives in Venezuela and whose major interest is Venezuelan freshwater fish. The University of Florida awarded his master's (1976) and doctorate (1990). He is founder and Director, Museum of Zoology, Biocentro-Universidad Nacional Experimental de Los Llanos, Venezuela.

Tarzan

Tarzan Chameleon *Calumma tarzan* Gehring et al., 2010

Named after Edgar Rice Burroughs' jungle hero, the subject since his creation in 1912 of many books, comics, and screen portrayals. Gehring et al. hope to draw attention to destruction of essential habitat by naming the species after the jungle superhero. Gehring said, "The Tarzan chamaeleon is going to use his celebrity name to promote protection for this last patch of forest."

Tasma

Tasma's Bevel-nosed Boa *Candoia paulsoni tasmai* **H. M. Smith** and **Tepedelen,** 2001

Budiyanto Tasma of Jakarta provided the describers with specimens and much local data. He breeds reptiles for export to collectors.

Tasman

Tasman's Girdled Lizard *Cordylus tasmani* Power, 1930

Tasmanian Leaf-toed Gecko *Hemidactylus tasmani* **Hewitt,** 1932

Father Kenneth Robert Tasman (1890–1968) was a Jesuit priest and missionary at Macheke, Mashonaland, Southern Rhodesia (Zimbabwe). He was a founding member of the Herpetological Association of Africa (1965), having been a member of the Herpetological Association of Rhodesia (founded 1957).

Tate

Tate's Neusticurus *Neusticurus tatei* **Burt** and Burt, 1931

Dr. George Henry Hamilton Tate (1894–1953) was a zoologist and ecologist particularly interested in marine mammals. He worked at the Museum of Comparative Zoology, Harvard, but traveled widely, collecting in Ecuador (1921–1924), Venezuela (1925–1928), and Australia (1952). He wrote many scientific papers (1920s–1950s). Five mammals and two birds are named after him.

Taunay

Taunay's Teiid *Colobodactylus taunayi* **Amaral,** 1933

Afonso d'Escragnolle Taunay (1876–1958) was the son of a Brazilian viscount when Brazil was still a monarchy. He graduated in Rio de Janeiro as a civil engineer (1900). He taught engineering in São Paulo (1904–1910). He was Director, Museum Paulista (1917–1939), and also Professor, Faculty of Philosophy, Science, and Art, Universidade de São Paulo (1934–1937). He was more interested in history than zoology and wrote an 11-volume history of the coffee industry in Brazil (1929–1941).

Tautbato

Bow-fingered gecko sp. *Cyrtodactylus tautbatorum* Welton et al., 2009

Named after the Tau't- Bato peoples of southern Palawan Island, Philippines.

Taylor, E. H.

Taylor's Lipinia *Lipinia auriculata* Taylor, 1917

Taylor's Worm Snake *Typhlops canlaonensis* Taylor, 1917

Taylor's Fringed Gecko *Luperosaurus joloensis* Taylor, 1918

Taylor's Short-legged Skink *Brachymeles vermis* Taylor, 1918

Taylor's Gecko *Gekko porosus* Taylor, 1922

Forest Skink sp. *Sphenomorphus taylori* **Burt,** 1930

Taylor's Side-blotched Lizard *Uta stansburiana taylori* **H. M. Smith,** 1935

Taylor's Spiny Lizard *Sceloporus edwardtaylori* H. M. Smith, 1936

Taylor's Ground Skink *Scincella silvicola* Taylor, 1937

Taylor's Red Forest Skink *Sphenomorphus assatus taylori* **Oliver,** 1937

Taylor's Blind Snake *Leptotyphlops nasalis* Taylor, 1940

Taylor's Coral Snake *Micrurus browni taylori* **Schmidt** and H. M. Smith, 1943 [Alt. Acapulco Coral Snake]

Taylor's Anole *Anolis taylori* Smith and Spieler, 1945

Taylor's Large-scaled Lizard *Ptychoglossus plicatus* Taylor, 1949

Taylor's Black-striped Snake *Coniophanes piceivittis taylori* **Hall,** 1951

Taylor's Cantil *Agkistrodon bilineatus taylori* **Burger** and **Robertson,** 1951

Taylor's Alligator Lizard *Gerrhonotus infernalis taylori* Tihen, 1954

Taylor's Snail-eater *Dipsas tenuissima* Taylor, 1954

Taylor's Tropical Racer *Mastigodryas sanguiventris* Taylor, 1954

Utah Milk Snake *Lampropeltis triangulum taylori* **Tanner** and Loomis, 1957

Taylor's Burrowing Snake *Pseudorabdion taylori* **Leviton** and **Brown,** 1959

Cuatrocienegas Slider *Trachemys taylori* Legler, 1960

Taylor's Bow-fingered Gecko *Cyrtodactylus quadrivirgatus* Taylor, 1962 [Alt. Four-striped Forest Gecko]

Taylor's Limbless Skink *Dibamus alfredi* Taylor, 1962

Taylor's Writhing Skink *Lygosoma frontoparietale* Taylor, 1962

Taylor's Blind Skink *Dibamus taylori* **Greer,** 1985

Taylor's Thailand Gecko *Gekko taylori* **Ota** and Nabhitabhata, 1991

Taylor's Tree Skink *Lankascincus taylori* Greer, 1991

Edward Taylor's Gecko *Cyrtodactylus edwardtaylori* Batuwita and Bahir, 2005

Dr. Edward Harrison Taylor (1889–1978) was a zoologist and herpetologist. He went to teach in a Philippine village school (1912), returning briefly to his university, Kansas, to finish his master's and doctorate (1926). He was Chief of the Division of Fisheries, Philippines (1916–1920). He was Head of the Zoology Department of the Philippines

(1923–1927), then worked at Kansas University (1927–1949), becoming Full Professor (1934). He was an early broadcaster, giving a series of 10 radio talks on herpetology (1932–1933). He took students on collecting trips to Mexico (1937–1948), traveling in "marginally reliable vehicles." He collected after retirement in Costa Rica (1949), Thailand, and Brazil and wrote 19 papers on Philippine herpetology (1915–1928). A mammal and an amphibian are named after him.

Taylor, R. H. R.

Taylor's Fat-tail Gecko *Hemitheconyx taylori* **Parker,** 1930
Taylor's House Gecko *Hemidactylus taylori* Parker, 1932
Taylor's Strange Agama *Xenagama taylori* Parker, 1935
Taylor's Long-tailed Lizard *Latastia taylori* Parker, 1942
Racer (snake) sp. *Coluber taylori* Parker, 1949
Taylor's Wolf Snake *Lycophidion taylori* **Broadley** and **Hughes,** 1993

Captain R. H. R. Taylor was an army officer and administrator in British Somaliland (Somalia). He was a member of the Anglo-Ethiopian Boundary Commission (1934). He returned to England in 1935, traveling through Ethiopia and Sudan. He co-wrote "The Lizards of British Somaliland" (1942).

Taylor, W. E.

Taylor's Ground Snake *Sonora (semiannulata) taylori* **Boulenger,** 1894 [Alt. Southern Texas Ground Snake; Syn. *Contia taylori*]

The original description says only that the holotype came from "W. Taylor Esq." This may well refer to Walter Edgar Taylor, who was an American ornithologist and herpetologist. He wrote "Catalogue of the Snakes of Nebraska" (1892).

Tayra

Volcan Tacana Centipede Snake *Tantilla tayrae* **Wilson,** 1983

Tayra Barbara Wilson is the describer's younger daughter. She is named after the Tayra, a species of Neotropical mustelid.

Tchernov

Tchernov's Chainling Snake *Micrelaps tchernovi* **Werner,** Babocsay, Carmely, and Thuna, 2006

Professor Dr. Eitan Tchernov (1935–2002) was an archeozoologist and paleontologist who co-founded the Department of Evolution, Systematics, and Ecology, Hebrew University, Jerusalem.

Teague

Northern Blind Snake *Leptotyphlops teaguei* Orejas-Miranda, 1964

Gerard Warden Teague (1885–1974) was a systematic ichthyologist and herpetologist who was British Vice Consul for Paraguay. He worked in South America for the Midland Uruguay Railway Company, being the company's General Manager (1937), based in Paysandú where he was President of his local golf club. He lived in Lisbon (1960–1961), then in Quebec, before returning to live in Montevideo until his death. He wrote "Plants of Central Paraguay" (1965).

Teale

Teale's Delma *Delma tealei* **Maryan,** Aplin, and Adams, 2007

Roy Teale is a zoologist and Research Associate, Western Australian Museum. He studied at the University of Western Australia (1986–1991). He is a Director of Biota Environmental Sciences, which is involved in the mining and metals industries at North Perth, Western Australia.

Tehuelche

Tree Iguana sp. *Liolaemus tehuelche* Abdala, 2003

This lizard is named after a people, the Tehuelche Indians.

Telfair

Telfair's Skink *Leiolopisma telfairi* Desjardin, 1831 [Alt. Round Island Skink]

Dr. Charles Telfair (1778–1833) was an Irish physician, naval surgeon, botanist, sugar planter, and, probably, rum smuggler. He lived in Mauritius but traveled widely in the Indian Ocean and further afield. He traveled in China (1826), where he acquired some banana plants that he sent to England. They were passed to the Duke of Devonshire, who successfully grew them in the glasshouses at Chatsworth. He imported Nile Crocodiles from Madagascar, where he spent time collecting, as Bennett reported in "Characters of a New Genus of Lemuridae, Presented by Mr. Telfair" (1832). A mammal and some plants are named after him.

Tello

Tello's Thread Snake *Leptotyphlops telloi* **Broadley** and **Watson,** 1976

Dr. José Luis Pessoa Lobão Tello is a herpetologist who has collected both reptiles and amphibians, notably in poorly known regions of southern Mozambique (1970s). He combined collecting with training wildlife technicians. He was also a leading philatelist in Mozambique, with an award-winning collection of stamps depicting

African animals. He co-wrote "Check List and Atlas of the Mammals of Mocambique" (1976).

Temminck

Alligator Snapping Turtle *Macrochelys temminckii* Harlan, 1835
Forest Skink sp. *Sphenomorphus temmincki* **Duméril** and **Bibron,** 1839

Coenraad Jacob Temminck (1778–1858) was an ornithologist, illustrator, and collector. He was the first Director of Nationaal Natuurhistorisch Museum, Leiden (1820–1858). He catalogued his father's extensive collection of birds. He wrote *Manuel d'ornithologie, ou Tableau systematique des oiseaux qui se trouvent en Europe* (1815). Nineteen birds and 13 mammals, plus other taxa, are named after him.

Tempel

Tempel's Chameleon *Chamaeleo tempeli* **Tornier,** 1899 [Alt. Tanzania Mountain Chameleon]

Dr. Max Ludwig Tempel (b. 1865), one of Tornier's friends, was a veterinary surgeon in Chemnitz.

Templeton

Templeton's Kukri Snake *Oligodon templetoni* **Günther,** 1862

Dr. Robert Templeton (1802–1892) was an Irish naturalist and entomologist. His father was the botanist John Templeton. He studied medicine at Edinburgh and served as an army surgeon (1833–1860). He was an Assistant Surgeon at Woolwich, London (1833), and Mauritius (1834). He visited Brazil (1835), then went to Colombo (Sri Lanka) and became a corresponding member of the Zoological Society of London. He was posted to Malta (1836), collecting also in Corfu and Albania, and became a corresponding member of the Entomological Society, London (1839). When in Colombo (1839–1851), he was promoted to Surgeon (1847). After service during the Crimean War he retired with the rank of Inspector-General of Hospitals. He obtained the holotype of this snake. A bird is named after him.

Tennent

Tennent's Leaf-nosed Lizard *Ceratophora tennentii* **Günther,** 1861 [Alt. Rhinoceros Agama]

Sir James Emerson Tennent (1804–1866) was an Irish lawyer, politician, and traveler. He was born James Emerson but added his wife's name to his own (1831). He was called to the Bar at Lincoln's Inn (1831) and became a Member of Parliament (1832). He was knighted (1845) and was Colonial Secretary of Ceylon (now Sri Lanka) (1845–1850). Changes he proposed to taxation were wildly unpopular and a cause of the Matale Rebellion (1848). He wrote *Ceylon, Physical, Historical and Topographical* (1859), which includes the first English use of the expression "rogue elephant."

Tepedelen

Tepedelen's Bevel-nosed Boa *Candoia carinata tepedeleni* **H. M. Smith** and **Chiszar,** 2001

Kumaran Tepedelen, who is a herpetologist and dealer in reptiles, lives in Boulder, Colorado. He worked with the senior describer.

Ternetz

Ternetz's Blind Snake *Liotyphlops ternetzii* **Boulenger,** 1896

Dr. Carl Ternetz (b. 1870) was an ichthyologist and naturalist who collected all over South America over a period of more than 30 years. Several fish are named after him.

Teruel

Teruel's Anole *Anolis terueli* **Navarro,** Fernandez, and **Garrido,** 2001

Rolando Teruel Ochoa is a Cuban arachnologist and biologist. A scorpion is named after him.

Thales de Lema

Ground Snake sp. *Atractus thalesdelemai* Passos, Fernandes, and Zanella, 2005

Professor Dr. Thales de Lema is a Brazilian herpetologist based at Pontifícia Universidade Católica do Rio Grande do Sul, Porto Alegre, where he originally graduated in biology, sciences, and mathematics (1958). He was variously at the Museum of Zoology, Universidade de São Paulo, and at Instituto Butantan, São Paulo (1956–1975). He undertook postdoctoral work at the University of Toronto, Canada (1993). He wrote "Redescription of *Apostolepis sanctaeritae* (Serpentes, Colubridae), and a Comparison with Related Species" (2002).

Thanh

Thanh's Reed Snake *Calamaria thanhi* Ziegler and Quyet, 2005

Professor Vu Ngoc Thanh is a Vietnamese zoologist and physician at the Zoological Museum and Department of Vertebrate Zoology, Faculty of Biology, University of Science, Vietnam National University, Hanoi. He was an army doctor during the Vietnam War. His major area of study is the primates of Vietnam. He co-wrote "Cyrtodactylus phongnhakebangensis sp.n., ein neuer Bogenfingergecko aus dem annamitischen Karstwaldmassiv, Vietnam" (2002).

Thayer

Spiny Lizard sp. *Sceloporus thayerii* **Baird** and **Girard,** 1852 [Syn. *Sceloporus undulatus hyacinthinus*]

Brigadier General Sylvanus Thayer (1785–1872) was an early proponent of engineering education for everyone in the USA and gave $30,000 to Dartmouth College (1867) to create the Thayer School of Engineering. He was commissioned into the Corps of Engineers as Lieutenant (1808). He directed the defense of Norfolk, Virginia, during the War of 1812 and was promoted to Major. He studied in Paris at L'École Polytechnique (1815–1817). He was Superintendent of West Point military academy (1817–1833). After quarreling with President Andrew Jackson, he resigned from West Point and served again with the Corps of Engineers (1833–1863), retiring as a Brigadier General.

Theobald

Theobald's Toad-headed Agama *Phrynocephalus theobaldi* **Blyth,** 1863

Theobald's Kukri Snake *Oligodon theobaldi* **Günther,** 1868

Theobald's Gecko *Cnemaspis sisparensis* Theobald, 1876

William Theobald (1829–1908), a naturalist and herpetologist, was Deputy Superintendent, Geological Survey of India. He is often said to have found a very old fossil skull of a human (1860) believed to be the oldest example in India, but this is an error due to a mixup of labels. Despite his herpetological knowledge, he once picked up a King Cobra by the tail thinking it another species. He was also interested in malacology and collaborated with Blanford and Godwin-Austen (1876–1883). He wrote *Descriptive Catalogue of the Reptiles of British India* (1876). A mammal is named after him.

Theresia

Theresia's Pacific Iguana *Microlophus theresiae* **Steindachner,** 1901

Princess Therese Charlotte Marianne Augusta von Bayern (1850–1925) was a botanist, zoologist, ethnologist, and author. This remarkable woman started traveling in Europe and North Africa (1871) and became fluent in 12 languages. She traveled incognito with a maximum of three attendants. She traveled in Mexico and the USA (1893) and in the Caribbean, Colombia, and Argentina via Ecuador and Peru (1898). She wrote *Over Mexican Seas* (1895).

Therezien

Perinet Leaf Chameleon *Brookesia therezieni* **Brygoo** and **Domergue,** 1970

Colubrid snake sp. *Liophidium therezieni* Domergue, 1984

Yves Thérézien was an expert on freshwater algae and a hydrobiologist at Muséum National d'Histoire Naturelle, Paris, where Brygoo and Domergue were his colleagues. He was a fisheries researcher with Centre Technique Forestier Tropical, Madagascar (1958); inspector for the Water and Forestry Service, Madagascar (1960–1966); and designated collector of botanical specimens for Service Forestier de Madagascar. He worked for Institut Nationale de la Recherche Agronomique, Guadeloupe (1979), and was on a Franco-Austrian hydrobiological expedition to the Lesser Antilles. He wrote several scientific papers. A fish is named after him.

Thiel

Thiel's Pygmy Chameleon *Brookesia thieli* **Brygoo** and **Domergue,** 1969 [Alt. Domergue's Leaf Chameleon]

Forest Water Snake *Liopholidophis thieli* Domergue, 1972

Jean Thiel is a French expert on tropical forests, especially hardwood trees. He worked and collected botanical specimens in Madagascar (1968–1971). By 1978 he was working in French Guiana. He wrote "Reconnaissance pratique des arbres sur pied de la forêt guyanaise" (1983).

Thierry

Thierry's Cylindrical Skink *Chalcides thierryi* **Tornier,** 1901

Gaston Thierry (1866–1904), judging by his name, should be French but was an Oberleutnant in the Imperial German Army. He was among those sent to Togoland (1896) to establish a series of bases to enforce German control over the country after the French and German governments had reached agreement on the border between Togoland and Dahomey. He left Togo (1899) and was killed by a poisoned arrow in Cameroon (1904) (then a German colony). A mammal is named after him.

Thirakhupt

Thirakhupt's Bent-toed Gecko *Cyrtodactylus thirakhupti* **Pauwels** et al., 2004

Professor Dr. Kumthorn Thirakhupt is a Thai herpetologist and zoologist, Department of Biology, Faculty of Science, Chulalongkorn University, Bangkok. He coauthored *Photographic Guide to Snakes and Other Reptiles of Thailand, Peninsular Malaysia, and Singapore* (1998).

Thollon

Thollon's African Water Snake *Grayia tholloni* **Mocquard,** 1897

François-Romain Thollon (1855–1896) was a French collector in the Congo and a member of the de Brazza mission in Gabon. Many taxa, including a bird and a mammal, are named after him.

Thomas, B.

Thomas' Mastigure *Uromastyx thomasi* **Parker,** 1930
Thomas' Semi-banded Racer *Coluber thomasi* Parker, 1931

Bertram Thomas (1892–1950) was a civil servant who served during WW1 in Belgium and then in Mesopotamia (Iraq) (1916–1918). He then became Political Officer (1918–1922) and Assistant British Representative in Transjordan (Jordan) (1922–1924). He was Finance Minister to the Sultan of Muscat and Oman (1925–1932) and went on a number of expeditions, being the first Westerner to cross the Rub' al Khali (1930–1931). He filmed his journey, creating a documentary film entitled *Crossing the Empty Quarter*. He wrote *Arabia Felix* (1932).

Thomas, F.

Skink sp. *Panaspis thomasi* **Tornier,** 1904

Felix Thomas, who was an engineer in British East Africa (Kenya), collected the holotype.

Thomas, R.

Thomas' Bachia *Bachia panoplia* Thomas, 1965
Thomas' Blind Snake *Leptotyphlops pyrites* Thomas, 1965
Thomas' Sauresia *Sauresia agasepsoides* Thomas, 1971
Thomas' Worm Snake *Typhlops hectus* Thomas, 1974
Thomas' Galliwasp *Celestus macrotus* Thomas and Hedges, 1989
Tree Iguana sp. *Liolaemus thomasi* **Laurent,** 1998

Dr. John Paul Richard Thomas (b. 1938), a zoologist and herpetologist, is Professor, Department of Biology, University of Puerto Rico. He gained his doctorate in zoology from Louisiana State University (1976). He collected amphibians and birds in Peru with Duellman for the University of Kansas' Natural History Museum (1971) and as a member of the Louisiana State University Museum of Zoology expedition to Peru (1974). He co-wrote "The Status of *Sphaerodactylus gilvitorques* Cope and of *Sphaerodactylus nigropunctatus* Gray (Sauria: Gekkonidae)." A mammal is named after him.

Thomas White

See **White, T.**

Thompson

Thompson's Least Gecko *Sphaerodactylus thompsoni* **Schwartz** and Franz, 1976
Thompson's Leaf-toed Gecko *Phyllodactylus thompsoni* Venegas et al., 2008

Dr. Fred Gilbert Thompson (b. 1934) is Curator of Nonmarine Malacology, Florida Museum of Natural History. He collected the Least Gecko holotype (1974).

Thurj

Crowned River Turtle *Hardella thurjii* **Gray,** 1831
[Alt. Brahminy River Turtle]

Like many early descriptions of species, this one is lacking in etymological details. Gray calls this turtle the Thurgy Terrapin, but it isn't clear to whom or what he is referring.

Thurston

Thurston's Worm Snake *Typhlops thurstoni* **Boettger,** 1890

Dr. Edgar Thurston (1855–1935) was an ethnographer, natural historian, and museologist who qualified as a physician in England (1877). He was Superintendent of the Government Museum, Madras (now Chennai), establishing the natural history and anthropology sections (1885–1910). He returned to England (1910) and eventually settled in Cornwall, where he was a noted plant collector (1915–1926). He mainly published on ethnography but did write a book on the amphibians of southern India. A number of other taxa, including a fish, are named after him.

Thwaites

Thwaites' Skink *Chalcidoseps thwaitesi* **Günther,** 1872

George Henry Kendrick Thwaites (1811–1882) was an accountant who studied botany in his spare time. From 1847 he lectured on botany at the School of Pharmacy and Bristol Medical School. He worked at the Botanical Garden, Peradeniya, Ceylon (now Sri Lanka) (1849–1880), first as Superintendent and then Director (1857). He wrote the first flora of Ceylon, *Enumeratio plantarum Zeylaniae* (1864).

Tien

Tien's Mountain Stream Snake *Opisthotropis daovantieni* **Orlov,** Darevsky, and **Murphy,** 1998

Dao Van Tien (1917–1995) was a Professor of Biology, National University of Hanoi, educated in Hanoi under the French colonial administration. He was a primatologist and is widely acknowledged as the father of his field in Vietnam, although he is probably best known for asserting his belief in the existence of "Forest Man"—a supposed primitive hominid reported from remote parts of Asia. A rodent is named after him.

Tilbury

Tilbury's Chameleon *Chamaeleo marsabitensis* Tilbury, 1991
Tilbury's Fringe-fingered Lizard *Acanthodactylus tilburyi* **Arnold,** 1986

Dr. Colin R. Tilbury is a herpetologist, zoologist, and

molecular biologist in the Department of Zoology, University of Stellenbosch, South Africa. He has collected in Saudi Arabia and wrote "An Annotated Checklist of Some of the Common Reptiles Occurring around Riyadh, Kingdom of Saudi Arabia" (1988).

Tillier

Tillier's Maquis Skink *Lioscincus tillieri* Ineich and **Sadlier,** 1991

Professor Dr. Simon Tillier is a paleontologist, zoologist, and malacalogist who works for Muséum National d'Histoire Naturelle, Paris. He was Curator of Land and Freshwater Mollusks (1976–1992) and coordinated the museum's research program on the flora and fauna of New Caledonia (1984–1992). He was Director of the museum's faculty in molecular systematics (1993–2002). His wife, Annie, works at the same molecular laboratory in Paris, and they both have taxa named after them.

Timon

Chameleon sp. *Furcifer timoni* Glaw, Köhler and Vences, 2009

Timon Robert Glaw (b. 2004) is, according to his father, Frank, more interested in dinosaurs than any other herp.

Tindall

Tindall's Worm Snake *Typhlops tindalli* **M. A. Smith,** 1943 [Alt. Nilgiri Hills Worm Snake]

Roger Tindall. Unfortunately, Smith gives no further details about him.

Tinkle

Raza Island Leaf-toed Gecko *Phyllodactylus tinklei* **Dixon,** 1966

Dr. Donald Ward Tinkle (1930–1980) was a herpetologist. His doctorate came from Tulane University (1956). He was Curator of Reptiles and Amphibians, University of Michigan Museum of Zoology (1965–1980), taking a sabbatical as Visiting Professor, Arizona State University (1972). He wrote "Ecology, Maturation and Reproduction of *Thamnophis sauritus proximus*" (1957).

Tiwari

Tiwari's Wolf Snake *Lycodon tiwarii* Biswas and Sanyal, 1965 [Alt. Andaman Wolf Snake]

Tiwari's Bronzeback *Dendrelaphis humayuni* **Tiwari** and Biswas, 1973

Dr. Krishna Kant Tiwari is a zoologist and carcinologist. He was a member of the Zoological Survey of India (1951), of which organization he become Joint Director (1980–1981). He is a retired Vice Chancellor, Jiwaji University, Gwalior. He co-wrote "Two New Reptiles from the Great Nicobar Islands" (1973).

Todd

Carlos Todd's Anole *Anolis carlostoddi* **Williams,** Praderio, and Gorzula, 1996

Carlos Todd is an environmentalist in Venezuela. The etymology for this anole says that he was "long active in conservation work [and] participated in the exploration of Chimantá Tepui."

Tokobajev

Kirghizia Even-fingered Gecko *Alsophylax tokobajevi* Yeriomtschenko and Szczerbak, 1984

Dr. Marat M. Tokobaev (b. 1932) is an entomologist and parasitologist. His name has several spellings; *tokobajevi* appears in the original description and so is used in the official binomial even though he spells his name Tokobaev. He became a corresponding member of the Russian Academy of Sciences (1977) and Director, Biology Institute, Frunze (Bishkek), Kirghiz Republic (1980).

Tolampy

Grandidier's Dwarf Gecko *Lygodactylus tolampyae* **Grandidier,** 1872

Grandidier's description of this gecko is a few lines long and has no mention of Tolampy. Amba-tolampy is a place in Madagascar, and the binomial may be derived from it.

Tolpan

Graceful Brown Snake sp. *Rhadinaea tolpanorum* Holm and Cruz, 1994

The Tolpan are a native people who live in Honduras.

Tornier

Tornier's Leaf-toed Gecko *Hemidactylus squamulatus* Tornier, 1896 [Alt. Nyika Gecko]

Tornier's Tortoise *Malacochersus tornieri* **Siebenrock,** 1903 [Alt. Pancake Tortoise]

Tornier's Cat Snake *Crotaphopeltis tornieri* **Werner,** 1908 [Alt. Werner's Cat Snake]

Dr. Gustav Tornier (1859–1938) was a zoologist, artist, anatomist, paleontologist, and taxonomist, Museum für Naturkunde Berlin. He is best remembered for his (incorrect) "crawling" restoration of the dinosaur *Diplodocus*. An amphibian is named after him.

Torre

Barbour's Least Gecko *Sphaerodactyulus torrei* **Barbour,** 1914

Professor Carlos de la Torre y la Huerta (1858–1950) was regarded as the foremost Cuban naturalist of his generation. He was closely associated with the Smithsonian, but he died some 10 years before Castro took power and Cuba became isolated from USA. He was a leading figure in the

Academia de Ciencias Medicas, Fisicas y Naturales de La Habana. A bird and an extinct mammal are named after him.

Torres

Torres-Mura's Dragon *Phrynosaura torresi* **Núñez** et al., 2003 [Syn. *Liolaemus torresi*]

Juan Carlos Torres-Mura is a biologist who is Curator of Zoology, Museo Nacional de Historia Natural, Santiago de Chile. His bachelor's degree (1984) and master's (1990) are from Universidad de Chile.

Tourneville

Sahara Agama *Trapelus tournevillei* **Lataste,** 1880

Albert Tourneville was a French herpetologist. He wrote *Étude sur les vipers* (1881).

Touzet

Dwarf Iguana sp. *Enyalioides touzeti* Torres-Carvajal, Almendáriz, Yúnez-Muños, and **Reyes,** 2008

Jean-Marc Touzet became Assistant Director, Lyon Zoo, France (2002). He lived in Ecuador (1977–1999), where he established Fundacion Herpetologica Gustavo Orces, Quito. He collected the holotype.

Tovell

Darwin Blind Snake *Ramphotyphlops tovelli* **Loveridge,** 1945 [Syn. *Austrotyphlops tovelli*]

Gunner Tovell (among many others) sent material to Loveridge during WW2 in response to Loveridge's *Reptiles of the Pacific World* (1945). One hundred thousand copies were printed for the armed forces, and servicemen collected in response.

Towns

Towns' Skink *Cyclodina townsi* Chapple et al., 2008

Dr. David Towns is a biologist, ecologist, and herpetologist in New Zealand's Department of Conservation, and President of the Society for Research on Amphibians and Reptiles in New Zealand.

Townsend

Townsend's Anole *Anolis townsendi* **Stejneger,** 1900

Townsend's Worm Lizard *Amphisbaena townsendi* Stegneger, 1911

Townsend's Least Gecko *Sphaerodactylus townsendi* **Grant,** 1931 [Alt. Townsend's Dwarf Sphaero]

Charles Haskins Townsend (1859–1944) was a zoologist who worked for the U.S. Fish Commission (1883–1902). He explored northern California (1883–1884) and the Kobuk River, Alaska (1885), and was an expert before the Russo-American fisheries arbitration at The Hague (1896). He was Director of the New York Aquarium (1902–1937) and was president of the American Fisheries Society (1912–1913). Among his publications is *Field Notes on the Mammals, Birds, and Reptiles of Northern California* (1887). A bird and a mammal are named after him.

Toyama

Iheya Ground Gecko *Goniurosaurus toyamai* **Grismer** et al., 1994

Miyako Grass Lizard *Takydromus toyamai* Takeda and **Ota,** 1996

Masanao Toyama is a Japanese herpetologist. He wrote "Distribution of the Genus *Eumeces* (Scincidae: Lacertilia) in the Ryukyu Archipelago" (1988).

Tracy

Halmahera Python *Morelia tracyae* **Harvey, Barker,** Ammerman, and Chippindale, 2000

Tracy M. Barker is an Australian python breeder. She co-wrote *Pythons of the World Volume I, Australia* (1994).

Tracy, C. R.

Tree Iguana sp. *Liolaemus dicktracyi* **Espinoza,** 2003

Dr. C. Richard "Dick" Tracy (b. 1943) is Professor of Biology, University of Nevada, Reno. His bachelor's (1966) and master's (1968) degrees in biology were awarded by California State University and his doctorate in zoology (1972) by the University of Wisconsin. He co-wrote "Thermal Biology, Metabolism, and Hibernation" in *The Biology of Reptiles* (1997).

Trape

Senegal Garter Snake *Elapsoidea trapei* Mane, 1999

Dr. Jean-François Trape (b. 1949) is a physician and herpetologist at Institut de Récherche pour le Dévelopment, Dakar, Senegal, where he conducts research into malaria. He studied medicine in Paris (1967–1973) and was a general practitioner (1974–1977). Since 1977 he has worked in tropical medicine in Congo, French Guiana, and Senegal. He co-wrote "Les serpents des environs de Bandafassi (Sénégal oriental)" (2004).

Trapido

Trapido's Brown Snake *Storeria dekayi temporalineata* Trapido, 1944

Professor Dr. Harold Trapido (1916–1991) was an entomologist, a herpetologist, and a physician who specialized in tropical medicine. He took all his degrees at Cornell, the last a doctorate in zoology (1942). During WW2 he served in the U.S. Army Air Corps and was posted to Panama (1944), where he was involved in the first field-testing of DDT in the control of malaria. He was in Sardinia (1952), then worked on the control of yellow fever in Panama (1953–1956). He then joined the Rockefeller Foundation and went to live in Poona (1957–1961) as adviser to the Indian Council of Medical Research, followed by writing up his Indian notes in Oxford (1962–1963). He was at Universidad del Valle, Cali, Colombia (1964–1970), becoming President of the Cali Cricket Club. He was head of the Department of Tropical Medicine and Medical Parasitology, Louisiana State University Medical School (1970–1884), retiring as Professor Emeritus. His interest in herpetology was focused on the systematics of snakes. He wrote *The Snakes of New Jersey* (1937).

Trefaut

Blind Snake sp. *Liotyphlops trefauti* Freire, Caramaschi, and Argolo, 2007

See **Rodrigues, M. T. U.**

Tregenza

Tregenza's Lizard *Liolaemus tregenzai* Pincheira-Donoso and Scolaro, 2007

Thomas "Tom" Tregenza is an evolutionary biologist who is Royal Society Research Fellow, Centre for Ecology and Conservation, University of Exeter, England. He co-wrote "Comparative Evidence for Strong Philogenetic Inertia in Precloacal Signalling Glands in a Species-Rich Lizard Clade" (2008).

Tremper

Eldama Ravine Chameleon *Chamaeleo tremperi* **Necas,** 1994

Ronald "Ron" L. Tremper is a herpetologist who was Curator of Reptiles, Chaffee Zoological Gardens, Fresno, California. He breeds geckos and other reptiles and co-authored *Herpetoculture of Leopard Geckos: Twenty-seven Generations of Living Art* (2005).

Treutler

Leaf-toed Gecko sp. *Hemidactylus treutleri* Mahony, 2009

Uli Treutler (1951–2006) was a highly regarded German-born herpetoculturist in Ireland. He died of cancer.

Trevelyan

Trevelyan's Earth Snake *Rhinophis trevelyanus* **Kelaart,** 1853

Kelaart gave no indication in his original description as to who Trevelyan was.

Tristram

Tristram's Spiny-footed Lizard *Acanthodactylus tristrami* **Günther,** 1864

Rev. Henry Baker Tristram (1822–1906) was Canon of Durham Cathedral and a traveler, archeologist, naturalist, and antiquarian. Despite being a churchman, he was an early supporter of Darwin. He traveled and explored as described in *A Journal of Travels in Palestine and the Great Sahara: Wanderings South of the Atlas Mountains* (1860). He wrote extremely interestingly on the indigenous peoples and their customs, and on the natural history of the region, including references to mammals, birds, reptiles, molluscs, and plants. Despite his early penchant for collecting with a gun, he was a Vice President of the Royal Society for the Protection of Birds (1904–1906). Eleven birds and a mammal are named after him.

Troost

Troost's Turtle *Trachemys scripta troostii* **Holbrook,** 1836 [Alt. Cumberland Slider]

Dr. Gerard Troost (1776–1850) was a Dutch physician, naturalist, mineralogist, and pharmacist. He studied at École des Mines, Paris. He arrived in the USA (1810) and settled in Philadelphia, where he was a founding member and first President of the Academy of Natural Sciences (1812). He joined the idealistic settlement of New Harmony (1826) but left to become Professor, University of Nashville, Tennessee (1827), and was also Tennessee's first official geologist (1831).

Troschel

Troschel's Pampas Snake *Phimorphis guainensis* Troschel, 1848

Dr. Franz Hermann "Fritz" Troschel (1810–1882) was a zoologist, malacologist, herpetologist, and ichthyologist. Universität Berlin awarded his doctorate (1834). He was Assistant to Lichtenstein at Museum für Naturkunde, Humboldt-Universität, Berlin (1840–1849), and became Professor of Zoology, Friedrich-Wilhelms-Universität Bonn (1850). A number of other taxa are named after him.

Tryon

Tryon's Velvet Gecko *Oedura tryoni* **De Vis,** 1884 [Alt. Southern Spotted Velvet Gecko]

Tryon's Water Skink *Eulamprus tryoni* **Longman,** 1918

Henry Tryon (1856–1943) was an English scientist. He abandoned medicine in favor of natural science, particularly botany and entomology. He collected in Sweden and New Zealand before going to Queensland (1882), where he first became an Honorary Assistant at the Queensland Museum and then was officially employed there (1883–1893), becoming Assistant Curator (1885). His extracurricular activities for government departments, such as investigating the rabbit menace for the government of New South Wales (1888–1889), brought him into conflict with his Director, and he left, becoming a government entomologist (1893–1925) and vegetable pathologist (1901).

Tschudi

Tschudi's Blind Snake *Leptotyphlops tesselatus* Tschudi, 1845

Tschudi's False Coral Snake *Oxyrhopus melanogenys* Tschudi, 1845

Tschudi's Lightbulb Lizard *Proctoporus pachyurus* Tschudi, 1845

Tschudi's Pacific Iguana *Microlophus thoracicus* Tschudi, 1845

Tschudi's Lizard *Placosoma cordylinum* Tschudi, 1847

Desert Coral Snake *Micrurus tschudii* **Jan,** 1858

Baron Dr. Johann Jacob von Tschudi (1818–1889) was a Swiss explorer, physician, diplomat, naturalist, zoologist, hunter, anthropologist, cultural historian, language researcher, and statesman who traveled in Brazil, Argentina, Chile, and Peru. He wrote *Untersuchungen uber die Fauna Peruana Ornithologie* (1844). Five birds and five mammals are named after him.

Tulkas

Iguanid lizard sp. *Liolaemus tulkas* Quinteros, Abdala, Gómez, and **Scrocchi,** 2008

In the mythology of J. R. R. Tolkien, Tulkas is one of the Ainur (demiurgic powers) who helped to shape Middle Earth. One of Tulkas' characteristics is the power to run faster than any other creature. *Liolaemus tulkas* is also very fast in short sprints.

Turner, H. J. A.

Malindi Centipede-eater *Aparallactus turneri* **Loveridge,** 1935

H. J. Allen Turner (1876–1953) was a British taxidermist who lived in Kenya (1909–1953). He collected birds in East Africa (mainly Kenya) (1915–1917). A bird is named after him.

Turner, J. A.

Turner's Thick-toed Gecko *Pachydactylus turneri* **Gray,** 1864

J. Aspinall Turner (1797–1867) was a British entomologist, cotton merchant, manufacturer, and Member of Parliament (1857–1865). He founded the Manchester Field Naturalist Club and was a member of the Royal Entomological Society. Gray's etymology says he names the gecko "in honour of J. Aspinall Turner, Esq., M.P., who has done so much to make known the zoology of Western Africa, and formed such a fine collection of insects."

Tuzet

Ambiky Chameleon *Furcifer tuzetae* **Brygoo,** Bourgat, and **Domergue,** 1972

Professor Odette Tuzet (1906–1976) was a biologist and parasitologist who worked at Banyuls, the University of Paris' laboratory on the Mediterranean. She became head of the Department of Invertebrate Zoology, Université Paul Valéry, Montpellier, France.

Tweedie

Tweedie's Mountain Reed Snake *Macrocalamus tweediei* **Lim,** 1963

Michael Wilmer Forbes Tweedie (1907–1993) was a herpetologist, ichthyologist, and malacologist who worked (1932–1971) at Raffles Museum (now the Singapore National Museum) as Assistant Curator (1932), becoming Director (1946). He wrote *Snakes of Malaya* (1953).

Tyler

Arthrosaura sp. *Arthrosaura tyleri* **Burt** and Burt, 1931

Sidney F. Tyler Jr. (d. 1937) was a historian and photographer. He was wealthy enough to support the American Museum of Natural History's Tyler Duida expedition (1928–1929) to the headwaters of the Orinoco River.

Tytler

Tytler's Mabuya *Mabuya tytleri* Tytler, 1868

Colonel Robert Christopher Tytler (1818–1872) was a soldier, naturalist, photographer, and collector. He served throughout India (1835–1864) and in Kabul and other parts of Afghanistan, interrupted only by two periods on sick leave in England. He and his wife, Harriet (1827–1907), were keen photographers and took around 500 large-format calotype negatives of scenes associated with the Indian Mutiny (1857). He was the third Superintendent of the convict settlement at Port Blair, part of the Andaman Islands Administration. He spent the last six months of his life in charge of the museum in Simla. According to Theobald's *Catalogue of Reptiles* in the Asiatic Society's Museum, Tytler described the mabuya (skink) and named it after himself—not the action of a gentleman, but we cannot corroborate this, having not found Tytler's own original description. Two birds are named after him.

Tzarewsky

Tzarewsky's Toadhead Agama *Phrynocephalus birulai* Tzarevsky 1927

Dr. Sergius F. Tzarevsky (sometimes spelled Tsarewsky, Zarevsky, or Czarevsky) was a herpetologist at the Russian Academy of Sciences, Leningrad (St. Petersburg), where he was Curator of Herpetology (1915–1929).

Tzotzil

Tzotzil Montane Pit-viper *Cerrophidion tzotzilorum* **Campbell,** 1985

Named after the Tzotzil people of the central mountainous region of Chiapas, Mexico, the direct descendants of the classic Maya civilization.

U

Ulikovski

Ulikovsi's Pacific Gecko *Gekko ulikovskii* **Darevsky** and **Orlov,** 1994 [Alt. Vietnam Golden Gecko]

Dimitri Ulikovsky is a Russian amateur herpetologist.

Ulmer

Dwarf/Reed Snake sp. *Calamaria ulmeri* Sackett, 1940

Frederick A. Ulmer Jr. (1892–1974) was a zoologist and Curator of Mammals at the Philadelphia Zoological Gardens. He was a member of the George Vanderbilt Sumatran expedition (1936–1939), during which this snake was collected. Primarily a mammalogist, he wrote "A Longevity Record for the Mindanao Tarsier" (1960).

Underwood

Underwood's Spectacled Tegu *Gymnophthalmus underwoodi* **Grant,** 1958

Underwood's Marked Gecko *Homonota underwoodi* **Kluge,** 1964

Thick-tailed Gecko genus *Underwoodisaurus* **Wermuth,** 1965

Underwood's Least Gecko *Sphaerodactylus underwoodi* **Schwartz,** 1968

Underwood's Mussurana *Clelia errabunda* Underwood, 1993

Underwood's Tree Snake *Dipsadoboa underwoodi* **Rasmussen,** 1993

Underwood's Bronzeback *Dendrelaphis underwoodi* Van Rooijen and **Vogel,** 2008

Dr. Garth Leon Underwood (1919–2002) was a British herpetologist and an Honorary Research Fellow at the Natural History Museum, London. His writings include "On the Classification and Evolution of Geckos" (1954).

UNSAAC

Lightbulb Lizard sp. *Proctoporus unsaacae* Doan and Castoe, 2003

This reptile is named after UNSAAC: Universidad Nacional de San Antonio Abad de Cusco, Peru.

Upton

Tree Iguana sp. *Liolaemus uptoni* Scolaro and **Cei,** 2006

Jorge Arturo Upton (d. 2003), an Argentine zoologist, was a friend and collaborator of the authors, who named this reptile in his memory.

Uribe

Uribe's False Cat-eyed Snake *Pseudoleptodeira uribei* Bautista and **H. M. Smith,** 1992

Dr. Zeferino Uribe-Peña (b. 1947) is, according to the etymology, "an eminent herpetologist on the faculty of the Universidad Nacional Autónoma de México." He was the first President of the Mexican Herpetological Society (1988–1990). He co-wrote, with Ramírez-Bautista (the senior describer) and Pérez-Ramos, "New Herpetological Records from Islands of the Gulf of California" (1989).

Ursini

Ursini's Viper *Vipera ursinii* Bonaparte, 1835 [Alt. Orsini's Viper, Meadow Viper, Steppe Viper]

See **Orsini.**

Uthmoeller

Uthmöller's Chameleon *Kinyongia uthmoelleri* **Müller,** 1938 [Alt. Müller's Leaf Chameleon, Hanang Hornless Chameleon]

Dr. Wolfgang Uthmöller was a herpetologist who discovered this chameleon while traveling in East Africa (1927–1936). He was attached to Zoologische Staatssammlung Munchen. He wrote *Schlangen wie ich sie sah* (1946).

Uzzell

Uzzell's Neusticurus *Neusticurus apodemus* Uzzell, 1966

Uzzell's Prionodactylus *Prionodactylus dicrus* Uzzell, 1973

Uzzell's Lizard *Darevskia uzzelli* **Darevsky** and Danielyan, 1977

Uzzell's Riolama *Riolama uzzelli* **Molina** and Señaris, 2003

Dr. Thomas Marshall Uzzell Jr. (b. 1932) is a herpetologist. His bachelor's (1953), master's (1958), and doctorate (1962) were awarded by the University of Michigan. He was an Instructor (1962) and Assistant Professor (1965–1967) at the University of Chicago. He was an Assistant Professor at Yale and first Assistant Curator of Herpetology, Peabody Museum of Natural History (1967–1974). He was appointed Adjunct Associate Professor of Biology, University of Pennsylvania (1974).

V

Vadon

Mossy Pygmy Leaf Chameleon *Brookesia vadoni* **Brygoo** and **Domergue,** 1968

Jean Pierre Léopold Vadon (1904–1970) was a naturalist. He served in the French army in Morocco (1924–1926), then became a schoolmaster (1927–1930). He developed an interest in entomology (1931) and obtained a post in Cameroon, transferring (1933) to teach at the European School, Tananarive, Madagascar. He taught (1934–1970) at Maroantsetra, being also French Consul (1963). He became a corresponding member of the Entomological Department, Muséum National d'Histoire Naturelle, Paris (1945). Eleven genera of insects are named after him.

Vaillant

Vaillant's Mabuya *Mabuya vaillantii* **Boulenger,** 1887

Somali Two-headed Snake *Micrelaps vaillanti* **Mocquard,** 1888

Colubrid snake sp. *Liophidium vaillanti* Mocquard, 1901

Professor Léon Louis Vaillant (1834–1914) was a zoologist, herpetologist, ichthyologist, and malacologist at Muséum National d'Histoire Naturelle, Paris. He was on several French naval expeditions (1880–1883). See also **Ludovic.**

Valenciennes

Short-tail Anole *Anolis valencienni* **Duméril** and **Bibron** 1837

Valenciennes' Rock Gecko *Pristurus crucifer* Valenciennes, 1861

Achille Valenciennes (1794–1865) was a zoologist, ichthyologist, and conchologist who worked as an Assistant, Muséum National d'Histoire Naturelle, Paris. He wrote, with Cuvier, *Histoire naturelle des poissons* (22 vols., 1828–1849).

Valentin

Valentin's Lizard *Darevskia valentini* **Boettger,** 1892

Dr. Jean Valentin (1868–1898) was a naturalist at Naturmuseum Senckenberg, Frankfurt, who went to Karabagh with Radde (1890).

Valeria B.

Smooth Earth Snake *Virginia valeriae* **Baird** and **Girard,** 1853

Valeria Biddle Blaney (1828–1900), who collected the holotype, was Baird's first cousin. She married Brigadier General Washington LaFayette Elliott (1858).

Valeria D.

Donoso's Steppe Iguana *Urostrophus valeriae* **Donoso-Barros,** 1966 [Syn. *Pristidactylus valeriae*]

Valeria is Donoso-Barros' fourth daughter.

Valerie

Valerie's Leaf Chameleon *Brookesia valerieae* **Raxworthy,** 1991 [Alt. Raxworthy's Leaf Chameleon]

Valerie M. Raxworthy supported Christopher Raxworthy's work in Madagascar.

Valverde

Valverde's Lizard *Algyroides marchi* Valverde, 1958 [Alt. Spanish Keeled Lizard]

Professor José Antonio Valverde (1926–2003) was a zoologist, ornithologist, environmentalist, and herpetologist who helped save what is now Doñana National Park. Among his publications is *Birds of the Spanish Sahara* (1957).

Van Dam

Van Dam's Girdled Lizard *Cordylus (warreni) vandami* **FitzSimons,** 1930

Van Dam's Round-headed Worm Lizard *Zygaspis vandami* FitzSimons, 1930

Gerhardus Petrus Frederick Van Dam (d. 1927) was a South African herpetologist who worked in the Transvaal (1920s). He also collected botanical specimens in South Africa and Mozambique with Vivian FitzSimons. He wrote "Description of New Species of *Zonurus*, and Notes on the Species of *Zonurus* Occurring in the Transvaal" (1921).

Van Denburgh

Southern Sagebrush Lizard *Sceloporus graciosus vandenburgianus* **Cope,** 1896

Van Denburgh's Ground Skink *Scincella formosensis* Van Denburgh, 1912

Van Denburgh's Rock Racer *Platyplacopus kuehnei* Van Denburgh, 1909

Monterey Ring-necked Snake *Diadophis punctatus vandenburgii* **Blanchard,** 1923

California Lyre Snake *Trimorphodon biscutatus vandenburghi* **Klauber,** 1924

Tsushima Smooth Skink *Scincella vandenburghi* **Schmidt,** 1927

Dr. John Van Denburgh (1872–1924) was a physician and herpetologist. He was one of the original students at Stanford, which awarded his zoology doctorate (1897). He qualified as a physician at Johns Hopkins (1902). He practiced medicine in San Francisco and was Curator of the Herpetological Collections, California Academy of

Sciences. The San Francisco earthquake (1906) destroyed the academy (see **Slevin**). He wrote *The Reptiles of Western North America* (1922). He died while on holiday in Hawaii.

Vanderhaege

Vanderhaege's Toad-Headed Turtle *Phrynops vanderhaegei* Bour, 1973 [Syn. *Mesoclemmys vanderhaegei*]

Maurice Vanderhaege is a cheloniophile and expert in breeding reptiles in terraria. He co-wrote *Guide du terrarium: Tous les conseils pour élever plus de 300 espèces.*

Vanderyst

Vanderyst's Worm Lizard *Monopeltis vanderysti* **Witte, 1922**

Father Hyacinth Julien Robert Vanderyst (1860–1934) was a missionary, anthropologist, agronomist, botanist and naturalist in the Belgian Congo (Zaire). He collected botanical specimens in Belgium, the Canary Islands, Morocco, Senegal, and Angola as well as the Congo (1891–1934). He wrote "La population préhistorique au Congo Belge" (1932).

Van Dijk

Van Dijk's Chitra *Chitra vandijki* **McCord** and **Pritchard,** 2002 [Alt. Myanmar Narrow-headed Softshell Turtle]

Dr. Peter Paul Van Dijk has been, since 2004, Director of the Tortoise and Freshwater Turtle Conservation Program, Center for Applied Biodiversity Science (part of Conservation International). He worked in Malaysia (1999–2002) analyzing trends in turtle farming and trading. He co-wrote *Turtles of the World, Volume 4: East and South Asia* (2006).

Van Heurn

Forest Skink sp. *Sphenomorphus vanheurni* **Brongersma,** 1942

Willem Cornelis Van Heurn (1887–1972) was a taxonomist, biologist, and preparator who was wealthy but chose to work. He sent most of his specimens to Nationaal Natuurhistorisch Museum, Leiden, and worked there as Assistant Curator for Fossil Mammals (1941–1945). He was in Surinam (1911), Simaloer (off Sumatra) (1913), and Dutch New Guinea (1920–1921). He then lived in the Dutch East Indies (mostly Java) (1921–1938), where he studied rat control on Java, Timor, and Flores and ran a sea research laboratory, was a schoolteacher, and headed the Botany Department, Netherlands Indies Medical School, Java. He returned to Holland in 1939. He liked to give his articles amusing titles, such as "Do Tits Lay Eggs Together as the Result of a Housing Shortage?" (1955). Among other taxa, a mammal is named after him.

Van Heygen

Van Heygen's Day Gecko *Phelsuma vanheygeni* **Lerner,** 2004

Bearded Chameleon sp. *Kinyongia vanheygeni* **Necas,** 2009

Dr. Emmanuel Van Heygen started keeping *Phelsuma* geckos in terraria in his bedroom while still a child in Belgium. Since 1992 he has been visiting Indian Ocean destinations where he can see these geckos in the wild. He lived in Canada (2000–2003). He is the brand manager for Exo Terra, a company that makes and markets paraphernalia for pet reptiles and amphibians. The company also sponsors the specialist magazine *Phelsuma*.

Van Kampen

Van Kampen's Gecko *Nactus vankampeni* **Brongersma,** 1933

Pieter Nicolaas Van Kampen (1878–1937) was a Dutch herpetologist and ichthyologist who was Professor of Zoology, Universiteit Leiden, until his retirement (1931). He wrote *The Amphibia of the Indo-Australian Archipelago* (1923).

Vanmeerhaeghe

Vanmeerhaeghés Pond Turtle *Mauremys leprosa vanmeerhaeghei* Bour and Jerome, 1999

The etymology says only that this turtle is dedicated to Bertrand Vanmeerhaeghe, and we have been unable to uncover more about him.

Van Son

Van Son's Thick-toed Gecko *Pachydactylus vansoni* **FitzSimons,** 1933

Dr. Georges Van Son (1898–1967) was an entomologist and botanist. He was born in Russia to a Dutch diplomat father and a Russian countess. They had only French as a common language, which became Georges' mother tongue. He was an Imperial Russian Navy cadet and visited China and Japan. During the Russian revolution his father was killed by a Bolshevik sniper, and the family was imprisoned until 1921, when the Dutch Embassy obtained their release and they emigrated to Holland to join the father's family. He worked at Nationaal Natuurhistorisch Museum, Leiden, from where he was recruited by Dr. A. J. T. Janse to work with his private entomological collection in Pretoria, South Africa (1923). He was employed at the Transvaal Museum (1925–1967), mainly working on butterflies. He was a pioneer in cultivating South African orchids and succulents. He was the expedition botanist and entomologist on the 1932 Vernay-Lang Kalahari expedition.

Vanzo

Vanzo's Whiptail *Cnemidophurus vanzoi* Baskin and **Williams,** 1966
Gallagher's Kentropyx *Kentropyx vanzoi* **Gallagher** and **Dixon,** 1980
Gymnophthalmid lizard genus *Vanzosaura* **Rodrigues,** 1991
Vanzo's Spectacled Tegu *Gymnophthalmus vanzoi* **Carvalho,** 1999

See **Vanzolini.**

Vanzolini

Vanzolini's Bachia *Bachia scolecoides* Vanzolini, 1961
Vanzolini's Worm Lizard *Amphisbaena vanzolinii* **Gans,** 1963
Vanzolini's Anotosaura *Anotosaura vanzolinia* **Dixon,** 1974
Vanzolini's Teiid *Colobodactylus dalcyanus* Vanzolini and Ramos, 1977
Vanzolini's Scaly-eyed Gecko *Lepidoblepharis heyerorum* Vanzolini, 1978
Vanzolini's Ground Snake *Liophis vanzolinii* Dixon, 1985
Vanzolini's Anole *Phenacosaurus vanzolinii* **Williams** et al., 1996
Bent-toed Gecko sp. *Gymnodactylus vanzolinii* Cassimiro and **Rodrigues,** 2009

Professor Dr. Paulo Emilio Vanzolini (b. 1924) is a zoologist and herpetologist, who worked at the American Museum of Natural History, New York (late 1970s–1980s). He was Director of the Zoological Museum, Universidade de São Paulo, Brazil (1962–1993)—compulsorily retired at age 70, but he still works there. He is also famous in Brazil as a composer of samba music. He wrote *Elementary Statistical Methods in Zoological Systematics* (1993). Two mammals are named after him. See also **Vanzo.**

Van Zyl

Namib Desert Gecko *Palmatogecko vanzyli* Stein and **Haacke,** 1966

Mr. and Mrs. Ben Van Zyl lived at Ohopoho, South-West Africa (Namibia), and were keen observers and investigators of anthropological and biological matters. They sent an unknown gecko to the museum in Windhoek, but it got lost in the post. The museum mounted an expedition (1965) specifically to investigate their report, and so this new genus and species was discovered. In the early 1960s Ben van Zyl was the "Bantu" Commissioner for the Kaokoveld, where a pass is named after him. The spelling in the binomial ought to be *vanzylorum*, as the gecko is named after them both.

Varcoa

Chinese Japalure *Japalura varcoae* **Boulenger,** 1918

Mrs. John Graham, née Varcoa, was the wife of the collector of the holotype. Boulenger received the specimen from Graham in 1914.

Vaucher

Andalusian Wall Lizard *Podarcis vaucheri* **Boulenger,** 1905

Henri Vaucher (1856–1910) was a Swiss botanist who worked in Morocco (1879–1910). A bird is named after him.

Vaueresell/Vauerocega

Sparse-scaled Forest Lizard *Adolfus vaueresellí* **Tornier,** 1902
Usambara Forest Snake *Buhoma vauerocegae* Tornier, 1902

Tornier gave no etymology in his descriptions, and we have been unable to trace the origins of these binomials.

Vautier

Brazilian Steppe Iguana *Urostrophus vautieri* **Duméril** and **Bibron,** 1837

Monsieur Vautier was a traveler in Brazil. The description states that he sent the first specimen of this lizard but adds no further useful comment. We think it may be Abel Félix Vautier (1794–1863), who was a French malacologist and a member of the French Parliament (1846). He had also been a shipowner who encouraged the masters of his vessels to bring back interesting specimens. He bequeathed his ornithological collection to the Caen museum.

Vedda/Veddha

Veddha's Blind Snake *Typhlops veddae* **Taylor,** 1947

Veddha is a Sinhala word from Sanskrit meaning "hunter with bow and arrow." It is also the name for the aboriginal inhabitants of Sri Lanka, a few of whom still live in its forested areas.

Velazquez

Central Peninsular Alligator Lizard *Elgaria velazquezi* **Grismer** and Hollingsworth, 2001

Victor Manuel Velázquez-Solis is a Mexican herpetologist with extensive knowledge of the herpetofauna of Baja California, where he and Grismer collected (1988) for the Department of Herpetology, San Diego Natural History Museum. He greatly helped the describers in their fieldwork. Hollingsworth collected the holotype (1997).

Vellard

Blind Snake sp. *Leptotyphlops vellardi* **Laurent,** 1984

Dr. Jehan Albert Vellard (1901–1996) was a Tunisian-born French physician, ethnologist, herpetologist, and arachnologist. He qualified as a physician (1924), then went to Brazil and worked at Instituto Butantan, São Paulo (1925–1929). He ran Museo Historia Natural del Paraguay (1930–1933). He traveled widely, always accompanied by his mother, and became caught up in the Chaco War between Paraguay and Bolivia (1932–1935). They once found an abandoned encampment in which was a coati, a pot of honey, and a one-year-old native girl. Whether they collected the coati is not reported, but they certainly ate the honey and adopted the girl, whom they called Marie-Yvonne; she was brought up under the strict regime of a French classical education and became an ethnologist. He was Director, Museo de Historia Natural, La Paz, Bolivia (1940–1943), and head of the Zoology Department, Universidad Nacional de Tucumán, Argentina. He founded the Herpetology Department, Natural History Museum, Universidad Nacional Mayor de San Marcos, Lima, Peru (1946), being Professor and Director there (1947–1956). Until 1966 he was also Director, Instituto Boliviano de Biologia de Altura, being interested in the physiological problems of adjusting to limited air at high altitudes and studying the native peoples of the high Andes. He retired to Buenos Aires (1966). He wrote *Civilisations des Andes* (1963).

Veloso

Tree Iguana sp. *Liolaemus velosoi* **Ortiz,** 1987

Professor Dr. Alberto Rafael Veloso-Martinez (b. 1940) is a herpetologist, Department of Ecology of the Sciences, Universidad de Chile, Santiago de Chile.

Vences

Vences' Chameleon *Calumna vencesi* Andreone et al., 2001

Dr. Miguel Vences (b. 1969) is a herpetologist whose doctorate was awarded by Friedrich-Wilhelms-Universität Bonn (2000). He did postdoctoral work at Muséum National d'Histoire Naturelle, Paris (2000–2001), and Universität Konstanz, Baden-Württemberg (2001–2002). He was Assistant Professor and head of the Vertebrate Section, Zoological Museum, Universiteit van Amsterdam (2002–2005). Since 2005 he has been Professor of Evolutionary Biology, Zoological Institute, Technische Universität Braunschweig, Germany. He co-authored *a Field Guide to the Amphibians and Reptiles of Madagascar* (3rd ed., 2007).

Venning

Venning's Keelback *Amphiesma venningi* **Wall,** 1910 [Alt. Chin Hills Keelback]

Brigadier Francis Esmond Wingate Venning (1882–1970) was born in Ceylon (now Sri Lanka) and educated at Sandhurst. He served in the Indian army (1902–1933). He was an ornithologist and oologist who collected in Burma (now Myanmar), northwest India (Pakistan), and Iraq. In retirement he was active in the Botany Section, Hampshire Field Club and Archaeological Society.

Vernay

Angola File Snake *Mehelya vernayi* **Bogert,** 1940

Arthur Stannard Vernay (1877–1960) was an English antiques dealer and philanthropist with a deep interest in natural history who lived in the USA. He was a Trustee of the American Museum of Natural History. Jointly with Colonel John C. Faunthorpe he financed six expeditions to Burma, India, and Thailand (1922–1928). He financed a British Museum collecting trip to Tunisia (1925). Having sold his business and all his collections and antiques, he retired to the Bahamas (1940).

Verreaux

Verreaux's Skink *Anomalopus verreauxii* **Duméril** and **Duméril,** 1851

Andamanese Giant Gecko *Gekko verreauxi* **Tytler,** 1864

Jean Baptiste Edouard Verreaux (1810–1868) and Jules Pierre Verreaux (1807–1873) were naturalists, collectors, and dealers. They both worked in China and South Africa's Cape Colony, where a third naturalist brother, Joseph Alexis Verreaux (d. 1868), lived. Jules was an ornithologist and plant collector for Muséum National d'Histoire Naturelle, Paris, which sent him to Australia (1842). He collected in Tasmania, New South Wales, and Queensland (1842–1850). He returned to France with a reported 115,000 items (1851). Earlier he helped Andrew Smith in founding the South African National Museum, Cape Town. The Verreaux family traded at Maison Verreaux, a huge emporium for feathers and stuffed birds. They were ambitious taxidermists and gained notoriety for having once attended the funeral of a tribal chief whose body they then disinterred, took to Cape Town, and stuffed. In 1888 the Catalán veterinarian Francisco Darder, then Curator of the Barcelona Zoo, purchased the "specimen" from one of the brothers' sons, Edouard Verreaux. This controversial exhibit was on show in Barcelona until the end of the 20th century, when the man's descendants demanded that it be returned for a decent burial. The skink is named after Jules. We are not sure after which brother the gecko is named. Eleven birds and two mammals are named after them.

Versteeg

Gymnophthalmid lizard sp. *Arthrosaura versteegii* Lidth De Jeude, 1904

Dr. Gerard Martinus Versteeg (1876–1943) qualified as a physician (1905) and went to the East Indies as an army surgeon. He went on an expedition to Surinam as a botanical collector and photographer (1904). He joined two expeditions to Dutch New Guinea (1907–1913). He was in charge of disease control in part of northern Java (1919–1923) and in the Health Department (1928–1931). He retired to Holland and was a medical administrator, Central Bureau for Statistics, The Hague (1931–1943).

Vesey-Fitzgerald

Vesey-Fitzgerald's Burrowing Skink *Janetaescincus veseyfitzgeraldi* **Parker,** 1948

Leslie Desmond Edward Foster Vesey-Fitzgerald (1909–1974) was an Irish zoologist, herpetologist, and environmentalist. He worked in the Seychelles (1938) and led an expedition to the Comoro Islands (1940). He undertook several wildlife surveys in the Abu Dhabi desert and collected many insects, amphibians, and reptiles. He was involved in locust control in Northern Rhodesia (Zambia) (1949–1964). He was employed by the Tanganyika (Tanzania) government as National Parks Officer (1965). His colleagues called him Vesey, and the Africans, Bwana Mungosi—"Mr. Skins," in reference to the boots he always wore. He wrote *East African Grasslands* (1973).

Victoria

Victoria Short-necked Turtle *Emydura victoriae* **Gray,** 1842 [Alt. Northern Red-faced Turtle]

Named after the Victoria River, Northern Territory, Australia.

Videla

Videla's Mountain Lizard *Phymaturus videlai* Scolaro and Pincheira-Donoso, 2010

Fernando Videla is an Argentine herpetologist at Unidad de Ecologia Animal, Mendoza, Argentina, who, for many years, accompanied José M. Cei in the field.

Vieillard

Vieillard's Chameleon Gecko *Eurydactylodes vieillardi* **Bavay,** 1869 [Alt. Bavay's Gecko]

Eugène Vieillard (1819–1896) was a French medical missionary, naturalist, and botanist who collected in New Caledonia, the Isle of Pines, and Tahiti (1861–1867).

Vilkinson

See **Wilkinson.**

Villa

Country Anole *Anolis villai* **Fitch** and **Henderson** 1976

Dr. Jaime Dolan Villa-Rivas (b. 1944) is a Nicaraguan biologist who was at the Department of Biology, University of Missouri. His doctorate was awarded by Cornell (1978). He wrote "Snakes of the Corn Islands, Caribbean Nicaragua" (1976).

Villiers

Villiers' Blind Snake *Rhinoleptus koniagui* Villiers, 1956

Dr. André Villiers (1915–1983) was an entomologist and herpetologist. He joined Muséum National d'Histoire Naturelle, Paris (1937), and was on several expeditions to Cameroon. He gained his doctorate (1943) and was requested (1945) to assume responsibility for entomology within France's African colonies. He worked in the Entomology Laboratory at Muséum National d'Histoire Naturelle, Paris (1956–1980). Among his publications is *Les serpents de l'ouest africain* (1950). In the binomial, *koniagui* refers to a West African language.

Vincent

Vincent's Least Gecko *Sphaerodactylus vincenti* **Boulenger,** 1891 [Alt. Central Lesser Antillean Sphaero]

Vincent's Sipo *Chironius vincenti* Boulenger, 1891 [Alt. St. Vincent Blacksnake]

These species are named after the island of St. Vincent.

Vinciguerra

Vinciguerra's Lipinia *Lipinia relicta* Vinciguerra, 1892

Vinciguerra's Writhing Skink *Lygosoma vinciguerrae* **Parker,** 1932

Dr. Decio Vinciguerra (1856–1934) was a zoologist, naturalist, and ichthyologist, Museo Civico di Storia Naturale di Genova (1883–1931). He was on the Italian expedition to Tierra del Fuego (1882) as the expedition's botanist and zoologist. He was (early 1890s) a Professor, Sapienza–Università di Roma, and Director, Acquario Romano, which was also used as a fish hatchery.

Vindum

Vindum's Bevel-nosed Boa *Candoia paulsoni vindumi* **H. M. Smith** and **Chizar,** 2001

Jens Verner Vindum (b. 1954) is the Senior Collections Manager, Herpetology Department, California Academy of Sciences. His bachelor's degree (1978) and master's (1983) were both awarded by San Francisco State University. He was a research collaborator at the Smithsonian's Department of Vertebrate Zoology, Amphibians, and Reptiles (2006–2009).

Vinson

Vinson's Gecko *Phelsuma vinsoni* **Mertens,** 1963 [Junior syn. of *P. ornata* **Gray,** 1825]

Joseph Lucien Jean Vinson (1906–1966), a zoologist from Mauritius, was Director, Mauritius Institute, and was awarded the O.B.E. (1963). The binomial is in the singular, but many sources state the animal is named after him and his son Jean-Michel. They wrote "The

Saurian Fauna of the Mascarene Islands" (published only 1969). A stag beetle is named after him.

Viquez

Víquez's Tropical Ground Snake *Trimetopon viquezi* **Dunn,** 1937

Dr. Carlos Víquez Segrada (b. 1890) collected the holotype. He wrote *Animales venenosos de Costa Rica* (1940).

Visser, G.

Visser's Shovel-snout *Prosymna visseri* **FitzSimons,** 1959

G. Visser was a South African from Cape Town who led and sponsored the expedition that collected this species.

Visser, J. D.

Gecko sp. *Pachydactylus visseri* **Bauer,** Lamb, and Branch, 2006

John D. Visser is a South African herpetologist and wildlife biologist. *Visseri* has a double meaning: it is both an eponym and a play on the name of the locality—the Fish (Vis) River valley—where the species can be found. He co-wrote *Snakes and Snakebite* (1978).

Vitt

Vitt's Ground Snake *Liophis vitti* **Dixon,** 2000

Dr. Laurie Joseph Vitt (b. 1945) is a biologist and herpetologist. After taking his bachelor's and master's degrees at Western Washington University, he moved to Arizona State University for his doctorate in biology (1976). After fieldwork in Brazil, he joined the University of California, Los Angeles, where he became a Full Professor. He moved to Oklahoma (1990), becoming Professor of Zoology and Curator of Reptiles, Sam Noble Oklahoma Museum of Natural History, Department of Zoology, University of Oklahoma. Among his works is "Shifting Paradigms: Herbivory and Body Size in Lizards" (2004).

Viv

Skink sp. *Lioscincus vivae* **Sadlier, Bauer, Whitaker,** and S. A. Smith, 2004

Vivienne "Viv" Whitaker, a New Zealander, is primarily a botanical collector but collected the holotype of this skink. She and her husband, Tony, a herpetologist, collected in the Philippines and New Guinea (2001).

Vlangali

Ching Hai Toadhead Agama *Phrynocephalus vlangalii* **Strauch,** 1876

Major General Aleksandr Georgiyevich Vlangali (1823–1908) was a traveler and diplomat who was the Russian Envoy to Peking (Beijing) about 1870. He was in Paris (1877) and was Russian Ambassador in Rome (1894). He wrote *Reise nach der Östlichen Kirgisen-Steppe* (1894).

Vogel

Vogel's Pit-viper *Trimeresurus venustus* Vogel, 1991 [Alt. Beautiful Pit-viper; Syn. *Cryptelytrops venustus*]

Vogel's Pit-viper *Viridovipera vogeli* David, Vidal, and **Pauwels,** 2001

Mountain Reed Snake sp. *Macrocalamus vogeli* David and Pauwels, 2005

Dr. Gernot Vogel (b. 1963) is a herpetologist and chemist. He is a member of the Society for Southeast Asian Herpetology, Heidelberg. One of his main interests is rugby football, and he is a Director of a German rugby club. He wrote *The Bronze-backed Snakes of Thailand* (1991). Two different species have at times been called "Vogel's Pit-viper."

Vogl

Savanna Side-necked Turtle *Podocnemis vogli* **Müller,** 1935 [Alt. Llanos Side-necked Turtle]

Father Cornelius Vogl (1884–1959) was a Benedictine priest. He was a missionary in Dar es Salaam, Tanzania (1910–1919), and in Venezuela (1925–1959). He collected botanical and herpetological specimens that he sent to the Natural History Museum, Munich (1933), and probably gave up active collecting in 1939.

Vogt

Vogt's Forest Dragon *Hypsilurus schoedei* Vogt, 1932

Theodor Vogt (b. 1881) was a German naturalist. He wrote *Reptilien und Amphibien aus Neu-Guinea* (1911).

Voris

Kharin's Sea Snake *Hydrophis vorisi* **Kharin,** 1984

Dibamid lizard sp. *Dibamus vorisi* **Das** and **Lim,** 2003

Dr. Harold Knight Voris (b. 1940), a herpetologist, is Curator and head of the Department of Zoology, Field Museum. He was awarded his bachelor's degree by Hanover College (1962) and his doctorate by the University of Chicago (1969).

Vosmaer

Saddle-backed Rodrigues Giant Tortoise *Cylindraspis vosmaeri* Schoepff, 1792 EXTINCT

Vosmaer's Writhing Skink *Lygosoma vosmaeri* **Gray,** 1839

Arnout Vosmaer (1720–1799) was the Curator of the Menagerie and the Museum of the Stadtholder (Dutch Head of State) from 1756. He left an unpublished autobiography, "Memorie tot het leven van Arnout Vosmaer." A bird is named after him.

Vosseler

Chameleon sp. *Kinyongia vosseleri* **Nieden,** 1913
[Formerly included in *K. fischeri*]

Professor Julius Vosseler (1861–1933) was a zoologist. He was in German East Africa (Tanzania) (1903–1909) and was Director of the Hamburg Zoo from 1910. A bird is named after him.

Vud

Bahamian Racer *Alsophis vudii* **Cope,** 1862

Dr. H. C. Wood Jr. was an expert on Myriapoda. It seems that *vudii* is a curious variation of *woodii*—perhaps a joke between Wood and Cope, who were friends.

Vulcan

Vulcan Lipinia *Lipinia vulcania* **Girard,** 1857
Centipede Snake sp. *Tantilla vulcani* **Campbell,** 1998

Means "of the volcano" rather than relating directly to the Roman god (or to the strategic bomber or an alien race in *Star Trek*).

Vulpinus

Western Fox Snake *Pantherophis vulpinus* **Baird** and **Girard,** 1853

Rev. Charles Fox (1815–1854) was an Episcopal minister. He emigrated from England (1836), arrived at Jackson, Michigan (1839), bought a farm, and built a church at Grosse Ile (1843). He moved to Ann Arbor (1853) to become first Professor of Theoretical and Practical Agriculture, University of Michigan. However, he never taught, as a fire destroyed his farm and he moved to Detroit, where he died of cholera. The scientific name seems to be an academic joke. Fox collected the holotype, but instead of naming the snake *foxi*, as might have been expected, the describers used the Latin *vulpinus* ("foxlike").

Vyner

Shelford's Skink *Lamprolepis vyneri* **Shelford,** 1905

Sir Charles Vyner deWindt Brooke (1874–1963) was Raja Muda of Sarawak and subsequently the Third and final White Rajah of Sarawak (1917–1946). He grew up in Sarawak but was educated in England and graduated from Cambridge, returning to Sarawak in 1911. A boom in oil and rubber gave him the revenue to modernize the local institutions and introduce (1924) a version of British law. He banned missionaries and fostered local traditions. He was in Australia (1942–1945) during the Japanese occupation but resumed as Rajah for a few months in 1946, then ceded Sarawak to Britain as a crown colony and retired to London.

W

Waanders

Bleeker's Kukri Snake *Oligodon waandersi* **Bleeker,** 1860

Henry Louis van Bloemen Waanders (1796–1851) was an ichthyologist and colonial administrator who was in Willemstad, Curacao (Dutch West Indies) (1831–1835), and at Semarang (Java) and Sumatra (1841–1845).

Wagler

Wagler's Blind Snake *Leptotyphlops albifrons* Wagler, 1824

Wagler's Puffing Snake *Pseustes sexcarinatus* Wagler, 1824

Wagler's Sipo *Chironius scurrulus* Wagler, 1824

Wagler's Snake *Waglerophis merremi* Wagler, 1824

Wagler's Worm Lizard *Amphisbaena vermicularis* Wagler, 1824

Wagler's Pit-viper *Tropidolaemus wagleri* **Boie,** 1827 [Alt. Temple Viper]

Sicilian Wall Lizard *Podarcis waglerianus* Gistel, 1868

Wagler's Ground Snake *Atractus wagleri* **Prado,** 1945

Johann Georg Wagler (1800–1832) was a herpetologist. He was Spix's assistant and, upon the former's death (1826), became the Director, Zoological Museum, Ludwig-Maximilians-Universität München, and continued working on their extensive Brazilian collections. In addition he worked on systematics of amphibians and reptiles in the museum. Wagler died from an accidentally self-inflicted gunshot wound while out hunting. He wrote the highly regarded *Monographia psittacorum*. Eight birds are named after him.

Wagner, J. A.

Ecuadorean Frog-eating Snake *Diaphorolepis wagneri* **Jan,** 1863

Johann Andreas Wagner (1797–1861) was a paleontologist, zoologist, and archeologist. He became Professor of Zoology and Assistant Curator, Zoological Museum of Ludwig-Maximilians-Universität München. He wrote *Diagnosen neuer Arten brasilischer Säugethiere* (1842). Five mammals are named after him.

Wagner, M.

Wagner's Viper *Vipera wagneri* **Nilson** and Andrén, 1984

Moritz Wagner (1813–1887) was a German traveler, naturalist, and geographer. He traveled through Persia (Iran), Georgia, and northern Iraq (1840s) and later through North and Central America and the Caribbean (1852–1855). He collected the holotype in 1846; it took 138 years for someone to realize that it was a new species. He committed suicide.

Wahlberg

Wahlberg's Snake-eyed Skink *Afroablepharus wahlbergi* **Andrew Smith,** 1849

Wahlberg's Velvet Gecko *Homopholis wahlbergii* Andrew Smith, 1849 [Specific name originally given as *walbergii*]

Wahlberg's Kalahari Gecko *Colopus wahlbergii* **Peters,** 1869

Wahlberg's Striped Skink *Trachylepis wahlbergii* Peters, 1870

Johan August Wahlberg (1810–1856) was a naturalist and collector. He studied chemistry and pharmacy at Uppsala Universitet (1829) and worked in a Stockholm chemist's shop while studying at the Forestry Institute. He traveled and collected in southern Africa (1838–1856), returned briefly to Sweden (1853), but was soon back in Africa at Walvis Bay (Namibia). He was exploring the headwaters of the Limpopo when a wounded elephant killed him. Three birds and a mammal are named after him.

Waite

Waite's Blind Snake *Ramphotyphlops waitii* **Boulenger,** 1895 [Syn. *Austrotyphlops waitii*]

Edgar Ravenswood Waite (1866–1928) was an English-born Australian zoologist and ichthyologist. After studying at Manchester University he worked at the Leeds Museum (1888–1892). He went to work at the Australian Museum, Sydney (1892), where he was Curator of Ichthyology (1893–1905). He was in New Zealand (1906–1914) as Curator of the Canterbury Museum, Christchurch. He was General Director of the South Australian Public Library, Museum, and Art Gallery (1914–1928). He wrote "Notes on Australian Typhlopidae" (1894). He died of malaria, contracted in New Guinea.

Wake

Wakes' Gecko *Cyrtodactylus wakeorum* **Bauer,** 2003

Dr. David Burton Wake (b. 1936) and his wife, Dr. Marvalee Hendricks Wake (b. 1939), are experts on salamanders and caecilians, respectively. His bachelor's degree (1958) was awarded by the Pacific Lutheran College, Tacoma, and both his master's (1960) and doctorate (1964) by the University of Southern California. He spent much of his career at the Museum of Vertebrate Zoology, University of California, Berkeley, including being Director of the Museum (1971–1998), Professor of Integrative Biology (1989), and Curator of Herpetology. He is now an Emeritus Professor. She joined the faculty of the University of California, Berkeley (1969), after her doctorate was awarded by the University of Southern California (1968). He wrote "Climate Change Implicated in Amphibian and Lizard Declines" (2007). She became

Chairman of the Department of Zoology and its successor, the Department of Integrative Biology. She was President of the American Society of Ichthyologists and Herpetologists (1983) and is now Professor of the Graduate School. See also **Dave Wake.**

Walker, J. J.

Walker's Lerista *Lerista walkeri* **Boulenger,** 1891
Flying Dragon sp. *Draco walkeri* Boulenger, 1891]

James John Walker (1851–1939) served in the Royal Navy and was an amateur natural historian and entomologist. He was on board HMS *Penguin* (1890–1892), which surveyed Australasian waters and the Pacific (1890–1907). He retired to Oxford (1904). He became President of the Entomological Society, London, and received an honorary degree from Oxford. During his naval career he sent many specimens of different taxa to the Natural History Museum, London. He wrote *The Natural History of the Oxford District* (1926). A mammal is named after him.

Walker, W. F.

Walker's Tree Iguana *Liolaemus walkeri* **Shreve,** 1938
Walker's Slender Snake *Tachymenis attenuata* Walker, 1945

Dr. Warren Franklin Walker Jr. (b. 1918) was a biologist and herpetologist who worked with K. P. Schmidt. His doctorate was awarded by Harvard. He taught at Oberlin College, Ohio (1947–1985), retiring as Emeritus Professor of Biology. He was an expert on reptile anatomy and locomotion. He co-wrote "Snakes of the Peruvian Coastal Region" (1943). An amphibian is named after him.

Wall

Sakishima Odd-tooth Snake *Dinodon rufozonatus walli* **Stejneger,** 1907 [Alt. Banded Red Snake]
Wall's Keelback *Amphiesma xenura* Wall, 1907
Wall's Krait *Bungarus sindanus walli* Wall, 1907 [Alt. Sind Krait]
Wall's Worm Snake *Typhlops oligolepis* Wall, 1909
Wall's Bronzeback *Dendrelaphis cyanochloris* Wall, 1921 [Alt. Blue Bronzeback Tree Snake]
Chitral Bow-foot Gecko *Cyrtopodion walli* **Ingoldby,** 1922 [Alt. Swat Stone Gecko; Syn. *Gymnodactylus walli, Tenuidactylus walli*]
Wall's Kukri Snake *Oligodon melanozonotus* Wall, 1922
Wall's Tawny Cat Snake *Boiga ochracea walli* **Stoliczka,** 1970
Wall's Hump-nosed Viper *Hypnale walli* **Gloyd,** 1977 [Alt. Gloyd's Hump-nosed Viper]
Wall's Sea Snake *Disteira walli* **Kharin,** 1989

Dr. Frank Wall (1868–1950), a physician and herpetologist, was born in Ceylon (now Sri Lanka). He qualified as a physician in the UK and worked for the Indian Medical Service (1893–1925). He was a member of the Bombay Natural History Society. He wrote *A Popular Treatise on the Common Indian Snakes* (installments, 1905–1919). By naming the krait after himself he committed a serious taxonomic faux pas.

Wallace

Bent-toed Gecko sp. *Cyrtodactylus wallacei* **Hayden** et al., 2008

Alfred Russel Wallace (1823–1913) was an English naturalist, evolutionary scientist, geographer, anthropologist, social critic and theorist, and a follower of the utopian socialist Robert Owen. He was one of the giants of Victorian science, with claims to be the father of zoogeography. He discovered and described the faunal discontinuity that now bears his name: Wallace's Line. This natural boundary runs between the islands of Bali and Lombok in the south and Borneo and Sulawesi in the north, and separates the Oriental and Australasian faunal regions. He started out as an apprentice surveyor, but his interest in natural history took over, and he went to Brazil on a self-sustaining natural history collecting expedition (1848–1852). On his way home, his ship caught fire and sank with all his specimens. He went to the Indonesian archipelago, where he covered 23,000 kilometers (14,000 miles) between 1862 and 1869, visiting every important island at least once and collecting 125,660 specimens. He was thinking along the same lines as Darwin and sent him his essay "On the Law Which Has Regulated the Introduction of New Species," which encapsulated his most profound theories on evolution, and later another essay, "On the Tendency of Varieties to Depart Indefinitely from the Original Type," presenting the theory of "survival of the fittest." Darwin and Lyell presented this essay, together with Darwin's own work, to the Linnean Society. Wallace's thinking spurred Darwin to encapsulate these ideas in *The Origin of Species.* The rest is history, but Wallace never has been so well known to the general public as Darwin. He wrote many books and papers, including *The Malay Archipelago* (1869), the most celebrated of all writings on Indonesia. Many taxa are named after him, including an amphibian, 2 mammals, and 12 birds.

Wallach

Nicobar Cat Snake *Boiga wallachi* **Das,** 1998

Dr. Van Wallach (b. 1947) is a herpetologist who was at the Harvard Museum of Comparative Zoology and who works with Das. They co-wrote "Scolecophidian Arboreality Revisited" (1998).

Warren, C. R.

Warren's Galliwasp *Diploglossus warreni* **Schwartz,** 1970

Tortuga Boa *Epicrates striatus warreni* **Sheplan** and Schwartz, 1974

C. Rhea Warren, who is a member of the University of Miami's Iron Arrow Society, collected specimens on the Ile de la Tortue, off Haiti (1968 -1970).

Warren, E.

Warren's Spinytail Lizard *Cordylus warreni* **Boulenger,** 1908 [Alt. Warren's Girdled Lizard; Syn. *Zonurus warreni*]

Professor Dr. Ernest Warren (1871–1945) was an English zoologist and first Director of the Natal Museum, Pietermaritzburg (1903–1935). He was also involved in the establishment of its university. He championed the establishment of national parks in Natal (1920s–1930s). He collected the lizard holotype.

Watson, C. F.

Watson's Asp *Atractaspis watsoni* **Boulenger,** 1908 [Alt. Watson's Mole Viper]

C. F. Watson. The etymology says the holotype was "presented by Mr. C. F. Watson," and nothing more.

Watson, E. Y.

Watson's Gecko *Cyrtopodion watsoni* Murray, 1892

Lieutenant E. Y. Watson of the 47th Regiment, Indian army, collected the gecko holotype at Quetta. He collected widely, particularly entomological specimens. He wrote *Hesperiidae Indicae: Being a Reprint of Descriptions of the Hesperiidae of India, Burma, and Ceylon* (1891).

Watson, H. E.

Lacertid lizard sp. *Mesalina watsonana* **Stoliczka,** 1872

H. E. Watson was Civil Officer at a "station" in Sakkar, India (now in Pakistan). Stoliczka wrote that Dr. Francis Day (q.v.) had been "energetically assisted by the Civil Officer at the station, Mr. H. E. Watson," while collecting reptiles in Sind.

Watts

Watts' Anole *Anolis wattsi* **Boulenger,** 1894

Dr. Sir Francis Watts was (1893) the government analyst and agricultural chemist for the Leeward Islands, based in Antigua, from where he sent specimens of reptiles to the Natural History Museum, London. He later became Commissioner of Agriculture, Imperial Department of Agriculture for the West Indies, and proposed that the Royal College of Agriculture and Institute of Tropical Research be located in Trinidad, becoming its first Principal.

Webb, R. G.

Sinaloan Mountain Kingsnake *Lampropeltis webbi* Bryson, **Dixon,** and Lazcano, 2005

Dr. Robert Gravem Webb (b. 1927) is a herpetologist and botanist. He received his doctorate in zoology from the University of Kansas (1960). He taught at the University of Texas, El Paso, but retired as Emeritus Professor (2005) and is now Co-Curator of Herpetology at the Laboratory of Environmental Biology there. He wrote "A New Kingsnake from Mexico, with Remarks on the *Mexicana* Group of the Genus *Lampropeltis*" (1961).

Webb, T. H.

San Diego Alligator Lizard *Elgaria multicarinata webbii* **Baird,** 1858

Dr. Thomas Hopkins Webb (1801–1866) was Secretary of the New England Emigrant Aid Company and Secretary of the U.S.-Mexican Boundary Survey (1848–1855), concurrently making collections of insects, fish, and reptiles. When the survey ran out of money, the party had to return under Mexican military protection as Comanches and Apaches were on the warpath. The survey team was attacked by Apaches, and one member was killed.

Weber

Weber's Pipe Snake *Anomochilus weberi* Lidth De Jeude, 1890

Weber's Thick-toed Gecko *Pachydactylus weberi* **Roux,** 1907

Weber's Sailfin Lizard *Hydrosaurus weberi* **Barbour,** 1911

Dr. Max Wilhelm Carl Weber van Bosse (1852–1937) was a German-Dutch physician and zoologist who was Director, Zoological Museum, Artis Amsterdam, from 1883 when he became a Dutch citizen. He did German military service, half the time as a doctor and half as a hussar. He combined the roles of watch-keeping officer, ship's doctor, and naturalist on a voyage (1881) in the *Willem Barents*, appropriately to the Barents Sea. His wife, Anna, was a botanist, and the Webers spent summers in Norway, where he dissected whales and she collected algae. They went to Sumatra, Java, Celebes, and Flores (1888) and to South Africa (1894). "Weber's Line," an important zoogeographical line between Sulawesi and the Moluccas, is often preferred over "Wallace's Line" (between Sulawesi and Borneo) as the dividing line between the Oriental and Australasian faunas. He co-wrote *The Fishes of the Indo-Australian Archipelago* (1911). Two mammals and a bird are named after him.

Webster

Webster's Anole *Anolis websteri* **Arnold,** 1980

Thomas Preston Webster III (1947–1975) was a student

at Harvard. E. E. Williams wrote of him, "He published extensively while he was a graduate student, although he never finished his Ph.D. He died in an auto accident in Montana just after he had gone there as an Assistant Professor."

Wegner

Wegner's Glass Lizard *Ophisaurus wegneri* **Mertens,** 1959

Dr. A. M. R. Wegner was a zoologist, entomologist, and collector for Museum Zoologicum Bogoriense, Indonesia. Before Indonesian independence, he was a professional collector and had a small private museum in Java's Tengger Mountains that was burned down (1947) by "extremists." He collected in Java and Borneo (1930s–1960s) and was on Ambon Island (1963). He sent over 10,000 specimens to the Field Museum (1963–1965). He published on entomology, and some insects are named after him.

Weidholz

Weidholz's Agama *Agama weidholzi* **Wettstein,** 1932 [Alt. Gambia Agama]

Alfred Weidholz was an Austrian banker and traveler who collected in West Africa, including fishes from the River Niger, for Schönbrunn Zoo in Vienna. An amphibian is named after him.

Weigel

Pygmy King Brown Snake *Pseudechis weigeli* Wellington and **Wells,** 1987

John Randall Weigel (b. 1955) dropped out of university and worked as a keeper and showman at the California Alligator Farm. He emigrated to Australia (1981) and since then has worked at the Australian Reptile Park, New South Wales, buying the business (1984) and becoming its Director in partnership with a local businessman. He co-founded the Reptile Keepers Association (1985), encouraging the keeping and breeding of native species. After the center and its collection were destroyed by fire, he collected replacements and established a captive breeding program. He milks 200 snakes every two weeks and has become the sole supplier of venom used in the production of antivenins in Australia. He was awarded Australia's highest civilian accolade, the Order of Australia, in recognition of his contributions to Australian tourism, herpetology, and antivenin production (2008). He wrote *Care of Australian Reptiles in Captivity* (1988). He discovered the snake and a toad, also named after him, during one of 20 expeditions to the Kimberly District.

Weiler

Black-tailed Green Tree Snake *Dipsadoboa weileri* Lindholm, 1905

Weiler's Gecko *Urocotyledon weileri* **Müller,** 1909

J. Weiler collected reptiles in Tanganyika (now Tanzania) and Cameroon, and gave the collection to Lorenz Müller.

Wellington

Wellington's Spiny-tailed Gecko *Strophurus wellingtonae* **Storr,** 1988 [Alt. Western Shield Spiny-tailed Gecko]

Betty D. Wellington of Mt. Helena was honored in the gecko's scientific name for her services to Western Australian natural history.

Wells

Wells' Death Adder *Acanthophis wellsi* Hoser, 1998

Richard C. Wells is a herpetologist who works closely with Ross Wellington. (See Introduction.)

Welwitsch

Angola Wedge-snouted Worm Lizard *Dalophia welwitschii* **Gray,** 1865

Dr. Friedrich Martin Josef Welwitsch (1806–1872) was an Austrian theater critic before fleeing to Portugal to escape the consequences of a youthful indiscretion. The Portuguese sent him to explore and collect botanical specimens in Angola, where over 12 years he accumulated some 5,000 specimens, many new to science. He was a proponent of the establishment of the Madeira botanical gardens. He caused an international quarrel by sending a large proportion of his collection to the Natural History Museum, London, instead of to Lisbon. The Portuguese took the view that as they had paid him, the collection belonged to them. The collection's duplicate specimens were split, so both museums got something out of it. Welwitsch died in London. Two mammals are named after him.

Werding

Flat-headed snake sp. *Xenopholis werdingorum* Jansen, Álvarez, and Köhler, 2009

The Werding family own Hacienda San Sebastián, Bolivia, where the holotype was collected. They provided great hospitality and broad logistic and financial support for the building of a biological research station on their property.

Wermuth

Wermuth's Anole *Anolis wermuthi* Köhler and Obermeier, 1998

Professor Dr. Heinz Wermuth (1918–2002) was a herpetologist. He graduated from Humboldt-Universität,

Berlin, in zoology, botany, paleontology, and chemistry. He became head of the Herpetological Section, Museum für Naturkunde Berlin (1952). The construction of the Berlin Wall (1961) created problems for him, as he lived in West Berlin and the museum was in East Berlin. He commuted across the wall but was caught bringing fake passports for museum colleagues to use to escape. He was lucky in that he only lost his job and was not arrested and sent to an East German prison. He was at Staatlisches Museum für Naturkunde, Stuttgart (1961–1983), first as Curator and retiring as head of the Division of Zoology. He lived in a small flat but kept a variety of reptiles as pets, including a caiman that was over 2.5 meters (8 feet) long. He co-wrote, with Robert Mertens, *Liste der Rezenten Amphibien und Reptilien, Testudines, Crocodylia, Rhynchocephalia* (1977).

Werner, F.

Usambara Centipede-eater *Aparallactus werneri* **Boulenger,** 1895
Werner's Gecko *Asaccus elisae* Werner, 1895 [Alt. Elisa's Leaf-toed Gecko]
Werner's Keelback *Helicops pictiventris* Werner, 1897
Werner's Tree Snake *Dipsadoboa werneri* Boulenger, 1897
Spatula-tooth Snake *Iguanognathus werneri* Boulenger, 1898
Werner's Three-horned Chameleon *Chamaeleo werneri* **Tornier,** 1899 [Alt. Uzungwe Three-horned Chameleon]
Werner's Gypsy Gecko *Hemiphyllodactylus harterti* Werner, 1900
Werner's Worm Lizard *Amphisbaena polygrammica* Werner, 1900
Werner's Worm Snake *Typhlops mutilatus* Werner, 1900 [Junior syn. of *Helminthophis flavoterminatus* Peters, 1857]
Colubrid snake sp. *Chamaelycus werneri* **Mocquard,** 1902
Werner's Cat Snake *Crotaphopeltis tornieri* Werner, 1908 [Alt. Tornier's Cat Snake]
Werner's Sipo *Chironius flavopictus* Werner, 1909
Werner's Thirst Snake *Dipsas maxillaris* Werner, 1909
Werner's Bush Anole *Polychrus femoralis* Werner, 1910
Werner's Coffee Snake *Ninia oxynota* Werner, 1910
Werner's Lipinia *Lipinia miangensis* Werner, 1910
Werner's Thick-toed Gecko *Pachydactylus serval* Werner, 1910 [Alt. Western Spotted Thick-toed Gecko]
Amami Odd-scaled Snake *Achalinus werneri* **Van Denburgh,** 1912
Werner's Agama *Agama sennariensis* Werner, 1914
Werner's Ground Snake *Atractus werneri* **Peracca,** 1914
Werner's False Coral Snake *Oxyrhopus leucomelas* Werner, 1916
Werner's Large-scaled Lizard *Ptychoglossus bicolor* Werner, 1916
Werner's Spider Gecko *Cyrtopodion gastropholis* Werner, 1917
Werner's Garter Snake *Elapsoidea laticincta* Werner, 1919 [Alt. Central African Garter Snake]
Werner's Diadem Snake *Spalerosophis dolichospilus* Werner, 1923
False Fer-de-Lance sp. *Xenodon werneri* Werner, 1924
Werner's Ornate Snake *Amphiesmoides ornaticeps* Werner, 1924
Werner's Thick-toed Gecko *Pachydactylus werneri* **Hewitt,** 1935

Professor Dr. Franz Josef Maria Werner (1867–1939) was an Austrian explorer, zoologist, and herpetologist who taught at Naturhistorisches Museum Wien. Steindachner, who was Director of the museum until 1919, disliked him and forbade him access to the herpetological collection there. Werner collected in North and East Africa, traveling to Egypt (1904) and the Sudan (1905), and made regular visits south to Uganda and west to Morocco up to the outbreak of WW1. He wrote much, including *Amphibian und Reptilien* (1910).

Werner, Y. L.

Negev Tortoise *Testudo werneri* Perälä, 2001 [Probably synonymous with *T. kleinmanni*]
Sinai Agama *Pseudotrapelus sinaitus werneri* Moravec, 2002

Dr. Yehuda Leopold Werner (b. 1931) was born in Germany but as a child emigrated to the Palestine Protectorate (Israel). He graduated in zoology (1956) from the Hebrew University, which awarded his doctorate (1961). He was an Assistant in the Department of Zoology (1953), becoming Curator of Amphibians and Reptiles (1973) and Director, Life Sciences Collections (1990). He was Professor, Alexander Silberman Institute of Life Sciences, Hebrew University, Jerusalem, and is now Emeritus. He wrote in English, Hebrew, German, Dutch, Hungarian, and Czech. His works include *A Guide to Our Reptiles and Amphibians* (1995).

Westermann

Westermann's Snake *Elachistodon westermanni* **Reinhardt,** 1863 [Alt. Indian Egg-eating Snake]

Geraldus Frederick Westermann (1807–1890) was a book dealer and pigeon fancier who became a zoologist. He was one of the "three W's" (Werlemann and Wijsmuller being the others) who in 1838 founded the Amsterdam Zoo, "Natura Artis Magistra," usually known as "Artis," and was its first Director (1838–1890). Two birds are named after him.

Westphal

Spiny Lizard sp. *Sceloporus westphalii* **Dugès,** 1877
[Syn. *S. dugesii intermedius*]

Westphal was a friend of Dugès, who made a mistake when naming this lizard. Only after he had named it *westphalii* did Dugès remember that he had already described the reptile as *Tropidolepis intermedius* (1869). Under the rules, the earlier naming takes precedence.

Wetmore

Wetmore's Ameiva *Ameiva wetmorei* **Stejneger,** 1913
[Alt. Blue-tailed Ground Lizard]
Wetmore's Pointed Snake *Uromacer frenatus wetmorei* **Cochran,** 1921
Diploglossid lizard genus *Wetmorena* Cochran, 1927
Gracile Anole *Anolis brevirostris wetmorei* Cochran, 1931
Wetmore's Tolucan Groundsnake *Conopsis lineata wetmorei* **H. M. Smith,** 1943

"Frank" Alexander Wetmore (1886–1978) was an American ornithologist and avian paleontologist. He was a bird taxidermist at the Denver Museum of Natural History, Colorado (1909). He was in Puerto Rico (1911) and later traveled throughout South America for two years, investigating bird migration between continents, while working for the U.S. Bureau of Biological Survey. He worked at the Smithsonian as Assistant Secretary (1925–1945), then Secretary (1945–1952). He conducted (1946–1966) an annual research program in Panama. He was President of the American Ornithologists' Union (1926–1929). He wrote *A Systematic Classification for the Birds of the World* (1930), wherein he devised the Wetmore Order, a sequence of bird classification that had widespread acceptance until very recently and is still in use. Numerous taxa, including 16 birds and 4 mammals, are named after him.

Wettstein

Wettstein's Viper *Vipera ursinii wettsteini* Knöpfler and **Sochurek,** 1955

Dr. Otto von Wettstein Ritter von Westersheim (1892–1967) was an Austrian zoologist who was Professor of Natural History, Naturhistorisches Museum Wien, where he was Curator of Herpetology (1920–1945). Originally interested in birds and mammals, he found that herpetology was the only department with a vacancy (1915). During WW2 he successfully kept the collection of tens of thousands of specimens, preserved in alcohol, safe in bunkers or in mines. He took over many other duties during that time, even continuing to publish the *Annals* of the museum (1941–1944). In 1945 the Allies officially barred him from the museum, so he retired to private life and worked for the Department of Forest Protection, studying insects and their parasites. He published 60 scientific papers on herpetology, and *Herpetologia Aegaea* was the result of field trips to Greece in 1934, 1935, 1942 (as part of a Wehrmacht biological research squad), and 1954.

Wetzel

South American Dwarf Gecko *Lygodactylus wetzeli* **H. M. Smith, Martin,** and **Swain,** 1977

Dr. Ralph M. Wetzel collected in Paraguay (1950) and then other areas of South America, sending his collections to the University of Connecticut, where he worked (1950–1983) until retiring as Professor of Zoology. He discovered (1972) the Chacoan Peccary as a living animal, previously known only from subfossil remains. He wrote "Systematics, Distribution, Ecology and Conservation of South American Edentates" (1982). A mammal is named after him.

Weyrauch

Argentine Blind Snake, *Leptotyphlops weyrauchi* Orejas-Miranda and **Zug,** 1964

Dr. Wolfgang Karl Weyrauch (1907–1970) was a German zoologist, malacologist, and ecologist who went to Peru (1938). He worked at the Museum of Universidad Nacional Mayor de San Marcos, Lima, and collected spiders and centipedes for the Department of Entomology (1940–1950). After 1950 his function was taken over by Hans-Wilhelm and Maria Koepke. His private collection was deposited later in Instituto Miguel Lillo, Tucumán, Argentina, where he probably resided.

Wheeler

Wheeler's Knob-tailed Gecko *Nephrurus wheeleri* **Loveridge,** 1932

Dr. William Morton Wheeler (1865–1937) was a biologist, entomologist, and myrmecologist. He was trained as an insect embryologist and became the leading expert on ants, particularly of Australia, where he led an expedition (1931). He was an assistant at Ward's Natural Science Establishment, Rochester (1884–1885). He was Director, Milwaukee Public Museum (1887–1890). His doctorate was from Clark University (1892). He was Professor, Department of Zoology, University of Chicago (1892–1899), with an interlude of research in Europe (1893–1894). He was Chairman, Department of Zoology, University of Texas, Austin (1899–1903), then Professor, Invertebrate Zoology, American Museum of Natural History, New York (1903–1908), and finally Professor at Harvard (1908–1937).

Whitaker, A. H.

Whitaker's Skink *Cyclodina whitakeri* **Hardy,** 1977
Whitaker's Sticky-toed Gecko *Hoplodactylus kahutarae* Whitaker, 1985

Anthony Hume Whitaker (b. 1944) is a New Zealand herpetologist. He co-wrote "A New Genus and Species of Live-Bearing Scincid Lizard (Reptilia: Scincidae) from New Caledonia" (2004).

Whitaker, R.

Whitaker's Sand Boa *Eryx whitakeri* **Das,** 1991

Romulus Whitaker (b. 1943) is an American-born Indian herpetologist and conservationist who has spent most of his life in India. He received his higher education in the USA, attending the University of Wyoming for a year but leaving without graduating. He became a merchant seaman and then was in the U.S. Army. He founded the Snake Park, Chennai, where the Irula people continue to practice their skills in snake collecting, nowadays to milk them of venom for the production of antivenins. With his former wife, Zai, and his sons, Nikhil and Samir, he established the Centre for Herpetology (also known as the Madras Crocodile Bank). He resigned as Director of the Crocodile Bank Trust (2001) but continues as a Trustee. He is currently (2009) in the process of setting up the Agumbe Rainforest Research Station. His documentary film about the King Cobra won him an Emmy award. He co-wrote *Snakes of India: The Field Guide* (2004).

White, E. W.

Argentine Marked Gecko *Homonota whitii* **Boulenger,** 1885

Ernest William White (1858–1884) lived in Argentina (1864–1882). He was a teenager when he visited London to meet naturalists but developed tuberculosis. He survived to return to Argentina and went to Mendoza in the Andes to recover. He collected on behalf of the Zoological Society, London, and sent the gecko holotype to Boulenger. He decided upon dentistry and studied (1882–1884) in Philadelphia but contracted typhus and died there. He wrote *Cameos from the Silver-Land; or, the Experiences of a Young Naturalist in the Argentine Republic* (1881). A bird is named after him.

White, J.

White's (Rock) Skink *Egernia whitii* **Lacépède,** 1804

John White (ca. 1756–1832) served in the Royal Navy (1781–1820), visiting India and the West Indies before 1788, when he was appointed Chief Naval Surgeon of the "First Fleet"— the 11 ships that in 1788 took the first 1,500 colonists to Australia, among them 778 convicts. Despite scurvy and dysentery only 34 people died on that voyage. White stayed in Australia until 1795. He served on various ships (1796–1799) and was Surgeon at the Sheerness Navy Yard (1799–1803) and at Chatham (1803–1820). A colorful character, he fought a duel with his Third Assistant, William Balmain, that left them both slightly wounded. He had three legitimate children but also had a son by a convict, and the child was brought up as part of his legitimate household. He was a keen naturalist and accompanied Governor Phillip on two explorations. He sent his wildlife notes, with many specimens, to a friend, Thomas Wilson. The notes were published as *Journal of a Voyage to New South Wales* (1790).

White, N.

White's Monitor *Varanus baritji* **King** and Horner, 1987 [Alt. Lemon-throated Monitor, Northern Ridge-tailed Monitor]

Dr. Neville White is a biological anthropologist who is an Associate Professor, Genetics Department, La Trobe University, Victoria, Australia. He served in the Australian Armed Forces in the Vietnam War. He has close links with, and has been incorporated into the kinship of, the Marralarrmirri clan at Yolngu, Arnhem Land, Australia, where he has researched annually since 1971. "White" in a northern Australian Aboriginal language is *baritj.*

White, T.

White's Nimble Gecko *Dierogekko thomaswhitei* **Bauer,** Jackman, **Sadlier,** and **Whitaker,** 2006

Dr. Thomas White. The etymology reads, "The specific epithet is a patronym honoring Dr. Thomas White, through whose generosity the automated sequencer used in our molecular phylogenetic analyses of the New Caledonian lizard fauna was obtained."

Whiteman

Whiteman's Anole *Anolis whitemani* **Williams,** 1963

Luc Whiteman was a collector in Haiti. He and George Whiteman (we assume a relation) collected for the Museum of Comparative Zoology (1961).

Whitten

Whittens' Rock Gecko *Cnemaspis whittenorum* **Das,** 2005

Dr. Anthony John Whitten (b. 1953) and his wife, Jane E. J. Whitten (b. 1954), are zoologists who work in Asia, specializing in Indonesian species. They were members of the Subdepartment of Veterinary Anatomy, Cambridge (1981), which awarded A. J. Whitten his doctorate (1980). He was an adviser for the Center of Environmental Studies, University of North Sumatra (1981–1983). He was employed by Dalhousie University, Halifax, Nova Scotia (1983–1985), but spent 10 of those 12 years on assignment in Indonesia. He is now (since 1995) employed as a wildlife biologist and Senior Biodiversity Specialist, Asia Technical Department, World Bank. A fish is named after them.

Whyte

Whyte's Water Snake *Lycodonomorphus whytii* **Boulenger,** 1897

Alexander Whyte (1834–1908) was a government naturalist in Nyasaland (Malawi) (1891–1897), where he collected extensively under the patronage of Sir Harry Johnston. Boulenger published an article entitled "A List of the Reptiles and Batrachians Collected in Northern Nyasaland by Mr. Alex Whyte, F.Z.S., and Presented to the British Museum by Sir Harry H. Johnston, K.C.B., with Descriptions of New Species" (1897). Three mammals and a bird are named after him.

Wied

Wied's Fathead Anole *Enyalius catenatus* Wied, 1821
Wied's Sipo *Chironius pyrrhopogon* Wied, 1824
Wied's Ground Snake *Liophis poecilogyrus* Wied, 1825
Wied's Keelback *Helicops carinicaudus* Wied, 1825
Wied's Blind Snake *Ramphotyphlops wiedii* **Peters,** 1867

See **Maximilian.**

Wiedersheim

Wiedersheim's Chameleon *Chamaeleo wiedersheimi* **Nieden** 1910 [Alt. Peacock Chameleon]

Robert Wiedersheim (1848–1923) was an anatomist who was originally trained in botany and zoology. He taught comparative anatomy at Albert-Ludwigs-Universität Freiburg (1876–1918). He published a list of 86 "vestigial organs" (1893) that he considered to have outlived their usefulness and thought the existence of such organs was a good indication that Darwin's theories on evolution were correct. Later research has shown that some of these organs are by no means redundant.

Wiegmann

Wiegmann's Agama *Otocryptis wiegmanni* **Wagler,** 1830 [Alt. Sri Lanka Kangaroo Lizard]
Wiegmann's Worm Lizard *Trogonophis wiegmanni* Kaup, 1830
Wiegmann's Horrible Spiny Lizard *Sceloporus horridus* Wiegmann, 1834
Wiegmann's Tree Lizard *Anisolepis undulatus* Wiegmann, 1834
Wiegmann's Tree Iguana *Liolaemus wiegmanni* **Dumeril** and **Bibron,** 1837
Wiegmann's Striped Gecko *Gonatodes vittatus* **Lichtenstein,** 1856

Arend Friedrich August Wiegmann (1802–1841) was a zoologist at what is now Humboldt-Universität, Berlin (1830–1841), having been Professor at Cologne (1828). He co-founded *Archiv für Naturgeschichte* (1835), which is otherwise known as "Wiegmann's Archive," a zoological periodical review. He died of tuberculosis.

Wiggins

Wiggins' Desert Night Lizard *Xantusia wigginsi* **Savage,** 1952
Gulf Coast Horned Lizard *Phrynosoma wigginsi* Montanucci, 2004

Dr. Ira Loren Wiggins (1899–1987) was a botanist and biologist. Stanford awarded him his doctorate, and he was on the staff there (1930–1964), rising from Assistant Professor of Botany to Emeritus Professor of Biology. He was also Director of the Stanford Natural History Museum. He collected mainly botanical specimens, in the Galapagos Islands and Mexico as well as California and Alaska, where he spent time in the 1950s as Director of the Arctic Research Laboratory, Point Barrow. He published on botany.

Wilcox

Chihuahuan Black-headed Snake *Tantilla wilcoxi* **Stejneger,** 1902

Dr. Timothy Erastus Wilcox (1840–1932) qualified as a physician (1864) and became an Assistant Surgeon in an artillery regiment of the Union army. At the end of the American Civil War he stayed on in the regular army until resigning (1868). He rejoined the army (1874) and served as Captain (1879), Major (1891), Lieutenant Colonel during the Spanish-American War (1898), and Colonel and Assistant Surgeon-General of the U.S. Army (1903), retiring as Brigadier General (1904). He collected the snake holotype shortly after the Battle of Wounded Knee (1892).

Wilder

Wilder's Blind Snake *Liotyphlops wilderi* **Garman,** 1883

Dr. Burt Green Wilder (1841–1925) was a comparative anatomist and naturalist. He was a student of Agassiz at Harvard, graduated as a physician (1862), served as a Surgeon in the Union army (1862–1865), and returned briefly to Harvard (1866). He was Professor of Neurology and Vertebrate Zoology at Cornell (1867–1910). He discovered (1862) that up to 140 meters (150 yards) of silk could be drawn from a living spider. This led him to investigate the habits of spiders and the qualities and usefulness of their silk.

Wilkins

Wilkins' Lerista *Lerista wilkinsi* **Parker,** 1926 [sometimes given as Wilkin's Lerista, in error]

Captain Sir George Hubert Wilkins (1888–1958) was an

Australian explorer and ornithologist who explored the Arctic (1913–1917) and the Antarctic (1920–1923). An official biography describes him as a war correspondent, polar explorer, naturalist, geographer, climatologist, aviator, author, balloonist, war hero, reporter, secret agent, submariner, and navigator. His exploits were legion, but to give one example: In 1928 he bought a surplus WW1 submarine for $1, renamed it *Nautilus* (perhaps he thought he was Captain Nemo), and attempted to cruise beneath the ice to the North Pole. Unfortunately the old ship broke down and the expedition failed. He was on Shackleton's last expedition (1922). He explored Northern Queensland and the Gulf of Carpentaria (1924–1925). He was on Lincoln Ellsworth's support ship when the aviator was stranded (1936) at Admiral Byrd's abandoned camp in Antarctica. His last expedition (1957) was to Antarctica, as a guest of Operation Deepfreeze. He died of a heart attack in the USA. His ashes were scattered at the North Pole.

Wilkinson

Texas Lyre Snake *Trimorphodon vilkinsonii* **Cope,** 1886

Edward Wilkinson (1846–1918) was an expert sheet-metal worker and a competent amateur naturalist. His personal collection, mostly from Mexico, is in a museum in Mansfield, Ohio, where he became Curator and Director upon its opening (1889).

Willard

Willard's Rattlesnake *Crotalus willardi* Meek, 1905 [Alt. Ridge-nosed Rattlesnake]

Professor Francis "Frank" Cottle Willard (1874–1930) was born in Germany while his parents were touring Europe. He graduated in Illinois (1896) and (until 1916) lived in Tombstone, Arizona, where he taught in the local school and worked in his uncle's general store. He was in partnership with another uncle in a sand and gravel company on Long Island, New York (1916–1930). He died in his office of a heart attack. The description says, "Named for Professor F. C. Willard, of Tombstone, AZ, its discoverer." Where the title "professor" comes from we do not know, but perhaps it was a nickname for a man known locally as an intellectual.

Willey

Blind Snake sp. *Ramphotyphlops willeyi* **Boulenger,** 1900

Dr. Arthur Willey (1867–1942) was a British-born Canadian zoologist who graduated from London University (1890). He taught at Columbia University, New York (1892–1894), resigning to work at Cambridge. He traveled to Lifu, Loyalty Islands (1896–1897), where he collected the holotype of this snake. He was Director, Colombo National Museum, Ceylon (now Sri Lanka) (1902–1910), and then was appointed Professor of Zoology, McGill University, Canada.

Williams, C. H.

Williams' Anole *Anolis williamsii* **Bocourt,** 1870

Charles H. Williams was the British Ambassador to the State of Bahia. At that period Bahia (Salvador) was an important administrative center even though it was no longer the capital of Brazil. He was a noted collector, especially of botanical specimens, being a regular correspondent (1863–1869) of Hooker. Sir Richard Francis Burton, who became British Consul in Damascus (1869), recorded escorting "an old Brazilian friend Mr Charles Williams of Bahia" on the road as far as Ramsah (1871). Bocourt's original text does not give more identifying information than "M [Monsieur] Williams," but as the holotype was from Bahia, we are confident that this is the man referred to.

Williams, E. E.

Williams' Dwarf Gecko *Lygodactylus williamsi* **Loveridge,** 1952 [Alt. Turquoise Dwarf Gecko]
Williams' Ground Snake *Liophis williamsi* **Roze,** 1958
Williams' Spiny-tailed Gecko *Strophurus williamsi* **Kluge,** 1963 [Alt. Eastern Spiny-tailed Gecko, Soft-spined Gecko]
Williams' Hinged Terrapin *Pelusios williamsi* **Laurent,** 1965 [Alt. Williams' Mud Turtle]
Williams' Tree Snake *Sibynomorphus williamsi* Carrillo de Espinoza, 1974
Williams' Anole *Anolis vaupesianus* Williams, 1982
Williams' Least Gecko *Sphaerodactylus williamsi* Thomas and Schwartz, 1983
Williams' Toadhead Turtle *Phrynops williamsi* Rhodin and **Mittermeier,** 1983
Williams' Scaly-eyed Gecko *Lepidoblepharis williamsi* **Ayala** and Serna, 1986
Williams' Worm Lizard *Cynisca williamsi* Gans, 1987
Williams' Lizard *Liolaemus williamsi* Laurent, 1992

Dr. Ernest Edward Williams (1914–1998) was a herpetologist, Museum of Comparative Zoology, Harvard, retiring as Professor Emeritus of both Biology and Zoology. He was a renowned expert on *Anolis* lizards, recognizing that this speciose genus offered insights into evolutionary and biogeographical studies. Among his publications is "The Ecology of Colonization as Seen in the Zoogeography of Anoline Lizards on Small Islands" (1969). See also **Eew** and **Ernest.**

Williams, K. L.

Williams' Canyon Lizard *Sceloporus merriami williamsi* **Lemos-Espinal, Chiszar,** and **H. M. Smith,** 2000

Dr. Kenneth Lee Williams (b. 1934) is a specialist in

Honduran and Mexican herpetology. He was on the 1962 University of Illinois expedition to Mexico. He taught at Louisiana State University, where he was awarded his doctorate (1970), and was Curator of the university's Museum of Zoology. He also taught biology at Northwestern State University of Louisiana. He co-wrote *Turtles and Lizards from Northern Mexico* (1960).

Williams-Mittermeier

Williams-Mittermeier Anole *Anolis williamsmittermeierorum* Poe and **Yañez-Miranda,** 2007

This lizard is named after E. E. Williams (q.v.) and Dr. Russell A. Mittermeier (b. 1949), a primatologist, herpetologist, and conservationist who has been President of Conservation International since 1989, before which he worked for the World Wildlife Fund–USA (1977–1988). His bachelor's degree is from Dartmouth College (1971) and his doctorate from Harvard (1977). He has been an Adjunct Professor at the State University of New York, Stony Brook, since 1978. He has undertaken fieldwork for over 30 years and discovered several new primate species. He has published over 300 paper, articles, and books. He co-authored *Wilderness: Earth's Last Wild Places* (2002). Two mammals are named after him.

Williamson, J. H.

Williamson's Kentropyx *Kentropyx williamsoni* **Ruthven,** 1929 [Junior syn. of *K. altamazonica* **Cope,** 1876]

Jesse H. Williamson collected the holotype. He took a bachelor's degree at the University of Indiana (1907). He led an expedition to Brazil for the Museum of Zoology, University of Michigan (1922).

Williamson, K. B.

Williamson's Reed Snake *Collorhabdium williamsoni* Smedley, 1931

Professor K. B. Williamson was an entomologist who investigated mosquitoes in Malaya (Peninsular Malaysia) (1920s–1930s) and wrote *Control of Rural Malaria by Natural Methods* (1935). He collected the snake holotype (1930), having "found an opportunity to collect a number of herpetological specimens."

Wills

Wills' Chameleon *Furcifer willsii* **Günther,** 1890

Rev. James Wills (1836–1898) was an English ornithologist and missionary in Madagascar (1870–1898). He wrote "Notes on Some Malagasy Birds Rarely Seen in the Interior" (1893) and made a collection of birds that was later sold to the Smithsonian.

Wilmara

False Coral Snake sp. *Pliocercus wilmarai* **H. M. Smith,** Perez-Higareda and **Chiszar,** 1996

William P. "Wil" Mara (b. 1966) is a herpetologist and author of about 80 books—fiction and nonfiction, on topics from herpetology to snowmobile racing—including *Venomous Snakes of the World* (1993).

Wilson, A. T.

Iranian Worm Snake *Typhlops wilsoni* **Wall,** 1908

Wilson's Dwarf Skink *Afroablepharus wilsoni* **Werner,** 1919

Colonel Sir Arnold Talbot Wilson (1884–1940) was a soldier and amateur naturalist. As Lieutenant in charge of a detachment of Bengal Lancers guarding the British Consulate in Ahwaz, Persia (Iran) (1904), he made a collection of snakes that he donated to the Bombay Natural History Society. He transferred to the Indian Political Department (1907) and became an Acting Consul (1912). He worked as Resident Director in the Persian Gulf of the Anglo-Persian Oil Company (British Petroleum) (1920–1932). He was both knighted and sacked (1932). He returned to England and was elected to Parliament (1933). He joined the RAF as a volunteer reservist, became an air gunner (while still an MP), and was killed in action over France. He wrote *S. W. Persia* (published posthumously, 1941).

Wilson, D. E.

Wilson's Blind Snake *Leptotyphlops wilsoni* Hahn, 1978

Dr. Don Ellis Wilson (b. 1944) is a zoologist who graduated from the University of Arizona (1955). His took his master's (1967) and doctorate (1970) at the University of New Mexico. He joined the U.S. Fish and Wildlife Service (1971), based at the Smithsonian, where since 2000 he has been Senior Scientist and Curator of Mammals. He was President of the Washington Biologists' Field Club (1993–1996). He co-wrote *The History of the Raccoons of the West Indies* (2002). Two mammals are named after him.

Wilson, H.

African Bighead Snake *Hypoptophis wilsonii* **Boulenger,** 1908

Rev. Henry Wilson was a missionary at Inkongo, Congo, in 1904. He presented the snake holotype. He was still at Inkongo in 1939, as he is known to have supplied samples of fish from the Sankuru River at that time.

Wilson, L. D.

Roatan Vine Snake *Oxybelis wilsoni* **Villa** and McCranie, 1995

Dr. Larry David Wilson (b. 1940) is a herpetologist at the

Department of Biology, Miami-Dade Community College, Florida. His main area of interest is Central America. He wrote *Snakes of Honduras* (1985).

Wilson, S. K.

Wilson's Spiny-tailed Gecko *Strophurus wilsoni* **Storr,** 1983

Stephen "Steve" Karl Wilson is an Australian naturalist, author, and photographer. He collected the gecko holotype. He wrote *A Complete Guide to Reptiles of Australia* (2004).

Wingate

Skink sp. *Trachylepis wingati* **Werner,** 1908

General Sir Francis Reginald Wingate (1861–1953) was an army officer and the first British Governor of Sudan (1899–1916). He was commissioned in the artillery (1860) and assigned to the Egyptian army (1883). He became Director of Egyptian Military Intelligence (1889). He fought several battles against the forces of al-Madhi, the nationalist "rebel." He became British High Commissioner for Egypt (1917). He freed Father Joseph Ohrwalder and others held captive by Mahdi forces and translated his narrative into English: *Ten Years Captivity in the Mahdi's Camp 1882–1892 from the Original Manuscript of Father Joseph Ohrwalder* (1892).

Winnecke

Winnecke's Two-pored Dragon *Diporiphora winneckei* Lucas and Frost, 1896

Charles George A. Winnecke (1856–1902) was an Australian explorer, surveyor, and botanist who joined the Government Survey Office (1873). He was a member of the North Eastern Exploring Expedition (1877–1881) surveying the South Australia–Queensland border. He led the Horn expedition to Central Australia (1894). He wrote *Journal of the Horn Scientific Exploring Expedition* (1897). The Winnecke Goldfields, north of Alice Springs, are named after him.

Wirot

Wirot's Pit-viper *Trimeresurus wiroti* Trutnau, 1981

See **Nutaphand.**

Wirshing

Puerto Rican Leaf-toed Gecko *Phyllodactylus wirshingi* Kerster and **H. M. Smith,** 1955

Juan A. "Tito" Wirshing (d. 1967) collected the gecko holotype in 1953. He seems to have been a wealthy man, as he kept the Isla Caja de Muerto as a private zoological reserve (1954–1967). He appears on the passenger list for the Great World Cruise of 1958 on board the Cunard liner *Caronia*.

Wislizenius

Longnosed Leopard Lizard *Gambelia wislizenii* **Baird** and **Girard,** 1852

Dr. Frederick Adolph Wislizenius (1810–1889) attended Friedrich-Schiller-Universität Jena, Georg-August-Universität Göttingen, and Eberhard Karls Universität Tübingen but received his degree as a physician from Universität Zürich. He had fled to Switzerland (1833) to avoid the political unrest then sweeping Germany. He arrived in the USA (1835) and moved to Illinois. He traveled through the West and the Rockies (1839), then practiced medicine near St. Louis (1840–1846). He was captured by the Mexican army (1846) and held prisoner near Chihuahua for six months. He was allowed out on parole and spent time collecting botanical specimens. Chihuahua was captured by Colonel Doniphan's troops (1847), and he joined that regiment as a military surgeon. He was in St. Louis during an epidemic of cholera (1848–1849). He visited Europe (1850–1851). He wrote *Memoir of a Tour to Northern Mexico* (1848).

Witte

Witte's Five-toed Skink *Leptosiaphos luberoensis* Witte, 1933

Witte's Worm Lizard *Monopeltis remaclei* Witte, 1933

Witte's Five-toed Skink *Leptosiaphos dewittei* **Loveridge,** 1934

Witte's Beaked Snake *Letheobia wittei* **Roux-Estève,** 1974

See **De Witte.** Two different skinks have become known by the same vernacular name, Witte's Five-toed Skink.

Wolf

Wolf's Forest Skink *Sphenomorphus wolfi* **Sternfeld,** 1918

Dr. Eugen Wolf was a member of the 1909 Hanseatische Südsee-Expedition, the year in which he became Director, Naturmuseum Senckenberg, Frankfurt. He wrote the expedition's travel report (published 1909–1911).

Wolter

Mountain Grass Lizard *Takydromus wolteri* **Fischer,** 1885

Karl Andreas Wolter was an amateur naturalist and a businessman from Hamburg who collected the holotype. He went from Shanghai to Chemulpo, Korea (1884), to establish a branch of E. Meyer and Co., a company of which he was a Director. He took over the company and renamed it Karl Wolter and Co. (1907), then returned with his family to Germany (1908).

Wolterstorff

Wolterstorff's Gecko *Urocotyledon wolterstorffi* **Tornier,** 1900

Dr. Willy Georg Wolterstorff (1864–1943) was a geologist

and herpetologist. An illness deprived him of his hearing and power of speech (1871), but he learned to lip-read. He was also myopic and so had a lonely childhood. He compensated by collecting and keeping amphibians; they remained a major lifelong interest. He joined Naturkundemuseums Magdeburg as an assistant (1891), retiring (1929) but continuing to work there until his death. The museum was totally destroyed by the RAF (1945), but his personal charge, 12,000 specimens of tailed amphibians in glass jars, survived the war in a cave, only to be destroyed later by arson.

Wombey

Wombey's Gecko *Diplodactylus wombeyi* **Storr,** 1978 [Alt. Pilbara Ground Gecko; Syn. *Lucasium wombeyi*]

John C. Wombey is a herpetologist who worked for the Australian National Wildlife Collection, Commonwealth Scientific and Industrial Research Organisation (1970–2001). Having retired as Curator of the collection, he continues to study Australian reptiles there as an Honorary Fellow. He co-wrote *List of Australian Vertebrates: A Reference with Conservation Status* (2006).

Wood, N. R.

Florida Scrub Lizard *Sceloporus woodi* **Stejneger,** 1918

Nelson R. Wood (d. 1920) was a taxidermist at the Smithsonian. He had a particular interest in birds, and he and Louis Agassiz were asked to give imitations of birdcalls at a convention of the American Ornithologists' Union. He collected the lizard holotype (1912).

Wood, W. C.

Wood's Anole *Anolis woodi* **Dunn,** 1940

W. C. Wood. When Dunn described this reptile he spoke of "the late" Mr. W. C. Wood. Dunn refers to Wood's Panama trip, so we wonder if Wood might have been an official in the Panama Canal Zone.

Woodford

Woodford's Scaly-toed Gecko *Lepidodactylus woodfordi* **Boulenger,** 1887

Woodford's Skink *Sphenomorphus woodfordi* Boulenger, 1887

Charles Morris Woodford (1852–1927) was an adventurer, naturalist, and philatelist and the Resident Commissioner, Solomon Islands Protectorate (1896–1914). He established the first postal service in the islands and issued their first stamps, personally franking the envelopes. He wrote *A Naturalist among the Headhunters* (1890), which is referred to in a letter by his friend, the novelist Jack London. Two birds and two mammals are named after him.

Wood-Mason

Woodmason's Earth Snake *Uropeltis woodmasoni* **Theobald,** 1876

Yellow-striped Kukri Snake *Oligodon woodmasoni* Sclater, 1891

Dr. James Wood-Mason (1846–1893) was a lepidopterist and specialist in marine animals who worked for the Indian Museum (1877). He made the first collection of molluscs from the Andaman and Nicobar islands (1872). He co-wrote *Natural History Notes from H.M. Indian Marine Survey Steamer "Investigator," Commander R.F. Hoskyn, R.N., Commanding* (1891). Woodmason Bay in the Andaman Islands is named after him.

Woolf

Mount Isa Death Adder *Acanthophis woolfi* Hoser, 1998

Paul Woolf is a herpetologist who, according to Hoser, has been unlawfully harassed by officialdom in the New South Wales and Queensland wildlife services.

Woosnam

Great Lakes Bush Viper *Atheris woosnami* **Boulenger,** 1906

Richard Bowen Woosnam (1880–1915) was on the British Museum scientific expedition with R. E. Dent to Bechuanaland (Botswana) (1906–1907). He was a game ranger and collector in Kenya. He was killed in action during WW1 at Gallipoli (1915). Two mammals and a bird are named after him.

Worontzow

Worontzow's Spotted Night Snake *Siphlophis worontzowi* **Prado,** 1940

C. Worontzow was an entomologist who collected the holotype. He worked as a technician, Parasitology Laboratory, Faculty of Medicine, Brazil (1937).

Worthington

Kenyan Horned Viper *Bitis worthingtoni* **Parker,** 1932

Dr. Edgar Barton Worthington (1905–2001), a British zoologist with a particular interest in fisheries and waterways, lived and worked in East Africa for many years. He wrote *A Development Plan for Uganda* (1949).

Wright, A. M. A. and A. H.

Cliff Tree Lizard *Urosaurus ornatus wrighti* **Schmidt,** 1921

Wright's Brown Snake *Storeria dekayi wrightorum* **Trapido,** 1944 [Alt. Midland Brown Snake]

The lizard is named after Dr. Albert Hazen Wright (1879–1970) and the snake after both him and his wife,

Anna Maria Allen Wright (1882–1964). Albert attended Cornell, where he was awarded his doctorate in vertebrate zoology (1908), and there met and married Anna (1910). He worked at Cornell (1908–1946), first as an Instructor, then as Assistant Professor (from 1915) and Professor (from 1925). Although the Wrights' main interest was amphibians, they jointly wrote *Handbook of Snakes of the United States and Canada* (1957).

Wright, C.

Wright's Dwarf Boa *Tropidophis wrighti* Stull, 1928

Charles Wright (1811–1885) was a botanist, teacher, and collector who explored the western USA for the Pacific Railroad Company. He sent a collection of plants to Professor Asa Gray at Harvard (1844). Gray helped him get places on various surveys and expeditions, including the Boundary Survey Commission, Texas. He was on the U.S. North Pacific exploring expedition (1853–1855) but was asked to leave at San Francisco before it was over, and he went alone to Nicaragua and thence to New York. He explored in Cuba (1856–1857). He went with a U.S. commission to Santo Domingo (1871). Between trips he spent time at the Gray Herbarium, Cambridge, Massachusetts. Wright probably collected the boa in the 1850s. A bird and many plants are named after him.

Wright, E. P.

Wright's Mabuya *Mabuya wrightii* **Boulenger,** 1887
[Alt. Wright's Skink; Syn. *Trachylepis wrightii*]

Dr. Edward Perceval Wright (1834–1910) was Professor of Zoology, Trinity College, Dublin. He was a physician, marine zoologist, botanist, and a naturalist who visited the Seychelles (1868). He became one of the earliest (1868) scientists to dredge in deep water, dredging in Setubal Bay, off Portugal, in depths of up to 900 meters (3,000 feet). He was a contributor to the *Challenger* expedition.

Wright, J. D.

Wright's Spenomorphus *Sphenomorphus wrighti* **Taylor,** 1925

John Dutton Wright (1866–1952) was an educator and philanthropist in New York and Santa Barbara, California. He provided financial support for Taylor's trip to Palawan. His specialty was the education of deaf children. He wrote *What the Mother of a Deaf Child Ought to Know* (1915).

Wright, J. S.

Wright's Short-legged Skink *Brachymeles wrighti* **Taylor,** 1925

John Suarez Wright was a naturalist from Santa Barbara, California. He accompanied Taylor on many collecting trips.

Wright, J. W.

Rio Huancabamba Leaf-toed Gecko *Phyllodactylus johnwrighti* **Dixon** and Huey, 1970

Dr. John William Wright (b. 1936) is a biologist who specializes in whiptail lizards. He studied under Professor Charles Herbert Lowe, University of Arizona, from where he received his doctorate (1965). He was Curator of Herpetology, Los Angeles Museum (1992).

Wu

Wu's Rock Agama *Laudakia wui* **Zhao,** 1998

Gangfu Wu (b. 1935) is a Chinese herpetologist who, like Zhao, works at the Department of Zoology, Chengdu Institute of Biology. Among his publication is, with Zhao and Inger, "Ecological and Geographic Distribution of the Amphibians of Sichuan, China" (1989).

Wucherer

Wucherer's Ground Snake *Xenopholis scalaris* Wucherer, 1861

Wucherer's Lizard-eating Snake *Elapomorphus wuchereri* **Gunther,** 1861

Wucherer's Worm Lizard *Leposternon wuchereri* **Peters,** 1879

Dr. Otto Edward Henry Wucherer (1820–1874) was a Portuguese-born German physician and herpetologist. He qualified at Eberhard Karls Universität Tübingen and practiced at St. Bartholomew's Hospital, London, and in Lisbon. He discovered the cause of the tropical disease elephantiasis. He left Europe and settled in Salvador Bahia, Brazil (1843). He wrote "Sobre a mordedura das cobras venenosas e seu tratamento" (1867).

Wynn

Wynn's Worm Snake *Typhlops castanotus* Wynn and **Leviton,** 1993

Addison Hartwell Wynn (b. 1955) is a herpetologist with the Division of Amphibians and Reptiles, the Smithsonian. With others he wrote "Apparent Triploidy in the Unisexual Brahminy Blind Snake, *Ramphotyphlops braminus*" (1987).

X

Xantus

Night Lizard genus *Xantusia* **Baird** 1858

Leaf-Toed Gecko *Phyllodactylus xanti* **Cope,** 1863

Louis Janos (John) Xantus de Vesey (1825–1894) was a Hungarian who worked for William Hammond, the collector. He fled Hungary (1848) and lived the USA (1855–1861). Xantus was a member of the Austro-Hungarian Empire expedition to Siam (Thailand), China, and Japan, collecting botanical and zoological specimens and investigating ethnography and applied arts (1868). Afterward he traveled independently, visiting Hong Kong, Manila, Singapore, Borneo, Java, and Sumatra. He returned to Hungary in 1870 with 155,644 specimens in 200 crates. He is renowned as a pathological liar (a 19th-century Baron Munchausen). "A poor but educated and ambitious man, he wrote grandiose accounts of his American exploits. They were published in Hungary where he became famous. His letters make Private Xantus sound like he was in charge. Despite the fact that he plagiarized other travel accounts of the American West, lied about himself, and always claimed to be superior to those around him, Xantus did great work for Baird and the Smithsonian. Xantus once had a photo taken of himself as a US Navy captain, which was published in Hungary. Xantus never even served in the Navy" (from Schoenman's introduction to Xantus' *Travels in Southern California*, translated from the Hungarian by Theodore Schoenman and Helen Benedek Schoenman, Detroit, Wayne State University Press, 1976). Many species, including six birds, are named after him.

Y

Yam

Lanyu Scaly-toed Gecko *Lepidodactylus yami* **Ota,** 1987

The name *yami* refers to Filipino immigrants now inhabiting Lanyu Island.

Yamagishi

Skink sp. *Sirenoscincus yamagishii* Sakata and **Hikida,** 2003

Dr. Satoshi Yamagishi is a zoologist and ornithologist who was Director-General of the Yamashina Institute for Ornithology, Japan (2007), and of the Department of Zoology, Graduate School of Science, Kyoto University. He has studied Madagascan fauna. Hikida specifically thanked Yamagishi for providing him opportunities to visit Thailand and Madagascar.

Yanez

Yanez's Lava Lizard *Microlophus yanezi* **Ortiz,** 1980 [Syn. *Tropidurus yanezi*]

Yanez's Tree Iguana *Liolaemus fabiani* Yanez and **Núñez,** 1989 [Alt. Fabian's Lizard]

José Lautaro Yáñez-Valenzuela (b. 1951) is a zoologist, herpetologist, and marine mammalogist at Museo Nacional de Historia Natural de Chile.

Yarrow

Yarrow's Spiny Lizard *Sceloporus jarrovii* **Cope,** 1875

Dr. Henry Crecy Yarrow (1840–1929) was an army surgeon who had studied in Philadelphia and Switzerland. He graduated as a physician (1861) and served with a cavalry regiment during the American Civil War. He was in Georgia (1866), helping to cope with a cholera outbreak, and in Baltimore (1871), where he met the naturalist Coues. He was surgeon and naturalist on Wheeler's expedition (1871–1876) exploring west of the 100th meridian, returning to become Professor of Dermatology, George Washington University. He was Honorary Curator, Department of Herpetology, the Smithsonian (1879–1889). He went on the reserve list (1908) but was recalled (1917) when the USA entered WW1. He wrote *Checklist of North American Reptilia and Amphibia, with Catalogue of Specimens in the United States National Museum* (1883).

Yasuma

Yasuma's Fringed Gecko *Luperosaurus yasumai* **Ota,** Sengoku, and **Hikida,** 1996

Dr. Shigeki Yasuma (b. 1944) is a zoologist. He works at the Hiraoka Environmental Science Laboratory and was seconded by the Japan International Cooperation Agency to advise in Borneo. He wrote *The Ryukyu Archipelago: Diversity of Biota and Geological History of the Islands* (2000).

Yav

Neotropical House Snake sp. *Thamnodynastes yavi* **Myers** and **Donnelly,** 1996

This snake is named after a mountain, Cerro Yavi.

Yerbury

Southern Leaf-toed Gecko *Hemidactylus yerburyi* **Anderson,** 1895

Lieutenant Colonel John William Yerbury (1847–1927) was an amateur entomologist and dipterist. He collected in Aden (1884–1896). He was in Rhodesia (Zambia/Zimbabwe) just before WW1. He wrote *Seashore Diptera* (1919).

Yonenaga

Yonenaga's Worm Snake *Typhlops yonenagae* **Rodrigues,** 1991

Dr. Yatiyo Yonenaga-Yassuda is a Brazilian biologist of Japanese descent who specializes in vertebrate genetics, particularly rodents. She took three degrees at Universidade de São Paulo, the last one a doctorate (1973), and became an Assistant Professor in the university's Biological Science Institute (1969). A mammal is named after her.

Yoshi

Yoshi's Bow-fingered Gecko *Cyrtodactylus yoshii* **Hikida,** 1990

Professor Dr. Ryozo Yoshii (d. ca. 1999) was a biologist and the Chief of the Entomological Section of the Forest Research Centre, Sabah, Malaysia (1978–1985). He later taught at Kyoto University. He was Vice President of the Speleological Society of Japan (1976–1977).

Youngson

Youngson's Ctenotus *Ctenotus youngsoni* **Storr,** 1975

William Kenneth Youngson is an Australian zoologist. He co-wrote *The Islands of the North-West Kimberley, Western Australia* (1978). A mammal is named after him.

Yuwono

Sulawesi Forest Turtle *Leucocephalon yuwonoi* **McCord, Iverson,** and Boeadi, 1995

Black-backed Monitor *Varanus yuwonoi* **Harvey** and **Barker,** 1998 [Alt. Tricolored Monitor]

Frank Bambang Yuwono (b. 1958) is a biologist, herpetologist, and breeder of small mammals and reptiles. After early education in Indonesia he studied environmental science and biology at Loyola Marymount University and the University of California, Los Angeles. He now lives in Melbourne.

Z

Zarudny

Zarudny's Skink *Eumeces schneiderii zarudnyi* **Nikolsky,** 1899

Zarudny's Worm Lizard *Diplometopon zarudnyi* Nikolsky, 1905

Nikolai Alekseyivich Zarudny (Zarudnyi) (1859–1919) was a zoologist, traveler, and ornithologist. He taught at the Military High School, Orenburg (1879–1892), and undertook five expeditions through the Trans-Caspian region (Turkmenistan). He taught natural history at the Pskov Military School (1892–1906) and made four journeys through Persia (Iran). He worked in Tashkent (1906). His extensive collections are now in the Zoological Museum, Russian Academy of Science. He wrote *Third Excursion over Eastern Persia (Horassan, Seistan and Persian Baluchistan) in 1900–1901* (1916). Many taxa, including a bird and two mammals, are named after him.

Zaw

Zaw's Wolf Snake *Lycodon zawi* **Slowinski** et al., 2001

Professor Khin Maung Zaw is Pro-Rector, University of Forestry, and Director, Nature and Wildlife Conservation Division, Forestry Department of Myanmar. He co-wrote *Developing a National Tiger Action Plan for the Union of Myanmar* (2006).

Zeledon

Zeledon's Earth Snake *Geophis zeledoni* **Taylor,** 1954

José Cástulo Zeledón (1846–1923) became an internationally known ornithologist and co-administrator of the drugstore Botica Francesa, one of Costa Rica's largest private companies. He co-founded Museo Nacional de Costa Rica. He studied under Baird and Ridgway at the Smithsonian, then returned to Costa Rica with Dr. William Gabb's expedition (1871) to explore the forest of Talamanca, where Zeledón made an important bird collection. He wrote *Catalogue of the Birds of Costa Rica* (1885). Among other taxa, three birds are named after him.

Zenker

Zenker's Worm Snake *Typhlops zenkeri* **Sternfeld,** 1908

Georg August Zenker (1855–1922) was a German botanist, ornithologist, and gardener who collected in Central Africa from ca. 1895. He had significant landholding around Bipindi. His collecting activities extended, apparently, even to human bones, which he disinterred. He made a particular study of "pygmies" and other native peoples. Two birds and three mammals are named after him.

Zetek

Zetek's Neckband Snake *Scaphiodontophis zeteki* **Dunn,** 1930 [Junior syn. of *S. annulatus* Duméril, Bibron, and Duméril, 1854]

Professor James Zetek (1886–1959) was an entomologist who graduated from the University of Illinois and was employed by the U.S. government in the Panama Canal Zone (1911–1953). He became Professor of Biology and Hygiene, National Institute of Panama (1916), and the first Director of the Smithsonian Tropical Research Institute, Barro Colorado Island, Panama Canal (1923). He wrote *Report on Reptiles from Barro Colorado Island* (1950). Two amphibians are named after him.

Zeus

Anole sp. *Anolis zeus* Köhler and McCranie, 2001

Zeus was supreme ruler of the gods in Greek mythology. The etymology states that the name "refers to the Cordillera Nombre de Dios" (the "Name of God" mountain range).

Zhao

Zhao's Pit-viper *Gloydius shedaoensis* Zhao, 1979

Japalure sp. *Japalura zhaoermii* Goa and Huo, 2002

Pit-viper genus *Zhaoermia* **Gumprecht** and Tillack, 2004

Sichuan Hot-spring Keelback *Thermophis zhaoermii* Guo, **Liu,** Feng, and He, 2008

Bow-fingered Gecko sp. *Cyrtodactylus zhaoermii* Shi and Zhao, 2010

Dr. Zhao Er-mi (b. 1930) is a zoologist and China's foremost herpetologist. He graduated from the University of Huaxi (1951). He has been Research Professor at Cornell and Visiting Professor at the University of California, Berkeley. He is an academician of the Chinese Academy of Sciences and is retired from the Chengdu Institute of Biology. The genus *Zhaoermia* was created in his honor, as *Ermia* had already been created for a genus of locusts. He wrote *Snakes of China* (2006).

Zhou

Zhou's Box Turtle *Cuora zhoui* **Zhao,** 1990

Zhou Jiufa is a herpetologist who started the first Chinese Turtle Museum, Nanjing. He wrote, with his daughter, Zhou Ting, *Chinese Chelonians Illustrated* (1992).

Zidok

Zidok's Ground Snake *Atractus zidoki* Gasc and **Rodrigues,** 1979

Zidok is a place in French Guiana.

Ziegler

Bent-toed Gecko sp. *Cyrtodactylus ziegleri* Nazarov, **Orlov,** Nguyen, and Ho, 2008

Ziegler's Tree Lizard *Pseudocalotes ziegleri* Hallermann et al., 2010

Dr. Thomas Ziegler (b. 1970) is a zoologist and Curator of the Aquarium, Cologne Zoo, and an expert on Vietnamese fauna.

Zimmer

Forest Skink sp. *Sphenomorphus zimmeri* Ahl, 1933

Dr. Carl Wilhelm Erich Zimmer (1873–1950) was a zoologist who was Director of the Zoological Museum and Professor of Zoology, Universität Berlin (1926–1943). Ahl, who described this skink, worked under Zimmer. Several crustaceans—his specialty—are named after him. He wrote *Anleitung zur Beobachtung der vogelwelt* (1917).

Zolio

Scortecci's Orange-tailed Lizard *Philochortus zolii* **Scortecci,** 1934

Signor Zolio was the President of the Italian Geographic Society, which supported the expedition to Libya during which the holotype was collected.

Zong

Zong's Odd-scaled Snake *Achalinus jinggangensis* Zong and Ma, 1983

Zong Yu (b. 1936) is a Chinese herpetologist, one of many authors involved in producing the *Fauna Sinica*, including *Reptilia Volume 3. Squamata: Serpentes* (1998).

Zug

Zug's River Cooter *Pseudemys gorzugi* Ward, 1984 [Alt. Rio Grande Cooter]

Zugs' Monitor *Varanus zugorum* **Böhme** and **Ziegler,** 2005

Bow-fingered Gecko sp. *Cyrtodactylus zugi* **Oliver** et al., 2008

Dr. George Robert Zug (b. 1938) joined the staff of the Smithsonian as Assistant Curator (1968), becoming Curator of Amphibians and Reptiles (1975) and then Curator Emeritus, Department of Systematic Biology–Vertebrate Zoology. His bachelor's degree was awarded by Albright College (1960), his master's by the University of Florida (1963), and his doctorate by the University of Michigan (1968). He co-wrote "Age and Growth in Olive Ridley Seaturtles *(Lepidochelys olivacea)* from the North-Central Pacific: A Skeletochronological Analysis" (2006). The monitor is named after Zug and his wife, Patricia.

Zugmayer

Zugmayer's Rock Agama *Stellio tarimensis* Zugmayer, 1909 [Junior syn. of *Laudakia stoliczkana* **Blanford,** 1875]

Zugmayer's Toadhead Agama *Phynocephalus erythrurus* Zugmayer, 1909

Professor Dr. Erich Zugmayer (1879–1938) was an Austrian explorer, zoologist, ichthyologist, and herpetologist at Zoologische Staatssammlung München. He visited Iceland (1902). He explored the area around Lake Urmia, Persia (Iran), and collected in Tibet, Ladakh, and Panggong Lake (1906 and 1911), Baluchistan, India (now in Pakistan). He published "Bericht über eine Reise in Westtibet" (1909).

Zulia

Zulia Toad-headed Turtle *Phrynops zuliae* **Pritchard** and Trebbau, 1984 [Syn. *Mesoclemmys zuliae*]

Zulia Skink *Mabuya zuliae* Miralles et al., 2009

Zulia is a province of Venezuela.

Zully

Tree Iguana sp. *Liolaemus zullyae* **Cei** and Scolaro, 1996

Mrs. Zully Ortega de Scolaro is presumably the junior author's wife (or mother).

Zuma

See **Montezuma.**

Zweifel

Zweifel's Ground Snake *Liophis reginae zweifeli* **Roze,** 1959

Zweifel's Whiptail *Aspidoscelis costatus zweifeli* **Duellman,** 1960

Zweifel's Leaf-toed Gecko *Phyllodactylus nocticolus zweifeli* **Dixon,** 1964

Zweifel's Helmet Skink *Tribolonotus annectens* Zweifel, 1966

Zweifel's Coral Snake *Micrurus distans zweifeli* Roze, 1967

Zweifel's Snail-eating Snake *Tropidodipsas zweifeli* **Liner** and **Wilson,** 1970 [Alt. Zweifel's Snail-sucker; Syn. *Sibon zweifeli*]

Zweifel's Sea Snake *Enhydrina zweifeli* **Kharin,** 1985

Dr. Richard George Zweifel (b. 1926) is a herpetologist and leading expert on Australian frogs. He joined the American Museum of Natural History (1954), served as Chairman, Department of Herpetology (1968–1980), retired as Curator Emeritus (1989), and now lives in Arizona, continuing his studies at the museum's Portal research station. Among his many publications, he co-wrote *Encyclopedia of Reptiles and Amphibians* (1998).

BIBLIOGRAPHY

American Museum Novitates. New York, 1921–.

Amphibia-Reptilia Leiden. 1980–.

Annales des sciences naturelles. Zoologie et paléontologie. Paris, 1864–1915.

Annals and Magazine of Natural History. London, 1841–1966.

Annals of the Transvaal Museum. Pretoria, 1911–.

Annuaire du Musée Zoologique de l'Academie Imperiale des Sciences. St. Petersburg, 1896–1932.

Beolens, B., and M. Watkins. 2003. *Whose Bird?* London, Christopher Helm / A. and C. Black.

Beolens, B., M. Watkins, and M. Grayson. 2009. *The Eponym Dictionary of Mammals*. Baltimore, Johns Hopkins University Press.

Bericht über die Senckenbergischer Naturforschende Gesellschaft. Frankfurt, 1869–1896.

Bocourt, M. F. 1870. "Description de quelques sauriens nouveaux originaires de l'Amérique méridionale." *Nouvelles archives du Museum d'Histoire Naturelle*, Paris, 2nd ser., 6.

Bonner zoologische Beiträge. Bonn, 1950–.

Boulenger, G. A. 1885–1887. *Catalogue of the Lizards in the British Museum (Nat. Hist.)*. Vols. 1–3. London.

Bulletin du Museum National d'Histoire Naturelle. Paris, 1895–1970.

Bulletin of the American Museum of Natural History. New York, 1881–.

Bulletin of the Museum of Comparative Zoology. Harvard University, Cambridge, Mass., 1863–.

Copeia. New York, 1913–.

Daudin, F. M. 1802–1803. *Histoire naturelle, générale et particulière des reptiles*. Vols. 4–6. Dufart, Paris.

Dictionary of National Biography. 1992. Oxford, Oxford University Press.

Duméril, A. M. C., and G. Bibron. 1839. *Erpétologie générale ou Histoire naturelle complète des reptiles*. Vol. 5. Paris, Roret / Fain et Thunot.

Duméril, A. M. C., G. Bibron, and A. H. A. Duméril. 1854. *Erpétologie générale ou Histoire naturelle complète des reptiles*. Vol. 7, pt. 1. Paris.

Dumerilia. Paris. 1994–.

Frank, N., and E. Ramus. 1995. *A Complete Guide to Scientific and Common Names of Reptiles and Amphibians of the World*. Pottsville, Pa., Ramus / NG Publishing.

Glaw, F. and M. Vences. 1994. *A Fieldguide to the Amphibians and Reptiles of Madagascar*. Cologne: Vences and Glaw.

Gotch, A. F. 1995. *Latin Names Explained—A Guide to the Classification of Reptiles, Birds. and Mammals*. London, Blandford Press.

Gray, J. E. 1849. *Catalogue of the Specimens of Snakes in the Collection of the British Museum*. London.

Hamadryad. Madras, 1976–2000.

Herpetologica. Lawrence, Kans., 1936–.

The Herpetological Journal. London, 1985–.

Herpetological Monographs. Austin, 1982–.

Herpetological Review. Lawrence, Kans., 1967–.

Jornal de Sciencias Mathematicas, Physicas e Naturaes. Lisbon, 1866–1910.

Journal of Herpetology. Lawrence, Kans., 1968–.

Journal of the Bombay Natural History Society. Bombay, 1886–.

Mémoires de l'Academie Impériale des Sciences. St. Petersburg, 1809–1830.

Memoirs of the Queensland Museum. Brisbane, 1912–.

Ménétries, E. 1832. *Catalogue raisonné des objets de zoologie recueillis dans un voyage au caucase et jusqu'aux frontières actuelles de la Perse*. St. Petersburg, L'Academie Imperiale des Sciences.

Mittellungen aus dem Naturhistorisches Museum. Hamburg, 1882–1908.

Mocquard, F. 1894. "Reptiles nouveaux ou insuffisamment connus de Madagascar." *Compte rendu sommaire de séances de la Société Philomathique de Paris*, 17.

Monatsberichte der Königlichen. Preussischen. Akademie der Wissenschaften zu Berlin. Berlin, 1856–1881.

Murray, J. A. *The Zoology of Beloochistan and Southern Afghanistan (Reptiles and Batrachia)*. Bombay Education Society Press, 1892.

Nouvelles archives du Museum de l'Histoire Naturelle. Paris, 1865–1914.

Philippine Journal of Science. Manila, 1906–1994.

Proceedings of the Zoological Society of London. London, 1831–1965.

Publicaciones ocasionales del Museo de Ciencias Naturales. Caracas, Venezuela.

Ramírez Leyton, G., and D. Pincheira Donoso. 2005. *Fauna del altiplano y desierto de Atacama*. Phrynosaura Ediciones.

Records of the Albany Museum. Grahamstown, 1903–1935.

Records of the Australian Museum. Sydney, 1890–.

Records of the South Australian Museum. Adelaide, 1918–2002.

Records of the Western Australian Museum. Perth, 1974–.

Rochebrune, A. T. de. 1884. *Faune de la Sénégambie. Reptiles*. Paris, Octave Doin.

Russian Journal of Herpetology. Moscow, 1994–.

Salamandra: Zeitschrift für Herpetologie und Terrarienkunde. Frankfurt am Main, 1965–.

Sitzungsberichte der Kaiserlichen Akademie der Wissensschaften. Vienna, 1850–1947.

Smith, H. M., D. Chiszar, K. Tepedelen, and F. van Breukelen. 2001. "A Revision of Bevelnosed Boas (*Candoia carinata* Complex) (Reptilia: Serpentes)." *Hamadryad* 26 (2).

Smith, M. A. 1926. *Monograph of the Sea Snakes (Hydrophiidae).* London, British Museum (Natural History).

———. 1943. *The Fauna of British India, Ceylon and Burma, Reptilia and Amphibia, Volume III, Serpentes.* London, Taylor and Francis.

Spawls, S., K. Howell, R. C. Drewes, and J. Ashe. 2001. *A Field Guide to the Reptiles of East Africa.* San Diego, Academic Press.

Sterling, Keir B., Richard P. Harmond, George A. Cevasco, and Lorne F. Harmond. 1997. *Biographical Dictionary of American and Canadian Naturalists and Environmentalists.* Westport, Conn., Greenwood Press.

Thomson, Keith S. 1995. *HMS Beagle—The Ship That Changed the Course of History.* New York, W. W. Norton and Co.

Titschack, E., ed. 1942. *Beiträge zur Fauna Perus.* Vol. 2. Hamburg.

Tropical Zoology. Florence, 1988–.

Vestnik zoologii. Kiev, 1967–.

Videnskabelige Meddelelser fra den Naturhistoriske Forening i Kjöbenhavn. Copenhagen, 1849–1912.

Wynne, Owen E. 1969. *Biographical Key—Names of Birds of the World—to Authors and Those Commemorated.* Privately published.

Zoologischer Anzeiger. Jena, 1878–.

Zootaxa. Auckland, 2001–.